NEW CANAAN LIBRARY
3 1457 00420 1088

P9-ELR-324

Edgers water damage
11/2/06

05/22/08-10

WITHDRAWN

New Canaan Library
151 Main Street
New Canaan, CT 06840

(203) 594-5000
www.newcanaanlibrary.org

JUN 0 5 2006

THE GREAT DELUGE

ALSO BY DOUGLAS BRINKLEY

Parish Priest: Father Michael McGivney and American Catholicism
(with Julie Fenster)

The Boys of Pointe du Hoc:
Ronald Reagan, D-Day, and the U.S. Army 2nd Ranger Battalion

Tour of Duty: John Kerry and the Vietnam War

Windblown World: The Journals of Jack Kerouac, 1947–1954 (editor)

Wheels for the World:
Henry Ford, His Company, and a Century of Progress, 1903–2003

The Mississippi and the Making of a Nation
(with Stephen E. Ambrose)

American Heritage History of the United States

The Western Paradox: The Bernard DeVoto Reader
(editor, with Patricia Limerick)

Rosa Parks

The Unfinished Presidency: Jimmy Carter's Journey Beyond the White House

John F. Kennedy and Europe (editor)

Rise to Globalism: American Foreign Policy Since 1939, Eighth Edition
(with Stephen E. Ambrose)

The Majic Bus: An American Odyssey

Dean Acheson: The Cold War Years, 1953–1971

Driven Patriot: The Life and Times of James Forrestal
(with Townsend Hoopes)

FDR and the Creation of the U.N.

THE GREAT DELUGE

Hurricane Katrina, New Orleans, and the Mississippi Gulf Coast

Douglas Brinkley

WILLIAM MORROW

An Imprint of HarperCollins*Publishers*

976.335
B

Page 686 serves as a continuation of this copyright page.

THE GREAT DELUGE. Copyright © 2006 by Douglas Brinkley. All rights reserved. Printed in the United States of America. No part of this book may be used or reproduced in any manner whatsoever without written permission except in the case of brief quotations embodied in critical articles and reviews. For information address HarperCollins Publishers, 10 East 53rd Street, New York, NY 10022.

HarperCollins books may be purchased for educational, business, or sales promotional use. For information please write: Special Markets Department, HarperCollins Publishers, 10 East 53rd Street, New York, NY 10022.

FIRST EDITION

Designed by Lovedog Studio

Printed on acid-free paper

Library of Congress Cataloging-in-Publication Data

Brinkley, Douglas.
The great deluge: Hurricane Katrina, New Orleans, and the Mississippi Gulf Coast / Douglas Brinkley.
p. cm.
Includes index.
ISBN-13: 978-0-06-112423-5
ISBN-10: 0-06-112423-0
1. Hurricane Katrina, 2005. 2. Hurricanes—Louisiana—New Orleans. 3. Disaster victims—Louisiana—New Orleans. 4. Disaster relief—Louisiana—New Orleans. 5. Emergency management—Government policy—United States. I. Title.

HV6362005.L8 B75 2006
976.3'35064—dc22

2006043338

06 07 08 09 10 WBC/RRD 10 9 8 7 6 5 4 3 2 1

For the U.S. Coast Guard first responders,
whose bravery was unparalleled,

and

Houston, Texas,
whose efficient openheartedness was breathtaking

Things are going to slide in all directions
Won't be nothing
Nothing you can measure anymore.

—Leonard Cohen, "The Future"

Deluge:

1. a. A great flood. b. A heavy downpour.
2. Something that overwhelms as if by a great flood

—American Heritage Dictionary

Oh, Oh Deep Water
Black, and cold like the night
I stand with arms wide open
I've run a twisted line
I'm a stranger
in the eyes of the Maker

I could not see
for the flood in my eyes
I could not feel
for the fear in my life
From across the Great Divide
In the distance I saw a light
I saw Jean Baptiste
Walking to me with the Maker

My body is bent and broken
by long and dangerous sleep
I can't work the Fields of Abraham
and turn my head away
I'm not a stranger
in the hands of the Maker

Brother George
Have you seen the homeless daughters
standing there
with broken wings
I have seen the flaming swords
there over East of Eden
burning in the eyes of the Maker

—Daniel Lanois, "The Maker" (Katrina adaptation)

Contents

Author's Note

> Strange how people who suffer together have stronger connections than people who are most content.
> I don't have any regrets, they can talk about me plenty when I'm gone.
> You always said people don't do what they believe in, they just do what's most convenient, then they repent.
> And I always said, "Hang on to me, baby, and let's hope that the roof stays on."
>
> —Bob Dylan and Sam Shepard, "Brownsville Girl"

It was a Ripley's "Believe It or Not!" moment. "Looking out the window of a fifteenth-floor condominium at One River Place on New Orleans's East Bank on August 29, 2005, I witnessed a stunning aberration. Just below me, the whitecapped Mississippi River was roaring backward—northward—due to Hurricane Katrina's wrath. Earlier that morning I had left my Uptown home and evacuated vertically to this supposed safe haven. Now, as I gazed at the churning river, my miscalculation was evident. A few minutes later I took my wife and two young children to the crowded lobby, where a sense of panic ensued. A lone generator flickered on and off, while children huddled around a small refrigerator where baby bottles were stored. Storm-phobic dogs paced back and forth, their hindlegs quaking. Outside you saw everything from trembling street signs to lost umbrellas flying by as the piercing rain whipped needles and knives. You could hear glass shattering all around, and see the nearby Hilton parking garage lose its supposedly hurricane-proof roof.

A day earlier, the name "Katrina" had conjured whimsical images of a Gaelic ballad or a Vegas cocktail waitress. A close friend of mine, in fact, was named Katrina. There was no menace in the echo. Perhaps if the storm had been named "Genghis Khan" or "Attila the Hun" or "Caligula," I would have fled. But now, as the raging Mississippi frothed with primal madness, gushing around Algiers Bend, ripping open the huge riverfront warehouses where Mardi Gras floats were stored, it was clear that Katrina was no mere hurricane or flood. It was destined to be known as "the Great Deluge" in the annals of American history.

Because Hurricane Katrina devastated 90,000 square miles along the heavily populated Gulf Coast, it's truly impossible to capture the morose terror endured by each and every survivor. Certainly the people trapped in attics, desperately punching holes in their rooftops with axes and hammers hoping to be rescued, cannot be forgotten; they've earned a special spot in our collective memory. Neither can the indomitable Coast Guard helicopter pilots and Louisiana Wildlife and Fisheries boatmen who hoisted stranded people out of the muck. But in the weeks after the hurricane first made landfall near Buras, Louisiana, on Monday, August 29, dozens of other horrific words and repellent images became part of our national discourse: feces contamination, storm surges, toxic soup, pervasive damage, highway triage, police suicide, FEMA's indifference, and so on.

The general feeling emanating from New Orleans—at first—was that the storm could have been worse; outside the region, this was interpreted to mean that it wasn't *too* bad. Once Katrina had passed, everybody patted themselves down, pleased that their vital organs were in place and that their pulse registered. Most New Orleanians who survived Katrina felt eerie, though, as if something ominous still lingered over the port city like a gothic shroud. Katrina had been a palpable monster, an alien beast that had gotten under the goose-fleshed skin of those who lived through the storm. Many of these non-evacuees felt vaguely ill that Monday evening. They sensed that something was horrifically wrong with their beloved city, something deeper than surface wounds. The ornate St. Louis Cathedral and the elegant Garden District mansions were only slightly battered, however, looking as opulent as ever. If there was a wave of post-storm optimism at first, it was because the pallid veil of Katrina had lifted, prematurely assuring residents in trickster fashion of their supposed safety. Unbeknownst at first to anyone but the Katrina victims within neighborhoods like the Lower Ninth Ward, Lakeview, Hollygrove, Gentilly, and New Orleans East, large areas of the city were gone—perhaps gone forever. With the exception of WWL radio, nearly all communication within the Greater New Orleans area had broken down. People were left to rely on their own firsthand experiences, hunches, and instincts. Virtually nobody in New Orleans knew what had happened just a few lonely blocks away.

As a historian I knew a wicked hurricane could alter world history. One, in fact, prohibited Kublai Khan from conquering Japan in 1274. Somewhere along the line I had learned in a biography of Mahatma Gandhi that during World War II over 35,000 Indians were killed by a major storm just south of Calcutta. Last year, when reading Ron Chernow's fine biography of Alexander Hamilton, I was reminded of how on August 31, 1772, a hurricane near the tiny island of Dominica crashed into the harbor, smashing all anchored boats and destroying the sugarcane crop. A seventeen-year-old Hamilton, traumatized by the storm, wrote an essay about it which was subsequently published in the *Royal Danish-American Gazette*, launching his career as a polemicist. "It seemed as if a total dissolution of nature was taking place," Hamilton said. "The roaring of the sea and wind, fiery meteors flying about it [sic] in the air, the prodigious glare of almost perpetual lightning, the crash of the falling house, and the ear-piercing shrieks of the distressed, were sufficient to strike astonishment into angels." I was also well aware that the great Galveston hurricane of 1900 killed between 8,000 and 10,000 people, and that when John F. Kennedy was president a tropical cyclone in East Pakistan annihilated over 22,000 citizens. Hurricanes mattered. I knew that much.

As to storms in my own times, I remember reading about the ravages of Hurricane Gilbert of 1988, Hurricane Hugo of 1989, and Hurricane Andrew of 1992 (which inflicted the costliest hurricane damage in U.S.

history—$44.9 billion). These hurricanes, however, were abstractions to me, dramatic *New York Times* accounts and NBC News visuals that I watched from safety. I never fully comprehended their power or their tricks—or their ability to unleash widespread devastation on all aspects of civil society. Hurricane Katrina was my adult-dosage wake-up call.

This book attempts to document what novelist Saul Bellow in *Ravelstein* called "the freaky improvisations of creatures under stress." Hurricane Katrina had created widespread anarchy. In helter-skelter situations, citizen-heroes like Reverend Willie Walker of Kenner; Mama D of New Orleans; Sara Roberts of Lake Charles, Louisiana; and Mayor Eddie Favre of Bay St. Louis, Mississippi, rose to the forefront among first responders and performed beyond expectations. In New Orleans an ex-con, a newspaper reporter, an animal-rights activist, a reggae dreamer, a states' rights advocate, an AIDS doctor, and a prep-school teacher became civic leaders while the New Orleans Police Department, with notable individual exceptions, was the face of fecklessness.

But this book is not just about New Orleans. Too often during the Great Deluge the national media focused on New Orleans rather than the Mississippi-Alabama Gulf Coast area. To a degree this was understandable. While the hurricane devastation was much worse in towns like Bay St. Louis (where I lived for eight years) and nearby Waveland (where I often wrote along the beach front), the New Orleans drama unfolded in an escalating way once the 17th Street Canal, London Avenue Canal, and Industrial Canal levees breached. As evenhandedly as possible, I try to flash back and forth between New Orleans, the Mississippi Gulf Coast, and southeastern Louisiana parishes such as St. Bernard and Plaquemines, with an occasional look at Alabama. An apology, however, must be offered at the outset. The magnitude of the Great Deluge was so great, the implications for the Gulf South region so mindbogglingly huge, it was impossible to tell what happened to every town or hamlet. Now that I'm finished with the text, and my publisher is preparing an index, I see that communities such as Pearlington, Moss Point, D'Iberville, Delacroix, Port Sulphur, Grand Isle, and Vinton aren't meaningfully included. This is regrettable. My hope in writing this book is to give historical dignity to the people and places that seized the moment, however surreal, and took charge.

If I gleaned one pertinent insight about human nature from writing this book, it's that love of geographical places is more all-encompassing than most of us imagine. When reading about the horror of the St. Bernard Parish "Wall of Water" or the Bay St. Louis "Lake Borgne Surge," you might think only a crazy person would rebuild there. Yet people are rebuilding and they are not crazy. I interviewed more than three hundred people, and none, not even those who lost everything they had, want to live anyplace else. They were born in Pascagoula or Ocean Springs or Belle Chasse, and they plan on dying there. It's their unflappable spirit, with private-sector and federal help, which guarantees that all of these devastated communities—even poor Chalmette, Louisiana—will be back.

Anybody reading the endnotes will see that I relied on newspaper reporters quite a bit in writing this narrative. What a truly superb job the *New York Times,* New Orleans *Times-Picayune, Washington Post*, *Los Angeles Times*, *Philadelphia Inquirer*, *Houston Chronicle, Chicago Tribune*, Baton Rouge *Advocate*, Biloxi *Sun Herald, USA Today, Wall Street Journal,* and other dailies did in covering Katrina. When FEMA and the White House were floundering, the media stepped into the fray with gutsy reporting and deep moral principle. They had some details wrong but, more important, they got the urgency exactly right. Likewise CNN, ABC News, Fox News Channel, NBC News, MSNBC, and National Public Radio helped me write this book more than the tiny note numbers assigned to specific quotations indicate. And the Associated Press—as always—did a wondrous job fanning out across the region to make sure no town was left behind.

It was literature and music that got me through writing this book. I kept thinking that it was too bad John Steinbeck wasn't around to document the New Orleans diaspora or George Orwell to sleep with refugees at the Reliant Center. Wouldn't it have been something if Hunter S. Thompson had been alive to vent his spleen at Homeland Security's Michael Chertoff or if Arthur Miller had written a drama about the daily dilemmas facing evacuees as they started new lives in FEMA trailer camps? But Joan Didion was alive to write *The Year of Magical Thinking*, which Knopf published just as the Katrina disaster unfolded, a memoir full of grief, loss, and derangement about the death of her husband and the serious illness of her daughter. "Confronted with sudden disaster we all focus on how unremarkable the circumstances were in which the unthinkable occurred," she wrote, "the clear blue sky from which the plane fell, the routine errand that ended on the shoulder with the cars in flames, the swings where the children were playing as usual when the rattlesnake struck from the ivy."

Because I was in New Orleans for Katrina, and later took part in boat rescues around Memorial Medical Center and Central City, a memoir of the Great Deluge was possible. Instead I chose to write a history of a single week in the summer of 2005: August 27 (Saturday) to September 3 (Saturday). Eventually I evacuated my family to Houston and interviewed people for days at the Astrodome. I have and always will have a highly personal stake in the Alabama, Mississippi, Louisiana, and Texas regions, which might color some of my judgments; my extensive eyewitness interviews helped me forge strong feelings about what went down on the plain of the Gulf South; but my job, as I saw it, was to chronicle the plight of others, not my own. And in these pages you'll read only prose in the past tense. The historical narrative stops on September 3, when the U.S. 82nd Airborne took control of New Orleans and evacuation buses finally arrived at the Convention Center. "That Saturday, the combination of troops and buses was like taking a great big wet blanket and dropping it over the city," Terry Ebbert, director of Homeland Security for New Orleans, later recalled. "By nightfall command and control had been stabilized. The crisis was over."

Perhaps, someday, I'll write a sequel titled *After the Deluge*, about how the Mississippi Coast, with casino gambling, came back, I hope—stronger than ever—and how Louisiana flourished once the rotten politicians and corrupt cops were flushed out of the system. About how New Orleans reimagined itself as the Afro-Caribbean capital of the world with a New South self-confidence and how Plaquemines Parish reclaimed its land from the sea in a saga as glorious as that of the province of Zeeland in the Netherlands. But maybe not. History is still being made in the Gulf South, and only a foolish booster would try to sell you a "Better Than Ever" paddle. Really, it's too early to say. Before the Gulf South can rebuild, it needs time to heal. While entrepreneurial zeal is the rightful engine of the rebuild, the Gulf South needs to take a breath, to pause, and to remember what transpired on August 29, 2005—the day it was so gravely wounded. History, in the end, is homage; it's about caring enough to set the record straight even if reliving the past is painful or disappointing. Buried history leads to rank defilement of the human spirit. Any politician involved with Katrina, who espouses the cliché that "the blame game" is unnecessary is probably harboring a chestful of guilt.

My hope is that this history, fast out of the gates, may serve as an opening effort in Katrina scholarship, with hundreds of other popular books and scholarly articles following suit. There is, in my opinion, no such thing as too many Great Deluge anecdotes and offerings. Only by remembering, and holding city, state, and federal government officials responsible for their actions, can a true Gulf South rebuild commence in the appropriate fashion. Meanwhile, I stand by something Christopher Columbus once wrote from the Caribbean: "I don't say it rained, because it was like another deluge."

Douglas Brinkley
March 17, 2006
New Orleans

Chapter One

IGNORING THE INEVITABLE

> More than once, a society has been seen to give way before the wind which is let loose upon mankind; history is full of the shipwrecks of nations and empires; manners, customs, laws, religions—and some fine day that unknown force, the hurricane, passes by and bears them all away.
>
> —Victor Hugo, *Les Misérables*

I

NO WIND WAS BLOWING when forty-four-year-old Laura Maloney arrived at the Louisiana Society for the Prevention of Cruelty to Animals (LSPCA) on Japonica Street in New Orleans's Ninth Ward. With the exception of some storefront windows plywooded-up and Mandich's Restaurant, which was closed, August 27 was, by and large, a fairly normal Saturday morning. In a building across the street from the Industrial Canal, Maloney's LSPCA staff had lots of work to do. Hurricane Katrina—a possible Category 5 storm—was headed toward New Orleans and the shelter had a total population of 263 stray pets, ranging from boxers to Heinz 57 mutts and Siamese cats. All of them had to be evacuated. "Each animal got its own digital picture shots," Maloney recalled. "We made sure each pet's paperwork was in order. And we IDed each collar;

we had a tracking system, in case any animal got separated from their paperwork."[1]

Maloney could have been a fashion model, with her long blond hair, perfect white teeth, and eyes that implied an internal kindness. The only problem was that she didn't care for high fashion; her passion was animals. Raised in Maryland, Maloney had earned her undergraduate degree at West Virginia University and her MBA at Tulane University. She had worked at the Philadelphia Zoo and New York's Central Park Zoo before landing employment at the Aquarium of the Americas near the French Quarter. She loved everything about New Orleans, except the way stray animals weren't properly cared for. Her husband, Don Maloney, also an animal enthusiast, was general curator of the Audubon Zoo, where he took care of everything from apes to zebras and every species in the alphabet in between. "Animals were a big part of our lives," she recalled. "We shared a deep appreciation for them."

Back in 1997 they had gone to the LSPCA together to adopt what Laura called "the muttiest dog we could." They succeeded in their quest. Tucked away in the back of a kennel was a black-tan German shepherd mix inflicted with chronic tics, heartworm, and a hip crack from what they assumed was an automobile accident. "She was on death row," Laura recalled. "About to be put down, so we adopted her. We named her Filé."

Maloney was hooked. She quit her job as assistant to the president of Freeport-McMoran, a huge New Orleans–based mineral exploration company, and took over the LSPCA as executive director. Many of New Orleans's nursing homes may have been a shambles, and the housing projects that populated the city in a state of ghastly disrepair, but under her tutelage, the Louisiana SPCA was run with the spic-and-span efficiency of a Swiss hospital. She wouldn't have it any other way. That Saturday morning, Maloney, dressed in blue jeans and a T-shirt, and her staff created an assembly-line approach to load all the animals into a pair of climate-controlled refrigerated trucks headed for Houston's SPCA on Portway Drive. Although the two animal shelters were independent agencies, they operated under the mission statement of the 140-year-old national organization: "Compassion and mercy for those who cannot speak for themselves."[2]

Transporting 263 dogs and cats was no small task, but there weren't any other options. "The Louisiana SPCA," according to its own stated policy, "evacuates its shelter for Category 3 hurricanes and above."[3]

At 5 A.M., the National Hurricane Center (NHC) had released an update from its headquarters in Miami. Advisory Number 16 on the tropical storm named Katrina affirmed that with sustained winds of 115 mph, the disturbance had already become a Category 3 hurricane and, moreover, that "some strengthening is forecast during the next twenty-four hours."[4] Katrina was still about 350 miles out in the Gulf of Mexico. It had ripped through Florida as a Category 1 hurricane two days before, leaving approximately 500,000 people without power. About eighteen inches of rain had fallen. Driving winds had torn doors off houses, bent trailers like horseshoes, sent sloops surfing onto front lawns, and chewed up industrial parks, coughing out plywood and shards. There were seven reported storm-related deaths from falling trees and other mishaps. Despite the horror, Floridians were hardened to hurricanes. In 2004 alone they had been hit with four of them. The state recovered quickly from Katrina's blow, with the lightning-fast help of the Federal Emergency Management Agency (FEMA), which trucked in water and ice, hospital supplies, and even microclips to properly tag dead bodies. But just because Florida had recovered quite quickly didn't mean that Katrina, still growing in fury, was through with the American coast. "It could be," meteorologist Christopher Sisko told the *New York Times*, "an extremely dangerous storm."[5] According to Advisory 16, in fact, forecasters expected Katrina to turn west-northwest, toward the city of New Orleans, during the weekend.

That was enough for the Louisiana SPCA, which brooked no discussion and no debate: with the announcement that a major hurricane was on the way, the preset plan went into motion. The two trucks arrived at the Japonica Street shelter. "We reached out to them and offered our shelter for the New Orleans animals," Kathy Boulte of the Houston SPCA recalled. "They arrived in Houston and later we all watched on television while the storm grew into a Category 5."[6] Laura Maloney had overseen the evacuation of her four dogs, all of the stray pets, and fifteen staff members. "If we had stayed at Japonica Street," Maloney recalled, "we'd have all been goners."[7]

Twenty miles to the west of New Orleans, near the town of Taft in St. Charles Parish, the Waterford 3 nuclear plant also heeded Saturday's warning. Relying on its own advance plan, the generating facility, owned by Entergy Corporation, which employed 14,000 people in the region, was required to shut down completely in the face of a Category 3 storm. Crews prepared diesel-fueled generators to supply the facility with enough current to maintain the reactor, as part of an emergency action level called an "unusual event" (the lowest of four emergency action levels). Because Waterford 3 had an output capacity of 1091 megawatts of electricity, the shutdown was certain to result in a blackout in southern Louisiana. But, rightly, Entergy wanted to make sure its employees—and the reactor itself—were safe. Entergy, New Orleans's only Fortune 500 company, performed according to plan.[8]

St. Charles Parish's president, Albert Laque, responded to the NHC advisory by issuing a *mandatory* evacuation order for all residents. A few other Louisiana parish presidents did the same that Saturday. Plaquemines Parish lies south of New Orleans, where it escorts the Mississippi River into the Gulf of Mexico. Locally, some refer to the parish as "the ragged boot heel of Louisiana." Suburban in the north and rural in the south, it had about 26,000 residents. Not surprisingly, it was dependent on the waters that surrounded it—commercial fishing being the main livelihood and sport fishing the biggest tourist draw. With this reputation, visitors to Plaquemines were often astonished to see rows and rows of citrus trees lining the Mississippi River for more than forty miles. Unlike the sandy dirt of Florida, the alluvial soil of Plaquemines was soft and rich, derived from the thirty-one states drained by the Mississippi. If you gave even a cursory glance at a map of Louisiana, you quickly saw that the parish was, even in normal times, half under water, vulnerable to massive flooding even if Katrina were a low-grade tropical storm. Nobody in Plaquemines ever really believed the levees would withstand a powerful hurricane. "After [Hurricane] Betsy [in 1965] these levees were designed for a Category 3," Sheriff Jeff Hingle of Plaquemines Parish told CNN that anxious weekend. "You're now looking at a Category 5. You're looking at a storm that is as strong as Camille [in 1969] was, but bigger than Betsy was size-wise. These levees will not hold the water back. So we're urging people to leave. You're looking at these levees having ten feet of water over the top of them easily."[9]

When Plaquemines Parish President Benny Rousselle declared a mandatory "Phase I" evacuation on August 27—that Saturday when there was still plenty of time to flee—parish employees fanned out on preappointed routes, picking up residents with special needs and busing them to state-run shelters in Shreveport, Alexandria, Houma, and Lafayette.[10] The parish not only knew *which* residents required special help, it knew exactly *where* they lived. "We were putting them on buses that Saturday morning and, you know what?" Rousselle recalled. "When we ran out of drivers, I went up to evacuees, determined which ones had valid driver's licenses and knew how to drive stick shift, and told them to bring folks north. They were tapped . . . deputized or whatever you want to call it. They were now official parish drivers. Out . . . out . . . out. I wanted everybody out of Plaquemines Parish. We were able to get our people out."[11]

II

Throughout all nine of Louisiana's coastal parishes,* only a *mandatory* evacuation drew the attention of storm-tested residents—anything less was inadequate. In New Orleans (Orleans Parish), no such mandatory evacuation order had *ever* been issued during the course of the city's 287-year history. This was not to imply that New Orleans had never before been in harm's way, the object of a natural disaster. In fact, it had never been out of peril. The city was founded in 1718 by the Sieur de Bienville, a French-Canadian nobleman who believed that the Mississippi River was in crying need of an ocean port. Having explored the delta region for almost twenty years, he had an intimate understanding of its alternating swamps and bayous (naturally occurring canals). With his brother, Pierre Le Moyne, Bienville had been one of the earliest Europeans to set eyes on Lake Pontchartrain: a 630-square-mile body of water at the edge of southeastern

* The nine coastal parishes are Cameron, Vermilion, Iberia, St. Mary, Terrebonne, Lafourche, Jefferson, Plaquemines, and St. Bernard.

Louisiana. Pontchartrain was part bay, in that it opened onto the Gulf of Mexico, and part lake, in that it was generally filled with freshwater. Native Americans called it Lake Oktwa-ta, or "Wide Water."[12]

For years, Bienville had carefully considered a site for the port he envisioned. With advice from local Native Americans, especially Bayougoulas and Choctaws with whom he had navigated the region in hand-hewn cypress pirogues, he selected a tract of relatively high ground along an eastward curve in the Mississippi River. Bienville calculated that it was about ten feet above sea level; it was probably not quite that high, but the site was distinctly above the surrounding waters in those early days. The Native Americans assured him that it was the driest spot in proximity to both the river and the lake. Bienville was easily convinced that this location—on "the most beautiful crescent of the river"—would grow into the commercial vortex of the Mississippi Valley.[13] In his view, the site was a great natural port, allowing access to the river, which lay along the southern edge, as well as Lake Pontchartrain, about three miles to the north by way of Bayou St. John.[14] Nor was Bienville dissuaded by the Native American name for the tract: Chinchuba (in Choctaw, that meant "alligator"). He immediately renamed it La Nouvelle Orléans in honor of Philippe II, Duke of Orléans, the prince regent of France. Soon immigrants began arriving from France and Canada.[15] "Unfortunately, the rude wooden huts and log cabins that arose among these mud streets," historian Jim Fraiser wrote, "seemed far more appropriate residences for bastards than for saints."[16]

During the first year, as Bienville was laying out the first roads in his new city with the help of sixty or seventy laborers, the Mississippi River swelled and put New Orleans, such as it was, under nearly a foot of water.[17] "We are working on New Orleans, with such diligence as the dearth of workmen will allow," he wrote to the French consul in June 1719. "I myself went to the spot to choose the best site. . . . All the ground for the site, except the borders which are drowned by floods, is very good and everything will grow there."[18] But the area's proclivity toward flooding worried Bienville. He immediately started work on the city's first levees, earthen berms that would protect the land from floodwater. On September 23 and 24, 1722, the so-called Great Hurricane slammed into New Orleans, bringing with it an eight- to ten-foot storm surge. The town, for all practical purposes, ceased to exist. The stubborn Bienville would have saved a great many people future trouble if only he had abandoned Chinchuba right then and moved New Orleans inland. Higher ground was his for the taking around present-day Baton Rouge, farther north along the Mississippi. Many people surrounding Bienville campaigned for just such a new, safer site. But there was an economic reason for keeping New Orleans where it was. In the eighteenth century, sailing a fully laden ship up a swift-flowing river such as the Mississippi was achingly difficult and, at times, utterly impossible. Sandbars were a particularly mettlesome navigation obstacle. For the sake of commerce, the Port of New Orleans stayed put, while Bienville—thinking himself very astute and a master of nature—redirected his men to build more levees between the town and the river.

No human will ever master the Mississippi River Delta (nor will the U.S. Army Corps of Engineers), as nearly three hundred years of flooding attest. Yet, starting with Bienville and continuing once the Louisiana Purchase became part of the United States in 1803, entrepreneurial delusion became a mind-set in the region. There was money to be made in the burgeoning river port, and that's what mattered most. Traders were bolstered by unflappable engineers. The river, they insisted, could be tamed with the proper dykes and levees. Charged with keeping the city dry, the U.S. government *did* build miles of levees. The levees were to be foolproof bulwarks against both the seasonal flooding of the river and storm surges from the sea, via Lake Pontchartrain. Sometimes, however, those levees didn't work: in 1849, during the Mexican-American War, the collapse of one near the Carrollton neighborhood of Uptown New Orleans sent a cascade of water surging through the Crescent City. Ultimately more than two hundred city blocks were inundated with floodwaters. For more than six weeks, New Orleans rotted, submerged underwater. Another brutal hurricane blew through southeastern Louisiana in 1893, killing more than two thousand. The Great Storm of 1909 had Category 4 hurricane-force winds, causing Lake Pontchartrain to flood along the South Shore, killing hundreds in the New Orleans area. In 1927, the city faced the wrath of the Great Mississippi Flood that devastated the Lower Mississippi Valley. Worried, New Orleans's civic leaders decided to stop flooding in the city by any means necessary. They actually detonated thirty tons of dynamite on the Caernarvon Levee, just fifteen miles downriver from the French Quarter, breaching the levee in poor black neighborhoods, in the interest of *protecting the rest of the city*. Ever since 1927, African Americans in New Orleans had been distrustful of the levee boards, believing that if the white gentry did it once, they'd do it again.

For New Orleans residents, peace of mind came from the system of levees and flood walls that protected the city from surrounding water, whether from the lake, river, sea—or, in the form of rain, the sky. If the Dutch reclaimed miles from the North Sea after the tragic 1953 flood, then, logic suggested, certainly the United States could engineer a safe levee system to protect Louisiana from the Gulf of Mexico. New Orleans levee walls were typically ten to fifteen feet high, forming the sides of the saucer in which New Orleans sits. The system of levees was overseen by the U.S. Army Corps of Engineers, which had thirteen hundred employees in the New Orleans office. The Corps of Engineers, the civil engineering arm of the federal government, traced its roots to 1775, when the Continental Congress hired experienced men specifically to take on building projects around Bunker Hill in Boston. Largely staffed by civilians, it had responsibility for domestic construction and maintenance jobs that sought to minimize flooding, drought, and erosion, among other natural phenomena. The New Orleans District of the Corps administered 30,000 square miles, all of southern Louisiana, the most complex section of the United States in terms of water control and, at the same time, the busiest, measured by shipping traffic. If engineering a reliable way to satisfy all the construction demands wasn't sensitive enough, the Corps also had to balance itself between the often conflicting interests of parochial Louisiana politicians and federal officials. What the national government ordered and what the locals insisted upon were nearly always different. The Corps itself was pushed and pulled just as much as any of the rivers or bayous that it oversaw. Because of the extraneous projects and inefficiencies that resulted, by 2005 overseeing a system of levees in New Orleans based on fifty-year-old engineering and suspect construction techniques was taking its toll. Meanwhile, as the politicians bickered (and the Corps dawdled), the threat of flooding in New Orleans constantly grew.

Through coastal erosion and man-made engineering mistakes, nearly one million acres of buffering wetlands in southern Louisiana disappeared between 1930 and 2005. Public awareness campaigns with names like Coasts 2050 and No Time to Lose were launched but did little good. Too many Americans saw these swamps and coastal marshes as wastelands. "The impact of losing their wetlands was overwhelming," explained Park Moore, assistant secretary of the Louisiana Department of Wildlife and Fisheries. "All the habitat for animals and invertebrates was disappearing along with a vital natural filter, which prevents pollution in the Gulf from toxic agents from oil and gas. Dredging killed the wetlands, which, in time, would leave Louisianans more vulnerable to hurricanes."[19]

Between 1932 and 2000, from the Atchafalaya Basin in the north to the Gulf of Mexico in the south, from Mississippi in the east to Texas in the west, Louisiana lost coastal marshland in twenty parishes, a swath roughly equivalent to the acreage of Delaware.[20] (By 2050 it will have lost another 700 square miles.) In 2005 the state was losing critical wetlands at the rate of one football field every thirty-eight minutes.[21] Mike Dunne, author of *America's Wetland: Louisiana's Vanishing Coast,* broke it down even more: "A tennis court every thirteen seconds slips under the water or is nibbled off the edge of one of the most ecologically sensitive regions of the nation and the world."[22] The reasons were many. As a result of the 1927 flood, the U.S. Army Corps of Engineers started channeling the Mississippi River, which had ugly, unintended consequences. Because the Mississippi was controlled by levees that kept it on one path, all of its sediment was deposited into the Gulf of Mexico, instead of spreading out along the coast, where it would build up in wetland areas. Another reason the coastline receded lay with the disruptive and long-lasting effect of massive construction projects. Oil companies installed miles of crisscrossing pipelines. Traditionally, oil was extracted from land or offshore, but new technologies allowed drilling in the coastal area known as America's Wetlands.

Meanwhile the government, under pressure from shipping interests, dredged channels, including the granddaddy of them all, the Gulf Intracoastal Waterway. It was built in the early 1900s, and eventually grew to connect the Rio Grande at Brownsville, Texas, with Apalachee Bay near Tallahassee, Florida. Along some stretches, the Intracoastal route followed natural waterways, but in the New Orleans section it was man-made, a cement-sided canal cut into the soft marshland. Blocking the natural ebb and flow of the water, the Intracoastal and other channels weakened the delicate ecology of the wetlands. Ongoing maintenance caused further damage. Together, industrial factors contributed to as much as 40 percent of Louisiana's total land loss.[23] "The amazing part was how little was done to expedite the wetlands reconstruction process, even after Hurricane Ivan in 2004," John W. Sutherlin, an environmental professor at the University of Louisiana at Monroe, wrote after Katrina. "The state and local government (with a noticeably absent federal government) seemed to stare reality in the face and pretend that the good times would just continue to roll."[24]

No less a conservation authority than Theodore Roosevelt, in fact, took a special interest in the Louisiana coast. He established Breton Island, off the eastern shore of the Louisiana "toe," as the second U.S. National Wildlife Refuge in 1904, insisting that it must be protected if near extinct bird species were to thrive again and New Orleans was to be spared massive flooding. He visited the Louisiana barrier islands for four days in 1915 as ex-president, enjoying the sight of dolphin schools and bird rookeries. As a Harvard-trained naturalist, he kept field notes about the brown pelicans, caspain terns, roseate spoonbills, and laughing gulls he encountered. He enjoyed watching the shrimp boats trawl and the egrets swoop down from the sky looking for minnows to eat. Roosevelt chronicled the trip in his 1916 travelogue, *A Booklover's Holiday in the Open*, and told readers about conservation conclusions he made upon visiting the pristine Gulf island. "Birds should be saved because of utilitarian reasons; and, moreover, they should be saved because of reasons unconnected with any return in dollars and cents," he wrote. "A grove of giant redwoods or sequoias should be kept just as we keep a great and beautiful cathedral."[25]

Over the decades, groups including the Sierra Club, Audubon Society, National Wetlands Research Center (Lafayette), and the organization America's Wetlands, tried to sound alarms. Scientists believed that miles of coastal wetlands could reduce hurricane storm surges by over three or four feet. Sometimes politicians listened. The Coastal Wetlands

Planning, Protection, and Restoration Act of 1990, known as the Breaux Act for its primary author, John Breaux, then a Democratic senator from Louisiana, brought ongoing funding to needed projects, but the modest appropriations were never in the billions of dollars that the dire situation called for. The Breaux Act gave $50 million annually (one-eighth of the money was provided by the states and the rest came from the federal government) for wetlands restoration projects. But the funding was done in piecemeal fashion, not project by project. It was not a comprehensive, holistic engineering effort. The historical record shows that Louisianans and certain responsible corporations, like Royal Dutch Shell, *did* care—but alas not enough. Task forces were created and blue-ribbon commissions forged. Gorgeous photography books were published showcasing shorebirds flying over the Chandeleur Islands, and baby alligators being hatched at the Rockefeller Wildlife Refuge. But for all the Save the Wetlands fanfare, these legislative gestures remained largely hollow.

Louisiana's ecosystem was dying, weakening every year with no miracle remedy in sight. Starting in the 1950s, millions of barrels of oil and trillions of cubic feet of natural gas were tapped from the Gulf of Mexico. Wetlands were lucrative. "Pork barrel projects for the oil-gas industry and Port of New Orleans always won the cash prizes," Louisiana Sierra Club representative Daryl Malek complained in late 2005. "The port lobby was unreal. There was not the same political push for wetlands. The politicians' view was: out of sight, out of mind."[26] Turbulent seawater was eating away the marshes and barrier islands—New Orleans's hurricane protection—but that was generally considered a "tomorrow problem" for the world's largest port system. Louisiana wetlands restoration was perceived as a squishy "green" issue, a flashpoint for environmental-minded citizens and tree huggers who needed to be placated even as the causc was denied sufficient funding. Massachusetts would not have tolerated Cape Cod's disappearing, nor would Virginia have let the Chesapeake Bay vanish without a fight. As Tulane University environmental law professor Oliver Houck framed the issue: "If Texas annexed fifty square miles of Louisiana's coast, we would go to war! Yet the political community doesn't even notice as it sinks away."[27]

Two million Louisiana residents, about 50 percent of the state's population, lived in coastal parishes, according to the 2000 census. Some people thought of coastal Louisiana as a watery marsh at best, a dumping zone at worst. They underestimated its importance to the business of seafood. In 2001 commercial fishing in Louisiana brought in $343 million dollars, about 27 percent of the U.S. fishing industry. Whether it was Gulf shrimp or blue crabs or Louisiana oysters, the coastal parishes of America's Wetlands were a seafood lover's nirvana (menhadan fishing alone was a major industry in the area). Louisiana's license plates may have read "Sportsman's Paradise"—but Louisiana wouldn't fit that description for long if the recreational fishing industry was decimated. And there was more to it than fish. As the Audubon Society liked to point out, more than 5 million migratory birds—70 percent of avians that migrate through the Mississippi Flyway—rested in Louisiana's coastal habitat.[28] To the frustration of modern ecologists, Theodore Roosevelt continued to be the only powerful politician to understand, or care, about the Louisiana coast. "Roosevelt was a president committed enough to his dream of conservation that he traveled to a distant group of barrier islands off the coast of Louisiana to see firsthand the affected land and wildlife," U.S. Senator Mary Landrieu said in May 2005. "Unfortunately, even with the efforts of conservation visionaries like Roosevelt, the past [century] has been one of continued coastal and wildlife losses."[29]

If America's Wetlands continued to vanish, New Orleans, which was fifty-five miles from the Gulf of Mexico, would soon be only twenty-five miles away. Given this precarious geographical reality, the city of New Orleans desperately needed the levees and flood walls provided by the U.S. Army Corps of Engineers. According to research scientists at the University of New Orleans, the city had been sinking at the rate of three feet per century, due in part to the destruction of barrier islands and coastal wetlands.[30] In addition, the ongoing effort to keep the city free of excess rainwater, through the use of massive pumps, had weakened the underlying geology, accelerating the process engineers call "subsidence." Basically subsidence was the sinking effect that had taken hold of coastal Louisiana through a number of contributing factors both natural (invasive flora-eating insect species, shifting subsurface faults), and man-made (dredging, canal digging, polluting). The net result was that instead of being slightly above sea level, as in Bienville's day, New Orleans had sunk to an average of six feet below the waterline, and as much as eleven feet in some parts of the Ninth Ward and Lakeview. In New Orleans East the elevation was eight feet below sea level.[31] The real threat to New Orleans came from Lake Pontchartrain, north of the city, and from the primordial swamps that bordered it on the east and the west. Remarkably the Mississippi wasn't much of a concern because the U.S. Army Corps of Engineers had built high-quality, sturdy levees along the river.[32] The levees along the lake, however, were known to be more unstable. You could tell that by just looking at them. They resembled the tall cement protective fences that safeguarded Los Angeles from forest fires, not from fourteen-foot storm surges. "For every 2.7 miles of marshes/swampland that disappeared, there was a corresponding increase of one foot of storm surge," Whitney Bank President and America's Wetlands spokesperson King Milling explained. "Storm surges that used to be in the neighborhood of ten to twelve feet could suddenly be eighteen to twenty feet. It's scary and real."[33]

In geographical terms, New Orleans was no more stable than a delicate saucer floating in a bowl of water. Any turbulence in the surrounding water is bound to flood the saucer. (Humorist Roy Blount Jr. preferred the analogy of the oyster, "the half-shell being the levees that keep Lake Pontchartrain and the Mississippi River from engulfing the city."[34]) In view of the city's close proximity to the Gulf of Mexico, turbulence in the surrounding area was inevitable. Most experts ranked a hurricane in New Orleans with an earthquake in California and a terrorist attack on New York as the gravest threats to the nation. To make matters worse, the escalating carbon dioxide in the Gulf of Mexico, some scientists believed, was causing the ferocity of hurricanes to increase. "In a paper published in *Nature* just a few weeks before Katrina struck, a researcher at the Massachusetts Institute of Technology reported that wind-speed measurements made by planes flying through tropical storms showed that the 'potential destructiveness' of such storms had 'increased markedly' since the nineteen-seventies," Elizabeth Kolbert wrote in *The New Yorker*, "right in line with rising sea-surface temperatures."[35]

III

Locals were certainly not ignorant of the looming danger of lethal hurricanes and storm surges jumping the levees, which formed the lip of the saucer. A steady stream of so-called doomsday treatises on the inevitability of a major hurricane laying waste to New Orleans were published in recent years. Pulitzer Prize–winning author John McPhee's 1989 book *The Control of Nature*, for example, explained in elegant detail how the Army Corps of Engineers' levees would ultimately prove futile against the wrath of Mother Nature.[36] A *Scientific American* article published in October 2001—"Drowning New Orleans," by Mark Fischetti—offered a truly grim prognosis for the city. "New Orleans is a disaster waiting to happen," Fischetti predicted. "The city lies below sea level, in a bowl bordered by levees that fend off Lake Pontchartrain to the north and the Mississippi River to the south and west. And because of a damning confluence of factors, the city is sinking further, putting it at increasing flood risk after even minor storms."[37] Then there was a landmark five-part series by John McQuaid and Mark Schleifstein published in June 2002 in the local *Times-Picayune*, titled "Washing Away." The reporters didn't mince words: the "Big One" would "turn the city and the east bank of Jefferson Parish into a lake as much as thirty feet deep, fouled with chemicals and waste from ruined septic systems, businesses, and homes. Such a flood could trap hundreds of thousands of people in buildings and in vehicles. At the same time, high winds and tornadoes would tear at everything left standing." The series also warned of the high probability of the levees breaking.[38]

Others joined the doomsday choir. In October 2004 *National Geographic* published "Gone with the Water," a frightening description of the havoc that a herculean Category 3, 4, or 5 hurricane would wreak on New Orleans. Written by Joel K. Bourne Jr., the article imagined a deadly Lake Pontchartrain storm surge thundering into New Orleans, "the car-less, the aged and infirmed, and those die-hard New Orleanians who look for any excuse to throw a party" left behind. Bourne's piece wasn't just descriptive writing—it was prophetic.[39]

In 2001, *Houston Chronicle* science writer Eric Berger attended an

American Meteorological Society meeting and first heard the startlingly dire endgame for New Orleans if a major hurricane ever struck. A trained astronomer with a master's degree in journalism, Berger started researching the gradual "sinking" of New Orleans. His frightening story appeared in the *Chronicle* on December 1, 2001; he didn't pull any punches. "In the face of an approaching storm, scientists say, the city's less-than-adequate evacuation routes would strand 250,000 people or more, and probably kill one of ten left behind as the city drowned under twenty feet of water," Berger wrote. "Thousands of refugees could land in Houston. Economically, the toll would be shattering. . . . The Big Easy might never recover."[40]

As Katrina drew a bead on New Orleans, Berger, at home in Houston, feared that his bleak sinkhole article was about to become a reality. "I knew that New Orleans dodged a bullet with Ivan in 2004," he recalled thinking as Katrina's menacing radar track appeared on his television screen. "They were lucky to survive a direct hit for as long as they did. And—although it's unfathomable—in the city proper they had established no Hurricane Command Center for police and city officials to cope with what was going to be a certain future reality. They had no communication network established in advance. Go figure."[41]

Historian John Barry's 1998 *Rising Tide* was almost mandatory reading for the college-educated class in the Crescent City.[42] In fact, it was the 2005 selection for One Book One New Orleans, an annual community-wide reading program encouraging all New Orleanians to read and discuss the same literary work. In exacting detail Barry explained the social, economic, racial, and political firestorm that accompanied the Great Mississippi Flood of 1927. *Rising Tide* was particularly tough on the U.S. Army Corps of Engineers and the so-called Carnival class, who worried more about Mardi Gras floats than flood protection. "The blame," Barry told the *Harvard University Gazette*, referring to New Orleans's vulnerability to hurricane flooding in modern times, "lies with the companies that have constructed gas and oil pipelines and shipping channels through coastal Louisiana, making the area more subject to erosion and providing less protection to the city."[43]

For decades, university professors all over Louisiana were desperately trying to get government officials to wake up to the dire threat. "Many scenarios had been run prior to Katrina years ago," Gregory Stone of Louisiana State University's Coastal Studies Institute told BBC World Service, "and with that type of storm, the scenarios should have been taken seriously. We'd been telling people all over the world that New Orleans was a nightmare waiting to happen."[44] But New Orleans had so many urban problems (e.g., inadequately funded schools, political corruption, and collapsing infrastructure) that it was near impossible to get locals to focus on barrier islands, wetlands rebuilding, and coastal erosion. Their view was, by and large, that the Louisiana coastal parishes had to lead the charge on those regional problems. It was a tall order to try to save the seventh-largest delta on earth when you weren't safe on your block. Billions would be needed for coastal erosion; what New Orleans needed was millions to pay better wages to teachers and the street cleaners who pick up rubbish after parades. Besides, people liked to believe they were safe behind their Category-3-proof levees. New Orleans was a fortress—so what if it was surrounded by a moat?

IV

The idea of categorizing hurricanes into five ascending classes originated in 1969, the brainchild of Herbert Saffir, a consulting engineer in Coral Gables, Florida, and Robert Simpson, then the director of the NHC. Their classification, known as the Saffir-Simpson Damage Potential Scale, was based on myriad criteria that indicate the likelihood of a storm surge. The dominant consideration, however, was wind speed. It grew out of the experience the two men had had with Camille, the 1969 hurricane that killed 172 people along the eighty-six-mile coast of the state of Mississippi (256 was the eventual death toll, including Alabama, Mississippi, Louisiana, and Virginia). According to the National Oceanic and Atmospheric Administration (NOAA), hurricane intensity on the Saffir-Simpson Scale ranged from Category 1, with winds of 75–94 mph ("No real damage to building structures. Damage primarily to unanchored mobile homes, shrubbery, and trees. Also, some coastal road flooding and minor pier damage."), through Category 3, with winds of 111–130 mph ("Some structural damage to small residences and utility buildings with a minor amount of curtain wall failures. Mobile homes are destroyed. Flooding near the coast destroys smaller structures with larger structures damaged by floating debris."), to the rare and monstrous Category 5, with winds in excess of 155 mph ("Complete roof failure on many residences and industrial buildings. Some complete building failures with small utility buildings blown over or away. Major damage to lower floors of all structures located less than fifteen feet above sea level and within 500 yards of the shoreline. Massive evacuation of residential areas on low ground within five to ten miles of the shoreline may be required.").[45] In recent years, hurricane ratings according to the Saffir-Simpson Scale have become the central factor in determining whether to evacuate in advance of a hurricane. Due to the scale's proven value in the United States, countries like Japan and Australia adopted it as well.[46]

New Orleans engineers learned to think in terms of the Saffir-Simpson Scale. With the protection of their system of levees and pumps, the city was considered just barely safe in the face of a Category 3 hurricane, specifically a fast-moving Category 3 storm. A slow-moving Category 3, crawling along at 7 mph or less, would be even more treacherous, drenching the coast with a greater amount of rain than a fast-moving storm. As for Category 4 or 5 hurricanes, the city was simply not prepared to withstand them. And yet they occurred with frightening regularity in the Gulf of Mexico. As the commonplace New Orleans refrain went, "It's not *if*, you know . . . it's *when*." The parlor game, come hurricane season, was to wish the cursed storm away, praying it would hit Panama City or Biloxi or Galveston, or a foreign country like Mexico or Nicaragua—anywhere but Louisiana's fragile 397 miles of shoreline.

While New Orleans residents often mused about the possibility of cataclysmic destruction, even if they were just whistling in the dark, their city remained starkly vulnerable. In 2005, it could not boast of a single shelter certified by the Red Cross. Not one. That nonprofit behemoth, based on E St. NW in Washington, D.C., refused to operate a facility in New Orleans (or Key West) because *none* of the commonly used emergency sites (schools, hospitals, gymnasiums, etc.) was much above sea level. The Red

Cross would open up shelters *outside* the flood zone, but not in the saucer. Meanwhile, New Orleanians didn't even protest as their flood-prone hometown became a dumping ground for toxic waste, the location of thirty-one Superfund sites.* The collective nonchalance in the face of such omens triggered novelist James Lee Burke to deem New Orleans "an insane asylum built on a sponge."[47] Novelists couldn't quite agree on the nature of New Orleans's problem—to wit, whether it was dysfunctional or just nonfunctional. Or as the forty-four-year-old New Orleans native Michael Lewis—author of best sellers like *Liar's Poker* and *The Next Next Thing*—wrote in the *New York Times Magazine* on October 9, 2005, "There's a fine line between stability and stagnation, and by the time I was born, New Orleans had already crossed it."[48]

Nobody in New Orleans, not even the hospitals, protested that the Red Cross had blacklisted their city. But in 2004, 270 officials from all levels of government *did* participate in a FEMA-funded, weeklong simulation of a Category 3 Hurricane striking New Orleans, a fake but very realistic storm called Hurricane Pam based on extensive computer models developed at Louisiana State University. The primary assumption of the Hurricane Pam exercise was that "Greater New Orleans is inundated with ten feet of water within the levee systems as a result of a Category 3 or higher hurricane."[49] The attendees learned that it would not be just water, in fact, but a "HAZMAT 'gumbo.'"† They heard that the total number of people left stranded in the toxic water "may approach 500,000" if residents didn't properly evacuate.[50] And they were further informed that a monstrous storm such as Pam would leave 30 million cubic yards of debris—not counting human remains—spread out over thirteen parishes in southeast Louisiana. But to most Louisiana officials, discussing Pam was akin to

* A Superfund site is a location, such as a warehouse or landfill, abandoned by previous owners that has been designated by the Environmental Protection Agency as hazardous and dangerous because of past dumping of chemical waste and other toxic materials. The Comprehensive Environmental Response, Compensation, and Liability Act (enacted in December 1980 in response to the Love Canal disaster) authorized the EPA to designate and then clean up such sites. Of 11,281 sites, 160 are in Louisiana.

† Hazmat is the apocalyptic jargon for "hazardous materials" or, more succinctly, poison.

reading *A Thousand and One Nights;* it was make-believe. "Before Katrina people looked at Pam exercises like kids do fire drills in school," Clancy Dubos, publisher of the local weekly *Gambit*, recalled. "They just weren't going to take it seriously until it happened."[51]

Having lived through Pam for a week, the 270 officials just went home. Art Jones, the head of the disaster recovery division of the Louisiana Office of Homeland Security and Emergency Preparedness, later explained that the days were well spent, though. They allowed for interaction between the emergency officials who would be working together if—that is, when—the major hurricane hit.[52] That sense of familiarity would not prove to be enough on Saturday, August 27, when the moment was upon them.

New Orleans did not possess a *realistic* hurricane plan, but it did have an evacuation plan. This document, "City of New Orleans Comprehensive Emergency Management Plan," was prepared by city officials and various consultants in 2000. A fourteen-page booklet, it devoted about a page and a half to evacuation, but backed away from substantive directives. "It must be understood that this Comprehensive Emergency Plan is an all-hazard plan," it explained. "Due to the sheer size and number of persons to be evacuated . . . specifically directed long-range planning and coordination of resources and responsibilities efforts must be undertaken." In other words, the booklet didn't have any novel answers or fresh approaches. But it did offer a series of clear-cut guidelines, which Mayor Ray Nagin seemed to ignore. After suggesting that "evacuation zones" based on probable storm flooding should be used as the basis of mass evacuation, it advised that such zones "will be developed pending further study."[53] They never were. Nothing further was done with the vague start offered by this emergency management plan. "The city of New Orleans followed virtually no aspect of its own emergency management plan in the disaster caused by Hurricane Katrina," Audrey Hudson and James G. Lakely of the *Washington Times* wrote. "New Orleans officially also failed to implement most federal guidelines."[54]

Apparently unimpressed by the emergency management plan, even though it was posted on the City Hall Web site (it was taken down shortly after Katrina), Mayor Nagin behaved in a hesitant, perplexing fashion. The plan was the collective wisdom of an entire generation of New Orleans political thinking, going back as far as those who had grappled with Hurricane Betsy in 1965. The plan instructed that when a serious hurricane approached, the city should evacuate seventy-two hours prior to the storm to give "approximately 100,000 citizens of New Orleans [who] do not have the means of personal transportation" enough time to leave. Mayor Nagin also ignored FEMA guidelines, which urged City Hall to "coordinate the use of school buses and drivers to support evacuation efforts." What neither the "Emergency Management Plan" authors nor FEMA's lawyers could have known was that Mayor Nagin didn't prioritize hurricane evacuation plans. With his city's schools floundering and the murder rate rising, he had other more immediate woes to grapple with.

V

At noon on Saturday, forty-nine-year-old Mayor C. Ray Nagin staged a press briefing at City Hall in New Orleans. Casually dressed, his shaved head shining in the media lights, Nagin strained to seem like a man in charge. A careful observer of human nature could detect, however, by his twitching neck and glazed eyes that he was already unnerved by the prospect of Katrina. A state of emergency had been declared by Louisiana's low-key governor, Kathleen Babineaux Blanco, on Friday, August 26, at 11:00 P.M., and the mayor was nervous. Reports from the NHC insisted that Katrina was growing in menace by the minute. "Although the track could change, forecasters believe Hurricane Katrina will affect New Orleans," Nagin said tepidly, scratching his trimmed goatee. "We may call for a voluntary evacuation later this afternoon or tomorrow afternoon."[55] In emergency preparation, the three levels of evacuation are voluntary, recommended, and mandatory.[56] Only the third carries real weight—and places the responsibility for evacuation on state and local government officials. A halting, soft-spoken Nagin said that he needed to talk with his lawyers about what his options were. By stopping short of making a citywide exodus mandatory, he was squandering precious time. Even as Nagin dawdled, the first Louisiana SPCA pets were en route to Houston.

As politicians go, Nagin was an energetic show horse, not a nuts-and-bolts workhorse. That very Saturday, in fact, New Orleans *Times-Picayune* columnist Chris Rose reported on the mayor's latest venture: acting. Just days before Katrina hit landfall in Florida, Nagin had made his film debut in an independently produced thriller called *Labou*. For five hours—in the thick of the tropical storm season—he hung around Gallier Hall, New Orleans's old city hall, rehearsing lines (on taxpayers' time). He had been cast to play a corrupt Louisiana mayor. "I thought I was just gonna show up, do a cameo, say my lines and get out of there," Nagin complained on the set. "And they only pay a buck fifty for this." Rose interpreted the gripe for the *Times-Picayune* readership, explaining that "a buck fifty" was "Ray Speak" for 150 dollars. New Orleans residents were already familiar with "Ray Speak," a confusing inversion of words and ideas, all gathered up in tortured syntax, typically producing a mixed message, but marketed to his constituents as candor. You might call it pandering with seemingly earnest zeal. The net effect of "Ray Speak" was to sell his inaction as a form of action. After leaving the set, a boastful Nagin called out, "Hollywood South, baby!"[57]

A native of New Orleans, Nagin had been interested in business at the start of his adult life, earning an accounting degree at what was then Tuskegee Institute and an MBA from Tulane University. He rose through the corporate ranks to become a vice president of the cable television company Cox Communications, overseeing operations in southeast Louisiana. He also made a reputation for himself as a part owner of the New Orleans Brass, the local minor-league hockey team. In 2002, at the age of forty-six, Nagin suddenly jumped into politics and entered the New Orleans mayoral race. He was an unknown candidate with no record whatsoever. Suddenly the leading Democrat in the race, he constructed his campaign around a pointedly pro-business, Republican-sounding platform. He soon became a darling of the *Times-Picayune* and the conservative business elite, an African American who was a virtual Chamber of Commerce cheerleader when it came to New Orleans's future.

Always deferential to whites, Nagin broke with the civil rights tradition of the city's black leadership. At the same time, he worked hard to distance himself from the left-leaning legacies of such previous Democratic mayors as Sidney Barthelemy, Moon Landrieu, and Marc Morial. "I took great pains to bring in every segment of the community," Nagin later said of his campaign. "I got attacked. I was called 'Ray Reagan,' and that white man in black skin stuff. I had a stigma that Ray Nagin does not care about black people."[58] He campaigned for mayor claiming that he would sell Louis Armstrong International Airport to make money and build a new city hall. (After his election, neither happened.) He spewed anticorruption jive, go ing on about purging all the "bad guys" from city government: no more cronyism under his administration; the iron heel of justice had arrived. He was going to weed out the villains. He was a reform candidate, baby.

Even as Nagin represented oil and gas millionaires, shipping magnates, blue bloods, and nouveau-riche developers, he was a buffer zone between New Orleans's blacks and whites. He marketed himself as an African American with whom upwardly mobile Texas-Louisiana whites would feel comfortable. The popular Bishop Paul Morton of the Greater St. Stephen Full Gospel Baptist Church spoke for many in the African-American community of New Orleans when he called Nagin "a white man in black skin."[59] That was harsh and unfair: Nagin was a midlevel corporate manager who believed he could charm voters and move them in the direction he wanted. Nagin's self-promotion—combined with "Ray Speak"—worked. In the 2002 mayoral election, he defeated a strong adversary, Richard Pennington. While Pennington was the chief of police from 1994 to 2002, the murder rate in New Orleans fell by 50 percent. Under Nagin's leadership it subsequently skyrocketed to ten times the national average. On Saturday, December 27, in fact, a local weekly, *Gambit,* reported that New Orleans was on pace to end 2005 with 316 murders (71 per 100,000 residents). In 2004, New Orleans was the number two murder capital of America.[60] When it came to good governance and fighting crime, the glamorous CEO mayor was, in Texas parlance, all hat, no cattle.

That Saturday, with Katrina brewing in the Gulf, Nagin grew concerned and huddled with his team of lawyers, bodyguards, developers, and sycophants. His most pressing worry? "Nagin said late Saturday that he's having his legal staff look into whether he can order a mandatory evacuation of the city," Bruce Nolan reported in the *Times-Picayune*, "a step he's been hesitant to do because of the potential liability on the part of the city for closing hotels and other businesses."[61] That hotels would be in a position to sue the city if the tourist trade were disrupted because he called for a mandatory evacuation—*that* was Nagin's worry. As a protector of businesses—particularly ones of national stature—he couldn't have that. He stalled around on Saturday, trying to verify his legal position. Meanwhile, personnel had to be marshaled, resources deployed, and plans initiated. Direct action was necessary on a dozen fronts and Nagin was hesitating like a schoolboy afraid to receive his report card. "The assumption that poor people would be trapped was met with inaction, when it should have been met with a determination to save as many as possible," editorialized the *Times-Picayune*. "The words 'mandatory evacuation' mean nothing when state and local officials won't or can't deploy the resources necessary to make the mandate stick."[62]

According to the office of the Attorney General of Louisiana, a mayor was fully authorized by the law to "direct and compel the evacuation of all or part of the population from any stricken or threatened area." Nagin's imagination, however, was sorely limited. He could envision lawsuits, but apparently not Wilma Jones, stuck at the Magnolia Development off Claiborne Avenue. Besides sending around twelve buses to poor neighborhoods in a haphazard pickup and drop-off fashion, Mayor Nagin implemented no comprehensive plan to evacuate vulnerable people. Debbie Este, for example, sat in her wheelchair in her one-story Arts Street home with her infirm sixty-eight-year-old mother and two teenage daughters. She had no idea what was coming. Nagin had no list of people with special needs like Robert Green, who awaited Katrina's fury in his Tennessee Street house in the Lower Ninth Ward with his seventy-three-year-old mother (who had Parkinson's disease), his mentally handicapped cousin, and his three granddaughters. Both Green's mother and his granddaughter, Shenae, would be dead before Katrina passed. "They keep telling me I'm gonna break down from it," he told CNN as he searched for his mother's body in October 2005. "I won't break down, and I can't break down, not until it's over."[63]

Obviously Mayor Nagin cared about these special needs residents; just not enough. He was all Booker T. Washington, pull yourself up by the bootstraps, with a touch of Hollywood showboating, for good measure.

There was little W. E. B. Du Bois in his repertoire. Furthermore, tourists didn't come to see the unemployed, of whom there were many, or the ill, the toothless, the elderly. Visitors came to gawk at nude shops along Bourbon Street and eat bread pudding at Commanders' Palace in the Garden District. Five-star hotels with spas as big as Canyon Ranch in Las Vegas—those were things Nagin cared deeply about. He had little interest in the 112,000 adult New Orleanians who didn't own cars. They were, in his mind-set, a secondary concern.[64] "We didn't do enough to help earmark the special needs people," Terry Ebbert of Homeland Security admitted. "We will next time."[65]

VI

Sometimes it was hard for City Hall or the Chamber of Commerce or the New Orleans Visitors Bureau to visualize the life of a poor person in New Orleans—someone like Tonya Brown of the Lower Ninth Ward. Brown was a single mother earning a meager wage deep-frying at a restaurant and buying used clothes at the thrift shop while praying that the Sewerage and Water Board of New Orleans wouldn't cut service to her household. Occasionally, the New Testament gave her genuine comfort. She was a proud Baptist. Most of the time, however, her love gave way to hurt. There were bad-luck days when only pain was real. Tears often swelled in her brown eyes, but she waved them off, not wanting her six kids exposed to weakness. For all of her can-do attitude, she was still and all a scared mother. Self-pity was a loser's mask and she wanted no part of it. Quiet strength, like that of Rosa Parks—that's what Brown prayed for, kneeling down in church. Sometimes, when dusk hovered over the Lower Ninth Ward as she talked to Jesus, her anxiety dissipated. She was a covenant woman. Twilight, she said, was the best time to ask and listen. Long before, she had stopped asking City Hall for anything. Too proud for welfare, she walked to work across the St. Claude Avenue Bridge most mornings, declining to pay a dollar to ride the bus. She was on a shoestring budget, just trying to make ends meet. She jokingly referred to these walks as "dead marches," daydreaming that someday she'd make it to a better life—where the humidity index was zero. When she saw the *Times-Picayune* headline on Saturday, "Katrina Ends Lull: Leaving N.O. on Edge," she knew the city would forsake the old and poor. Mayor Nagin—"Chromedome," as she preferred to call him—had long since abandoned her people. The truth was obvious: her family would have to weather Katrina out in the Lower Ninth and pray that the Industrial Canal didn't break or that the Mississippi River levee didn't top. She had confidence in the U.S. Army Corps of Engineers; she believed the levee was built to endure hurricanes. Meanwhile, there were no evacuation buses coming for them. On the eve of Katrina, thousands of the poor, like Tonya Brown, were left to their own resources and God's blessing.[66]

But it wasn't just the poor who decided to stay in the New Orleans bowl. One lifelong Louisianan who never even thought about evacuating was Jackie Clarkson. Her entire life was centered on Algiers, the West Bank community on the opposite side of the Mississippi from the French Quarter. Algiers was part of New Orleans and yet a world apart. Having grown up there in the 1950s and 1960s, Clarkson knew it well. Her Algiers was a kind of Mayberry by the river. Grandma would rock on the porch, fried chicken was served on Sunday, and the ferryboat at Algiers Point brought them to the French Quarter's great open markets. "We were very geographically removed from the rest of the city," Clarkson said, "and we're buffered by Jefferson Parish and Plaquemines Parish, so we were both New Orleanians and West Bankers."[67]

During the early Cold War years, Algiers started getting a lot of U.S. military personnel assigned to the district. Algiers Naval Base became a mainstay. Unlike the more aristocratic Garden District, Algiers opened its arms to sailors and oil executives, and as a result an entwined middle-class community developed. If you moved to Algiers, no matter where you came from, there was a 90 percent chance you wouldn't leave. It was in this friendly environment that Clarkson and her husband, Arthur, raised five daughters (including the actress Patricia Clarkson). To bring in extra money, Clarkson became a real estate broker, and did exceedingly well. She was also a Girl Scout leader, Little League coach, Sunday school teacher, PTA regular, and member of the foundation board of the University of New Orleans. Simply put, Clarkson was a combination supermom and busybody.

Then in 1989, she made the plunge into electoral politics, running for the New Orleans City Council from District C, which contained Algiers, the French Quarter, Faubourg Marigny, and Bywater. They were all neighborhoods above sea level. People in them weren't sure whether a flood in the rest of the city would be a danger or merely an inconvenience. But when they elected Clarkson every four years from 1990 to 2002, it wasn't because she was a levee expert. She had heart—that's what they admired about her. With a Lucille Ball hairstyle, conservative suits (almost always red), and a belief that all U.S. soldiers were the epitome of goodness, Clarkson was a throwback figure. A self-described Truman Democrat, she cared whether a little Bywater park had a bench or whether an Algiers basketball hoop had a net. Voters elected her over and over again because they knew when an inconvenience like a Caribbean hurricane blew into town, she would be looking out for the citizens of District C.

The same West Bank pluck imbued middle-class New Orleans hospitality workers like Kathy A. Lawes-Reed of Algiers. A beloved waitress at the J. W. Marriott Hotel on Canal Street, she sometimes sang Italian opera or Billie Holiday blues to dinner guests. She lived on her tips, along with a small private income. The hotel would need Marriott employees like Lawes-Reed to tend to the tourists stuck in town during the coming hurricane. She didn't want to let her employer down. She loved Marriott. There were 38,000 hotel rooms in New Orleans on the eve of Katrina, and somebody had to address their needs. "I didn't have any money," she explained, "and my employers need me. So I stayed. New Orleans is a tourist town and I just couldn't leave the tourists to face Katrina on their own."[68]

Tourists, indeed, helped fuel the economic engine of New Orleans. Long before New York City and Hollywood dominated entertainment in the United States, before Las Vegas took up casinos, and before San Francisco began to pride itself on its gourmet restaurants, New Orleans was the capital of all three: music, gambling, dining, and anything else that held out the promise of a good time. In *Gone with the Wind*, Rhett Butler took Scarlett to New Orleans for their honeymoon, and while that was only a novel, Margaret Mitchell wrote with pinpoint accuracy about the sensibilities of the mid-nineteenth-century South. New Orleans was just the sophisticated setting for a rich and racy couple after the Civil War to impress each other. "New Orleans," Mitchell wrote, "was such a strange, glamorous place and Scarlett enjoyed it with the headlong pleasure of a pardoned life prisoner."[69]

During the twentieth century, New Orleans remained much the same, in spirit. It did not compete with hurly-burly Chicago, as the nation's capital of practical business and entrepreneurialism. It didn't pretend to intellectual or moral standards, as did Brahmin Boston. And it wasn't sports crazy, as was St. Louis, just 1,039 miles up the Mississippi River. Only in the latter pursuit did the city advance in the last third of the century, by attracting professional teams, including the football Saints and basketball Hornets. The 1960s, when the Saints arrived in town, were a time when, not coincidentally, sports were marketed as entertainment. New Orleans on the eve of Katrina still clung to its odd, old reputation as a roguish city saturated with what *Newsweek* called "indolent charm."[70] It had some of the best and oldest restaurants in America, like Antoine's, Galatoire's, and Arnaud's, all housed in nineteenth-century buildings.[71] It was the capital of the American roistering night. Whatever New Orleans's inherent attractions, it wasn't, however, a good place to do business. In 1960 it had a population of 600,000, larger by far than Houston or Atlanta. On the eve of Katrina, New Orleans had shrunk to about 460,000, and was home to only one Fortune 500 company (Entergy). By contrast, Atlanta had sixteen, Charlotte seven, and Memphis four. Farther up the Mississippi, Minneapolis–St. Paul—the so-called Twin Cities—had eighteen Fortune 500 companies between them.

New Orleans simply didn't draw business investment the way other American cities did and it didn't spawn entrepreneurs, either. The city—the anti-Seattle—missed the high-tech boom of the 1980s and 1990s entirely. In business, New Orleans rested on its natural livelihood in shipping lines, particularly Central Gulf, Lykes, and Delta Steamship. Driving along the Mississippi just above New Orleans, you couldn't help but see the humongous Avondale Shipyard, an industrial beehive of cranes and hard hats. After World War II, when engineers devised a way to build offshore oil rigs, New Orleans became a center for the moving and refining of oil. It was a world of cables, welding, barges, and pipelines that turned parts of the city into wild, roughneck territory. Through the years, there was nothing upstanding or elite about the general standard of business in New Orleans. Except for the addition of legal gambling in the city during the 1980s, the city was unprogressive and, in some odd way, seemed to like it that way. The old families with their old money hung on to their old power. Tourists were always welcomed and encouraged to leave.

Up until the Second World War, New Orleans was content within its own borders. That is, suburbs were, by and large, late in coming to the Crescent City; much of the surrounding terrain was too marshy to support tract housing developments. In the 1950s, however, much acreage in St. Bernard Parish, to the east, was drained and Levittown-like developments sprang up in Meraux, Chalmette, and Arabi. The aluminum siding and sheet metal industries flourished. Also, in the 1950s, the long-standing dream of building a bridge across the gaping width of Lake Pontchartrain became a reality. With the completion of the Lake Ponchartrain Causeway in 1956—the longest bridge in the world, at twenty-four miles—Jefferson Parish, just to the west of New Orleans, was connected with the north shore. The days of ferry transportation were largely over. Suburban communities sprang up in and around Covington and Mandeville on the far side of the lake, which had once been a piney summer refuge for New Orleans families. With the bridge, though, the largest growth of all on the north shore occurred in Slidell, a city of many developments, built around recreation and some industry. Another factor, as always in New Orleans, was race. Black empowerment growing out of the civil rights movement of the 1960s unfortunately led to massive "white flight" from New Orleans, causing suburbs like Slidell, Covington, and Metairie to burgeon. In the late 1960s, the area on the eastern side of the Industrial Canal, known as New Orleans East—touted as a "city inside a city"—was the place for upwardly mobile whites to be. With its fine walking path along the Lake Pontchartrain levee and new state-of-the-art shopping mall, New Orleans East was marketed by realtors as cookie-cutter modernity at its white-bread best.

By and large, though, New Orleans retained its upper-class, stately neighborhoods of the Garden District and Uptown. Both were located along the crescent that the Mississippi formed around the original settlement. The

French Quarter, which was in the very crook of the crescent, was also home to many of the city's wealthy elite. While mansions in other cities were often abandoned to Elks Lodges or medical clinics, in New Orleans they remained in private hands, the buildings well-maintained and the gardens fresh. The vibrant upper-class neighborhoods reflected the mood of the gentry inside. The blue bloods believed in a structured social life in New Orleans, where debutante balls were anything but passé, and membership in a Mardi Gras krewe was considered serious civic business. Being a member of the Boston Club was a matter of true prestige in New Orleans. Neither as independent as New York society, nor as determinedly casual as that of Los Angeles, New Orleans held out the prospect of participation in soirees and balls that few other cities could offer in the new millennium. It was fun to be part of the conscripted social swirl, as every Sunday the *Times-Picayune* ran a photography section featuring socialites smiling at fund-raisers or champagne parties.

Tourists constituted a second social tier in New Orleans. For a city of only about 460,000 to host 10 million visitors each year is bound to give it another side, however transitory.[72] The tourists were hosted, in a sense, by the 84,000 people employed in the hotels, restaurants, and convention attractions. On a monetary basis, they contributed $5 billion to the city's revenues. More than that, though, they gave the city a swagger—and a reason to accentuate its traditional tendency toward hedonism. New Orleans was not a spruced-up, clean-cut city, but one in which the saloons stayed open all night, drive-in daiquiri shops were pervasive, gay culture was embraced, and raunchy strip joints were located in the very center of town. In the New Orleans that tourism helped to perpetuate, drunken debauchery was just part of the bohemian-anarchistic-libertarian mix of an international port city that tourists were urged to embrace, in one pleasurable pursuit or another. It was an "adult entertainment" mecca. Eccentricity was embraced as a virtue. Tourists also helped give a reason for the French Quarter (or Vieux Carré) to maintain its beautiful nineteenth-century buildings, clad in wrought iron. The new construction in the adjacent business district gave hardly a nod to the old style of French influence. Quite the contrary, the central business district, home to monoliths like Entergy's headquarters and the Shell building, was lined with the same type of postmodern skyscrapers found in Houston or Atlanta. Around the rest of the city, however, enough vintage buildings were left untouched, usually by economic stagnation rather than historical appreciation, that New Orleans had fast become a favorite location for Hollywood movies. Filmmakers could find practically any sort of backdrop in New Orleans, from Spanish colonial mansions and Creole cottages to American Greek revival, art deco European, and postmodern ranches. And then there was Mardi Gras, the annual cash cow: in 2003, for example, the city spent $4.7 million on parades and got a return of $21.6 million.

Jazz was born in New Orleans and was cultivated in the Storyville neighborhood, where more than two thousand prostitutes worked in legal brothels from 1897 to 1917. The lively brothels served as the first jazz workshops for young, soon-to-be legends such as Jelly Roll Morton, Tony Jackson, Clarence Williams, and Joe "King" Oliver. Louis Armstrong, who grew up in what was then called the Back o' Town neighborhood, was exposed to these lively halls and to the rough and violent underside of the city. Armstrong captured the character of the city in his impassioned improvisational trumpeting and his gentle, deep melodic voice. With songs like "West End Blues," Armstrong was an icon by the 1940s and served as an informal ambassador of New Orleans all over the globe. The tradition has lived on: into the twenty-first century, live jazz was performed daily and nightly in the Crescent City, on street corners and in music halls—from the old Dixieland style of the Preservation Hall Jazz Band to the hiphop-infused jams of the Rebirth Brass Band, Dirty Dozen Brass Band, and the Soul Rebels.

The jazz personality of New Orleans, as distilled by the tourist segment of the city, was also the backdrop in a figurative sense for great literature. It was an irresistible setting for a story, with its crisscrossing impulses of age-old gentility and ribald nightlife. Most Americans were familiar with Stanley Kowalski screaming "Stella!" on Elysian Fields Avenue and enforcing the Napoleonic Code in Tennessee Williams's steamy *A Streetcar Named Desire*. Then there was Ignatius Reilly pushing his hot dog cart around the French Quarter and epitomizing the eccentric spirit of the city in John Kennedy Toole's Pulitzer Prize–winning novel *A Confederacy of Dunces*. A favorite among the elite was Walker Percy's lazy, lilting narrative of the Uptown life of Binx Bolling in *The Moviegoer*. In addition to hosting these unforgettable fictional characters, the real-life Crescent City had bred and inspired countless writers: Nobel Prize winner William Faulkner wrote his first novel, *Soldier's Pay*, while living in an apartment on Pirate's Alley (and according to local legend, he first decided to become a writer when, on a business trip to the city, he had a chance afternoon cocktail with *Winesburg, Ohio* author Sherwood Anderson in the Napoleon House bar). Poet and novelist Charles Bukowski wrote *Crucifix in a Deathhand* while living in New Orleans, and today you can read his graffiti in the sidewalk pavement of Faubourg Marigny.

More recently, Anne Rice, who lived in the Garden District, set her vampire tales, *Interview with the Vampire* and *The Queen of the Damned* among them, in the gothic city. Her First Street house was a popular tourist destination. Rice embraced the belief that time seemed to pass more slowly in New Orleans, and that people laughed more often at absurdity. "I was born in the city and lived there for many years," wrote Rice in a September 2005 *New York Times* op-ed piece. "It shaped who and what I am. Never have I experienced a place where people knew more about love, about family, about loyalty, and about getting along than the people of New Orleans. It is perhaps their very gentleness that gives them their endurance."[73]

The third segment of New Orleans was not necessarily composed of residents. In fact, the people who dominated it might well live on the other side of the globe. The shipping industry was generally hidden from the tourist areas and certainly did not impinge on the mansion districts, yet little money was spent on capital improvements to the city without this industry's tacit permission. It controlled the boards that controlled, however loosely, the future of New Orleans. The reason was simple, if shortsighted. Without shipping and its younger sibling, oil extraction, New Orleans would lose its economic base. Tourism may have brought in $5 billion, but ocean shipping alone contributed $13 billion. The port of New Orleans had three harboring areas. The smallest of them lay on the Industrial Canal, which leads from Lake Pontchartrain to the Mississippi River. Next, there were cruise ship docks along the Mississippi near the Convention Center. By far, though, the main portion of the port was located just a little way upriver at the Nashville Avenue wharf. In all, 107,000 people were employed in port activities.[74] That figure didn't even take into account the six major railroads and more than thirty-eight national trucking companies that used New Orleans as a regional hub. For all the activity, though, the city was more of a colony than a capital.

For a while after World War II, it looked as though America's domestic oil companies might make New Orleans their capital, but in the 1990s, those companies with a large white-collar presence in the city began moving to Texas or Oklahoma. In 2003, the number of oil-industry workers was 9,125, down about 45 percent from 1990. When ExxonMobil announced in 2003 that it was pulling the last of its operations out of New Orleans, Mayor Ray Nagin made a grandstand announcement that City Hall would give a $25,000 bonus to any engineer who would forsake the move and stay on with another New Orleans company. Over a year later, those few who took him up on the offer were still waiting for the check. One, Gary Wilson, called it "really just an empty political promise."[75] The relocation of Big Oil, however, put shipping in an even stronger position in New Orleans. While it had setbacks, including losing the position of number one in coffee imports to New York City in 2004, the port was secure as long as the American Midwest, served by the inexpensive barge traffic of the Mississippi, continued to require products from overseas, and to export them as well.

The city that Big Shipping wanted New Orleans to be was, however, not the same one that most of the residents would have wanted—if they had the choice. With power on the side of shipping, federal moneys earmarked for levee improvement and other public safety projects often turned into harbor upgrades, canal dredging, and other initiatives that actually improved the chances for a hurricane to do even more damage.

The fourth segment of the city of New Orleans—after the mansion districts, the tourism segment, and the shipping industry—was the population of poor people. New Orleans had a higher proportion of people living below the poverty line (27.9 percent in 1999) than similar-sized cities like Tucson, Arizona (18.4 percent), Kansas City, Missouri (14.3 percent), or Portland, Oregon (13.1 percent).[76] In New Orleans, blue-blood tradition and big business had pushed the poor off to the side and kept them hidden.

Once whites left for the suburbs, public schools in New Orleans became an abomination.

Intertwined with the city's poverty rate was its racial composition. African Americans constituted 67.3 percent of the population, whites 28.1 percent. Those two groups alone accounted for 95.4 percent of the population, leaving a small number of Hispanics, particularly Hondurans, and Asians, especially Vietnamese. New Orleans was much more distinctly divided along racial lines than most cities.

Another characteristic of the New Orleans population worth noting on the eve of Katrina was the proportion of the elderly. To compare cities of similar size, 11.7 percent of the New Orleans population was sixty-five or older, as opposed to Austin at 6.7 or Boston at 10.4.[77] New Orleans poverty figures had to be juxtaposed against its proportion of elderly, which was above average, and its noticeably large percentage of African Americans. Many social historians have traced the lively cultural personality of New Orleans and its jazz music to that disproportionately large percentage of the city's population, but in the face of a natural disaster, the world of the poor in New Orleans, particularly that of the poor black, and the poor elderly, couldn't be romanticized in a Harry Connick Jr. croon or a Kermit Ruffins horn riff. For these overlapping groups, New Orleans was built on a kind of isolation from the mainstream. The core of the city, built geopolitically in concentric circles around the mansion residents, the tourism world, and the shipping industry, rarely took the poverty-stricken neighborhoods into account. Poor neighborhoods occupied the majority of New Orleans's 181 square miles, stretching mile after mile with dilapidated houses and people on the porches, people with nowhere to go. Bifurcated, if not officially segregated, New Orleans gave the impression of a city that didn't even know itself. In metaphor, the city was a legendary beauty, but one that had refused to look in the mirror for a long, long time. Selling the world on the historic stage set that was so much of picturesque New Orleans, the city seemed not to care about its other decaying side. Citizens enjoyed being dubbed "the Capital of the Caribbean" and "the City That Time Forgot." But such a laissez-faire attitude toward civic improvement was loaded with unhappy consequences.

After nearly three hundred years, New Orleans had arranged its existence around those four segments, yet there was something hauntingly temporary about the arrangement. Always determined to squeeze the last drop out of midnight, New Orleans never looked to the future, except as a means of keeping things, right or wrong, the same. Something had to make New Orleans look hard in the mirror. Unfortunately, the one power on earth that could make it rise to see itself clearly was coming on fast, in the form of a Category 3 hurricane named Katrina.

VII

On Saturday afternoon, Mayor Nagin endangered the welfare of the poor and elderly as well as the tourists—and in the end, the city—by holding legal discussions about the impact of an evacuation on the hotel trade. At best, the mayor was being extremely shortsighted, exhibiting childlike leadership in an adult hour. He was thinking about the hotel owners' profits, not the hotel guests' lives. If New Orleans flooded, they would all be stuck in Radissons and Days Inns and Sheratons, with no electricity and no escape route, their lives in grave danger. Instead of worrying about having egg on his face if Hurricane Katrina *didn't* slam New Orleans, he should have put the long-term interests of the city's $5 billion tourism industry ahead of his short-term (and unfounded) fear of lawsuits for lost business. He had the power. There was time—not much, but enough to ferry all those out-of-towners right back out of town. Instead, he gambled with their lives, too. "The only thing I can say to the stranded tourists is I hope they have a hotel room and it's at least on the third floor up," Nagin said. "Unfortunately, unless they can rent a car to get out of town, which I doubt they can at this point, they're probably in the position of riding out the storm."[78]

On Saturday, in other words, Nagin was in the process of squandering what *New York Times* columnist Maureen Dowd in a different context called "the golden hour," a term that originated in combat medicine to describe the time when "acting fast may save those in jeopardy."[79] The mayor's irresponsible attitude toward the hurricane stood in sharp contrast to the ghastly reports that many people on the Gulf Coast were seeing on their televisions. The Weather Channel, while it may be prone to a certain hysteria at the first sign of a sprinkle in South Carolina or a few gusts in Montana, was rightly forceful on the subject of Katrina, devoting most of its coverage on Saturday to plotting the storm and repeating advisories from the NHC. Early on, when Katrina had yet to grow from a tropical storm to a hurricane, longtime weather photographer Jim Reed wrote to his TV network clients with an educated opinion: "Katrina may strike the southeast coast of Florida as a Cat 1, but we're becoming increasingly concerned she may become a major hurricane once in the Gulf."[80] Outsiders seemed to see more clearly than City Hall the danger facing metropolitan New Orleans. People who watched WDSU-TV news or listened to WWL radio, bypassing the hesitant drone from the mayor, were the ones who packed their cars and fled to higher ground. Without question Nagin should have ordered a mandatory evacuation twenty-four hours earlier than he did. "I'm not sure Nagin learned a whole lot from Ivan or Pam," Clancy Dubos recalled. "You would think that he would have but he just learned nothing."[81]

For example, Julie Silvers, an abstract painter from the suburb of Old Metairie, was uncertain what to do, hearing so many mixed messages from City Hall. Her husband was in Rhode Island on a business trip, so she bore the sole responsibility for her eleven-year-old daughter, as well as their black Labrador retriever. "I thought the storm was headed toward the Florida panhandle," Silvers recalled. "But I called my friend Lawrence Chehardy, assessor for Jefferson Parish. Unlike Nagin, he was direct. 'Get out of town now,' he told me. After hearing his voice, I loaded up my Infiniti, got on the I-10, and headed to Houston."[82]

At New Orleans's City Hall on Saturday, Mayor Nagin and Police Superintendent Eddie Compass—close friends since boyhood—presided over a closed-door briefing on the hurricane. Jasmine Haralson, chief of staff for a city councilman, was at the meeting. She later said that those present were not at all worried about the impending storm, believing it would veer off before it reached New Orleans. She described the meeting, overall, as "routine."[83] With Katrina edging closer by the hour, she could not have chosen a more harrowing word.

Chapter Two

SHOUTS AND WHISPERS

> Sleep at noon, window blind rattle and bang. Pay no mind. Door go jump like somebody coming: let him come. Tin roof drumming: drum away—she's drummed before.
>
> —Archibald MacLeish, "Hurricane"

I

IN WASHINGTON, D.C., MICHAEL D. Brown, the director of the Federal Emergency Management Agency, received a briefing on Saturday from the National Hurricane Center on the severity of Hurricane Katrina and the likelihood that it would indeed make a direct hit on New Orleans. Like Nagin, Brown responded by letting the day pass. He didn't send emergency-response management teams to the region, normally a reflex action for a FEMA director in the face of potential problems. He didn't send hundreds of buses to the periphery of the Gulf Coast, within easy post-storm striking distance. He sent two public affairs officials and waited to see what would happen. "When FEMA finally did show up, everybody was angry because all they had was a Web site and a flyer," Senator Mark Pryor (D-Arkansas) told the Select Bipartisan Committee to Investigate the Preparation for and Response to Hurricane Katrina. "They didn't have any real resources that they could give."

Because the advance plans in Louisiana were nebulous, where they existed at all, hour after hour on Saturday evening was devoted to telephone calls between officials, none of whom seemed to be moving at the same speed. Governor Blanco stepped up to try to advance the preparations. She had proclaimed a state of emergency on the statewide level the day before.[1] On Saturday, she wrote to President George W. Bush, requesting that he declare a federal state of emergency in southeastern Louisiana.[2] The letter was a formality, written according to a prescribed text by which governors exercised emergency provisions in the federal-state relationship. Blanco even neglected to omit prompts suggested by the federal government for such letters. On the second page, Blanco's letter ran, "I request Direct Federal assistance for work and services to save lives and protect property. (a) List any reasons State and local government cannot perform or contract for performance (if applicable). (b) Specify the type of assistance requested." On the latter subject, Blanco filled in: "I am specifically requesting emergency protective measures, direct Federal Assistance, Individual and Household Program (IHP) assistance, Special Needs Program assistance, and debris removal."[3]

Blanco's requests were as boilerplate as the rest of her letter. The Individual and Household Program and Special Needs assistance both referred to FEMA programs that provided grant money to those displaced by a disaster. It was right that she included them in her pro forma list. Both, however, were to be activated well after Katrina passed. So was debris removal. A suddenly overwhelmed Blanco failed to indicate that the region needed federal help with transportation in advance of the storm, and rescue boats immediately thereafter. She failed to fully abide by Fleet Admiral Chester W. Nimitz's famous 1945 warning to future leaders on the grave perils of hesitation: "Preoccupation with the job at hand, or a desire not to disturb the skipper, should never result in disregard of a rapidly falling barometer."[4]

If Blanco's message to Bush had been an emphatic letter or frantic telephone call, and not merely a legal form—if it had actually communicated what wasn't happening in Louisiana (i.e., evacuation)—various U.S. government agencies might have mobilized more quickly. Just as New Orleans wasn't properly communicating with Baton Rouge, Baton Rouge wasn't properly communicating to Washington, D.C. There was a chain of failures. "The federal government does not have the authority to intervene in a state emergency without the request of a governor," Bob Williams, a Washington State legislator from the district most devastated by the eruption of Mount St. Helens in 1980, wrote in the *Wall Street Journal*, helping readers understand post-Katrina relief. "President Bush declared an emergency prior to Katrina hitting New Orleans, so the only action needed for Federal assistance was for Governor Blanco to request the specific type of assistance she needed. She failed to send a timely request for specific aid. In addition, unlike the governors of New York, Oklahoma, and California in past disasters, Governor Blanco failed to take charge of the situation and ensure that the state emergency operation facility was in constant contact with Mayor Nagin and FEMA."[5] Blanco did send a request on Saturday, two days too late. Besides late timing, it was not much of a letter and not much of a list.

President Bush, who was vacationing at his 1,583-acre ranch in Crawford, Texas, responded in turn to the governor's form letter. In a legally correct fashion, he complied with her request for federal assistance, authorizing FEMA "to coordinate all disaster relief efforts which have the purpose of alleviating the hardship and suffering caused by the emergency on the local population."[6] Unfortunately, FEMA Director Brown wasn't entirely convinced of the urgency. After receiving notification of the president's action, he released a statement that didn't even mention the importance of evacuation for Gulf Coast residents. "There's still time to take action now," his Saturday afternoon statement read, "but you must be prepared and take shelter and other emergency precautions immediately."[7]

Governor Blanco wasn't as passive as the hours went by. She attended local news conferences in both Orleans and Jefferson parishes that Saturday, encouraged New Orleanians to go door-to-door to persuade neighbors to flee, and held a conference call with Louisiana officials (including sixty-five legislators) in the coastal parishes trying to coordinate last-minute programs. "There was certainly a sense of urgency about the situation," Blanco's communications officer, Bob Mann, explained. "We knew this was perhaps the Big One. This was an urgent situation. I think she communicated that pretty well."[8]

One person who really catapulted Governor Blanco into action mode that Saturday was Cedric Richmond, the president of the black caucus in the Louisiana legislature. He spent the entire weekend telling everybody in New Orleans East, part of the Ninth Ward, to "get the hell out." Only thirty-one years old, he had grown up in NOE, throwing rocks into Lake Ponchartrain as a kid, and later hanging out at the video arcade at Lake Forest Plaza. Richmond was a cautious lawyer and workaholic; his great indulgence was eating baby back ribs dripping in fat at the City Club once a week. On Saturday morning he had attended a Little League game in Gorretti Playground, with about eight or nine hundred people in attendance. "It was incredible," Richmond recalled. "Because the mayor's warning was so soft, nobody was taking Katrina seriously. Baseball. That's what they were up to. So that night I went from barroom to barroom saying, 'Y'all need to go.' "[9]

When Richmond told Blanco that afternoon about the blasé attitude at the ball game, the governor grew alarmed. Telephoning her assistant chief of staff, Johnny Anderson, she requested that all African-American ministers in below-sea-level areas dedicate their Sunday sermons to the need to evacuate at once. They would be called "pray and pack" sessions.[10] "She really tried to help," Richmond recalled. "But Nagin just ignored everything. He should have called a mandatory evacuation earlier; the governor was having to do his job."[11]

II

To those driving around New Orleans that afternoon, the sky pale and sunless, it was clear that the business community was taking Katrina seriously. All seven of the city's Starbucks coffeehouses closed early. The massive Wal-Mart on Tchoupitoulas Street locked its doors. Gas stations started shutting down their pumps. ATM machines were empty. Minimarkets sold out of Spam and Planters peanuts—survival snacks. The Audubon Zoo began safeguarding gorillas and bears. The aquarium exterminated its piranhas, worried that if they got loose they'd breed in the Mississippi. The Whitney Bank had not only closed, it had evacuated computers and files to Chicago. Delta Air Lines, in a wrongheaded corporate bungle, canceled all flights into or out of Louis Armstrong International Airport as of 1 P.M., leaving hundreds stranded in New Orleans (by contrast, Continental Airlines evacuated people up until the last possible minute on Sunday). Production was suspended on several film projects, including *The Last Time* with Michael Keaton, *The Reaping* with Hilary Swank, and *Vampire Bats* starring Lucy Lawless. The Hollywood stars and crews left town.[12]

Tulane University was holding its orientation weekend, when incoming freshmen are squired around campus and Mom and Dad get to see just what a $32,000 tuition check meant. Normally, it is an exciting event for everyone. The approach of Katrina, however, forced Tulane President Scott Cowen to make a wrenching decision. Even though it wasn't good public relations at the time, he did the right thing and officially closed the campus at 5 P.M. The school encouraged all students, parents, staff members, and faculty to leave the city for safety. Most other colleges in the vicinity—including Xavier, Loyola, and the University of New Orleans—did the same. "We closed Saturday so our people could board up and get out of town fast," Nick Mueller, director of the National D-Day Museum, recalled. "Having lived through Camille, I knew Katrina was going to be an ordeal."[13]

Most of New Orleans's political bigwigs, both past and present, congregated that Saturday in the Lawless Memorial Chapel at Dillard University for the funeral of Clarence Barney Jr., the longtime leader of the local Urban League. As the Associated Press reported, the funeral was a who's who of Louisiana politicians. The Landrieu clan was there in force: Senator Mary Landrieu, Lieutenant Governor Mitch Landrieu, and the family patriarch and former New Orleans mayor Moon Landrieu. Those in attendance couldn't help but wonder what Ray Nagin was doing at the Dillard chapel when the poor and elderly needed to be evacuated out of the bowl. They thought, "Ray has it under control," or else "he wouldn't be hanging around."[14]

Hanging over the funeral, like a dark shroud, was the specter of the Big One. While eulogies paid homage to Barney's twenty-five years as the executive director of the Urban League in Greater New Orleans, virtually everybody in attendance was distracted by the storm. Everyone knew the "bowl" analogy. Was New Orleans going to fill up? Was the Great Deluge just around the river bend? Even as Bible passages from Luke and Isaiah were being read in the chapel, mourners could hear the sound of plywood being hammered over building windows and traffic helicopters flying overhead. Amid the prayers, trepidation was the collective sentiment; Nagin gave out handshakes and hugs, seemingly in a calm and carefree mood, until he made an early exit to get back to City Hall. "In a surreal way it seemed like almost a funeral for the city, or at least an era in the city," Jacques Morial, the brother of the former mayor, recalled. "Another interesting thing about the funeral was everybody was on edge, because something else was in the air. Usually, after a funeral service, people hang around and mingle and visit for a long time. That really wasn't the case on that Saturday. After three hours of services, people bee-lined it out of town right afterward."[15]

During the service, a staffer came up to the pew where Lieutenant Governor Landrieu sat and whispered in his ear that Governor Blanco and Mayor Nagin needed him at a City Hall meeting. "I didn't go," Landrieu recalled, "because I didn't want to leave the funeral early." After the funeral at 2:30 P.M. he immediately drove home to Octavia Street, in the Uptown neighborhood, and prepared to evacuate his five children. "My wife, Cheryl, and I had a discussion about when we should be leaving," he recalled. "I wanted to leave soon. She didn't want to leave, so we compromised and got up Sunday at 6:30 A.M. and drove to Baton Rouge. Dropped my kids off and then went to the EOC."

The Emergency Operations Center, built in 2002, featured the latest radio, computer, and Web-based communication systems for emergency management. Once Governor Blanco activated the EOC, at 7:30 A.M. Saturday, it was where state, parish, and city emergency directors met with military and FEMA officials, along with the governor's staff and scores of others. That Saturday afternoon Mitch Landrieu went to the EOC and began monitoring the storm. He was particularly worried about a Katrina surge damaging Plaquemines and St. Bernard parishes. The high-tech "command table" was in an underground overview room, which resembled NASA's mission control during the Apollo moon shots of the late 1960s. A large electronic map of Louisiana was on a ten-foot wall. Just studying the map while the NHC was telephoning in cyclonic data was disconcerting to Landrieu. You didn't have to be a meteorologist to recognize that 145 mph winds blowing water up the various rivers, canals, and waterways of Louisiana was a recipe for disaster. Wisely, Landrieu, without informing the understandably busy Governor Blanco, established pre-Katrina cell phone contact with government and emergency officials in most of the coastal parishes. His message was essentially twofold: keep evacuating the residents and then let him know where the officials were going to hunker down. If Katrina hit hard, and communications vanished, at least Landrieu would know exactly where the Coast Guard, state police, and Louisiana National Guard were congregated. They would be the state's first responders. He could get frank assessments from them on where further evacuations via helicopter and boat needed to take place. "Now I'm in a bit of a difficult situation because my role in the EOC is to be passive and to pay attention and to basically just watch," Landrieu said. "It's Governor Blanco's show and the governor as the commander in chief dictates what everybody is supposed to do, at the advice and consent of [Louisiana National Guard General Bennet C.] Landreneau and [Colonel Henry] Whitehorn and her cabinet secretaries and other people who are basically in charge. So basically there was no role for the lieutenant governor at EOC except educate yourself before the storm. That's what I did."

City Councilman Oliver Thomas, a hulking, soft-spoken, lighthearted Creole with a face full of freckles, had also made an appearance at Barney's funeral, but it was a brief one. With Katrina coming, he was on a mission to evacuate his boyhood stomping grounds, the Lower Ninth Ward, a tight-knit largely African-American community nestled along the Mississippi River across the St. Claude Avenue Bridge on the way to St. Bernard Parish. Growing up there in the 1960s, Thomas had the river for a backyard, whether he was sprint-racing along the levee, poling for the fattest catfish, or hunting for wild pigs behind the Florida Avenue Bridge. Because whites had dynamited the levees back in 1927, causing the Lower Ninth Ward to flood, African Americans in the two-square-mile neighborhood didn't trust government authority to take care of them. "We had trouble with law enforcement," Thomas recalled. "Even when we were little kids, the wrong deputy would show up and shoot over our heads and run us back just for sport."[16]

Basketball was Thomas's ticket out of the Lower Ninth. After graduating from Clark High School, he was recruited as a forward to the University of Wisconsin-Parkside and given a scholarship. Uncomfortable in Wisconsin's cold climate, Thomas transferred to the College of Santa Fe in New Mexico. Surrounded by the Sangre de Cristo Mountains and finding art galleries even down gravel roads, Thomas flourished. He got involved in drama and studying Native American painters such as R.C. Gorman. Watching the famous dramatization of Erskine Caldwell's *Tobacco Road* was to him a revelatory experience, poverty transformed into art. "I became an artsy person," Thomas explained. "Native American art helped me understand my plight; they were always expressing their plight and victories on canvas. I said that I was *never* moving back to New Orleans. To me New Orleans was a place you could socialize but not advance."

In 1985, after living in Los Angeles, New York, and Boston, Thomas did come back. His mother was ill and he wanted to take care of her. Quick on his feet, he got a job at Hughes Aircraft, dabbled in sales, and eventually found his way working as an assistant to veteran City Councilman Jim Singleton. "Jim was a real hard-edged military person at the time," Thomas recalled. "A lot of community issues, issues for women, gays, and lesbians—he was hard on those issues. And I was always of the opinion that everyone mattered. Even if you don't agree with them, they're human beings. So I think I helped Jim be more tolerant, to understand people."

Entering politics, Thomas became a spokesperson for the underdog, New Orleans's tireless proponent of affordable housing for the poor and a lobbyist for all things advantageous to the Lower Ninth. Urban planning became his bailiwick and he spearheaded an effort to refurbish historic homes in the Lower Garden District, transforming the neighborhood from skid row to middle-class prosperity.

But it was the Lower Ninth that had Thomas worried those last days of August. It was New Orleans's orphan neighborhood, separated geographically from the rest of the city by the Industrial Canal. Historically, the Lower

Ninth was grossly neglected. When the rest of New Orleans had running water and sewers, it had open drainage canals. When every other ward had paved streets, it had dirt roads. After Katrina, *Newsweek* did a story delineating the Lower Ninth's chronic urban problems, deeming the area fifteen times more crime-ridden than New York City.[17] Suspicion of the "white power structure" was part of the neighborhood's DNA, and in the Family Tree Restaurant on St. Claude Avenue on the eve of Katrina, Lower Ninthers were talking about how rich honkies were going to dynamite the levees (a repeat of 1927) and turn Reynes Street into a river. Ironically, even though the neighborhood was troubled, 60 percent of the residents were home owners, compared with 46 percent in the rest of New Orleans. "Washington D.C., Baton Rouge, the city leaders, they just don't have any idea of the history of that all-black community," Thomas said. "They've been pissed on before. They're used to it. They survived it. And that little property down there, it may not be much but it's theirs. People took pride in owning those little shacks."

Realizing the inherent stubbornness of the Lower Ninth, Thomas spent Saturday going door-to-door like some college-age canvasser, telling everybody to evacuate. He wasn't just playing politician. Thomas himself had been a hurricane survivor. Back in 1965, when Betsy came to town, Oliver's parents had refused to evacuate, and lived to regret their recalcitrance. "I told everybody that as a child, the scariest thing in my life was Hurricane Betsy," Thomas said. "You listen to the wind coming through the cracks in your house, the breaking windows, the howling sound." Betsy flooded out the Thomas home; they were stranded, in fact, for a day on their rooftop and had to be saved by the Coast Guard. Forty years later, as a city councilman, Thomas was in a position to make sure Lower Ninth folks like his parents, people afraid to leave the one thing they owned, dropped their antagonism toward municipal warnings.

At one juncture on Saturday, while Thomas was on Claiborne Avenue, he spotted Frank Watson, an old friend, walking down the sidewalk. "Where are you going?" Thomas shouted at him. "I'm goin' home," Watson replied. "Why are you still here?" Thomas shot back. A smile broke out on Watson's face and he shook his head in disbelief; his childhood buddy was acting like an old grandma worrywart. "Oliver, come on now," he said. "I ain't goin' nowhere. Y'all always talkin' about leaving and every time we leave and just turn around and come right back."

Such stubbornness was typical of everybody Thomas knew in the Lower Ninth Ward; they were like mules with no sleep. "Ah, cut it out," Thomas said. "Come on now, Frank, I don't feel good about this one."

Watson never evacuated, and his conversation with Councilman Thomas was the last he was known to have. He was never found after the storm.

III

While the worlds of business, academics, and politics were responding to the latest news, some were strangely isolated from it. Louisiana, and especially New Orleans, like much of America, was divided by class, race, and neighborhood. The rich, for the most part white, were living in Lakeview, Mid-City, Uptown, the Garden District, the Warehouse District, and in the French Quarter. The poor were mostly African American, living predominantly in New Orleans East, the Lower Ninth Ward, Central City, Hollygrove, and Tremé. But there was crossover in all neighborhoods. At least, that is the way it looked at first sight. The truth was far more complex. There were many varieties of African Americans, some of whom were the light-skinned Creoles, who had been the political leaders of the Seventh, Eighth, and Ninth Wards for decades. The poorest New Orleanians were country folk from the delta, who had been coming to the city looking for jobs since the woeful days of the Great Depression. They were the people of the housing projects; by and large, they didn't trust whites or cops.

The prisonlike projects of New Orleans were beehives for drug trafficking. Hoodlums armed with AK-47s could make up to $50,000 a week dealing crack cocaine. (The marijuana trade had moved primarily out to the white suburbs.) The drugs would flow into New Orleans via I-10 and I-12. The projects became a virtual farmers' market of drugs. The NOPD often turned a blind eye to the "rock dealers," making sure they received a cash cut for their enabling services. The drug suppliers were sometimes Iranians, called Talibans by local African Americans. In some cases, the suppliers were Vietnamese, as were those at the convenience store beside the St. Thomas project, where crack cocaine was sold in bubble gum wrappers. All over the projects five- or ten-dollar bags of heroin were readily available. The rivalry for control of the illegal drug industry turned New Orleans's housing projects into killing fields. Every week the *Times-Picayune* would list the names of the dead, casualties in urban gang warfare, with the phrase "gunshot wound" closing the cases on autopsy reports.

Even outside of the projects, there was an imprisoned quality to life for poor blacks. Housing was relatively inexpensive in New Orleans, with an average monthly rent of $488, but it tended to be flimsy wood-frame construction that would be considered substandard in other cities. Small houses, barely more than shanties, and "shotgun" two-family structures were common in the poorer neighborhoods. Those neighborhoods were insulated from mainstream life in a way that made being poor in New Orleans a special hardship. "New Orleans has a 40 percent literacy rate; over 50 percent of black ninth graders won't graduate in four years," Michael Eric Dyson explained in *Come Hell or High Water,* his post-Katrina study of race relations in America. "Louisiana expends an average of $4,724 per student and has the third-lowest rank for teacher salaries in the nation. The black dropout rates are high and nearly 50,000 students cut class every day. When they are done with school, many young black males end up at Angola Prison, a correctional facility located on a former plantation where inmates still perform manual farm labor, and where 90 percent of them eventually die. New Orleans's employment picture is equally gloomy, since industry long ago deserted the city, leaving in its place a service economy that caters to tourists and that thrives on low-paying, transient, and unstable jobs."[18]

Although New Orleans has been branded "jazz capital" (rightfully so), in truth, on the eve of Katrina, one would be hard-pressed to hear Wynton Marsalis or Irvin Mayfield CDs blaring out of federally subsidized housing. Jazz had become establishment music except for the brass band phenomenon. Times had changed. Louis Armstrong had fled New Orleans long ago, living his last decades in Queens, New York, where he was buried in July 1971. But hiphop and rap were flourishing, angry lyrics being shouted out of Magnolia and St. Bernard and Iberville. Drum machines and turntables had replaced trumpets and trombones as the instruments of choice. Master P, raised in the Calliope Projects, was a true hiphop hero to African-American youths, admired both for such outrageous hits as "99 Ways to Die" and for being CEO and founder of No Limit Records. Then there was B.G., the Baby Gangsta, who at age eleven signed a major recording deal with Cash Money Records. His first album, *True Story,* released in 1993, graphically depicted life in a crime-ridden, racist New Orleans where schools were like prisons. He was accurately depicting his reality, no matter how uncomfortable his lyrics might make some listeners. Terius Gray, a.k.a. Juvenile, was a pure product of the Magnolia Projects—or "Wild Magnolia" as he called it. His first band was UTP, three letters he has tattooed on his stomach. "UTP," he explained, "was like a coalition of rappers." His hiphop lyrics lambasted New Orleans's racism and classism in a searing, no-holds-barred, in-your-face fashion. Growing up in the Hollygrove section of New Orleans, Lil' Wayne (Dwayne Carter), also known as Weezy F Baby, Birdman Junior, and Raw Tunes, wrote dozens of gangsta rap hits while still in his teens. To understand the African-American youth culture of New Orleans during August 2005, put aside CDs like *The Magic Hour* and *Half Past Autumn Suite* and listen to the bleak, thuggish, violent inner-city lyrics of Lil' Wayne in such songs as "Shooter," where he raps: "So many doubt cuz I come from the south / But when I open my mouth, all bullets come out."[19]

Hiphop-infused gang members were frequently arrested in New Orleans. They were usually let go before trial. In August 2005, the Metropolitan Crime Commission completed an extensive analysis of the arrest-to-conviction success between the NOPD and the District Attorney's office, headed by Eddie Jordan. During the preceding year, 2004, the NOPD had made 114,000 arrests. Only 17,004 of those arrests, however, were for state offenses, and the breakdown of that number was about 60 percent felonies and 40 percent misdemeanors. In any case, only 7 percent of those 17,004 people who were arrested were actually sent to jail. The NOPD, in other words, was arresting a lot of criminals, but very few were being convicted. Most arrestees walked in the front door of the city jail and then, after signing a couple of forms, walked out the back door. "So the reason I think that we have a high crime rate is that police measured success by arrests," explained Rafael C. Goyeneche III, president of the commission, a watchdog group dedicated to keeping repeat offenders off the streets and the NOPD honest, "and not how many of the arrests resulted in incarceration or incapacitation."[20]

On the eve of Katrina, New Orleans was a city of 460,000; the 114,000 arrests made there in 2005 reflected a rate of one for every four adult citizens had been arrested by the NOPD. Quite understandably, many citizens felt the NOPD was overzealous in arrests and underzealous in follow-up. New Orleans's juries often didn't trust the arresting NOPD officers' credibility. Often DA Jordan had to drop cases because the arresting officer simply refused to be a witness. Many bad guys—drug kingpins or child molesters—were let go on technicalities because of sloppy NOPD paperwork. To put it simply, the NOPD was high on street action and low on desk work. In 2005 only 12 percent of the people arrested for homicide were convicted, which meant that 88 percent of those arrested were free to walk the streets. Murderers were wandering around New Orleans as Katrina approached, in large measure because the NOPD didn't know how to work the judicial system.

But there were able officers in the department on the eve of Katrina, cops who were determined to clean up the corruption. "You have to be by the book," Warren Riley said late in 2005, after becoming superintendent, explaining the need for a new era of public integrity in the NOPD. "That's the bottom line. I think that accountability is vital, swift punishment is absolutely necessary. Even within an organization, like the law enforcement organization, the officers need to fear the administration. They need to believe that if they don't do what they're supposed to do, they're going to suffer the consequences . . . termination, prosecution, whatever."[21]

IV

At forty-six Warren Riley, as husky as a Russian weight lifter, with the facial features of a puffy Hank Aaron, was in many respects an old-fashioned community cop. The youngest of five children, he was the first person in his family to earn a college degree. Combating racism was a reality the Riley family faced head-on. His father, Sam, who hailed from Wilson, Louisiana, never let verbal slights by whites get under his skin. His mother, Selma, however, was a child of a woman raped by a white man. "My mother was half-white," Riley explained. "Her mother was an African-American female who worked in a grocery store in Plaquemines. The story that my mother told was that the white Irishman store owner, when his wife left, raped her, at least one time. So my mother was a very fair-complexioned lady, long straight hair."[22]

Riley's father worked three jobs as a maintenance man to keep the family together. A stern disciplinarian, he made Warren take responsibility for household chores. As a student at Booker T. Washington High School from 1974 to 1978, Warren seldom encountered racism, although he admits white gangs and black gangs sometimes fought on Walmsey Avenue. Ironically, it was a white NOPD officer named Tom Fierce who influenced Riley's decision to become a cop. A true community officer, Fierce would bring football helmets and pads to give to the poorer African-American kids in the playground. This impressed Riley immensely. The police were the good guys, especially Fierce. Riley's sports hero was also white: quarterback Johnny Unitas of the Baltimore Colts. "Something about the crew cut," he later reflected on Unitas, "the clean-cut look he had." Riley's favorite subject was history, particularly battlefield heroics ranging in time from the Crusades to World War II, but he also enjoyed TV detective shows. He recognized that men in uniform garnered society's respect. "My mother was a maid somewhere and my father went to pick her up and it was this big house," Riley said. "That was the first time I saw a community where I realized there was something really different between these homes and my community. But what drove me was I was just the kind of person who wanted the house and the picket fence and the two kids and the car. So, becoming a police officer, initially for me, it was just a steppingstone. It was a job, an opportunity to get a car, to get an apartment."

Oddly, Riley would claim that his inaugural encounter with racism came when he joined the NOPD in 1979. "It was clearly an internally segregated system back then," Riley explained. "You'd go to roll call, blacks would sit on one side of the room, whites would sit on the other side of the room. It was very seldom that blacks and whites rode together in the Sixth District, where I was working. . . . We merely existed in our own worlds." Back then, Riley said white officers harbored a death wish for their black colleagues. Over twenty-five years later, he remembered the chilling words he heard from a couple of different white officers: "Don't call for any help 'cause nobody's coming."[23]

From 1984 to 1989, the NOPD had Riley working vice in the St. Bernard Housing Development in the Ninth Ward. He became an undercover narcotics agent, staying out of Uptown, where folks might recognize him. It was known as the "teas and blues" circuit—a homemade concoction of the barbiturate Talwin (teas) and antihistamines (blues), which, when injected, hyped abusers up, making them violent. Unlike heroin, which tended to make junkies lethargic and easygoing, or marijuana, which the police largely ignored, crack cocaine was something that vice wanted to get off the streets. Eight balls (eight ounces) of crack cocaine were selling for $400. The NOPD needed somebody young to become an undercover drug dealer, and Riley was in his late twenties. "I was set up in an apartment and I'm selling drugs back to the drug dealers," Riley recalled.

There was no question that Riley was a tough, uncompromising narc. He believed in swift punishment for drug dealers, whom he deemed scum, just like those who were committing theft, burglarizing homes, and even murdering people. "I was never looking at it in a humanitarian way, so to speak, like 'these guys need a break,'" Riley said. "I grew up in a pretty rough neighborhood. The big thing then was sniffing glue. . . . I had two outstanding parents. That's what made the difference in my life. So there was no regret. Never did I think about what was going to happen to these guys [once arrested]."

On the eve of Katrina, Riley was living in Algiers. He had run against Marlon Gusman for Orleans Parish criminal sheriff in the previous fall and had lost. He was, however, deputy chief of police of the NOPD. Riley was determined to stick out the storm, to start rescuing people if necessary when it passed. With media training by the FBI, he was, in police jargon, a "command presence." To his credit, Riley tried to be proactive about preparing the department for the impending storm. But as number two, there was only so much he could do. Worried that the NOPD had antiquated radio communication that would have made Marconi jeer, he put out a search for satellite phones. He contacted the National Guard, requesting that they put five boats in each district station so his men could get around if it flooded. But the Guard nixed the idea, preferring to keep them *all* at Jackson Barracks, its headquarters. Located on St. Claude Avenue, nestled on 100 acres, Jackson Barracks was created in 1834 to offer logistical assistance to far-flung Mississippi River Valley forts. A campus of fine brick buildings with white columns, the facility had been used earlier in 2005 by film director Steven Zaillian as a set for his remake of *All the King's Men.* This was where the Louisiana National Guard was going to make its stand during Katrina. Unfortunately that pre-Katrina weekend, feeling safe from the hurricane, the Guard was bringing rescue equipment like boats and high-water vehicles *to* the compound, not out of it. As fate would have it, Jackson Barracks got wiped out. Good-bye, trucks.

Riley also helped set up an emergency operations center on the ninth floor of City Hall. While it would be headed by Assistant Police Chief Danny Lawless, a whole beehive of representatives—from the state police, harbor police, National Guard, Coast Guard, Air National Guard, Homeland Security, fire department—would be ensconced there. For some reason it never seemed to dawn on any of these official representatives that they were in the bowl, that if City Hall flooded there would be no command center. When asked if this kind of consolidation was smart, Riley stiffened, claiming it made perfect sense. "Every storm we'd been through," he said, "that command center at City Hall had been sufficient." For all his street smarts, Riley, like most city officials, ignoring the Pam exercises and the five-part *Times-Picayune* doomsday series, felt like a man on top of his game. He was going to ride out Katrina at police headquarters on 715 Broad Street with his men—the men who were his brothers of the shield.

In most cities, first-rate police officers garnered the respect of its citizens. But not in New Orleans. For residents, the problems of the projects were intertwined with the attitude of the police. But the housing projects at least possessed the virtue of being located in the inner city. Hardworking families living there could walk to work or ride the trolley or the bus.

Many project residents, therefore, didn't own a car. Truth be told, they couldn't afford one. When the St. Thomas, Desire, and Florida projects had been condemned, as well as the Melpomene project in Central City, new housing for the evicted needed to be found. Two bad solutions were hatched by City Hall, both contributing mightily to the unfolding Katrina tragedy. The first blunder in social engineering occurred in 2000, when those families living in the defunct St. Thomas project were shipped to live at the St. Bernard project in the Seventh Ward. It soon became known in the black community as "St. Bathomas." What the merger accomplished was an increase in violence, as gangs fought over drug turf. This mistake of ghettoizing thousands in "St. Bathomas" contributed to New Orleans's skyrocketing murder rate. The other unhelpful local-state solution was to relocate residents of other projects to New Orleans East, a long ten-mile drive from downtown, in areas like Little Woods and Michould Boulevard. This was where Cedric Richmond saw the last Little League game of the New Orleans summer. Packed into flimsy apartment complexes, ugly condos, and tiny houses near Lake Pontchartrain, these residents had no reliable public transportation. Carjacking became epidemic in New Orleans East. Many of those new to the neighborhood had worked in what blacks called the "servant industry," toiling as hotel maids, parking attendants, or domestic help for well-to-do whites. It was an honest living. Suddenly, with their relocation, they had no easy way to get to work downtown. Singer Aaron Neville, a resident of New Orleans East, gave the failed relocation a name, the "Outer City Blues," and even wrote an unrecorded song about it.

These powerless city poor were what sociologist Michael Harrington once called "the Other America"—those living in desperate poverty, living on minimum wage or welfare checks, hidden from the view of the mainstream, and often denied basic services, like proper sewage, reliable electricity, or decent schools. On any given day, you could encounter them redeeming aluminum cans at Walgreens on St. Charles or holding cardboard signs asking for money around Lee Circle. They didn't hear about Katrina on television, for a simple reason: they didn't own a set. Even if they did hear about the storm, they didn't have the money to leave. They had no credit cards with which to rent a car and reserve a motel room in

Dallas, Memphis, Little Rock, or Baton Rouge. Poorly educated, and often illiterate, they couldn't figure out what all the evacuation commotion was about. With no driver's license or other form of identification, some were afraid the NOPD would arrest them at city-run shelters or handcuff them for hitchhiking on I-10.

V

At 4 P.M. on Saturday, the Louisiana State Police turned over all lanes to outward traffic on four New Orleans interstate highways. The metro area's two toll roads, the Crescent City Connection and the Lake Pontchartrain Causeway, were now free. Called "contra-flow," the redirected traffic represented the one plan that the state had worked out in enough detail to operate effectively in the face of Katrina. Governor Blanco oversaw the creation of the surprisingly complex contra-flow plan after the bottleneck traffic debacle caused by the approach of Hurricane Ivan in September 2004. Later, when preparing to testify before a congressional committee, Blanco offered a defense of her contra-flow plan. She rightly pointed out that her plan had been designed in collaboration with appropriate parish leaders and that, as bad as Katrina was, it "would have been far worse if the initial evacuation had not been so efficient and safe."[24]

Without question, Blanco's contra-flow plan saved lives. All of New Orleans's hospitals, for example, started evacuating patients—those they could move—in a reliably easy fashion. At every hospital, supervisors decided not to move critically ill patients; Charity Hospital, for example, the oldest continuously operating hospital in the country, had fifty beds occupied in its intensive care unit.[25] A group of doctors, nurses, respiratory therapists, and other staff were staying behind to take care of them. "I was assigned as teaching physician for the infectious diseases unit on the ninth floor of the hospital," Ruth Berggren later wrote in *The New England Journal of Medicine*. "There were eighteen patients in the unit, of whom four had active tuberculosis and thirteen had opportunistic infections related to HIV and AIDS. We also had a boarder from surgery—with a complicated gunshot wound and vascular access problems."[26]

Even with contra-flow, however, traffic moved at a snail's pace, and by late Saturday afternoon, it was virtually impossible to reserve a motel room in towns as far north in Louisiana as Alexandria, Monroe, and Shreveport. Prophecies of bad weather for the Gulf South area had reached a saturation point. Still, cars were making hasty dashes about, drivers looking for the last flashlight batteries and bottled water in each vulnerable parish.

One person who was extremely worried that City Hall wasn't recognizing the devastation a Category 3, 4, or 5 hurricane would wreak was the outspoken Nick Felton, president of the New Orleans Firefighters and captain of Engine Company 21. Six feet tall, with salt-and-pepper hair, Felton was a twenty-two-year veteran, a hardboiled, no-nonsense professional, the kind who would have rushed up the fateful stairs of the World Trade Center on September 11, 2001. He was appalled that the New Orleans Fire Department didn't own a single boat. At a ninth-floor EOC meeting he spoke bluntly about the Big One, telling Superintendent of Fire Charles Parent and Deputy CAO Cynthia Sullivan-Lear that he felt firemen needed water reserves, compensation for working the storm, family protection, and food supplies. If flooding occurred, and natural gas pipelines broke, they were going to be putting out raging infernos for a week. When he was finished with his demands, Sullivan-Lear—with not even an iota of tact—told Felton he was being "an alarmist." Felton couldn't believe she was copping a lackadaisical attitude when he was in crisis mode. "They just couldn't comprehend that this was it," Felton said. "I kept saying, 'Aren't you guys watching the Weather Channel?' I had been through Camille and Betsy and this damn thing, at least on the Weather Channel, looked far worse."[27]

An hour after the contra-flow changes were put into effect, at 5 P.M., Governor Blanco arrived in New Orleans for a press conference with Mayor Nagin. They were visibly uncomfortable with each other, a crackle of tension in the air. They jointly announced that a *voluntary* evacuation order had been issued for the city in anticipation of Katrina. The year before, Ivan had been on a similar path, heading straight for New Orleans, but it swerved eastward at the last minute, missing the city. So, as the illogic went, Katrina would probably do the same thing. Hope had become Nagin's main evacuation strategy. "We strongly advise citizens," the mayor said, "to leave at this time." In the ears of longtime New Orleanians, his cautionary words ringed of the chance for yet another hurricane reprieve. "We want everyone to not panic," Nagin continued, "but to take this very seriously. Every projection still has it hitting New Orleans in some form or fashion."[28]

In some form or fashion—a New Orleans resident could take that to mean a lot of things, most of which were far removed from the pessimism of the National Hurricane Center's latest bulletin. Issued that afternoon, Advisory 18 warned that "there remains a chance that Katrina could become a Category 5 hurricane before landfall."[29] That should have been Nagin's lead. It was, at least, in the *Times-Picayune*'s early Sunday edition (which came out Saturday night). The paper ran a big-block headline: "Katrina Takes Aim."[30] The chain of responsibility for urban evacuation, highly debated after Katrina, was really quite simple. The pecking order, according to protocol, was (a) the mayor; (b) the New Orleans director of Homeland Security (a political appointee of the mayor, who reports to the mayor); (c) the governor; (d) the secretary of Homeland Security; and (e) the U.S. President. Both Nagin and Blanco, even before Katrina hit, recognizing they were unable to cope with the impending doom, were already chastising the federal government for New Orleans's rank unpreparedness. The blame game had begun even before Katrina made landfall.

VI

Unlike New Orleans's hurricane evacuation strategy, tracking hurricanes *was* the responsibility of the federal government. Toward that end, Congress had authorized the U.S. Weather Bureau in March 1870. Based in Washington, D.C., it was soon the hub of twenty-four far-flung observatories, connected by telegraph and constantly monitoring meteorological conditions in order to help communities prepare for such climatic curses as heat waves, blizzards, and hurricanes. By 1899, the bureau had opened a hurricane-forecasting center on the island of Jamaica. When the bureau, a component of the Commerce Department, had its centennial, it was renamed the National Weather Service.

During tropical storm season, August to October, it was the National Hurricane Center, a unit of the National Weather Service, that always took the lead. Based on the campus of Florida Atlantic University, the modern NHC was founded in 1955 and put into the specific business of storm prediction. Everyone seemed to be fascinated by how the center collected its cyclonic data: the NHC grew into a popular tourist attraction, with busloads of schoolchildren, among many others, taking trips to western Dade County to see hydrogen-filled weather balloons launched twice daily and to hear about the Doppler Radar Network, which covers much of the Gulf of Mexico.

When a powerful weather disturbance first blipped on the radar screen, NHC responded in dramatic fashion, sending manned twin-jet Gulfstream IV-SP aircraft straight into the eye of the storm to bring back detailed reading on wind force and barometric pressure. If the depression was a tropical storm (i.e., sustained wind speeds between 39 and 73 mph), the NHC alerted governmental agencies, including FEMA and the Department of Defense, regarding the potential threat. The media and civil defense authorities in affected areas were also contacted. Using the Saffir-Simpson Scale, meteorologists could quickly communicate the severity of a storm. Typically, word that any hurricane ranked at Category 3 or above aroused intense interest. Only in the rarest instances was a personal appeal also required.

Late in the day on Saturday, August 27, Max Mayfield, the mild-mannered, bespectacled scientist who had been the director of the NHC since 2000, grew concerned at the lack of activity in advance of Katrina. As he put it later, he wanted to do "everything that I could do" to warn the country that the Gulf Coast—and New Orleans in particular—was in grave danger. Mayfield made scores of telephone calls, in addition to participating in a teleconference with Michael Brown of FEMA, as he tried to convince officials at all levels that even though Katrina was still out in the Gulf and could go anywhere, his data showed that the hurricane was on a direct track to New Orleans. In case his meteorological explanations were not persuasive enough, he used language anyone could understand. "This is really scary," Mayfield insisted. "The guidance we get and common sense and experience suggest this storm is not done strengthening."[31]

What concerned Governor Blanco on Saturday evening was that Mayor Nagin, for whatever reason, just wasn't taking Katrina seriously enough. When she was back at EOC in Baton Rouge, Mayfield telephoned her. They had developed a warm, special friendship over the years. Once again, he wanted the governor to know that Katrina was barreling Louisiana's way and that he "was sorry." His voice was maudlin, full of trepidation, so Blanco conveyed to him that she understood. "Thank you, Max," she said. "But you need to talk to Ray Nagin." A frustrated Mayfield said, "I've been trying to talk to him, but I can't reach him." An exasperated Blanco, sympathizing with Mayfield, said, "I've got his cell number. Give me your number and I'll call him." So Blanco tracked Nagin down. He was at a restaurant with his wife. "So, I gave Nagin Mayfield's number," she said. "I put them in touch."[32]

Virtually every local official in the Hurricane Belt states—except Ray Nagin—knew of Max Mayfield. He was a critical figure in the most important role of government, protecting people from danger. The mayor later recalled, "I got a call from the head of the hurricane center, Max somebody . . . and he said, 'Mr. Mayor, I've never seen a storm like this. I've never seen conditions like this.' "[33]

Nagin claimed that he ordered the *mandatory* evacuation of his city after speaking with Mayfield on Saturday night, but there is no evidence that he acted then. He certainly did not make any public announcement. One of the others who received Mayfield's message was Walter Maestri, emergency director of Jefferson Parish, which abuts New Orleans on the south. A longtime hurricane watcher, he immediately called Jeff Smith, the deputy director of the Louisiana Department of Homeland Security and Emergency Preparedness. Maestri impatiently asked if Mayfield had called Smith yet. "He said, yes, he had received the call," Maestri recalled later. "So I said, 'Then you know what he's sharing?' And he says, 'Yes, but the storm right now'—and I said, 'Please, please. You've indicated you don't know Max. Let me tell you. When he calls you like that, he's telling you you need to be ready, be prepared.' "[34]

On CNN's *Anderson Cooper 360,* on January 20, 2006, Nagin, to his credit, stopped passing the blame and took personal responsibility for not properly preparing New Orleans for Katrina. "There's . . . things that I would do totally different now," he told Cooper. "I wish I had talked to Max Mayfield earlier, number one . . . so the possibility of a mandatory evacuation would have been done 24 hours earlier . . . [w]hen I got that call, and he was so emphatic and so passionate, we had never—this city had never done a mandatory evacuation in its history. I immediately called my city attorney and said, look, in the morning, I don't care what you have to do. Figure out a way for us to do this. I wish I had done that earlier."[35]

For all of the weak, confused, and bureaucratic messages from government officials, there were many who heard the NHC warnings and decided on evacuation pronto. Elizabeth Daigle, of the New Orleans suburb of Metairie, was determined to leave town on Saturday evening. At forty-four, she had heard warnings before and had even survived one major hurricane, 1965's Betsy, which killed sixty-five people in New Orleans. "We just have a bad feeling about this one," she explained, "We just don't know. That's what's scary."[36] Likewise, Garden District resident Janine Butscher, originally from Oxfordshire, England, was planning to stay in New Orleans for Katrina. Then she woke up on Saturday and talked to her next-door neighbor, who worked for Schlumberger, an oil services company specializing in geophysical data collection and analysis. "He said his company had been notified to get out. That this was the Big One. He scared me so much that I grabbed my two-year-old daughter and drove to Houston in my flannel pajamas."[37] Real estate broker Judy Oudt, famous locally for selling Garden District mansions, had planned to stay in her Lee Circle condominium and ride out the storm. But when she visited her local pharmacy, she saw hordes of semipanicked locals filling their shopping carts with hurricane provisions. The line to pay was twenty people long. "Hell, if this many people are freaking out," she said, "so will I."[38] She walked out of the pharmacy with no purchases and headed to Seaside, Florida.

It took others just a little longer than Daigle, Butscher, and Oudt to get out of town. Andrew Travers, a graduate student in history at Tulane, for example, spent his Saturday evening at Pat O'Brien's French Quarter bar, downing "hurricanes," a potent rum–fruit juice concoction created in the early 1940s by reckless revelers waiting out a ferocious tropical storm. (According to New Orleans chef Emeril Lagasse, the best recipe for a "hurricane" was 2 ounces light rum, 2 ounces dark rum, 2 ounces grenadine, 1 ounce orange juice, 1 ounce sour mix, 1 teaspoon of sugar, and orange wedges for garnish.)[39] O'Brien's was packed that night. It was a giant "hurricane party" with cocktails being pounded back by rebels determined to booze and boogie their way through the natural disaster. (In August 1969, a group of young people drank hurricanes at a beachside party in Pass Christian, Mississippi, daring Camille to make landfall. It did, and twenty of them drowned.) When Travers sobered up the following morning, he looked at CNN on television and learned that Katrina seemed to be developing into a Category 5 storm. "They were rehashing the doomsday scenario we'd been talking about for years," Travers recalled, "a direct hit and broken levees and an American Atlantis." He telephoned his girlfriend, who had stockpiled water and nonperishables in her apartment and invited a handful of friends to stay through the storm. She wisely called off the would-be survivors party, saying, "It would have been fun, but yeah, we gotta go." They fled town in a Honda Civic coupe, with two dogs crammed in the backseat.[40]

Parties all over town were ending early on Saturday. Andy Ambrose, the son of the late historian Stephen E. Ambrose, was supposed to be celebrating his forty-second birthday that afternoon. A soulful rhythm-and-blues vocalist, Andy planned to party at the Maple Leaf Bar on Oak Street, a smoky music club that showcased New Orleans's finest talent. The bar was holding a midsummer Mardi Gras party featuring the legendary George Porter, founder of the Meters, the seminal New Orleans funk band that influenced everybody from Phish to Widespread Panic. Ambrose decided to forgo his party, though. Instead, he picked up a hammer and spent his birthday boarding up his neighbors' windows on St. John Court in Mid-City, on the Carrollton Avenue side of the bayou. "Saturday evening, people who said they were going to stay put for the storm started to have second thoughts," Ambrose recalled. "By midnight it was ominous. The town was desolate: my wife and I decided to skedaddle. At 3 A.M. we drove to Columbia, Mississippi, and got a room."[41]

On Saturday morning U.S. Senator David Vitter, a Jefferson Parish resident, woke up to learn that a major hurricane was headed toward Louisiana. He just stared at the television with his wife, Wendy, and shook his head. He had one prevailing thought: Welcome to the Big One. "We both said let's not kid around," Vitter recalled. "We're leaving." While his wife started packing the minivan, Vitter called his brother Al, a mathematics professor at Tulane, and his sister Martha in Atlanta. They were worried about their mother. She lived on Vincennes Place in Uptown, which was susceptible to flooding. Moreover, she had an enlarged heart. Around-the-clock medical attention—or the possibility of it—was necessary. A decision was made to evacuate her by airplane to Atlanta, ASAP. With that plan in motion, David Vitter headed up Highway 61 to Memphis, dropped his family off at a relative's house, and then flew back down to Louisiana. He was going to remain available in Baton Rouge, much like Mitch Landrieu, at the EOC. Vitter would sleep in a dormitory room provided by the state police and take part in the relief efforts. That Saturday afternoon, however, Vitter couldn't help but wonder why Mayor Nagin wasn't calling for the mandatory evacuation of the city. "Hindsight is twenty-twenty," Vitter later said. "But it was clear to me very early Saturday that, yes, it should have been ordered."[42]

Ronald Mack Jr. of the Seventh Ward thought all those people evacuating on I-10 and I-55 were fools. Everybody knew the Big One always veered east when it approached the mouth of the Mississippi—everyone, that is, except his wife. "I was going to suck it up," he later recalled. "But she starting nagging me. She made us leave for Houston."[43] Officials could count themselves successful in the cases of Ambrose, Vitter, Travers, Mack, and hundreds of thousands like them. Last-minute evacuation was preferable to no evacuation at all.

VII

But there were many who heard the dire predictions for Katrina and decided to stay put. Kenny Bourque, a twenty-two-year-old bartender from the French Quarter, counted himself among their ranks. The only precaution he took, as he brazenly defied meteorologists, was purchasing a life preserver for his dog—just in case of a flood.[44] He adhered to the spiritual lament that goes, "Bury me down in New Orleans/so I can spend eternity aboveground/you can flood this town/but you can't shut the party down/ain't no drownin' the spirit." Street performer Gaetano Zarzana, full of aperçus, told the *Houston Chronicle* he was going to "have fun and watch God's fury" and "hang out in Johnny White's bar on Bourbon Street and watch the flood come up."[45] Then there was Michael Barnett, who decided to hole up in his office on Poydras Street in the central business district, stating on Saturday that he was determined to keep an hourly blog of Katrina. "Like when P. Diddy sang, 'we ain't goin' no where,'" he logged at 8:19 A.M. "Come on with it then, storm. Bring me what ya got. Let's see who wins."[46] Officials could count themselves powerless in the cases of defiant citizens like Barnett, Bourque, and Zarzana.

Thousands of healthy, well-informed citizens simply made the personal decision that they didn't want to leave. Some were traffic-phobic while others believed such natural forces as hurricanes were in the Lord's hands. A few thousand of the unmovable were gamblers, long ago courting risk like a lover. There was a Good Samaritan contingent who wanted to keep an eye on their neighborhoods, and the doctors and nurses who just couldn't leave. There were thrill seekers, tarot-card readers, and professional squatters. A parochial pride informed some New Orleanians' decision to stay; this was their town, by God, and they weren't going to abandon it in its time of peril. An unusually high number of boat owners stayed, convinced that—worst-case scenario—if their houses flooded, they could just sail away until the water receded. Some people saw Katrina as a chance to hole up and get long-overdue chores done, write delinquent thank-you notes, and file the tax extensions they'd been putting off. Taken all together, these people were a breed unto themselves. As the city emptied out, and New Orleans felt like a forlorn tomb, they had unknowingly volunteered to be first responders in the worst natural disaster in modern U.S. history. A large part of the civic burden of saving the poor, infirm, elderly, and confused in Katrina's wake would fall on their vigorous shoulders.

It was getting late for everyone in southeastern Louisiana. Ninety-year-old Amantine Marie Verdin, a member of the Pointe-au-Chien Indian tribe, lived in St. Bernard Parish with her mentally handicapped son Xavier. Another son, Herbert, and other relatives lived nearby. As the storm approached, Herbert's daughter, Monique Michelle Verdin, drove down from her home in Baton Rouge to check on them all. "When I got to my grandmother's house that Saturday," Monique said, "I found her frying fish and cooking shrimp étouffée. My father, Xavier, and my first cousin were all at the house living like it was just another day." Monique took a box of her grandmother's keepsakes, and wrote out a list of storm precautions. When she left, though, nothing had much changed. "Clothes were drying on the line outside," she said. "Xavier was sitting on the porch."[47]

In other arenas, there was only confusion and, under the circumstances, that was an outrage. Benjamin Johnson, a U.S. Marine from 1977 to 1987, was employed as a security guard. "The biggest mistake in New Orleans history was Nagin's not calling a mandatory evacuation on Thursday or Friday, at the latest," Johnson declared. "My view was that if it wasn't mandatory it can't be a bad storm. I did sneak out under the wire on Sunday. It took me eleven hours to drive from New Orleans to Baton Rouge. But the people in the projects, those I knew, kept saying, 'Katrina ain't nothing. They ain't even askin' us to leave.' "[48]

For others, evacuating in the face of Katrina was not entirely a matter of impetus or money, but of finding help collecting their treasured belongings. Some simply refused to abandon their dogs or cats. The very sick were afraid to be disconnected from their oxygen supply, respirators, or dialysis machines. In the case of the elderly, many suffered from dementia or chronic fatigue, and barely knew what was going on around them. In effect, City Hall deemed automobile drivers the first-class citizens. If you didn't have a car, you were second-class. As Saturday came to a close, city officials were despicable, ignoring the car-less, not just because they didn't help such people evacuate, but because they didn't even know who they were. As far as Mayor Nagin was concerned, it seemed the down-and-outers were an inconvenience to City Hall—pure and simple. There wasn't much he could do for their welfare so late in the game. And, in many cases, he was right.

Stone Phillips of NBC's *Dateline*—among other journalists—would hound Nagin for his pre-Katrina blunders. He asked Nagin why Regional Transit Authority buses sat idle on Chickasaw Street. Why was the fleet of yellow buses padlocked away on Metropolitan Street? Why weren't all the buses used to evacuate large numbers of folks? Why was nothing in New Orleans mobilized? Why weren't National Guard troops in proper post-hurricane position? Why wasn't there a high-tech hurricane command center? Why weren't rescue helicopters and evacuation buses standing by on the periphery of the storm, ready to swoop down and rescue Superdome evacuees and the poor when Katrina passed? All Nagin, skirting any personal responsibility, could meekly answer to Stone was "I don't know." . . . Those were questions for someone else.[49]

Nearly five months after Katrina, however, under stinging criticism from U.S. senators and congressmen, Nagin admitted guilt, in hesitant fashion, for failing to evacuate his city's buses before Katrina made landfall. "If I had to do it again," he told CNN, "I would probably go to the school board, cut a cooperative endeavor agreement with them, move all the city-controlled buses to another section of the state probably up north, so that they're readily available, and we will just deal with the driver issue later."[50]

VIII

That Saturday Joe Donchess, executive director of the Louisiana Nursing Home Association, was extremely worried. An Ohio native and a 1975 graduate of Southern University Law School in Baton Rouge, Donchess became the state's leading voice on issues pertaining to health planning. Ensconced at the EOC in Baton Rouge, Donchess had been closely monitoring the sixty to seventy nursing homes likely to be affected by Katrina. Electricity blackouts were a certainty. That meant elderly patients would be disconnected from life-support machines, alone in the clammy darkness. Evacuating elderly and disabled people on beds and in wheelchairs took time—a couple of days. It was, in fact, a logistical ordeal for them. Every two hours or so on Saturday, Donchess at least e-mailed the nursing homes, updating them on Katrina. There was one major stumbling block regarding New Orleans, and it made him edgy. "Because Mayor Nagin refused to call a *mandatory* evacuation, the nursing homes didn't feel compelled to evacuate," Donchess explained. "It was not my job to tell homes whether to leave, it was up to the mayor. . . . Nagin and other Orleans Parish officials were dilatory in not calling for a mandatory evacuation earlier. I know, for sure, that twenty-one facilities would have evacuated on Saturday if he had called it. That would have been just enough time for buses to properly bring the patients out of harm's way."[51]

The dilatory response in New Orleans, however, was not shared by surrounding parishes. In St. Bernard Parish, for example, a mandatory evacuation was declared on Saturday; nevertheless, there were nursing homes that chose to ride out the storm, with tragic results. Donchess was in a difficult position. As he updated the nursing homes about millibars and storm surge predictions, he felt like a nag. Every time he heard back that a nursing home was evacuating, he cheered. By late Sunday twenty-one of the homes had evacuated while around forty hunkered down and sheltered in situ. One reason a number of nursing homes didn't evacuate was bad memories of Hurricane Ivan. Back in September 2004, when Ivan was in the Gulf, many of these homes had buses pick up their patients and then transport them on I-10. The bumper-to-bumper traffic—and the uncomfortable fact that it took eight hours just to get from New Orleans to Baton Rouge—caused two elderly patients to die in transit. An eighty-six-year-old woman also died of a heart attack when she was evacuated from a nursing home during Hurricane George in 1998.[52] The word in the state nursing-home world was that the pre-storm commotion was potentially harder on seniors than the hurricane itself.

Another problem facing the Nursing Home Association was a shortage of nurses and caregivers. Even in good times the more than three hundred nursing homes in Louisiana were understaffed; when news of Katrina's path broke, many essential medical assistants left the state for higher ground.[53] And finally, transportation was hard to find, and getting more scarce by the hour. "We really needed busing help, and just didn't get it," Donchess recalled. "The state office of emergency preparedness didn't listen to our needs. They thought that because we were an association we surely knew how to evacuate all the nursing homes. That wasn't the case. We needed a mandatory evacuation, called earlier, and buses to help us move thousands of patients."[54]

Although Donchess was correct that city, parish, and state officials needed to help homes evacuate, the ultimate responsibility lay with the individual homes. At single-story St. Rita's Nursing Home near Poydras, Louisiana, Coroner Bryan Bertucci pleaded with owner Mabel Mangano to close the facility. The NHC was predicting 140 mph winds and a twenty-foot storm surge for St. Bernard Parish. A Category 3 or 4 storm was nothing to mess around with. "I told her I had two buses and two drivers who could evacuate all seventy of her residents and take them anywhere she wanted to go," Bertucci recalled. "She told me, 'I have five nurses and a generator, and we're going to stay here.'"[55]

According to the *Times-Picayune*, terrible misjudgment was nothing new at St. Rita's. Back in 1999 the home had been cited twice for endangering the lives of residents and was denied U.S. government funding for more than forty days for failing to rectify the malfeasance. The nursing home didn't properly stop patients' infections from spreading. Under constant regulatory heat, St. Rita's did, in November 2004, finally meet health inspectors' basic standards. But it was not, by any stretch of the imagination, a good place for your grandparents to spend their twilight years. As Katrina approached, the business practices of Mabel Mangano and her family were putting patients' lives at great risk. "They had a duty as a standard of care to people who could not care for themselves," Louisiana Attorney General Charles Foti later said of the Manganos. "If you or I decided we are going to stay, we do it of our own free will. . . . The people at the nursing home don't have that chance."[56]

Most of the elderly in New Orleans, however, couldn't even afford to be cared for in a home like St. Rita's. Fully one-quarter of the families in New Orleans lived on a per capita annual income of $15,000 or less.[57] Even worse, many of the elderly had no family. They were all alone against the storm. And August 31—the last day of the month—was when social security and welfare checks were handed out. Many poor, elderly people just weren't going to evacuate without that check.

When it came to hurricane evacuation, there was nothing new or novel about the poor or elderly having a harder time fleeing than the rich. Zora Neale Hurston, in her brilliant 1937 novel *Their Eyes Were Watching God*, captured the helpless attitude African Americans around Lake Okechobee had about an evacuation during the Great Depression. On the evening before a mammoth hurricane was supposed to slam South Florida, Janie Crawford and her boyfriend, Tea Cake, decide to stay at their rickety fishing camp, despite the fact that it was extremely vulnerable to storm surge. "Everybody was talking about it that night," Hurston wrote. "But nobody was worried. You couldn't have a hurricane when you're making seven or eight dollars a day." Just as in New Orleans on August 27, 2005, Hurston's economically depressed characters believed they were protected by flood walls and levees. "The folks in the quarters and the people in the big houses further around the shore feared the big lake and wondered," Hurston wrote. "The people felt uncomfortable but safe because there were the seawalls to chain the senseless monster in his bed. The folks let the people do the thinking. If the castles thought themselves secure, the cabins needn't worry."[58]

IX

Even on Saturday at 5 P.M., with the highways leading out of New Orleans crammed with drivers escaping the coming hurricane, nothing much was moving out of poor neighborhoods like the Carver-Desire section of New Orleans East, Lower Ninth, and Tremé. A few people were leaving, but operable buses were a rare sight. And hardly anyone was arriving to help—with some very notable exceptions. Driving into New Orleans on Saturday against the contra-flow traffic was thirty-nine-year-old Willie Walker, senior pastor of Noah's Ark Missionary Baptist Church, located on South Saratoga Street, about a mile west of the Superdome. The area surrounding his church was a depressingly blighted Central City neighborhood, the domain of dope dealers, garbage heaps, and a high crime rate. Many residents relied on their monthly food stamps just to survive. But the neighborhood also had a rich civil rights history. In 1957, in fact, the Southern Christian Leadership Conference was founded in Central City at the Mount Zion Baptist Church, with Martin Luther King Jr. holding court.

Reverend Willie, as his parishioners called him, was a true man of God—not the bombastic Bible-thumping kind, but a coolheaded servant of the poor. Born in 1966 at Charity Hospital, Walker was the son of a Marine sergeant who served during the Vietnam war. He may have inherited his exterior toughness from his leatherneck father. But, as in too many African-American households, that father abandoned his family; Walker was only four at the time.

Walker's mother, by contrast, was devoted to her four children, always ready to cook up a pot of okra gumbo or to read Psalms out loud. She would have them pray to Jesus as a family, hands clasped and eyes closed. Mrs. Walker worked as a secretary for the New Orleans public schools. Often the gospel records of Reverend James Cleveland and Mahalia Jackson blared out at full volume from their old-time phonograph. Although the Walker kids were all baptized in Noah's Ark Missionary Baptist Church, he and his two sisters and brother were raised as Catholics, attending St. Raymond on Paris Avenue.

Money was tight in the Walker household, but the children never complained. They thanked God for every meal. At age eight, Walker started working for Aquarius Janitorial Service, mopping up dirty floors from the Shell Building on Poydras Street to Jimmy's Bar on Willow Street. "I was always looking for a dollar," he recalls of his youth. "I just wanted to sustain some kind of income." When he turned thirteen, however, a near-death experience changed his life. One evening in October 1979, Walker was crunched in the back of his stepfather's little burgundy Vega hatchback as it cruised down General de Gaulle Boulevard near Lake Pontchartrain. Suddenly, there was a collision. On impact, Walker flew out of the car, landing on the roadway, a tire running over his limp body. Because New Orleans's ambulance drivers were on strike, it took over two hours for medics to arrive. Walker was drifting in and out of consciousness; all he remembers about the tragedy is his mother holding his head, wailing, "Lord, don't take my child . . . Lord, don't take my child" over and over again. "They didn't know whether I was dead or alive," he recalled of the accident. "I felt my soul leave my body headed to some beautiful bright light."

Eventually an ambulance arrived and Walker was rushed to Jo Ellen Smith Hospital. His face was swollen like a giant strawberry and his teeth cracked. A medic poured peroxide over his head as if baptizing him on the run. His hair turned reddish. His sobbing mother, he remembered, kept chanting "Take me instead, Lord. Oh God, take me instead." Desperately, he wanted to comfort her. Somehow, he found the strength to softly say, "It's all right. I'm back."[59]

Through a demanding combination of hourly prayer, intense willpower, and physical therapy, Walker did come back, forever changed. A favorite with the young ladies and a charismatic leader of all, he had a bright New Orleans future, a wide-open canvas in which an ambitious African American with drive could perhaps help break the cycle of poverty. But he had these haunting dreams—visions, really, with Jesus Christ talking to him. It was unnerving. He felt that he had been tapped, that making money (lots of money), his old ambition, was somehow corrupt. Like many people who have had near-death experiences, he knew there was something better than being rich. He decided to become a pastor for the destitute. Hence, in 1998, he took over Noah's Ark, hoping to give hope to AIDS patients, heroin addicts, and down-and-outers. His new pulpit was located in a small stucco building with a wooden cross on its tiny steeple. On the day he accepted leadership of the 145-member congregation, he made a sacred vow to himself: Never would he shove Jesus down people's throats like some Elmer Gantry wannabe. He cringed at the very notion of Holy Rollerism, but by his caring, loving ways, by his direct actions, he aspired to show the lost the way out of earthly hell and into heavenly salvation. No matter who you were, no matter how broken down or unlucky you were, Walker, who called everybody "dude," would offer you a hug.

Although keeping Noah's Ark open was always a financial challenge, Reverend Willie lived a charmed existence. He was madly in love with his elegant wife, Veronica, who sang mournful renditions of "I Don't Worry Tomorrow" and "Thank You Lord" in church. Together they were raising three children in a house near Kenner, not far from the airport. It was a relatively safe suburban environment in which to raise kids. Every day, however, no matter what the weather, Reverend Willie headed into the Magnolia housing project near his church to minister to the needy. Like a door-to-door canvasser, he would wander around and check up on the dudes, particularly those living a marginal existence. If the Morrises were having a marital spat, he tried to ease the tension. If "Big Boy" was selling

OxyContin for two dollars a pop to teenagers, Walker turned beat cop, confiscating the bottles and flushing the contents down the toilet. Every time he saw a young man scratching, taking on a fidgety persona, he intervened, instructing him on how to get off crack cocaine. Sometimes in Central City, however, all he could do was cry. On bad days it was a torturous ghetto that sometimes seemed too hard for love. Like when his cousin Wanda Morgan was raped in a vacant lot across from Estelle J. Wilson Funeral Home; the attackers smashed her head with a brick, causing her to bleed to death. "It was horrible," Walker recalled. "But I didn't give up on the neighborhood."[60] Whether you liked Reverend Willie or not, he was a fixture around South Saratoga Street, and even his detractors admitted that he "walked the walk."

Reverend Willie hustled around Central City at dusk on Saturday, August 27, trying to broker rides out of New Orleans for those without cars. He knew before he started that there was no way of saving everyone. The city would have to send buses, but so far, City Hall hadn't said anything about organizing convoys to help those still left in the city. He had heard that FEMA had prepositioned buses at Camp Beauregard in Alexandria, Louisiana, ready to evacuate stranded New Orleans after the storm; true, it was only a rumor, but the mirage of such a fleet made him feel better. Reverend Willie was frustrated and frantic as night fell, with no help in sight and so much left to be done; he said prayers over and over again as an incantation against evil.[61]

By then, the 263 pets from the LSPCA on Japonica Street had already reached Houston and were warm, dry, and safe in their temporary home.

Chapter Three

STORM VS. SHORELINE

> Day breaks through the flying wrack, over the infinite heaving of the sea, over the low land made vast with desolation. It is a spectral dawn: a wan light, like the light of a dying sun.
>
> —Lafcadio Hearn, *Chita: A Memory of Last Island*

I

JOSEPH CONRAD, ONE OF the greatest writers of sea tales, offered an unforgettably vivid depiction of a storm's fury in his 1903 novel *Typhoon*, with headstrong Captain MacWhirr guiding his battered steamer *Nan-Shan* through a China Sea tropical storm against the advice of his chief mate. "It was something formidable and swift, like the sudden smashing of a vial of wrath," wrote Conrad of the typhoon's first wind and rain. "It seemed to explode all round the ship with an overpowering concussion and a rush of great waters, as if an immense dam had been blown up to windward. In an instant the men lost touch of each other. This is the disintegrating power of a great wind: it isolates one from one's kind. An earthquake, a landslip, an avalanche, overtake a man incidentally, as it were—without passion. A furious gale attacks him like a personal enemy, tries to grasp his limbs, fastens upon his mind, seeks to rout his very spirit out of him."[1]

Ships usually sail around a hurricane, but inevitably, they sometimes get caught. Airplanes have more of a chance, even when the pilot, as in the case of Captain MacWhirr, chooses to take a storm head-on. At 3 A.M. on Sunday morning, a 1965-vintage WC-130 Turboprop airplane was flying from the Mississippi Gulf Coast straight into the swirling skies of Hurricane Katrina, 8,500 feet above the Gulf of Mexico. Meanwhile, the storm itself was continuing to press toward New Orleans at about 10 mph. At a distance of 300 miles, it was just over a day away. The WC-130 had taken off from Keesler Air Force Base, located near Biloxi. Home to the second-largest hospital facility in the Air Force, Keesler boasted an array of high-tech training facilities, including wide-band maintenance, ground radio, and cryptography. At any given time, the base—once host to the famous Tuskegee Airmen of World War II—employed approximately 4,700 military personnel. Most important, on the weekend of August 27–28, 2005, Keesler was the base camp of the Fifty-third Weather Reconnaissance Squadron—the "Hurricane Hunters."

While Navy and NASA planes also flew research missions into Katrina's eye wall, it was the small fleet from Keesler that gathered data for the National Hurricane Center.[2] Formally created in 1944, the Hurricane Hunters were charged with gathering firsthand information on tropical storms, including readings on barometric pressure, sustained wind speeds, and upper-level circulation patterns. Such information was essential in the effort to warn the American public about a storm's fury *before* landfall. The dramatic flights were typically bumpy as they burst through thunderclouds, but they were safe for the propeller-driven planes and the brave individuals who sent back meteorological data from them.

When the WC-130 known as *Miss Piggy* took off from Keesler at 2:30 A.M., Katrina was rated as a Category 3 storm, with winds of 115 mph surrounding a barometric pressure of 945 millibars (the lower the measurement, the more potent the storm's energy). Just after four in the morning *Miss Piggy* was nearing the eye, having rocked and rattled its way through the coiled air of a monstrous hurricane, that was almost 500 miles in diameter. While the eye was an area of perfect calm, it was protected by a wall composed of the strongest winds in the storm. *Miss Piggy,* crammed with equipment, took readings as it flew. The winds tightening around the hurricane eye were clocked at 200 mph.[3] The combination of low pressure and high winds made Katrina the most ferocious storm that any of the sixteen crewmen onboard had ever experienced. "We had no idea," admitted Michael Kelly, lead scientist of *Miss Piggy,* "how devastating it was going to be."[4]

Moreover, with a barometric pressure reading of 915 mb, Katrina was one of the most powerful storms ever recorded in U.S. history. It had joined the ranks of the Labor Day Hurricane of 1935 in the Florida Keys (winds nearly 200 mph, pressure 892 mb), 1969's Hurricane Camille in Mississippi (winds nearly 190 mph, pressure 909 mb), and 1992's Hurricane Andrew in Dade County, Florida (winds at 165 mph, 922 mb). Since hurricane records were first kept in 1851, those three were the only Category 5 hurricanes known to have hit the United States.[5] A number of Category 4 hurricanes had almost reached 5 status: New Orleans (1915), Florida Keys (1919), Miami–Pensacola, Florida (1926), Lake Okeechobee (1928), Donna (1960), and Carla (1961).[6] Ernest Hemingway noted the ravages of the Labor Day 1935 storm firsthand and wrote, "Indian Key absolutely swept clean, not a blade of grass, and over the high center of it were scattered live conchs that came in with the sea, craw fish, and dead morays. The whole bottom of the sea blew up over it."[7]

A nonmeteorological indicator of the magnitude of a hurricane was the death toll. The U.S. historical winner in this grisly contest was the Great Galveston Hurricane of September 8–9, 1900. It was a Category 4 hurricane, with 140 mph winds, 9 inches of rain, and a 16-foot storm surge. Galveston, then Texas's largest city, was decimated even though the Gulf community had constructed a 17-foot-high flood wall. Somewhere between 8,000 and 12,000 people died in the Storm, as it became known. Clara Barton, founder of the American Red Cross, rushed to the Gulf South to nurse the maimed. Stories abounded about burials at sea and mass cremations in the Great Galveston Hurricane, which originated near the Cape Verde Islands off Africa. Nobody knew who or what to blame. The impact of the hurricane was gruesome, as has been related in such fine narratives as Erik Larson's *Isaac's Storm: A Man, a Time, and the Deadliest Hurricane in History* (1999) and Isaac Cline's *Storms, Floods and Sunshine* (1945). As late as 2005, deadly stories of the Storm were still told in communities around Galveston Bay. Fifteen nuns and one hundred children at the Catholic Orphanage Asylum, for example, had all drowned. At St. Mary's Infirmary, ninety-two patients perished when the entire building crumbled like a house of cards. The death and devastation in Galveston, in fact, was so extensive that Texans largely abandoned the Gulf South maritime hub, moving to higher ground in the then small city of Houston located fifty-one miles from the sea.

No one knows precisely how a storm graduates from a tropical depression (a localized decrease in the barometric pressure) to a tropical storm (an unleashing of the energy drawn toward the depression) and then on to an outright hurricane. A hurricane is a unique phenomenon in nature, a veritable machine for drawing energy out of warm tropical waters and forcing it into the atmosphere, by way of circling winds. The warmer the ocean waters, the more powerful the hurricane can become, leading many observers to conclude that the rash of major hurricanes seen in the Caribbean in the decade after 1995 must have had some relation to global warming. As Herman Melville had perceptively written in *Moby-Dick*, "warmest climes nurse the cruelest fangs."[8] The temperature in the world's oceans did indeed increase by a full degree F in the 1990s, according to the National Oceanic and Atmospheric Administration (NOAA). The evidence suggests that this increase is due to global warming—the process by which man-made pollutants diminish the upper atmosphere's ability to filter sunlight. At the time of Katrina, according to *Time* magazine, the water in the Gulf of Mexico was actually 5 degrees F higher than usual, which triggered high levels of rainfall and in effect made the Gulf into "a veritable hurricane refueling station."[9] Many scientists at the time of Katrina were unconvinced that global warming had anything to do with the increased frequency and ferocity of hurricanes. But global warming—if the theory was true—would undoubtedly make bad hurricanes worse.

Professor William Gray, one of the most influential storm trackers in recent times, contends that several factors were responsible for the glut of storms in the last decade. Gray heads the Tropical Meteorology Project at Colorado State University in Fort Collins. While it may seem incongruous for a world-renowned hurricane expert to be hidden away in the Rocky Mountains, Gray, a specialist in the Pacific and Indian oceans, might be smart to live ten thousand feet above sea level and far from any beach. In any case, he had gained widespread attention with the accuracy of the annual forecasts he issued, with the help of his associate Philip Klotzbach and other team members. For 2005, Gray projected that there would be twenty named storms, twice the average. That estimate, which seemed so extravagant before the season began, was ultimately short: there were thirty named storms in 2005. Furthermore, Gray predicted that six major hurricanes would develop in the Atlantic, three times the average (there were indeed six)[10] and said that there was a 77 percent chance that one would strike the United States.[11]

According to Gray, the profusion of storms was related to cyclical factors, including rainfall rates in Africa and the absence of a warming trend, such as an El Niño prevailing wind, in addition to the unusually warm temperatures of the Atlantic Ocean, and the lower surface pressures. These factors converged, as they had in past centuries, to produce the unusual frequency of storms. "Between 1995 and 2003, we had thirty-two major Atlantic Basin hurricanes and of those thirty-two, only three hit the U.S.," Gray explained in an interview. "The long-term average is about one in three. So, we were very lucky then. Now, in the last two years, that hasn't been the case. We've tended to have this ridge off the U.S. coast that has driven a lot of these storms westward until they recurved in the longitude of the U.S."[12] This was precisely the path that Katrina was taking, putting it on a collision course with New Orleans.

Soon after *Miss Piggy* relayed its eye-wall data, including Katrina's barometric pressure reading of only 915 mb, another Hurricane Hunter plane broke through to the eye and registered an even lower reading: 902 mb.[13] That was the fourth-lowest atmospheric pressure ever recorded in an Atlantic storm. Katrina had the potential to grow into something even more fearsome than 1969's Camille, which was still the stuff of legend along the Mississippi-Alabama coast. But no one could imagine just what that might mean, in actuality. Katrina was a Category 5 storm at dawn on Sunday, but there were some signs that it would not be able to sustain its force for long. "We cheered each millibar of weakening," the Hurricane Hunters later posted on their Web site, "but it was with heavy hearts, because we knew it was still going to be very terrible. The day before Katrina struck, our crews were flying over the 'birdfoot,' where the Mississippi River extends into the Gulf of Mexico. With the towering anvil clouds of Katrina's outflow menacing in the background the region looked so vulnerable."[14]

Any hurricane acts like a vacuum as it drives across the ocean, literally pulling water up into its grasp. When the hurricane hits land, all that vacuumed-up water is let loose, in a great wave known as the storm surge. In the case of Katrina, several factors combined to make it a practically perfect water vacuum: it had an irresistibly low-pressure core, it was huge, and it was flowing over the unusually warm water present in the Gulf of Mexico during the last week of August. Playing off the title of Sebastian Junger's nonfiction thriller, Katrina was, in popular parlance, another Perfect Storm. Meteorologists all over America, on radio and TV, started using that tag, a euphemism for Big Disaster Soon to Come. Whatever it was, there was no way to stop it.

II

Most people along the Gulf basin, especially those who remembered Camille, knew that the U.S. government couldn't stop a hurricane. They had tried to between 1961 and 1983 by seeding clouds with silver iodide crystals as part of Project Storm Fury. But FEMA *could* help communities recover from one. And they were aware that some of the worst hurricane damage came from the surge, not the wind. A hurricane packing Katrina's punch could flood out any edifice within a mile or more of the beachfront. People with the Gulf of Mexico as a front yard lived with one sobering certainty: a hurricane would be kicking on their door someday soon. *Very* soon. No magical silver iodide crystals, like fairy dust, could scare the brute away. In fact, one reason the Gulf Coast had developed at a slower pace than other coastal areas in America was that it was considered too hurricane-prone to support primary residences. Second homes were another matter. During most of the twentieth century, wealthy and sporting Southerners built cottages along the coast, most on stilts, assuming that if a storm wiped a beach house away, it wouldn't be a great tragedy (if no one was home at the time, of course).

In the 1970s—a time when hurricane activity was at a cyclical low—development of permanent homes accelerated. In fact, the Mississippi Gulf Coast (or Redneck Riviera, as it was popularly called) became highly urbanized from one end to the other. "For the Coast, these are the best of times," wrote Philip D. Hearn in *Hurricane Camille: Monster Storm of the Gulf Coast* in early 2005. "New U.S. census figures show that nearly 90,000 people, pursuing warm climate, job opportunities, and southern hospitality, moved into Mississippi's three coastal counties and three adjacent communities immediately north between 1995 and 2000."

All over the United States, home owners were drawn to property near open water. Between 1980 and 2003, in fact, America's coastal population increased by 28 percent.[15] And in keeping with that trend, the scenic Mississippi coast was too tempting to bypass. Given the harsh potential for hurricane damage, however, insurance companies that were writing coverage there couldn't handle the risk of storms *and* of subsequent flooding. This probably ought to have sent a message that the coastline was simply too dangerous for permanent development. At some point, risk of natural disaster *should* be a consideration—and insurance companies are in a good position to determine just where that point lies. Apparently, though, the idea that people can't live wherever they want seemed restrictive and antithetical to American liberties and, so, starting in 1968, the federal government offered policies for those in high-risk areas. That invited the development boom along the coast. Even then, many residents didn't think ahead—and didn't take out flood coverage. Many poorer Hurricane Belt residents simply couldn't afford any home insurance. They had no choice but to accept the woeful old line: "Nothing, baby, is guaranteed." At any rate, by Sunday, August 28, it was too late to make long-term provisions. It was practically too late for anything except to quickly pack the car and flee. And if you didn't happen to have a car, then in most communities, you were in a very bad way.

As those who were still left along the Alabama-Mississippi-Louisiana coast on Sunday morning woke up, they heard the terrifying news from the NHC: Katrina had grown into a Category 5 storm. Incredibly, Governor Haley Barbour of Mississippi had yet to declare a mandatory evacuation. Governor Bob Riley of Alabama, which was farther east of the anticipated storm's track, was also biding his time. They may have been waiting to see where the storm would track, but they already knew it was coming and that should have spurred them to firmer action. For those stranded in New Orleans with no means of escape, the Superdome, which Mayor Nagin deemed a "shelter of last resort," opened at 8 A.M. Some nine hundred Louisiana National Guardsmen were on duty there, and no one passed through the doors without a weapons search by the Guard. That slowed the process and within an hour, a long line snaked around the huge stadium, as if the New Orleans Saints were playing the Atlanta Falcons and only one door allowed admittance as kickoff approached. "Only a fool would have waited in such a line," longtime resident Terry Jenson recalled. "And I was one of them."[16]

As the predominant shelter the city, the Superdome was not enough. Back during Hurricane Georges in 1998, Mayor Marc Morial had also opened a shelter of last resort in New Orleans East on the third floor of the old Dillard's department store. Morial was worried about the car-less in NOE—of whom there were many. They had an above-sea-level place to go. Making it downtown to the Superdome without transportation, he decided, was unrealistic. But under Nagin, the car-less were left stranded in NOE, as seemingly undisturbed as death. "He screwed up real bad," Morial later said. "What else can you say?"[17]

Reverend Willie Walker was on his way into his church, Noah's Ark, in Central City early Sunday morning. Every week, on his way into town from his house in suburban Kenner, he picked up a parishioner named Diane Johnson at her home on Tricou (pronounced "Tree-cue") Street in the Lower Ninth Ward. Johnson was not a well woman, suffering with sickle-cell anemia, a genetic disease that caused the body to be starved for oxygen. She had a husband to take care of and three grown children to monitor, one of whom was in prison. Cognizant that there was no cure for adults with sickle-cell anemia, Johnson accepted the fact that she would not live forever, or anything like it. Reverend Willie's preaching was the best kind of medicine for her. That and a gnarled ancient oak tree across the street. "My favorite thing to do," she said, "is sit on my porch and look at that beautiful tree." In the car with Reverend Walker on the way to Noah's Ark, Johnson could not engage in her usual gossip about the superettes and washaterias of the Lower Ninth. With Katrina nearing the Louisiana coastline, Reverend Walker, refusing small talk, pleaded with Johnson to evacuate her low-lying home on Tricou Street. "Over and over again," she recalled, "he kept telling me not to stay. Oh, he wouldn't stop. He offered to buy me a motel room or he would find me a hospital that would take me in. But I said, 'No, no, no.'"[18]

At 10 A.M., Advisory 23 on Katrina confirmed that "reports from an Air Force Hurricane Hunter aircraft indicate that the maximum sustained winds have increased to near 175 m.p.h. . . . With higher gusts. Katrina is a potentially catastrophic Category Five hurricane." Moving at 12 mph, the storm was now 225 miles from reaching shore. That put landfall within twenty-four hours. Most ominous of all was the amount of water that Katrina was predicted to bring along, according to this advisory: "Coastal storm surge flooding of 18 to 22 feet above normal tide levels . . . locally as high as 28 feet along with large and dangerous battering waves . . . can be expected near and to the east of where the center makes landfall."[19]

For those who understood anything about meteorology, advisories such as this had calamitous implications. Few people in the general population, however, could visualize 175 mph winds, and fewer still knew what barometric pressure was. Many just assumed that staying indoors—"riding out the storm," to use the popular metaphor—would be enough to keep them safe. And so, for all of the warnings issued by the NHC, New Orleans still seemed to be lagging in its response to Katrina on Sunday morning. At 10:11 A.M., a warning far stronger than even Advisory 23 was sent to media outlets and government officials. It seemed to slap the sleepy city awake and give it a horrible glimpse of the future. There was very little science in it, only savage imagery like some "white paper" kin of Dante's *Inferno:*

DEVASTATING DAMAGE EXPECTED. . . .

HURRICANE KATRINA . . . A MOST POWERFUL HURRICANE WITH UNPRECEDENTED STRENGTH . . . RIVALING THE INTENSITY OF HURRICANE CAMILLE OF 1969. MOST OF THE AREA WILL BE UNINHABITABLE FOR WEEKS . . . PERHAPS LONGER.

AT LEAST ONE HALF OF WELL-CONSTRUCTED HOMES WILL HAVE ROOF AND WALL FAILURE. ALL GABLED ROOFS WILL FAIL . . . LEAVING THOSE HOMES SEVERELY DAMAGED OR DESTROYED.

THE MAJORITY OF INDUSTRIAL BUILDINGS WILL BECOME NON FUNCTIONAL. PARTIAL TO COMPLETE WALL AND ROOF FAILURE IS EXPECTED. ALL WOOD FRAMED LOW RISING APARTMENT BUILDINGS WILL BE DESTROYED. CONCRETE BLOCK LOW RISE APARTMENTS WILL SUSTAIN MAJOR DAMAGE . . . INCLUDING SOME WALL AND ROOF FAILURE.

HIGH RISE OFFICE AND APARTMENT BUILDINGS WILL SWAY DANGEROUSLY . . . A FEW TO THE POINT OF TOTAL COLLAPSE. ALL WINDOWS WILL BLOW OUT.

AIRBORNE DEBRIS WILL BE WIDESPREAD . . . AND MAY INCLUDE HEAVY ITEMS SUCH AS HOUSEHOLD APPLIANCES AND EVEN LIGHT VEHICLES. SPORT UTILITY VEHICLES AND LIGHT TRUCKS WILL BE MOVED. THE BLOWN DEBRIS WILL CREATE ADDITIONAL DESTRUCTION. PERSONS . . . PETS . . . AND LIVESTOCK EXPOSED TO THE WINDS WILL FACE CERTAIN DEATH IF STRUCK.

POWER OUTAGES WILL LAST FOR WEEKS . . . AS MOST POWER POLES WILL BE DOWN AND TRANSFORMERS DESTROYED. WATER SHORTAGES WILL MAKE HUMAN SUFFERING INCREDIBLE BY MODERN STANDARDS.

THE VAST MAJORITY OF NATIVE TREES WILL BE SNAPPED OR UPROOTED. ONLY THE HEARTIEST WILL REMAIN STANDING . . . BUT BE TOTALLY DEFOLIATED. FEW CROPS WILL REMAIN. LIVESTOCK LEFT EXPOSED TO THE WINDS WILL BE KILLED.

AN INLAND HURRICANE WIND WARNING IS ISSUED WHEN SUSTAINED WINDS NEAR HURRICANE FORCE . . . OR FREQUENT GUSTS AT OR ABOVE HURRICANE FORCE . . . ARE CERTAIN WITHIN THE NEXT 12 TO 24 HOURS.

ONCE TROPICAL STORM AND HURRICANE FORCE WINDS ONSET . . . DO NOT VENTURE OUTSIDE![20]

Such alarmist passion just didn't enter into weather bulletins. At least, it never did before. Katrina was clearly a beast of a different stripe. Brian Williams, the anchor of NBC's *Nightly News*, recalled the confusion that the advisory caused at his network. "A weather bulletin arrived on my Blackberry," he wrote about a week after the storm, referring to the 10:11 advisory, "along with a strong caveat from our New York producers. The wording and contents were so incendiary that our folks were concerned that it wasn't real . . . either a bogus dispatch or a rogue piece of text. I filed a live report by phone for *Nightly News* (after an exchange with New York about the contents of the bulletin) and very cautiously couched the information. Later, we learned it was real, every word of it."[21]

The advisory had been composed by meteorologist Robert Ricks, a forecaster at the NWS office in Slidell, Louisiana. He was the Paul Revere of Hurricane Katrina. For his wording, he relied on NWS's collection of clarified phrases for hurricane bulletins, none of which had ever been used before.[22] But Ricks, who had grown up in New Orleans's Lower Ninth Ward, determined that the time was right to break with 135 years of NWS techno-jargon, and create an apocalyptic picture that anyone could understand. "I happened to be on the shift," Ricks told Williams a couple of weeks later about his foreboding bulletin. "I happened to pull the trigger. It just happened to be me that day. . . . I would much rather have been wrong in this one. I would much rather be taking the heat for crying wolf."[23] In any case, his sense of urgency succeeded in putting a scare, a very necessary scare, into the hopeful population of the Gulf Coast.

Following on the heels of the NWS bulletin, Professor Ivor Van Heerden, director of the Louisiana State University Hurricane Center, also went public with a dire warning. The December 26, 2004, Indian Ocean tsunami, which devastated vast areas of South and Southeast Asia, was very much on the minds of most Gulf Coast residents, and Van Heerden seized upon the stark analogy. "This has the potential to be as disastrous as the Asian Tsunami," Van Heerden advised. "Tens of thousands of people could lose their lives. We could witness total destruction of New Orleans as we know it."[24] Analyzing data with the same computer model used for the Hurricane Pam exercise, the LSU Hurricane Center raised the ugly specter of mass flooding, predicting that the storm surge could reach 16 feet, swelling the Mississippi River–Gulf Outlet and topping levees in Chalmette and New Orleans East. As reported in the *Times-Picayune,* the computer model also suggested that terrible flooding could occur in the Ninth Ward, Mid-City, and Kenner. The north shore of Lake Pontchartrain—the area below I-12—could also be victimized by storm surges, putting such towns as Slidell, Madisonville, Lacombe, and Mandeville at serious risk.[25]

While New Orleans was poised for a direct hit from Katrina, the coast in Mississippi was drawn into the picture more and more as the weekend progressed and the storm was seen to be curving to the northeast.

The Mississippi Gulf Coast had been settled at the same time as New Orleans. In fact, Biloxi (pronounced "B-luxi" by locals) was the original supply post for Sieur de Bienville and his contemporaries, making it the oldest city in the region. Biloxi was located in Harrison County, which formed the middle of the Mississippi coast. It was near the end of a ten-mile-long peninsula running parallel to the mainland. Both the city and nearby Keesler Air Force Base lay between the Gulf and Biloxi Bay. Not surprisingly, Biloxi, with its salubrious climate, was a busy fishing port throughout the centuries. The sight of shrimpers bringing in their fresh catch was as commonplace as that of cornhusking in Iowa or Nebraska. On the ocean, not far from the spot where the peninsula splits away from the mainland, lay Gulfport, Biloxi's twin city. Both became major gambling centers after the state legalized dockside casinos in 1990, and before long glitzy Las Vegas–style emporiums were built along the 80-mile coastal region, a string of high-voltage neon lights on a pseudo-Nevada strip called "Casino Row." The casinos had become the economic engine for Harrison County. Only a few days before Katrina made landfall, in fact, Biloxi was preparing to open a new Hard Rock Café Casino.

To the east was Jackson County, home to Pascagoula, which had a coastline developed around palm-tree-lined bays replete with fishing fleets, golf courses, amusement parks, boutiques, and restaurants. Redeveloped from the rubble left by Hurricane Camille, the coast was steadily transformed into a tourist and retirement region. On the Fourth of July, Confederate flags were often flown along the beach, amplifying the area's Dixieland reputation.

The western section of the Mississippi Coast was occupied by Hancock County, which had its own claim to the picturesque St. Louis Bay. The name of the town on its western shore was inverted: Bay St. Louis. Long used as a retreat for rich families from New Orleans, Bay St. Louis had the look of a cozy New England fishing town, with clapboard houses, a long pier, and vintage storefronts in the streets nearest the bay. During the last third of the twentieth century, Bay St. Louis grew into a suburb of New Orleans, with commuters traveling fifty-one miles to work from fishing camps, tract houses, old-style mansions, and town houses. Only one gambling hall was in Bay St. Louis, Casino Magic; and on any given night, you could see the Neville Brothers, Jerry Lee Lewis, Don Rickles, or other top-tier performers in its theater. Bay St. Louis and other towns in Hancock County had also become home to several major plants associated with the aerospace industry, including the Stennis Space Center, where NASA tested shuttle engines.

On the eastern shore of St. Louis Bay was the village of Pass Christian. Although connected to Bay St. Louis by a bridge taking State Route 90 across the bay, Pass Christian was a world away, in large part because it was just beyond the range for commuters to New Orleans. As a result, it had retained an upper-class aura, with many substantial houses and a settled atmosphere.

The entire Mississippi Coast, from Bay St. Louis in the west to Pascagoula in the east, was scored by I-10, constructed about five miles from the beach. The six-lane highway that went from Jacksonville, Florida, to Santa Monica, California, was also a dividing line, separating the old-money sophistication and new economic prosperity of the Gulf Coast from the underdeveloped, rural atmosphere of much of the rest of Mississippi, where cotton, soybeans, and catfish farms still reigned economically supreme. That was where the Bible Belt took over, fanning out to encompass the so-called Deep South. The Gulf Coast, however, felt like an extension of New Orleans (and Louisiana) and, historically, that was how it developed. Most noticeably, the coastline population was heavily Roman Catholic, attending nineteenth-century cathedrals Our Lady of the Gulf Catholic Church in Bay St. Louis and Blessed Virgin Mary Cathedral in Biloxi, both known for their architectural dignity. North of I-10, however, the Catholic influence was scant. And of course, the threat of storms was most dire in the area south of the interstate, along that sandy urban sprawl of strip malls, souvenir shops, and outlet stores. But the recalcitrant settlers of the developing Mississippi Gulf Coast were not the first to ignore nature's threat. "The ancient Mayans learned by experience to build their cities inland, away from the reaches of hurricanes and their storm surges," Massachusetts Institute of Technology professor Kerry Emanuel wrote in *Divine Winds*, "yet despite centuries of hurricane disasters, our society continues to disregard collective experience and invite future tragedy by building more and more structures in surge-prone coastal regions. Short memory is usually cited as the main reason for this irrational behavior; after all, very few of those at risk have experienced a major hurricane."[26]

That Sunday afternoon, many Mississippians could be seen standing quietly on the beach, staring out into the Gulf of Mexico. There were no dolphins at play. The noisy seagulls were all gone. Their annoying *aack* . . . *aack* was not heard and was suddenly missed. The silence from the sky was ominous. These residents would soon join the seagulls in flight to the north, somewhere up around Tupelo or Meridian. The sea was the mysterious natural god in one's life; you swam in its cooling waters, foraged food out of its abundant reefs, and meditated during its sunrises and sunsets. For children it was a playground, for adults it was a place to get away from the modern rat race. Saltwater breezes, coupled with the surging surf, were often powerful tonics, agents of the lulling and pleasant kind. But pre-Katrina beach lovers who thought a Category 5 hurricane was aiming straight at them weren't admiring the cloud formations or acrobatic terns or salty air. As survivors of Camille and inheritors of its legacy, they knew, instinctively, that their lives were about to change dramatically. Come dawn, some of their fellow beachfront colonists, those too sick or stubborn to flee, would be dead. All the post-Camille casino building would be washed away. From then on, everything along the Mississippi Gulf Coast would be PK (pre-Katrina) or AK (after-Katrina). The sea had once again lost its mystery and magnetic allure . . . all that was left was dread of the coming storm. When these residents turned their back on the Gulf of Mexico, they were closing all the previous chapters in their lives for a strange new world. Their comfort zone was about to vanish. Their communities were about to be obliterated. Their self-reliance was about to be tested. "I just came down to see what it's like before," thirty-four-year-old Joe Moffett of Gulfport said, "because I know what it's going to look like after."[27] That's what he thought.

III

One Gulf Coast entrepreneur who decided to evacuate at the last minute was Greg Iverson, owner of the Fire Dog Saloon in Bay St. Louis. A native of Minnesota, the fifty-six-year-old Iverson ran a bar that was open—legally—around the clock. It was created around a firehouse motif, its walls adorned with vintage nozzles, fire hydrants, and weatherproof gear. Dalmatian posters and gimcracks were everywhere. The Fire Dog's centerpiece was a pair of pool tables. Nearby was a jukebox heavy on 1970s classic rock from groups like the Grateful Dead and The Band. Bleu cheeseburgers and fried pickles were the house specialties. Late at night dealers and croupiers from the nearby Casino Magic would file in, eager for a nightcap after their shift. "We catered a lot to the gambling industry," Iverson explained. "And tried to be a town hub."[28]

On Saturday evening at six Iverson had decided to close the bar. The weather forecast looked bleak. He told his waitresses and bartenders to board up the saloon and then leave town. He was, after all, in the hospitality (not the mortuary) business. But Iverson didn't take his own advice. Instead, he went to Henderson Point, an exposed sliver of land jutting into the Gulf, to sleep in his own luxurious condo. In 1969 Hurricane Camille had virtually wiped out the scenic Henderson Point village. Iverson wanted to forget that hard historical fact. "But at five in the morning a friend called. 'Get up and get out of there,' he said," Iverson recalled. "I said 'What?' I actually rolled back to sleep! He called me back a half an hour later and said, 'I'm not kidding, it's a Category 5.' So Sue Belchner, my business partner, and I leapt up, turned on the TV, and holy cow! We started loading the car."[29]

On the way out of Bay St. Louis, Iverson and Belchner stopped by the Fire Dog to pick up a few files and tell the manager, Nick Breazeale, he needed to evacuate. A stubborn, recalcitrant Gulf Coaster, Breazeale was, in some warped way, looking forward to the hurricane. High winds offered a change of pace. He was something of a beloved character around the bay, a native New Orleanian who had spent four years in the Navy during the 1960s. Short and stocky with blond hair turned gray, he was as much a fixture at the bar as the statues of Dalmatian dogs. For fifteen years he had owned Nick's Catfish House, a popular seafood eatery on Route 90, and had developed a local reputation as a good cook. Refusing to evacuate the coast for Katrina, Breazeale had decided to hole up on a bed in the Fire Dog's back business office. Whether the storm was a Category 2 or 5 didn't interest him in the least. He believed in the structural integrity of the concrete building, which had survived not only Hurricane Camille, but gonzo journalist Hunter S. Thompson, who frequented the joint whenever he was in the area, drinking Chivas Regal and using the alias "Ray." In addition, the Fire Dog was situated twenty-five feet above sea level. That seemed high enough for Breazeale, who prided himself in knowing the Gulf of Mexico like the back of his hand.[30]

Just as Breazeale was settling in, however, happy to both witness Katrina and get some rest, Iverson came barging into the bar. "Nick," he shouted, pointing his finger only inches from Breazeale's face. "Get out of Bay St. Louis! Now! You're gonna die in this one!" Nick merely scoffed. Clearly his Minnesotan boss was overreacting, a panicky "snowbird" unaccustomed to Mississippi's coastal ways. With a burst of macho pride, Nick said, "I'm staying right here." As Iverson walked out the front door around 8 P.M. preparing to drive all night to Florida (he ended up in Columbus, Georgia), his parting words were: "Nick, this one's gonna kill you, I'm telling ya." Robert Ricks's terrifying NWS warning was being read on the radio as Iverson drove out of town. He knew evacuation was the right thing.

Like Laura Maloney of the LSPCA, many Mississippians truly worried about evacuating the dogs and cats of the Gulf Coast as Katrina approached. One of them was veterinarian Dr. Charlie West. He was an amiable man, thirty-eight years old, with darting eyes and a pronounced jaw. West had grown up in Waveland in the midst of a five-acre menagerie of goats, ducks, geese, horses, dogs, and cats. His parents were, quite simply, animal lovers. "I learned to get along with animals," he said, "better than people." In the Deep South, animal rights often weren't given much credence. Groups like PETA or the SPCA were laughed at in most conservative Gulf Coast communities, their employees viewed as New Orleans lefties. Although he himself was a conservative, that was not the case with Dr. West. An advocate of spay-neuter programs in Hancock County, he worked closely with the Friends of the Animals shelter and the Humane Society. But his primary business concern was running Pethaven Veterinary Hospital in Bay St. Louis. "We were located a couple of miles from the coast," West recalled, "but I still put my kennels high off the ground, fearful of flooding if another Camille ever came around."[31]

On the weekend before Katrina, Dr. West evacuated most of the animals from his clinic. Meanwhile, he had rented three rooms at the Ramada Inn in Diamondhead, on the north side of I-10, for his wife, two kids, and their cats, Big Orange Kitty and Lucy. He also had his Jack Russell terrier with him. A few of his other pets, however, including a bloodhound (Flash), a springer spaniel (Brandy), and a gray tabby (Puss), were left in the clinic. They weren't alone. A basset hound (Marty) whose owner was in Utah and a little black kitten West had adopted August 27, the Saturday before Katrina, were also left in the clinic. In addition, one of his employees boarded her two dogs at Pethaven, believing it the safest place for them.

IV

Only moments before the NWS Advisory 23 was released on Sunday morning, Mayor C. Ray Nagin broke with a long tradition. At 10 A.M., he issued orders for the first mandatory evacuation in the history of the city of New Orleans. "I wish I had better news," he said, "but we're facing the storm most of us have feared. This is very serious. This is going to be an unprecedented event. . . . I want to emphasize, the first choice of every citizen should be to leave the city." The mayor advised those who couldn't find a way out of New Orleans to make their way to the Superdome. Regional Transit Authority (RTA) buses would be picking people up at twelve locations, he said, to take them to the shelter. He asked citizens to check on their relatives and neighbors, in advance of the crisis. "This is an opportunity," he said, "for us to come together in a way we've never done before."[32] Governor Kathleen Blanco and other state government officials were also at the press conference. Blanco even relayed a message from President Bush, expressing his concern about the storm and his hope that the mandatory evacuation order would be heeded. While she didn't mention it at the news conference, she had also "told President Bush we would need all the help we could get."[33]

As Katrina grew in fury, the sixty-two-year-old Governor Blanco looked ashen-faced, even though she had applied circles of rouge to her cheeks. A native of Coteau in the state's Cajun country, Blanco was not a typical Louisiana politician. Plump and schoolmarmish, with a sad hang-dog cast to her face, Blanco became known equally for her bedrock decency and fairly charmless courtesy. She was also shrewd. Whatever Blanco was, however, she was a far cry from the populist swagger of roguish former governors such as Huey P. Long and Edwin Edwards. She always dressed impeccably, often with a double strand of pearls draped around her neck and a Louisiana state seal pin on her lapel. A teacher in her younger days, she had quit the education profession to raise six children. After working as a door-to-door canvasser in a 1983 political campaign, she decided to run for the state legislature and won. She later served two terms as lieutenant governor, becoming known for her successful effort to increase Louisiana's tourist trade. On the campaign trail she was tireless. Her fast-talking, jovial husband, Raymond, was her political guru; he was known throughout Louisiana as "Coach," because he used to be head of the football team at the University of Louisiana–Lafayette. His mere presence in the governor's mansion, constructed by Huey Long during the Great Depression, added an air of old-time politics to everything his wife did. He was a good ole boy who got business done with a nod and a wink. Both were gun enthusiasts. They liked boating in swamps. She was the queen bee and Raymond was her kingfish. With Kathleen Blanco, each appearance was less a matter of working the crowd than of pausing with each individual for an extra-long couple of seconds, always making eye contact. As it turned out, she needed every one of those voters when she decided to run for governor in 2003.

Running against Piyush "Bobby" Jindal, a personable man of South Asian descent, Blanco counted on her own Cajun districts in the south and west of the state. Jindal, a former health policy advisor to the Bush administration, figured to run strong in his native Baton Rouge and in suburban precincts all over the state. The city that would swing the election was New Orleans, and it was there that Blanco received a rude shock: during the campaign Mayor Nagin, a fellow Democrat, endorsed her opponent, the Republican Bobby Jindal.[34] It was the beginning of a political feud between Blanco and Nagin that would have a profound effect on the lack of city-state coordination during Katrina. When asked on March 15, 2006–more than six months after Katrina—whether he even reconsidered the controversial endorsement, Nagin remained unflinching. "No," he said. "The only thing I regret is he didn't win."[35]

On election day, to the surprise of the *Baton Rouge Advocate,* Blanco won New Orleans by a wide margin, and managed a 52–48 percent victory over Jindal, who was gracious in defeat.[36] Blanco became Louisiana's first female governor and, perhaps even more significantly, one of the few Democrats to hold a governorship in the South. "People underestimate her as this nice grandmother figure," Troy Herbert, a Democratic state assemblyman from Jeanerette, said. "But she can be very tough."[37] Although Blanco was an uninspired speaker, her first year as governor was surprisingly productive. She spearheaded efforts to attract new businesses to Louisiana and attacked widespread political corruption. Nobody could question her integrity: she was a profoundly honest grandmother of seven.[38] But nobody ever accused her of being a decisive executive, either. As Katrina neared, tasks such as immediately mobilizing Louisiana's National Guard in staging areas in Gonzales or Covington seemed to elude her. Unfortunately, it was her deficiencies, not her strengths, that seemed to describe her public persona on Sunday. As the *Financial Times* later noted, the listless Blanco looked "more like a woman at a funeral than a pillar of support."[39]

By the time Nagin and Blanco announced the "mandatory" evacuation, it was unfortunately not enforceable. Approximately one-fifth of New Orleans's 460,000 residents were still in the city, and a similar proportion were left in each of the surrounding parishes (approximately

900,000 people lived in these suburbs). Orleans Parish, home to the city of New Orleans, was the last of them to call for a mandatory evacuation. Nobody at City Hall, however, knew what "mandatory" meant. Certainly the New Orleans Police Department wasn't in a position to arrest all those who disobeyed the directive. As a matter of fact, dozens of officers, ignoring the police oath, had already taken flight, in a serious dereliction of duty. Nor did the city offer much in the way of assistance to those who had no way to evacuate. But, at long last, Mayor Nagin *had* finally called for a mandatory evacuation, lawsuits be damned.

V

At 11:30 A.M. on Sunday, Reverend Willie was in the pulpit at Noah's Ark, Diane Johnson hanging on to his every word. There was a tinge of bitterness in his voice as he delivered his sermon. He was not bitter that only about twenty-five or thirty people attended the service. On the contrary, he was angry that anyone at all was left to attend church in New Orleans that morning. He thought about Job 37:9 ("Out of the south cometh the whirlwind"). Immediately after the service, he renewed his campaign to move his parishioners and their neighbors to safety. A handful were able to get car rides out of town. For the rest, Reverend Willie started looking for buses. What he soon learned angered him to no end. The plan to bus the needy to safety was essentially a ruse. "My goal was to get folks in the city and school buses that I thought were ferrying everybody out of town," Walker recalled. "But don't get me started, dude. When I got down to Magnolia, there were no buses in sight. Where are the RTA and school buses? I kept asking. Why are they telling folks the Superdome was a shelter of last resort, instead of shipping them out of town? The answer I got from City Hall folks was that Nagin had bigger things to worry about. Dude, what could be bigger than people?"[40] There were buses to at least take the poor to the Superdome.

When his sermon ended and the parishioners left, Reverend Willie Walker, with Diane Johnson riding her Pronto at his side, froze on the sidewalk. The sparrows that nested near Noah's Ark Missionary Baptist Church were chattering and hopping about erratically, sensing the approaching atmospheric convulsion. They were fear-crazed harbingers of doom. They fluttered on the mausoleums in Lafayette Cemetery #2, making a panicky racket. An empty feeling swept over Reverend Willie. He wanted to head home to Kenner, where his wife and children were anxiously waiting. First, however, he took Diane Johnson back to Tricou Street, giving her all of the water and food that he had purchased that morning at Winn-Dixie. It was a strange good-bye. He told her that he would pray for her throughout the storm. His conscience, however, nagged at him, making him wish he had done more. At home, he worked the telephones, trying to find out why buses were deemed inoperable and how Mayor Nagin could be so cavalier about it. He received the same answer on all fronts, essentially "get lost."

As Katrina approached, the RTA had approximately 360 buses available. As bloggers on nola.com have since noted, each could hold up to sixty people. Therefore, RTA's buses could have ferried almost 22,000 people out of New Orleans on each trip. And that does not even count school buses and contracted private buses, which also might have been pulled into evacuation service. The RTA buses were supposed to pick up poor and elderly residents at a dozen checkpoints in order to bring them to safety. But the service was first erratic and by midafternoon practically nonexistent—a horrific oversight on the part of City Hall. "Though more than 100,000 residents had no way to get out of the city on their own, New Orleans had no real evacuation plan," Evan Thomas of *Newsweek* concluded, "save to tell people to go to the Superdome and wait for buses."[41] Mayor Nagin, in a ridiculous counter to Thomas's charge, conjured up the lamest retort imaginable. "Get people to higher ground and have the feds and the state airlift supplies to them," Nagin told the *Wall Street Journal*, "that was the plan, man."[42]

Why not bus people out of the bowl? Where did the breakdown occur? Some people blamed the bus drivers, saying that they abandoned their jobs, choosing instead to evacuate their own families. The truth was more complicated, pointing to the need for leadership in any crisis. Out of ten full-time bus operators interviewed for this book—an admittedly unscientific sampling—eight insisted that they would have stayed and evacuated the poor and elderly if City Hall had given the order forcefully. Most of New Orleans's bus drivers felt little allegiance to RTA because, under Nagin's regime, no work agreement was in force, and as a result, bus drivers received low wages. "We worked almost half a year without a contract and then Katrina came," operator Oliver Armstrong recalled. "One reason Nagin was afraid to put us to work that Saturday or Sunday is that he never had us under contract. He could have gotten FEMA on the line and said 'Let's pay them and let's evacuate as many people as we can.' "[43]

Another oversight involved having no signs clearly marking the sites where evacuation buses would pick people up. By contrast, in Miami Beach, Florida, all bus stops post a huge notice instructing citizens how to get a free ride out of town in case of emergency. Other cities in the hurricane belt did much the same thing. But New Orleans had no such uniform contingency. After Katrina Mayor Nagin often complained that he had no money with which to do anything. How much money would a couple of hundred metal signs have cost?

Amtrak trains could also have been a tremendous asset in moving people out on Sunday, without adding to the traffic already clogging the highways. Had requests been made with even a little advance notice, trains could have been stacked in New Orleans railyards, and used to transport residents very quickly out of the bowl. Arrangements weren't made, though. Moreover, when Amtrak officials repeatedly tried to offer seven hundred seats on an unscheduled train being used to move equipment out on Sunday, Mayor Nagin's office would not accept their telephone calls. "We offered the city the opportunity to take evacuees out of harm's way," Amtrak spokesman Cliff Black complained. "The city declined." As the *Washington Post* reported, the "ghost train" left New Orleans, headed for high ground in McComb, Mississippi, at 8:30 P.M. Sunday without evacuees, just hundreds of empty seats.[44] "They took out five trains from the New Orleans station, leaving empty," Armstrong, the RTA driver, later lamented. "Why wouldn't he [Nagin] put people on those trains, on those buses, and get them out of here? Those poor people couldn't go anywhere, they couldn't afford to leave. And what about the sick? I mean, you're stocking them in the Superdome, you know that it's gonna be bad."[45]

VI

There were other pre-Katrina problems besides buses. Blanco had been elected in 2003, Nagin in 2002. Two of the key players in protecting New Orleans from the approaching hurricane were neophytes when it came to dealing with the U.S. Coast Guard or the Louisiana National Guard. That needn't have been a liability, necessarily, but it did mean that the city, without someone experienced in the art of politics, needed a person who would rise to the occasion on the strength of the moment. Sunday did not lend any hope of that. Blanco was going through the motions, but not much more. She did, however, try to look like a politician in charge. Nagin wasn't even managing that much. Primping for cameras wasn't the same as having an emergency blueprint for New Orleans in the desk drawer ready to be implemented when the Big One arrived. Too much planning had been done at the last minute or not at all. Nagin lacked grace under pressure. To the contrary, he got rattled too easily, and preferred winging it to sustained analysis. With Nagin in the mayor's office and the Big One seemingly on the way, New Orleans was in deep trouble. As former mayor Marc Morial, head of the National Urban League, explained, Nagin didn't have a real disaster plan "because *he* was the disaster."[46]

When Nagin made his announcement at ten that morning, offering up the ill-equipped Superdome as a refuge of last resort, former city planning director Collette Creppell moaned. Along with her husband and three kids, she was then in the process of evacuating. She worried that the traffic was going to be unbearable, with yahoos pulling over on the median to chug beer, kibitz, or smoke cigarettes—or all three at once. To some extent, she was right. Cars that ran out of gas gave rubberneckers something to pause for—and that made for bumper-to-bumper traffic for miles on end. The two-hour drive from New Orleans to Baton Rouge was taking five to seven hours. "We were tuned into WWL 87.8, like everybody else, waiting for the mayor's press conference," Creppell recalled of that morning's trek. "Everyone pretty much anticipated that there would be a mandatory evacuation announcement. My husband was driving, the kids in the back, the dog in the way back, a picnic basket and some clothes for two days. We were incredulous that the mayor had waited until the last moment before he announced a mandatory evacuation. The mayor, the city government, knew how many people didn't have cars—why not get buses shuttling in and out as soon as possible? He hadn't even finished his first sentence about 'You need to leave,' when he added, 'However, if you can't leave, we will take care of you.' What! The city had ignored hurricane planning and the mayor knew that people had to be evacuated out. We will take care of you, if you stay, was a misleading mixed message."[47]

Creppell was dismayed as she tried to understand Mayor Nagin's thinking. She knew he had participated in the Hurricane Pam simulation in 2004, when hurricane and flooding experts explained to him that if the Big One hit, lakefront houses would take seven feet of water, snakes would appear, and packs of wild dogs would become a menace. Because New Orleans had numerous sewarage treatment plants, the entire city could fill up with fecal matter. They told Nagin it would smell like a combination of putrid fruit and a country outhouse. There was going to be no electricity for weeks, maybe months. There would be no safe water or plumbing. "Maybe Nagin had visualized it to the point that it was all a hallucination, something that couldn't really happen," Creppell went on. "Connecting the dots between the enormity of that storm brewing in the Gulf and his very incomplete evacuation plans must have been trying. He was stuck with very incomplete plans, ones he never prioritized to complete. In truth, he never had the money, couldn't find it, to do a real plan. We're talking about an inexperienced man. I just think he never had to weather any big crisis before. But sending people to the Superdome with few provisions? . . . He knew better."[48]

New Orleans's famous enclosed football stadium—the Superdome—was not designed to be a long-term campground. It was an air-conditioned downtown bubble designed by internationally renowned architect Arthur Davis. Because the NFL season was about to begin, the arena floor was laid with bright-green artificial turf, decorated with the Saints black-and-gold fleur-de-lis. Nagin announced that this city landmark was open to people who would otherwise be stranded in the hurricane. But he added the admonition that the Superdome would most likely be out of power for days, if not weeks, in the aftermath of the storm. He suggested that anyone resorting to the Superdome bring enough food and water for four days. His discouraging words were, as Creppell noted, a confusingly mixed message. Nonetheless, residents arrived in droves to take shelter at the stadium, waiting in the sweltering 93-degree heat to go through weapons screening at the door.

Many of the evacuees were ill or handicapped, while others had small children—very few were easily capable of a two-hour wait, standing in the late-summer heat.[49] "I was going to the Superdome and then I saw the two-mile line," said Tony Peterson, who lived in the French Quarter. "I figure if I'm going to die, I'm going to die with cold beer and my best buds."[50] Most of the bars in the neighborhood, however, were closed and boarded up. A notable exception was Johnny White's, which had a sixteen-year-old policy of never closing, staying open round the clock 365 days a year. (Actually, it closed once for all of two hours, fourteen years earlier when the original owner died.) It was, as the *San Francisco Chronicle* described it, a "gum-stuck-under-the-counter joint." But it had heart—taps full of heart. "I'm not going anywhere," bartender Larry Hirst later told the *Chronicle*. "This is home. They're going to have to carry me out in a body bag."[51]

The crew at Johnny White's was not alone. A sign in one French Quarter window read, "Go Home Miss Thing." Another just said, "Beat It." There were also religious slogans on plywood, like "God Save Us" and "Left for the Devil's Storm." Some residents thought that staying at home with close friends and weathering Katrina would produce a bonding experience. Blues guitar wizard Walter "Wolfman" Washington, who lived near the New Orleans Fairgrounds racetrack in the eastern part of the city, explained his rationale for staying: "I said, it'll be cool. We're gonna hang."[52]

This "gonna hang" attitude was shared by the legendary French Quarter performer Chris Owens, the most flamboyant diva ever to sashay down Bourbon Street. In 1967 she had purchased a sturdy building on St. Louis Street, which became her entertainment castle. Dripping with diamonds, and always wearing gorgeous sequined gowns, Owens was New Orleans's version of a Vegas glamour queen. Tourists flocked from all over the world to hear her sexy versions of "Hot Legs" or "My Heart Belongs to Daddy." With the help of cosmetic surgery, Owens just never seemed to age. Even though her friend Jefferson Parish Sheriff Harry Lee had demanded that she evacuate, Owens decided to stay. Her escort, Mark Davidson (a combination bodyguard and cook), was staying with her, and, well, it would be romantic in a blacked-out French Quarter, candlelight flickering from balcony windows. In just a few months, in fact, she was going to be the newest inductee into the New Orleans Musical Legends Park, her statue slated to be installed next to those of Fats Domino, Al Hirt, and Pete Fountain. "I'd been through hurricanes before," she said. "I just wasn't afraid of it. Coming from West Texas, I was stubborn. St. Louis and Bourbon was my home."[53]

Another native of New Orleans who stayed was fifty-one-year-old Ivory Clark. Short, stocky, and blessed with a hundred-watt Louis Armstrong smile, Clark was a tireless worker in any endeavor, frowning on slackers. He never soured, never lost his good cheer. Although he moonlighted as a chauffeur or gardener, his primary work was cooking. Clark was a superb chef, trained in the old-style, butter-heavy tradition of New Orleans cuisine. Breakfast with Clark was an incredible indulgence of fluffy potato pancakes, fried eggs, pink grapefruit, and boudin sausage. "Cooking starts at home," he'd say. "If you can't do it in the home skillet, you can't do it at all." He had worked his way up in the restaurant business from a start as a dishwasher, through stints as a prep cook and line cook, before finally becoming recognized as a chef in 1984. A customer needed only to name a Gulf shrimp or crawfish dish, and Clark could whip it up with the flair of a master. His specialties were shrimp étouffée and barbecue shrimp. "You learn your little tricks," Clark said of cooking. "And if you're smart, you don't share them with too many people."[54]

When Katrina set its sights on New Orleans, Clark was working in a restaurant at the Pan-American Life Insurance Building on Poydras Street in the central business district. "So it was hard for me to evacuate," Clark recalled. "I had work, plus a family who couldn't leave due to old age." Family was everything to Ivory Clark. As one of nine children, he had grown up in a household that was close-knit, despite the fact that his father was a truck driver and was often gone for weeks at a time. At a time when every extra dollar was devoted to paying off the mortgage on the house, Clark's mother brought in her own income working as a butcher.

They were good, loving parents, and a work ethic was instilled in their children. Ivory was particularly receptive.

On Sunday, August 28, as Katrina approached, Clark helped fretful neighbors board up their homes and then picked up his mother-in-law, Sedona Green of New Orleans East. Her house was only a few blocks from Lake Pontchartrain and he feared that she'd be flooded out if the levee topped. At ninety-one, however, and in frail health, Mrs. Green wouldn't necessarily survive a hot and tedious road trip to East Texas or North Mississippi. At the Clark house, Mrs. Green joined other family members dependent on Ivory Clark: his wife, Donna; two teenage children, Gerald and Jeriel; one aunt; and a niece. On hearing that Katrina had uncoiled into a Category 5 hurricane, Ivory adopted a more drastic plan. "Better a little late than not at all," he recalled. "I made up my mind that we had to leave our [three-bedroom] house on Edinburgh Street and evacuate to a motel. I got everybody together and all of us crammed into my car."[55] As Donna put it, "Ivory really got us all moving fast."[56]

VII

At 12 noon (EST) that Sunday, Michael Brown of FEMA convened a videoconference from Washington that included officials in emergency management from Florida, Alabama, Mississippi, Louisiana, and Texas. Weather experts were online. And so were President Bush from Crawford, Texas, and Secretary Chertoff, from a different location in Washington, D.C. The purpose of the meeting was to verify that all those concerned with leading the recovery effort were briefed with the same hard facts and that they were all on the same page as the Katrina storm neared the Gulf Coast. For that reason, it was extremely significant that Bush and Chertoff were among those participating in the videoconference. Katrina was being taken seriously by everyone in government. In obtaining the participation of his bosses, Mike Brown had done well.

The meeting, which would last about forty-five minutes, was generally tense in atmosphere, void of frivolity. Early on, Brown introduced Max Mayfield, who appeared from his office at the National Hurricane Center near Miami. Mayfield said, "I don't have any good news here at all today."[57] Showing a series of four slides, Mayfield compared Katrina to previous killer hurricanes. Occasionally veering into atmosphere descriptions, he didn't keep to as steady a course in his talk as the hurricane was unfortunately taking on the map, but his predictions were specific. Referring to storm surges as "valleys" in the water, Mayfield said, "You know, there's a very complex system of levees there in the New Orleans area. Some of the valleys that we see [in computer models of the hurricane's path]—and I'm sure that all of these areas [pointing to southeastern Louisiana] are already going under water out near the mouth of the Mississippi River. The colors that you see here show inundation over the land areas. One of the valleys here in Lake Pontchartrain, we've got on our forecast track, if it maintains its intensity, about twelve and a half feet of storm surge in the lake. The big question is going to be: will that top some of the levees?"[58]

The next two speakers were a researcher at the Hydranet Prediction Center and a hydrologist from the National Hurricane Center. After hearing from them, Brown said very quickly, "At this time, I'd like to go to Crawford, Texas. Ladies and gentlemen, I'd like to introduce the President of the United States."

With that, President Bush, sitting in a small, paneled conference room at his ranch, extended greetings to the group. Deputy Chief of Staff Joe Hagin was the only other one at the table with him. President Bush's tone was oddly relaxed and genial, coming in the wake of such sobering predictions of inundation. In six short paragraphs, for example, Bush used the word "folks" nine times. Sometimes they were "good" folks, sometimes "local" folks. Once, they were "FEMA" folks. Overall, he apparently wanted to be encouraging, in the vague terms of a pregame pep talk. The most pointed remark Bush made was, "I want to assure the folks at the state level that we are fully prepared to not only help you during the storm, but we will move in whatever resources and assets we have at our disposal after the storm to help you deal with the loss of property."[59]

The President did not ask any questions of the assembled experts or attempt to ascertain the level of preparedness. Indeed, Bush was cued into the videoconference only to extend his verbal pat on the back. He didn't remain for the whole meeting, having made a commitment to address the nation at 11:30 A.M. (CST).[60] Even if he was listening only at the beginning, though, he couldn't have failed to recognize the magnitude of the impending storm. Later, when observers wondered how he could have been oblivious to the potential for breaching in the levees, his defenders pointed out that Mayfield had only warned that the levees might "top." He hadn't said anything about their breaching. As Haley Barbour would later explain on the television show *Hardball* with Chris Matthews, "Dr. Mayfield said that the levees may be topped. That is, the waves may be high enough that they would go over the top. He didn't say anything about the levees being breached."[61] It was true that Mayfield hadn't uttered the word "breached." But he did speak of inundation, and that ought to have been enough to indicate the deadly nature of the threat.

Mike Brown, for his part, led the meeting in a businesslike way, expressing his concern about the use of the Superdome as a shelter. "As you may or may not know," Brown said, "the Superdome is about twelve feet below sea level . . . and I am also concerned about that roof." He made an even better point in worrying that no other shelters in or outside New Orleans had been designated for evacuees. "My gut tells me," Brown concluded, that the coming hurricane "is a bad one and a big one." His intentions were good, but something was already falling through the cracks. At the end of the meeting, Secretary Chertoff spoke up for the first time to underscore the point that if Brown needed anything from the components of the Department of Homeland Security, he need only speak up. "Secondly," Chertoff said, "are there any DOD [Department of Defense] assets that might be available? Have we reached out to them, and have we, I guess, made any kind of arrangement in case we need some additional help from them?"

"We have DOD assets over here at the EOC [Emergency Operations Center]," Brown replied. "They are fulled engaged, and we are having those discussions with them now." The Department of Defense, however, was anything but fully engaged at that point.

Three basic types of military assistance were available in times of disaster. The first consisted of the National Guard, the part-time army that was organized by the state. During domestic service, each National Guard organization was under the command of its governor. (When a National

Guard unit was called to serve overseas, it moved into the control of the federal government.) The second was the Coast Guard, a full-time force that served domestically and was part of the Homeland Security department. The third form of military assistance consisted of America's standing army, which was part of the Department of Defense. According to the National Response Plan, the DOD had to be asked to engage in domestic relief operations. Typically, the request was expected to come from the Department of Homeland Security, although the President could also direct DOD's involvement in a disaster. The course described in the National Response Plan called for the secretary of the Department of Homeland Security to declare a particular event an Incident of National Significance. At that point, DOD would assign a Defense Coordinating Officer, who would move troops and matériel to a region as needed.

On Sunday afternoon, when Michael Brown said that he was in consultation with human "assets" from the DOD, it wasn't a very impressive fact. Secretary Chertoff surely knew that whoever those assets were, they were in no position to bring in troops or equipment. That would happen only with the designation of a Defense Coordinating Officer by DOD. And that designation wouldn't occur without the declaration of an Incident of National Significance—by Michael Chertoff. So it was that in a congressional report issued September 19, 2005, on the DOD's response to Katrina, the authors could only make the damning comment, "It is not yet clear when DHS/FEMA first requested DOD assistance or what was specifically requested."[62]

At 11:30 A.M. (CST), on August 28, President Bush delivered a short, artless speech from Crawford, mainly concerned with the progress of elections in Iraq. However, he prefaced his remarks with thoughts on the impending storm along the Gulf Coast, garnered from his teleconference, which was still in progress.

> This morning I spoke with FEMA Undersecretary Mike Brown and emergency management teams not only at the federal level but at the state level about the—Hurricane Katrina. I've also spoken to Governor Blanco of Louisiana, Governor Barbour of Mississippi, Governor Bush of Florida, and Governor Riley of Alabama. I want to thank all the folks at the federal level and the state level and the local level

> who have taken this storm seriously. I appreciate the efforts of the governors to prepare their citizenry for this upcoming storm.
>
> Yesterday, I signed a disaster declaration for the state of Louisiana, and this morning I signed a disaster declaration for the state of Mississippi. These declarations will allow federal agencies to coordinate all disaster relief efforts with state and local officials. We will do everything in our power to help the people in the communities affected by this storm.
>
> Hurricane Katrina is now designated a category five hurricane. We cannot stress enough the danger this hurricane poses to Gulf Coast communities. I urge all citizens to put their own safety and the safety of their families first by moving to safe ground. Please listen carefully to instructions provided by state and local officials.[63]

For anyone along the Gulf Coast who heard the President's comments, it was practically too late to prepare. People like Wolfman Washington who decided to stay may have been only postponing the misery Katrina would bring. Those who heeded the President's words and tried to leave New Orleans on Sunday afternoon were in the midst of their misery. A last-minute evacuation of America's thirty-fifth-largest city was fraught with problems, with or without the relief of contra-flow. Those on the highways were the modern-day equivalent of the Joads, the Dust Bowlers who escaped Oklahoma in an old touring car during the Great Depression. The Road of Flight made famous in John Steinbeck's novel *The Grapes of Wrath* was Route 66; its counterpart in advance of Katrina was I-10. Both led to Southern California. Each, in its time of crisis, was paved with equal parts hope and despair.

Jake Calamusa was one of the evacuees who decided to drive out of the low-lying city on Sunday. He had his eighty-five-year-old mother with him. "It seemed from the very beginning of the week," he said, explaining why he waited so long, "we were given 100 percent probability that the storm was not coming here at all. And then all of a sudden, we're faced with the fact that we have this Cat 5 that's going to hit us head-on."[64] Likewise John Bongard, a seventy-seven-year-old from New Orleans, made a last-minute escape, driving his white Oldsmobile northwest toward Baton Rouge. In his haste, he packed only a pillow, a towel, a few blankets, some documents, and a bottle of iced tea. Long before he arrived, he realized that he had no idea why he'd taken what he did.[65] Jake Calamusa hadn't been as lucky. He couldn't drive out of the doomed city, even if he wanted to. "We tried to get on the expressway," he said, "but it was total gridlock."[66] He took his mother to a sturdy old administration building at Tulane University to hunker down for the storm.

Ivory Clark and his family were still driving around New Orleans as the afternoon went on. Clark's vehicle was a blue Dodge Intrepid, not quite big enough for seven people—but his family made do. For about an hour, Clark cruised up and down thoroughfares in New Orleans, looking for a vacancy in a hotel or motel. Eventually, he found a room at the Grand Palace Hotel, a frayed 212-room high-rise at the junction of Claiborne Avenue and Canal Street. Although the lobby was seedy and the Clarks' room on the fifth floor smelled of stale cigarette smoke, the family felt blessed to have found a safe harbor at the last minute. As Ivory stretched out on a worn old chair, he comforted himself that his family was safe. And well provided for: he had brought along Popeye's fried chicken for dinner, along with Gatorade, Twinkies, Slim Jims, Cracker Jack, Chips Ahoy, and other snacks. It wasn't the gourmet fare he would have made in his own kitchen, but it was food. As the patriarch, he had done his job in protecting his family. Contented, he dozed off in an old chair.[67]

Ben and Sarah Jaffe had good reason not to speed out of New Orleans in advance of Katrina. They were members of the family that owned and operated Preservation Hall, the temple of New Orleans jazz located on St. Peter Street in the French Quarter. On Sunday, while Ben was running around the city getting last-minute supplies, Sarah visited Preservation Hall, removing valuable Noel Rockmore paintings and old promotion posters from the venerable stone walls. Convinced that the building was secure, the Jaffes then went looking for their veteran jazz musicians, many of whom were well along in years. Such aging musicians were, to the Jaffes' way of thinking, all part of the Preservation Hall fraternity. "We were particularly worried about Marvin Kimbel," Sarah recalled. "He's a banjo player who performed with Preservation Hall [Jazz Band] up until five years ago, when he had a stroke."[68]

Unable to perform anymore, Kimbel was bedridden in his home Uptown on Calhoun Street, with his wife, Lillian, who was ninety-four and blind. "They had live-in nurses, but were still calling their own shots," Sarah Jaffe said. "We went up to their house on Sunday afternoon and spent several hours convincing Mrs. Kimbel to get out of town and to get Mr. Kimbel out. They just didn't want to leave. They've seen a lot of hurricanes come and go. But we finally convinced their nurses to pack them up and put them in a car bound for Baton Rouge. We gave them $400. The nurse was like, 'I can't drive to Baton Rouge. I'm afraid of heights.' And we're like, 'What do you mean?' She says, 'I can't get on the interstate.' So she took the long River Road route to Baton Rouge. It took them twelve hours, but they made it! We had a room ready for them at the Marriott Courtyard."[69]

During the day on Sunday, President Bush responded to requests from Governor Haley Barbour of Mississippi and Alabama Governor Bob Riley, declaring emergencies in both states.[70] Barbour and Riley were both Republicans; and Barbour, a native of Yazoo City, Mississippi, was a longtime Bush crony. He was, in fact, a fixture in the Republican hierarchy. He had served nearly two years in the Reagan White House as director of the Office of Political Affairs. From 1993 to January 1997, Barbour was chairman of the Republican National Committee. During his tenure, Republicans won control of both houses of Congress for the first time in forty years. It was the greatest midterm majority sweep of the twentieth century. Gregarious, stubborn, and possessed with enough backroad grittiness to be an Ole Miss football coach, the silver-haired Barbour understood in the pre-Katrina hours that hurricanes—as dreaded and as horrific as they were—could also produce long-range economic benefits for a region. The important post-Katrina goals, from his perspective, would be to induce the federal government to pony up for disaster relief and then use the storm as an opening by which he might deregulate industries. Barbour couldn't help but see such possibilities. Nothing about him smacked of the quitter. He was, pure and simple, a master angler.

Still, Barbour was no more forceful than Governor Blanco in next-door Louisiana in utilizing his own mandated powers. To some extent, he was hampered from taking the drastic action required by the fact that 37 percent of Mississippi's National Guard were stationed in Iraq. As it was, Barbour called up only 850 troops in anticipation of Katrina. Like Nagin in New Orleans, he also waited until Sunday to declare a mandatory evacuation.[71]

Governor Riley, making sure that Alabama's fifty-three-mile coastline wasn't neglected, also declared a state of emergency. He had asked President Bush to issue an "expedited major disaster declaration" in six Alabama counties. Between 1959 and 1999, in fact, Alabama experienced forty-four hurricanes, with low-lying peninsulas, spits, and lagoons often getting hammered. Riley was particularly worried about the more than 56,000 Alabama residents who lived south of I-10 in such vulnerable areas as Dauphin Island and Bayou La Batre, places that, it had been calculated, got brushed or hit by a hurricane every 3.53 years.[72] Mobile was also a concern because 129,000 people lived in the downtown area, many along the waterfront. "The primary threat posed by Hurricane Katrina is going to be flooding from storm surge," Governor Riley told the Associated Press. "So those being evacuated just need to make sure they reach higher ground."[73]

VIII

The staff of the Hancock Medical Center in Bay St. Louis, Mississippi, didn't wait for any official directive on the subject of Katrina. On Sunday afternoon, administrators there held a "Code White" meeting. Normally just an exercise, a Code White was an emergency drill staged in anticipation of a terrorist attack or a natural disaster. The hospital's doctors, nurses, and administrators convened in the room designated as the emergency operations center. The chief administrator, CEO Hal Leftwich, gave everybody a choice, not unlike the one handed down at the Alamo in 1836: stay or go. Some of the hospital's employees had to choose the latter, citing responsibilities to children or elderly parents. Most chose to stay, though, including emergency room physicians Sean Applewood, Ronnie Ali, Jeff Giddens, and Fredro Knight. "That Code White meeting is forever buried in my memory," said Janet McQueen, the hospital's public relations director. "It was heartbreaking to see how many of our doctors and nurses were going to help those hurt by Katrina."[74]

Since the hospital was located a good two miles from the shore and sat 27 feet above sea level, flooding from the storm surge was not a serious consideration. With the NHC reports that winds might be as high as 180 mph, the greatest fear was that the roof might blow off the three-story building. The hospital thus began a lockdown period shortly after the Code White meeting. Under Leftwich's direction, the staff spent Sunday afternoon moving all of the thirty-three remaining patients from the ground floor onto the upper two floors. "We discharged as many patients as we could, telling them to go home and evacuate while they still could," Leftwich later recalled from his office. "The ambulance service came and got our ICU patients and transferred them to a hospital in Hattiesburg."[75]

Thirty-four-year-old Fredro Knight had arrived at the Hancock Medical Center at around three in the afternoon to work the overnight shift. He was a first-rate ER doctor who received his degree at the University of Chicago. Quick-witted, deeply caring, and good-natured in the extreme, Knight had been sought after by recruiters in New York, Los Angeles, and every major hospital in between. But he was a Gulf South man, born and raised in Mobile. He had done his undergraduate work at Xavier University in New Orleans, fallen in love with the city, and decided to make its West Bank neighborhood of Algiers his home. His mother, however, had recently suffered a severe stroke and he didn't like being too far away from her bedside in Mobile. So he split the geographical distance in half: Bay St. Louis was a long hour's drive from both New Orleans and Mobile. "Although the storm wasn't due to hit until Monday morning, I knew that, when I got off Katrina would be hitting," he recalled. "I knew I wasn't going to be able to leave. Then I was scheduled for Monday night and Tuesday night, so those were my regular scheduled shifts. The ER was going to be open, so I came from Algiers prepared to stay. I brought a little overnight bag."[76]

Nothing about Katrina had Knight worried that evening. He didn't rattle easily. As a child he had been an extreme sports fanatic. He raced bicycles, jumped off roofs, and in high school lettered in varsity baseball, football, and basketball from his sophomore year onward. Fear was seldom his companion. A dead ringer for a thin Denzel Washington, Knight was the star quarterback at his high school. He received football scholarships from five or six colleges, including Tulane University. Unfortunately, during his senior year, while playing third base, he dove for a ball and broke a rib, which pierced his kidney. "I was in real pain, peeing blood all the time," he recalled. "I was only seventeen years old and needed surgery. My mother refused to let me play college sports."

Since age nine, Knight had visited emergency rooms numerous times, getting stitched up for one thing or another. In fact, his pediatrician, Fay Roberts, had become like a second mom. "I knew that when I went to see Fay she'd make me feel better," he recalled. "So all I associated with medicine was making people feel better. That was my young image of what a physician did—and that is why I became a doctor."[77]

As for hurricanes, he'd experienced them with stark regularity since childhood. They'd blow in and out of Mobile, whooshing objects about, and then it was all over. In the four years Knight had worked at Hancock Medical Center, there had been six or seven lockdowns. These had become routine. Whatever ER doctors were on the schedule at the witching hour were locked in. "You'd come in, bring an overnight bag, the storm may hit, there's a rumbling through the actual course of the hurricane, then a day or two later, you'd go home," Knight said. "Our hospital was well built and during most storms all we lost was a roof shingle or two."[78]

Two of the nurses who worked regularly with Knight, Sydney Saucier and Angie Gambino, drove to Hancock Medical Center together on Sunday with carry-on bags. They were in Gambino's fire-engine-red Dodge Durango, a four-wheel-drive SUV, which they thought would allow them to maneuver unimpeded in high water. As they headed to the hospital on Sunday for the lockdown, they checked off all the personal items they had brought: change of clothes, blankets, pillows, flashlights, lots of snacks, and chocolate. They laughed a lot and they were well fortified. "When we got to the hospital, we tried to get everybody bedded down," Saucier recalled. "We tried to assign rooms for the staff to sleep. You have staff from all departments, nursing, respiratory, radiology, housekeepers. . . . So you try to keep track of everybody. The patients that we had on the second floor, we thought it would be a good idea to move them downstairs because we didn't know how the power would do with the strong winds."[79]

Forty-seven-year-old Saucier grew up in Ethiopia, where her father worked as a TWA pilot. She spent her teenage years in Biloxi. With blond hair cut in a short pixie style and an endearing Diane Keaton–like twitter, she made everybody in a room feel at ease. There wasn't a mean bone in her body. She had taken her associate's degree at Jefferson Davis Community College in Biloxi, and then earned a master's degree at the University of Phoenix in New Orleans. At the time of Katrina she was medical unit manager, overseeing sixteen employees and fifty-seven beds, often dealing with cardiacs, diabetics, and pneumonias. "Basically, Sunday night, we tried to get everybody settled down," Saucier recalled. "And we did. We were in good spirits."

Saucier's buddy, Angie Gambino, was everybody's best friend. She could be gentle or bawdy, supportive or wild, reassuring or a hellion. But she was always, whatever the mood of the moment, trying to lift morale. "We had five ICU patients left on Sunday," she recalled. "Nobody was on a ventilator. There were post-op surgery patients and we had a lovely lady about to give birth. But we had six doctors and plenty of staff. I wasn't too worried. We could handle the situation."[80]

Bay St. Louis had established an EOC at the city and county civil defense center, but all of the sick or wounded in Hancock County would end up at their hospital. For the most part, Sunday was a pretty uneventful night for the two nurses. They kept looking out the windows to make sure the streetlights were on—they were. They'd check the Weather Channel regularly. They'd call their husbands to learn the newest Katrina coordinates. They made sure coastal friends were all right. "The wind was kind of blustery," Saucier recalled. "There was mist and rain, so it would kind of swirl, like a Winnie the Pooh cartoon. But we all, including Dr. Knight, got some rest. We knew Monday was going to be a long, unpredictable day."[81]

IX

At 4 P.M. Sunday, the first rains of the hurricane began to fall in Louisiana. By that point, those left in the danger zone had only two choices: stay at home or split the Katrina threat in half by way of a partial evacuation.

Local officials in the parishes surrounding New Orleans felt pretty good about the contra-flow evacuation as the bands of rain started blowing across coastal Louisiana. In Jefferson Parish, as many as 65 percent of residents were thought to have left. In New Orleans, the proportion was even higher, approaching 80 percent. Mayor Nagin went to his own refuge—a room on the twenty-seventh floor at the Hyatt Regency on Poydras Street—congratulating himself that under his aegis, the city boasted the highest evacuation rate in its long and hurricane-prone history. Governor Blanco's contra-flow had worked; Nagin could ride on those coattails. His mantra, in fact, became "80 percent." He had been a baseball player at Tuskegee and knew that batting .800 was phenomenal—Ted Williams or Willie Mays never dreamed of such an average.

But what about the 20 percent who were left behind because they didn't own a vehicle or were sick or were waiting for their checks? Or those in nursing homes? Obviously, Nagin didn't think they were City Hall's responsibility. All he could do was shrug about the folks who didn't flee. *"Que será, será,"* as they say in Spanish. At any rate, it had been a hard Sunday. What was most astonishing, however, was that Nagin had abandoned City Hall, where the official Emergency Operations Center had been created on the ninth floor, for the supposed safety of a high-rise hotel. His excuse for jumping ship was weak. "I remember us coming over here [City Hall] first and my security guards were a little concerned about the building," Nagin recalled. "The swaying of the building. So we decided to go over to the Hyatt." Nagin's rationale was nonsensical. Why leave a nine-story swaying building for a twenty-seven-story one? Obviously the Hyatt was going to sway more than City Hall.[82] Apparently City Hall was a suitable place for the police, RTA officials, hospital administrators, National Guard, et al., to gather and work in unison, but not the all-important Mayor Nagin. His police force was already in disarray, many having abandoned their posts. Nagin also failed to position New Orleans's mobile command center, a retrofitted eighteen-wheeler, in a safe, out-of-the-bowl place during the storm. Nagin, in fact, never explained what happened to the vehicle or why it was never put to immediate use after Katrina.

Colonel Terry Ebbert, the director of New Orleans's Office of Homeland Security and Public Affairs, ran the EOC at City Hall as the mayor used the Hyatt as his base. A decorated veteran of the Vietnam war, Ebbert had been on active duty in the Marine Corps for thirty-five years. Officially, he was in charge of public safety, whether the threat was criminal, terrorist, or natural. "The superintendents of the Police and Fire departments and the director of the Office of Emergency Preparedness report directly to me," Ebbert said on assuming his post in 2003. "I report directly to the mayor."[83] When asked whether City Hall was structurally safe, Ebbert said yes, unable to explain Nagin's insistence that for security reasons he couldn't hunker down with the rest of his team.

At 6 P.M., with the winds rising, a curfew ordered by Mayor Nagin had taken effect. By then it was like the last hour on Christmas Eve—too late to do any more. The big day was coming, ready or not.

Chapter Four

THE WINDS COME TO LOUISIANA

I am the Rider of the wind,
The Stirrer of the storm;
The hurricane I left behind
Is yet with lightning warm;
To speed to thee, o'er shore and sea
I swept upon the blast.

—Lord Byron, *Manfred,* act I, scene I

I

Joaquin "Tony" Zumbado was relaxing at home in Homestead, Florida, early on Sunday when his telephone rang. He'd been expecting a call and wondered if this was it. For twenty-eight years Zumbado had been freelancing for NBC News as a videographer. Muscular and fearless by nature, Zumbado was a "hurricane jock" who had made a specialty of capturing tropical storms on film, putting himself in harm's way to get the most stunning footage imaginable. The fifty-one-year-old Zumbado, who wore wide-framed yellow-tinted glasses, had been born in Cuba but was raised in Miami. Somewhere in his polite demeanor, there was a streak of machismo, which was perhaps evident in his choice of pets: two rottweilers named Katie and Lulu. Zumbado, his wife, Lliam, and their three daughters had a family garage band—like the Partridge Family. They called themselves the Rockweilers. Tony was essentially a homebody, until the first palm fronds started to quake. Then he was all business.

During the previous week, Zumbado had been on assignment, shooting film of Katrina as it blew across the metropolitan Miami-Dade area. He was impressed by the city, state, and federal response. Governor Jeb Bush—the President's brother—and the emergency agencies on the scene executed a nearly flawless evacuation of hurricane-threatened areas. "Ever since Andrew, Florida has really gotten their act together as far as getting people evacuated, refusing to leave the disadvantaged behind," Zumbado recalled in an October 2005 interview. "They figured out which way the storm was going and they started mobilizing people and you saw supplies pour in, depending on where the storm was going, and they had staging areas where trucks meet and prepare to distribute relief aid. And the Red Cross and FEMA were very hands-on, helping people within hours after the storm. It was impressive."[1]

With Katrina rumbling its way north in the Gulf of Mexico, taking aim at the Mississippi-Louisiana coast, NBC News producers called Zumbado on August 28, wanting him to take his cameras to New Orleans and film both the storm's fury and its aftermath. The water in the Gulf of Mexico had risen above 85 degrees, a perfect incubator for hurricanes to flourish. "Of course I said yes," Zumbado recalled. "My soundman was to be Josh Holm. He was only twenty-one years old and this was his first hurricane." Well over six feet tall, Holm—an incurable fan of the Florida Panthers hockey team and of country singer Alan Jackson—was a native of Minnesota who had been drawn to Florida's warm weather. He originally worked in an automobile business, but switched into sound work through his friendship with Zumbado. "Tony kept his motor coach in the storage facility where I worked," Holm later recalled from his home in Pembroke Pines, Florida. "We became friends. His line of work was daring, but I knew he'd never put me in unreasonable danger."[2] Together they drove across the Florida Panhandle in a fully equipped, forty-foot-long white

Hallmark motor coach—which slept six and had a kitchen, den, and bathroom—with a Ford Econoline 350 van hitched on the back. What surprised Zumbado, as he drove west on I-10, was that he didn't see any relief trucks heading into New Orleans. It was the first red flag signaling that something was terribly wrong. Nor were there power-and-light company vehicles on the road. "I kept saying to Josh, 'Wow, I wonder where they're at. Maybe they're up north. Maybe they're in Georgia and they're going to come down into New Orleans.' The point is, after twenty-eight years of doing this, I was stunned not to see the support teams I was expecting to see. It made me nervous."

Once they got to Louisiana, Zumbado and Holm headed for Gonzales, fifty-seven miles northwest of New Orleans, where they stashed their motor coach in a Pontiac lot on high ground, void of nearby trees. It would stay hidden there in reserve, just in case the city flooded and they needed a dry, mobile place to live. They then drove the van to New Orleans. "That's when another red flag appeared," Zumbado recalled. "There were no police officers visually on duty in New Orleans when we arrived, no law enforcement trying to tell you not to come into the city or trying to check your IDs. It was weird."[3]

The lightning rod of NBC's New Orleans effort was fifty-two-year-old Heather Allan, based at the Los Angeles bureau. Born in South Africa and educated at the University of Witwatersrand in Johannesburg, Allan had spent most of her twenty-five years at NBC in war zones, disaster areas, genocide villages, and violent cities. Allan was NBC's troubleshooter, or as she put it, "queen of the shitholes." When Beirut was being bombed to smithereens, and 241 U.S. servicemen died in a 1983 terrorist attack, she was on the scene. When the Marcos dynasty was overthrown in the Philippines, and Corazon Aquino ushered in a new democratic era, Allan was there. Whether it was setting up a bureau in Kuwait during Desert Storm, bringing camera crews to Rwanda to film the piles of human skulls, or witnessing California's execution of Tookie Williams, Allan was, as Humphrey Bogart might say, "one tough broad." She had covered Ethiopia's famine, Somalia's civil war, and a toxic gas crisis in Cameroon. "When they have a big breaking story, I step in," Allan explained. "It's not always bleak, though. I've set up bureaus for the Olympics in Sydney, Salt Lake City, and Athens."

Make no mistake about it, Allan was a force of nature who didn't suffer fools gladly, but tolerated all *good* people. Her medium-length blond hair and rosy complexion belied her extraordinary sense of authority. Allan's casual manner (she often wore shorts and sandals while mothering her NBC family) was misleading. Recognizing early on that New Orleans was likely to be the epicenter of Katrina, Allan put together a hurricane team of freelancers and company technicians. As Zumbado and Holm drove from Florida, a number of her favorite reporters—Brian Williams, Carl Cantania, and Martin Savage—were on their way to Louisiana from Washington, D.C., and New York. She had booked a dozen rooms for her people at the Ritz-Carlton on Canal Street starting Saturday. Everything was off-kilter in New Orleans, however—including the hotel clocks, which were running behind time. Call it bad karma or impending doom, but the *feel* wasn't right; there was a beast already in New Orleans before the first raindrop fell. "We spent all of Sunday at the Superdome," Allan recalled. "It was truly bizarre. Nobody knew what was going on. We were ordered not to park our satellite trucks in the back ramp of the Superdome because it would flood, because our blue mobile was too heavy. But I saw huge National Guard trucks go up the ramp. Nobody was in charge of anything. It was all just a disorganized mess."[4]

When Zumbado arrived at the Ritz, he assumed that he and Holm would be practically the only guests. In Florida, when a hurricane was on the way, the first thing local officials usually did was evacuate all of the tourists—that obvious gesture, after all, took only a smidgen of common sense. "When we got to the Ritz, it was packed with people and, oh boy, did another red flag pop up," Zumbado recalled. "Normally, all these hotels are closed because obviously, everyone shuts down their businesses; they board up and they get out of town. Usually, the only way NBC can get us a safe haven is if we beg a manager to let us in, and then only after we agree, or we sign a legal document saying, that we're responsible for our own being." At the Ritz's registration desk, the manager said, "We don't know if we have a room for you anymore." Despite having reservations, Zumbado had been bumped, but that wasn't what irked him. Incredulous, he shot back: "What do you mean, you don't have a room? There's a hurricane coming. Didn't people evacuate?" To this, the manager replied, "No, a lot of people are coming in here and staying here." All Zumbado could do was walk away in bewilderment.

Zumbado was enough of a newsman to realize that there was a big story in the fact that the Ritz—and other hotels in New Orleans—were *sold out.* Never before in modern American hurricane history, to his knowledge, had a below-sea-level city—or any city—allowed citizens and tourists to stay in the face of a monster hurricane. "So we started videotaping this strange scene of people packing into the Ritz, scattered everywhere on every floor, in the lobby and in the hallways and in the staircases, up and down on every level," Zumbado recalled of his Sunday, August 28, arrival. "But the manager and police jumped in front of us and said we couldn't film. It was like they didn't want people in the outside world to know what was going on." The Ritz-Carlton, known for serene luxury, no doubt didn't want NBC News making a record of its descent into chaos.

Zumbado and Holm went to scout neighborhoods, doing rounds in the van and getting the lay of the land. They realized that the Ninth Ward was vulnerable to flooding from three sources: the Industrial Canal, the Intracoastal Waterway, and the Mississippi River–Gulf Outlet. They visited all the most vulnerable spots along the Industrial Canal levee in the Ninth Ward. Zumbado wondered why New Orleans's protection system—unlike that of the Netherlands, which had built a first-rate levee-dike system after a huge flood in 1953 and upgraded it continually ever since—seemed to have such shoddy-looking levees. Occasionally they saw a lone ambulance or a solitary bus on the desolate streets, but, as Zumbado put it, "nothing felt organized." Back at the Ritz, fifteen or twenty police officers were sitting in the lobby—one of them told the NBC freelancers where to stash their van and where they could go for the best chance to avoid flooding, warning them that the Ritz parking garage would probably be underwater. "Then what are you guys doing here?" Zumbado asked the officer. "You won't be able to rescue people; you'll be trapped here at the hotel. Why not take off for the periphery of the city, keep your vehicles safe, then come back in after the storm?" All the officer could say was "Well, this is where they have us staging."[5]

Zumbado noticed a family standing in front of the Ritz with a pirogue—a type of canoe—strapped to the top of their car. Thinking ahead, he arranged to rent the vessel for the duration. "We felt good," Holm recalled. "If we got stuck anywhere, at least we had a boat."[6]

Although Governor Blanco later was to be criticized for leaving Louisiana unprepared for Hurricane Katrina, she had authorized the adjutant general of the Louisiana National Guard, General Bennett Landreneau, to mobilize 2,000 soldiers and airmen. Eventually that number reached 4,000, and, in hindsight, Governor Blanco should have called for the maximum number from the outset. Her Department of Social Services had properly identified shelters and Red Cross facilities in Baton Rouge, Alexandria, Monroe, and other areas. It wasn't her fault that the Red Cross refused to go into New Orleans.[7] Even Marsha Evans, head of the Red Cross during Katrina, blamed City Hall and FEMA for the posthurricane debacle. "Louisiana had a plan," Evans said. "It's New Orleans and FEMA that really didn't have much of one."[8]

One measure Governor Blanco took—above all others, one that proved exceedingly wise—was ordering the Louisiana Department of Wildlife and Fisheries (LDWF) to pre-position more than 200 vessels at regional locations outside the coastal parishes (and Orleans Parish). As Katrina approached, LDWF agents from Shreveport, Monroe, Ferriday, and Alexandria congregated at Woodworth, a town just south of Alexandria. If widespread flooding occurred, Governor Blanco wanted to make sure LDWF agents and boats were poised for search and rescue missions. "I greatly appreciate the authority granted by Governor Blanco," Dwight Landreneau, secretary of LDWF, later wrote in *Louisiana Conservationist*, "to assign all LDWF assets to the critical search and rescue mission."[9] As LDWF Lieutenant Colonel Keith LaCaze later noted, the governor told his organization that if flooding occurred, their mission was to "get people out of the water."[10]

Some New Orleans Police Department officers, like forty-nine-year-old Tim Bayard of Vice and Narcotics, roamed the streets until the very last second on Sunday. He was operating out of the 1700 Moss Street building. In a frenzy of last-minute improvisation, Commander Bayard, who had grown up in the Ninth Ward, tried to place police cars out of harm's way. He was fuming mad because Police Chief Eddie Compass, whom he couldn't locate, didn't understand that New Orleans was going to flood—it was guaranteed. Bayard was a longtime NOPD veteran who, when he joined the force in 1976, was too young to even purchase a gun; his father, a fireman, had to buy it for him. "I lived through Betsy, so I knew where it was going to flood," Bayard recalled. "My dad was on the fire department for Betsy. So I remembered going with my dad and my brother and cutting relatives' roofs and everything else. So what was going to change now? Nothing's changed in the city." The entire NOPD had only five emergency boats, two of which Bayard said didn't work. "We kept the boats right across the street from HQ, in the shed," Bayard said. "Captain [Robert] Norton was trying to get them up and seaworthy. But it was futile."

Realizing that most New Orleans streets would flood, Bayard moved Narcotics to the Maison du Puy Hotel and Vice to the Marriott. He divided them up in case one division became immobilized in floodwater. All evening, however, he was cursing Mayor Nagin for keeping all the Regional Transit Authority rescue buses inside the bowl. "What Nagin did killed me," he said. "On Chickasaw Street, where we kept some of our cars, we knew it would flood. Our cars were by the bus barn for the RTA. Next to the Desire projects. That's the Ninth Ward. The school buses were three blocks away from there, on Metropolitan Street. If you read the hurricane evacuation plan, those RTA buses should have been taking people to the evacuation center, to the Superdome, sixty hours prior to the storm's arrival and the school buses should have been used. Then, as the storm approached, they could have evacuated people out of town and been on high ground."[11]

II

If you looked around downtown New Orleans Sunday you would have seen media trucks and reporters from everywhere. Fox News had sent its first tier of reporters—Shepherd Smith, Geraldo Rivera, and Jeff Goldblatt—to cover Katrina. At thirty-six, Goldblatt was a broadcasting veteran, having started his career right after graduating from Colgate University in 1991. He had been an itinerant reporter, holding jobs in Myrtle Beach, Wilmington

(North Carolina), Richmond, Washington, D.C., New York, and Miami before being hired by Fox in 1999. Handsome, with wavy sandy brown hair and a noticeable scar on his cheek, he was based in Chicago. His wife was seven months' pregnant with their second child. Fox News was notorious for operating on a shoestring, making sure its reporters were resourceful, not extravagant. On this assignment Goldblatt's first frustration was getting nothing better than a cheap rental car at the airport, because, as he put it, "there were lines out the wazoo."

Eventually, he got a worn-out SUV and headed straight for Jackson Square in the French Quarter to report on how ridiculous it was that people were partying on Bourbon Street with Katrina on the way. Earlier in the summer, when he had covered Dennis, Floridians treated the oncoming hurricane with respect. The French Quarter revelers he encountered, by contrast, seemed to be thumbing their noses at the sky. Not that hurricanes couldn't be comical. Back in September 2003, in fact, Goldblatt became a one-day national joke when his pants almost fell down while he was covering Hurricane Isabel in Virginia Beach; comedians Jay Leno and Jimmy Kimmel played the clip on their late-night shows. In New Orleans, Goldblatt saw scores of people milling about as if waiting for a throned Rex to appear waving from a gaudy Mardi Gras float. "It was all eerie and strange," Goldblatt recalled. "All these people were staying in New Orleans. On Sunday I went to the Superdome and it was filling up. That was incredible to me. During Ivan, with a few thousand people in there, it was my understanding that they ripped the seats out afterward, that there was theft in the Superdome. But they were back at it again."[12]

Goldblatt was also amazed at how young the National Guard soldiers at the Superdome were. Oliver Wendell Holmes once said you know you're getting old when the cops look like kids—that Sunday Goldblatt felt old. The African-American teenagers trying to enter the Superdome were forced to cough up their personal possessions, the Louisiana National Guard confiscating ice picks, knives, razor blades, guns, and bullets. Gaggles of people were obviously drunk, acting as if the entire evacuation were a tailgate party. The mantra of the moment was "Let me in . . . let me in." Eyeing a few women patiently waiting for their turn to enter, Goldblatt went over to talk. To his surprise one of them calmly said, "Look, it's our own fault that we're here. I commend the city for at least giving me this opportunity to have a place of refuge."[13]

That night, Goldblatt, his cameraman, Robert Lee, and his soundman, Mark Jeter, holed up at the W Hotel on Poydras Street in the central business district ten blocks from the Superdome. The W was a glass tower, and the Fox News team prudently congregated on the fourth floor. By 3 A.M. on Monday, they could feel the building sway and hear thuds as roofing material was ripped off. At this juncture Goldblatt was relying on his satellite phone to bring images of stormy New Orleans to a national audience; the W's dish couldn't transmit under such stiff winds. "We couldn't go out and give great coverage," Goldblatt explained. "But all the people who left their Gulf Coast towns, who were now in Georgia or Texas, want to see their hometown on TV. It's understandable. So we showed what we could of New Orleans. Not until midmorning Monday was it safe to travel about."[14]

The most respected broadcaster in Louisiana as Katrina approached was WWL radio's Garland Robinette, whose reassuring lilt everybody was listening to as the hurricane approached. Reared near the small Cajun community of Des Allemands, behind a Humble Oil swamp camp, Garland understood rank poverty. Feisty and independent-minded, with large wire-rimmed glasses and a Ronald Reagan smile, Robinette had been kicked out of Nicholls State University for punching the golf coach. He had a temper. He flunked out of the University of Southwest Louisiana. "Then I managed to flunk out at LSU," he said. "I wasn't a Rhodes scholar, I was a gravel scholar." Unable to use college as a fig leaf for avoiding U.S. military service, Robinette was sent to Vietnam, serving as a Swift boat officer in the Navy. Like so many in his generation, Robinette had experiences in the Mekong Delta from 1968 to 1969 that changed his outlook on life. Never again would he fully trust the government. Wounded twice, he received two Purple Hearts. Only you couldn't get him to talk about the destruction he saw; it was too painful. Mention Vietnam and Garland was apt to get up and walk out of the room.

Once back stateside, Robinette got a job as a janitor. Every day he also did an agricultural report on a small Louisiana radio station. Not for a Bourbon Street second did he think he had a career in broadcasting. But he enjoyed doing the "Soy Bean Report." And with his clear, soothing voice, Robinette was bound to be discovered. His break came in 1970, when the news director at WWL-TV, in desperation, offered him an anchor position. "I got my job as a part-time anchor at WWL," Robinette recalled, "because the main anchor got drunk on the air, and they needed immediate fill-ins. They put me on temporarily."

It didn't take long for Robinette to be discovered again, this time by a New York agent who thought he had national potential. An audition was set up at CBS News in New York, where he was introduced to Walter Cronkite and Barbra Streisand, who was in the building filming a prime-time musical special. But New York didn't appeal to Robinette—he felt out of his league. His great passions were Louisiana women and portrait painting, and he missed the back bayous, where life passed at a slower pace. So he quit the national news and returned to the CBS affiliate in New Orleans in 1971, where he felt his reporting might just make a difference. For the next fifteen years, he was the most outspoken local advocate of coastal restoration, warning New Orleanians that destroying Louisiana's fragile ecosystem was a death warrant for their city. He made hard-hitting documentaries for WWL such as *New Orleans: The Sinking City* and *The Dying Louisiana Wetlands*. And he hammered away at local companies like Freeport-McMoran for dumping waste products into Louisiana's waters and exploiting developing nations like Indonesia. He loved to stick it to firms that had oil spills and strip-mining accidents. "Much of my environmental awareness made my documentaries a bit one-sided," Robinette recalled. "Eventually I learned there was gray. For instance, Greenpeace would arrive at Canal Street hanging plastic signs on fences with copper wiring. They were hypocrites. Corporations weren't the bad guys per se; we as a society had to change our priorities."

Eventually, Robinette, the most beloved TV anchor in New Orleans's history, got out of the media business, tired of being "Ted Baxter" as he put it, referring to the airheaded anchor on *The Mary Tyler Moore Show*. He went to work for the enemy—Freeport-McMoran—for twelve years, finding ways the company could improve its environmental record. Environmentalists shouted "Judas," believing Robinette had betrayed the green movement. Ornery as always, and tired of having bosses, he quit in 1998 and formed his own crisis management/communications company, Planet Communications. "Done, done, done," Robinette recalled of being an on-air reporter. "Never did I want to get back into that racket. TV was humiliating. I went into the studio to read words, had a lot of hair, so I kept the job."

Everything was on the upswing for Robinette in the summer of 2005. He was married to Nancy, his fourth wife, and they had an eight-year-old daughter. His elegant portraits—including ones he painted of Pope John Paul II and Pete Fountain—were selling for top dollar nationally, and the only media people he chose to watch or listen to were Don Imus and Jon Stewart. "The rest," he said, "just didn't interest me." However, he got a call from his friend David Tyree, who had a popular talk-radio show on WWL with bad news: Tyree was suffering from prostate cancer and wanted Garland to fill in while he went through chemotherapy. "We had a confidential understanding that I'd do it temporarily," Robinette recalled. "But the second he was able to return, it was back to painting for me."

So it was that on Sunday, August 28, Robinette headed to work at 1450 Poydras Street, a high-rise that housed the WWL studio. The station had been broadcasting on a clear channel since 1922, so it was certain to have a lot of listeners for Katrina updates. He stopped at CC's coffee shop on the corner of Magazine Street and Jefferson Avenue. Standing in front of CC's, he got goose bumps. Robinette recalled, "It was Vietnam all over again. I looked up. There were no green parrots in the palm trees. I looked down the street, not a stray cat." Right before combat in Vietnam all the animals instinctively disappeared, particularly the birds—the jungle just became quiet. Although he was a veteran of numerous hurricanes, this was the first time the pre-storm silence unnerved him. He headed back home and took every drawing of his wife and daughter off the walls. "I'd already evacuated my family to Natchez," Robinette said. "Now I raced to the station—I had to be on the air in a few minutes—bringing a handful of my art inside with me."

As he settled in front of the microphone that evening, Garland Robinette didn't mince words. He told his audience about the vanishing parrots, and about his experience when the birds stopped chirping in Vietnam. Then, in uncharacteristic fashion for a professed libertarian, he blurted out, "You're going to think I'm stone-cold crazy. But the birds are gone. I know the powers that be say not to panic. I'm telling you, panic, worry, run. The birds are gone. Get out of town! Now! Don't stay! Leave! Save yourself while you can. Go . . . go . . . go!"[15]

III

As the national and local media focused on the impending hurricane, twenty-one-year-old Cody Nicholas was in a 125-foot utility boat that had been operating out of Port Fourchon, Louisiana. He had just evacuated the last rig workers from Apache oil platforms 205 and 206 in the South Tem block area of the Gulf of Mexico. Most of the oil companies had already evacuated workers. Chevron Texaco had pulled out 2,100 employees on Saturday while Royal Dutch Shell did the same for 1,000 of their own. Nicholas was doing the final roundup on Sunday for his company. The Louisiana Offshore Oil Port, which held the distinction of being the biggest oil-import facility in America, had halted the off-loading of vessels in preparation for the storm.[16] Ferrying back and forth between Fourchon and the platforms had made Nicholas's stomach turn. The water had been beyond choppy. Eight- to ten-foot waves, watery spires, caused the utility boat to bob up and down like an especially nauseating roller coaster. "We were getting our asses kicked," Nicholas later recalled. "By the time we brought our last platform workers to safety in Fourchon the winds were too high for them to be helicoptered out. But the oil companies, I believe, had vans and stuff so they could go, barely outracing the storm."[17]

A heavyset boatman from Mississippi, Nicholas sported a thin moustache and a face full of peach fuzz. For the first time in his young life, the Gulf of Mexico had turned madcap and violent on him, heaving with a sob and a sigh. The unmistakable advent of Katrina had left him truly startled. Now that his crew had locked down the oil platforms, they were racing to dodge Katrina themselves, fleeing full throttle up Atchafalaya Bay. These escapologists planned to ride out the storm in the petroleum port of Morgan City, Louisiana, far enough west to be out of harm's way. To ease the fears, the boat captain allowed Nicholas to play George Strait and Garth Brooks songs on the vessel's CD player, but the raging wind made the pop music hard to hear. "By the time we hit Morgan City we were getting steady fifteens," Nicholas claimed, referring to the height of the waves. "We docked, double-tied the old utility boat on every bet, and hoped for the best. We were going to have to sleep on the boat in the water. We knew we'd have to hang on for dear life."[18] By then, Katrina was bulldozing to the Louisiana coast. It was the hour when veteran boatmen rose to the occasion by clinging to shore.

Another riverman, Jimmy Duckworth, had already taken refuge at home on Old Metairie Road in New Orleans. Some New Orleanians believed that no one knew the byzantine waterways of Louisiana better than Jimmy Duckworth. Because he was going to leave active duty with the U.S. Coast Guard Reserve in a few days, he was going to sit Katrina out. Duckworth's great-grandfather had started the family business, Jimmy Duckworth Tires, in the late 1920s. Until he was twelve, Jimmy lived a block from the Mississippi in Jefferson Parish. Outside of business, Duckworth men had a primal passion for hunting and fishing in Plaquemines and St. Bernard parishes (and the Mississippi marshes). Their eighteen-foot Glaspar boat was essentially their second home. Gregarious and extremely well-mannered, Jimmy Duckworth befriended shrimpers in Pointe à la Hache, caught redfish in Black Bay, reeled in speckled trout at Triple Pass, and crabbed off the Lake Pontchartrain seawall. He became a wetlands rat, able to pilot a motorboat all over the confusing marshlands of coastal Louisiana with the ease of a latter-day Jean Lafitte.

When Duckworth entered Louisiana State University in Baton Rouge in 1974, he was already known by the hardworking watermen of the state. "I also belonged to Ship 46, a troop for sea scouts sponsored by St. Charles Presbyterian, and later got my Coast Guard license through a lot of sea time that I had gained working on some of the longer sea scout vessels," Duckworth recalled. "I then went to work downtown, at the foot of Canal Street, at nights for Streckfus Steamship Company, which owned the excursion boats."[19]

Although Duckworth admired all the various men who plied the mighty Mississippi, he became especially enamored of the Coast Guard—enough to earn a commercial captain's license in 1977. They were, to his mind, the best watermen of all—and they saved lives. In January 1983 he joined the Coast Guard Reserve and became a search and rescue coxswain. Operating a variety of boats for the Coast Guard was a childhood fantasy come true for Duckworth. He loved exploring the entire Mississippi River–Gulf Outlet (MRGO) and other state waterways. But as the years went by he grew saddened that his favorite childhood fishing holes were disappearing. Take, for example, Lake Athanasio, a gulfside body that was incorporated into the man-made MRGO in the 1960s. As a kid Duckworth had fished the drop-off in the lake. But dredging and the construction of oil pipelines allowed Lake Athanasio to erode. The subsequent intrusion of salt water into the MRGO, and into the wetlands along the entire coast, according to Duckworth, "just plain killed a lot of old oak trees and beautiful marshes." It pained him to watch his beloved marshes disappear.

From 1987 to 1990 Duckworth spent weekends around Venice in Plaquemines Parish, leading Coast Guard search and rescue missions. The first town upstream on the Mississippi, Venice was a strange outpost for fishermen and oil-platform rig hands. It was also the gateway community for the Delta National Wildlife Refuge, where a dazzling array of animal species coexisted with oil derricks. What Duckworth liked best about Venice was pulling his Coast Guard boat up to inspect the fresh catch of the local fishermen; on shore, he also enjoyed chatting with geologists, petrochemical engineers, and roughnecks. There was a "last chance" mentality that permeated Venice, reminiscent of the venerable trading posts that used to dot the Wild West.

After Venice, Duckworth's next assignment was at Station New Canal, the old lighthouse on the New Harbor Canal across from the Southern Yacht Club. Then he was assigned to the Coast Guard base at the Industrial Canal lock along St. Claude Avenue. In 1990, as an ensign, he was reassigned to the Marine Safety Office, located on the seventh floor of the Tidewater Building on Canal Street, working for the captain of the Port of New Orleans. The Coast Guard moved talented mariners like Duckworth around frequently so they would learn the waterway system surrounding Greater New Orleans like the backs of their hands. He knew where the Jefferson Parish canals along Lake Pontchartrain were. He knew how the locks at Empire, Algiers, the Industrial Canal, Harvey, and Port Allen (crucial to the Intracoastal Waterway System, moving east to west) worked. He didn't study nautical maps in a classroom—he experienced the waters in a boat. Through it all, Duckworth agreed with Mark Twain, who recalled how, when he was an apprentice river pilot in 1857, his teacher explained that even if he recognized every detail of the 1,039 miles of river between New Orleans and St. Louis, he might have to "learn it all again in a different way every twenty-four hours." This was especially true in Louisiana, with its ever-changing marine topography. Duckworth also embraced something Harnett T. Kane wrote in *The Bayous of Louisiana:* "It is a place that seems often unable to make up its mind whether it will be earth or water, and so it compromises."[20]

Besides memorizing the water topography of the greater New Orleans area, Duckworth participated in port operations, foreign vessels inspections, marine investigation and inspection, and emergency rescue; and in June 2001, he was promoted to the rank of lieutenant commander. He was a crack rifle shot, participating in national Coast Guard competitions. All the time Duckworth was in the Coast Guard Reserve, his real job was running Duckworth Tires and Repairs on Old Metairie Road. Besides raising two daughters and getting divorced along the way, he also found the time to build a fully operable replica of a World War II Higgins boat, the locally produced craft used in the D-Day landing in World War II (the boat went on display in 2000, when the National D-Day Museum opened in New Orleans).

The terrorist bombings of September 11, 2001, dramatically changed Duckworth's life. Because 20 percent of all imports and exports in the United States went through the Port of New Orleans, he was called to active duty. His father had to run the tire shop on his own. Duckworth was put in charge of port security, and in the months leading up to Katrina he was chief of port operations. A wide range of new security concerns had arisen since 9/11: if a boat full of explosives rammed into the Mississippi River Bridge in New Orleans, for example, it would block barge traffic for months, probably causing a severe economic recession in America. Duckworth also worried about hurricanes, which meteorologists were predicting would be increasing in menace.

In the runup to Hurricane Ivan in 2004, Duckworth had come up with a plan placing the Coast Guard's staging center in Carville. A small contingent of Coast Guard officers and the captain would wait out the storm in this town sixty-seven miles from New Orleans. On the evening Ivan approached, Duckworth's superior, Captain Frank Paskewich, didn't like being hunkered down, away from potential trouble spots. What, the captain wondered, if the Industrial Canal levee broke or MRGO overtopped? You would need massive search and rescue operations operating the second the winds died down. In the middle of the night as Ivan was approaching, Paskewich contacted Duckworth over the two-way radio. He wasn't satisfied with the plan: "If this storm kicks our ass, we'll be of no help rescuing desperate people. We'll be stuck here with not enough equipment and manpower. I'm not happy with this command post. Next time we've got to set up elsewhere."

Just as Governor Blanco fixed contra-flow after Hurricane Ivan, Paskewich intended to fix the Coast Guard's search and rescue approach before the next hurricane season. Paskewich was a spark plug who knew his job from the ground up. A graduate of the Coast Guard Academy, he was a consummate officer. Under his watchful eye, no aspect of marine safety went unchecked, be it responding to a ruptured pipeline or a damaged rudder, or a ship that had run aground. Hurricane Ivan had just barely missed New Orleans, but the Coast Guard had to learn from the near hit. Paskewich wanted a new command post, one that was bigger and better and would run like clockwork.

They chose Alexandria—a city along the Red River in central Louisiana—because there was a convention center that was conveniently located next to a few hotels. "The idea was that we could walk to and sleep in the hotel," Duckworth recalled. "We could work twelve hours off, twelve hours on." On Thursday, August 25, as Katrina approached the tip of Florida, Coast Guard Sector New Orleans moved its operations to Alexandria. By Friday, August 26, all of the New Orleans Sector Coast Guard functions were established there.

That Friday, back in New Orleans, Duckworth had grabbed a quick lunch with Paskewich at Coast Guard headquarters. Duckworth assured his captain that the Alexandria incident command post was fully staffed, with good accommodations for about 250 coasties. His hurricane preparation work finished, Duckworth was slated to leave active duty on September 1. He intended to see Katrina out at his family business, in the company of his girlfriend, Shelley Ford, the manager of a pediatrician's office. "I told the captain, 'Let the new guys get the experience of doing this.' I did Ivan and set up Alexandria, and I thought it would be great for the new guys to get some hands-on, not thinking that the levees would actually break."

Captain Paskewich agreed. Over the years he had developed a great fondness for Duckworth. So as Paskewich drove to Alexandria, Duckworth returned to his Old Metairie tire shop to board up the windows of his apartment, which was packed with World War II history books. He was a great fan of Stephen Ambrose's *Citizen Soldier,* a true believer in the fact that everyday Americans won World War II. He long knew it would be the same in a hurricane. And he knew from reading historian Samuel Eliot Morison's magisterial *History of United States Naval Operations in World War II* that in December 1944 Admiral William "Bull" Halsey lost 700 men in the Pacific Theater because of a typhoon.

As Sunday night turned into Monday morning, Duckworth reconsidered his decision to wait the storm out in his cozy wood-paneled apartment. Echoing in his ears were the words of his old friend Jesse St. Amant, who directed emergency operations in Plaquemines Parish. "Do not stay in New Orleans for a Cat 4 or 5 direct hit," St. Amant had told him over the years whenever the subject of the Big One came up. The two men had worked together on port security, pollution control, and hurricane response. "I knew from Jesse's point of view how bad things could potentially become," Duckworth recalled. "As I watched television and saw the eye wall move toward New Orleans, unswervingly, I thought about what Jesse said. Shelley and I just got in my truck and left."[21]

As Duckworth and Ford left for Baton Rouge at 2 A.M. Monday, Katrina was approaching the mouth of the Mississippi River near Venice. Driving west on the elevated section of I-10, they experienced hurricane-force winds. Pine trees were falling on the road right in front of them, turning I-10 into an obstacle course. Wave tops were breaking over the bridge. "We almost didn't make it out of town," Duckworth lamented. "It's the dumbest thing I ever did in my life, to wait that long to leave town. We were the only people on the highway. It was terrifying." Eventually, they made it to a friend's house in Baton Rouge. Duckworth's story illustrates just how confusing the day had been. With all his knowledge and experience, he made a poor decision. What could be expected of average landlubbers all over the region?

IV

Sunday evenings in Bay St. Louis, Mississippi, were usually tranquil times for Mayor Eddie Favre, a cousin of Brett Favre, the Green Bay Packers' quarterback. Entering his fifth term as mayor, he was a local politician who could do no wrong. Blessed with a breezy, slyly humorous demeanor, Favre loved to socialize. With his discolored moustache always in need of a trim and a predilection for laughing at his own jokes, the roly-poly Favre was the Falstaffian charmer of Bay St. Louis. He ran the beach community with one hand clenched into a fist and the other holding a Budweiser. There were very few Jimmy Buffett songs he didn't know by heart. Late at night he could often be found at the Casino Magic lounge or the Firedog Saloon, telling stories about catching record-size swordfish or about surviving Camille. On Sunday, August 28, he was a worried man. "Katrina had already come across the tip of Florida and was headed straight at us," he recalled. "I personally drove around all night, telling everybody to get out of town. I made sure all the backup generators in Bay St. Louis were working, even those at the water wells. I refused to sleep. None of my police officers slept either. We patrolled every street, stayed in the field until the very last moment."

By 9 P.M. on Sunday, a heavy rain was pelting the coasts of Mississippi and southeastern Louisiana.[22] Favre's windshield wipers were of little use—the downpour was blinding. Roads were washed out. Traffic lights swung back and forth like out-of-control pendulums. Favre felt unmoored. An eerie darkness loomed over Beach Boulevard, the main coastal drag in town, as angry clouds moved as quickly as if they were being played at fast-forward speed on a DVD. A green metal street sign suddenly blew in front of Favre's car—a close call.

Katrina was roaring toward land, with the eye less than 100 miles away. It was still a Category 5 hurricane, with sustained winds of 161 mph, but it was starting to weaken. That was probably inevitable, since it had been raging at maximum intensity for almost twenty-four hours. The larger waves generated when Katrina was at its strongest tended to be cooler than those on which it had built to a fever pitch. Because lower water temperatures contributed less energy to the rotating winds, Katrina was starting to fade slightly as it approached the Chandeleur Islands Area (CIA), the easternmost barrier islands of the Mississippi River Deltaic Plain. Encompassing Chandeleur, Curlew, Grand Gosier, and Breton, along with some flyspeck islets and seagrass meadows to their west, the CIA had shrunk in size in recent years as a result of coastal erosion.[23] They failed to serve as effective speed bumps to slow down Katrina's storm surge.

Favre drove by the Hancock Medical Center and checked inside. Everything seemed to be all right. "I was worried about surge because those islands had shrunk in recent years," Favre said. "But I wasn't particularly worried about Hancock Medical Center. They were far from the Gulf. I knew they wouldn't flood."[24]

At 2 A.M., the National Hurricane Center downgraded Katrina to a Category 4 hurricane.[25] For those waiting in its path, however, the distinction hardly mattered any longer. By the time a hurricane made landfall and its peculiar characteristics met those of the shoreline, five mere categories no longer described the potential misery index. Katrina packed a brutal punch even before it arrived *en force* along Louisiana's porous shore. At 3 A.M., a buoy monitored by the National Oceanic and Atmospheric Administration registered frightening new information. Located fifty miles east of the mouth of the Mississippi River, it reported waves that were forty feet in height—and growing.[26]

Turbulent water in the Gulf combined with heavy rain to fill Lake Pontchartrain to the very brim. The four wide drainage canals leading from the lake into New Orleans—17th Street, Industrial, Orleans Avenue, and London Avenue—were likewise full and still rising. The Big One—well, actually, the Medium One—had arrived. At 3 A.M., even before Katrina came to shore with its storm surge in tow, water broke through the concrete flood wall separating the 17th Street Canal from the city. The turbulence in the normally placid canal inevitably found the weakest part of the wall and burst through.[27] The breach wasn't large at first, but it grew with the impact of the hurricane. All of the lakefront area—full of restaurants, condominiums, marinas, and homes—was doomed. The specter of massive flooding was upon the city.

Because so much of New Orleans and the adjacent parishes was below sea level, rainwater would not drain off anyway. In New Orleans, the Sewerage and Water Board had twenty-two pumping stations in 2005 to collect and disperse the rain. Most was sent into Lake Pontchartrain, by way of one of two canals, or underground pipes. Some went into the Mississippi River. The stations were manned twenty-four hours a day; most of them had been installed around 1915. While that suggests they might be outdated, this was not the case. "The machinery has a certain antique splendor about it," wrote Brian Hayes, a specialist in infrastructure, "but the pumps are not museum pieces; they are still among the most powerful in the world."[28]

Nearby Jefferson Parish lay as much as five feet below sea level and depended on gravity to collect rainwater at forty-seven stations, containing 130 pumps. The staffing was not continuous at all pump stations, as in New Orleans; instead, pump operators left some houses unguarded, while they remained on standby status. When all the stations in the greater New Orleans area were manned, they had the capacity to pump 16 million gallons of water per minute. "Pump everything down," Walter Maestri, Jefferson Parish's emergency management chief, told his pump-station personnel. "Get as much of the water out without sucking in the walls of the canal."[29]

When the 17th Street Canal breached early on Monday morning, the New Orleans pumping stations were operational, and the pump-station mechanics on duty worked heroically to keep up with the frenetic, stark-raving flow. It was only late in the day on Monday, when four breached canals made New Orleans essentially a sill for Lake Pontchartrain, that the alarmed mechanics were told to turn off the great pumps and leave their stations. In Jefferson Parish, however, the flooding resulted from two sources: the hurricane rainwater and the breach of the 17th Street Canal. The system of pumping stations could have remained ahead of the rainwater and alleviated most of the flooding from the canal. Doing so would not have been futile, as it would have been in New Orleans. Unfortunately, the parish had pumping stations, but no mechanics to operate them. "All this panic is going on," remembered Brian Baudoin, a pump operator, of his colleagues on the Sunday night before the storm. "They wanted to get out. Well, I wanted to get out, too!"[30] Jefferson Parish's president, Aaron Broussard, citing a 1998 "doomsday" plan prepared to meet the contingency of a major hurricane, ordered the mechanics to evacuate on Sunday night—before the storm struck. Even beyond his obligation to follow the doomsday plan, Broussard later claimed that he stood by his decision. "You could easily be sentencing someone to death by staying at their post," he said. "That is illogical, unreasonable and we will never do that."[31]

It was also, however, illogical to evacuate essential personnel to a site 110 miles away. That was the case with the pump operators, who ended up in the town of Mount Hermon, near the Mississippi line. Amid the conditions that prevailed on Monday, it took them hours to return to Jefferson Parish at 7 P.M. and assume their posts. During those nine hours, thousands of homes in Lakeview and Metairie that might have remained dry were destroyed. More than that, some people drowned, when the pumps might have saved them from the deadly flood. In the aftermath of Katrina, the long absence of the pump mechanics would be the most incendiary issue in Jefferson Parish; a serious citizen effort was made to recall President Broussard. He promised that in the future, the mechanics would remain near their pumps, in safe houses constructed to withstand even a Category 5 hurricane. He was in a situation typical of the human response to Katrina: he hadn't been wrong, but he hadn't been right either.

When criticized later by Jefferson Parish Councilman-at-Large John Young for "pulling the operators," an irate Broussard exploded in disgust. "It will be a rainy day in hell . . . before I tell a pump operator that they will stay in any structure . . . during a Category 4 or 5 storm," Broussard said. "You say you're sensitive to life. I say you're full of shit."[32]

V

Incredibly, Tony Zumbado and Josh Holm were still out after midnight on Monday, driving around New Orleans in the wee morning hours, despite the danger. They knew that when the electricity went out, New Orleans would become a pre-Edison world, with only the benefit of flashlights and battery-operated generators. With Heather Allan as their logistics commander, however, they weren't worried. Allan was desperately trying to persuade the National Guard to allow Zumbado to park the van at the Superdome, the so-called shelter of last resort. Usually the news media get full cooperation in such matters. After all, the footage Zumbado and others took could influence U.S. Coast Guard, FEMA, and the Red Cross on how best to help out. All Allan was looking to do was embed NBC News at the Superdome, but she received no cooperation. "Yes," Zumbado recalled. "That was yet another red flag."[33]

Zumbado and Holm continued to patrol New Orleans, their van vibrating. Fast-food signs up on high poles spun around like weather vanes, out of control, and then crashed to the pavement. Palm trees shook like pom-poms, and streetlights popped out of the ground. Large chunks of sheet metal became the local equivalent of ghostly tumbleweed. Eventually, as the winds became unbearable, they returned to the Ritz and started filming rainwater on Canal Street. "It wasn't that unusual of a storm," Zumbado recalled. "Hard rain blowing in sheets, but nothing out of the ordinary."[34]

But the rain started showering down harder in pins and daggers, a deafening spectacle. Jennifer Broome, a reporter with WOIA, the NBC affiliate in San Antonio, arrived in Louisiana late Sunday night to cover the storm. She and her crew checked in at a hotel in Metairie, just west of New Orleans. A short time later, she watched as Katrina ripped the city apart, the rain piercing. "By 4:30 A.M.," she later wrote, "the winds are picking up and we're already standing in a foot of water. . . . We're already into tropical storm force winds, probably around 50 to 60 mph. Power is shut off at some point and the hotel generators turn on. It's surreal. The streets are starting to turn into rivers and the water keeps coming."[35]

Katrina was gargantuan in sheer size, 460 miles in diameter, but it was the eye wall that contained the strongest winds and promised the greatest destruction. As of early Monday morning, the storm was headed almost due north, straight toward New Orleans. Just before dawn, however, its track "wobbled," in the words of one meteorologist. In the words of another, the direction developed a "wrinkle." In any case, the cataclysmic storm began to curve eastward. Residents of Mississippi's Gulf Coast, who had thought that they would be on the near periphery, were now in the crosshairs. "I called my kids and told them to hit the road, to get to Florida," thirty-eight-year-old Waveland, Mississippi, Police Dispatcher Judy Frank recalled. "Dread came over me."[36]

The hurricane made landfall at three spots as it skirted the uneven Gulf Coast on its eastward curve. It was deemed a Category 4 storm by NHC at the time, but in December, after reassessing data, NHC downgraded it to a strong Category 3. The first place to feel its fury was just to the south of the hamlet of Buras, Louisiana, which sits near the mouth of the Mississippi in Plaquemines Parish, sixty-three miles southeast of New Orleans. Katrina struck there at 6:10 A.M., slashing the fishing community with winds of 161 mph. As far as is known, no one was left behind in Buras. All 3,348 people from Buras and its sister village, Triumph, had evacuated. By doing so, they saved their lives, for virtually all 1,146 households were flattened like pancakes; livestock and wildlife drowned en masse. The evacuated people of Buras and Triumph hunkered down in North Louisiana or East Texas or West Tennessee, listening to the National Weather Service reports like horse gamblers during the home stretch. What they heard was nothing short of sickening. *Chicago Tribune* reporter Tim Jones later accurately called the Buras area a "climatological war zone."[37]

As the hurricane moved, it was dropping rain at the rate of an inch an hour.[38] Fifty-eight-year-old Lee Walker lived in the small town of Poydras, about midway between Buras and New Orleans. At the end of a quiet road called Saro Lane, he had a small white house that suited him perfectly—he was a bachelor with a painful back condition. "My neighbors left a day or two before the hurricane," he said. "You know, when you got money, you can do that. When you cain't, you just get left." When Katrina came to Saro Lane, it brought a flood of water—the storm surge that pushed the seawater right up the Mississippi and over the river's banks. It came so fast at dawn Monday morning that Walker had only just enough time to gather his three small dogs and race upstairs to the attic.[39] For the time being, they were safe, but very much alone, cut off from the rest of the world, as the storm raged outside.

With 911 unresponsive in most Louisiana communities, Garland Robinette of WWL radio found himself in the unlikely position of being an emergency clearinghouse center. Desperate pleas were logged from Gentilly, where a roof blew off, and from the Lower Ninth Ward, where "I need a boat" was a common refrain. A woman called in and screamed, "We have a two-year-old. I think we're going to drown." The calls came ringing in from Tremé and Chalmette, Slidell and Metairie—everywhere. "A lot of callers were identifying locations," Robinette recalled. "We were trying to help, but what could we do? We didn't have boats. But people called into WWL just to say they were all right, that they were alive. Relatives and friends hear that. So, yes, we were providing a social service."

"Spooky" is the best word to describe what it was like for Robinette to be broadcasting from the fifth floor of the pinkish building near the Superdome, the high-rise shaking like a plucked tuning fork. The winds were hellacious and then *pow!* the studio's large plate-glass window blew outward as Robinette clutched his microphone, his life flashing before him. A jet-engine-like windstream came blasting through the building and almost sucked him out. Everything around him—papers, books, furniture, tapes—went flying into the morning sky. Behind the studio booth window, engineers, call screeners, and a few station stragglers were struggling to recover. Barely missing a beat, the team quickly regrouped in a large closet. Now they were stuck in the building. A group decision was made to keep Robinette broadcasting, shifting his operations to the closet, where the howling wind wouldn't be heard. "When the window blew in, they gave me a new microphone on a stand with wheels," he said, "and we moved down the hall into the closet."

As Robinette told his listeners, the hurricane was drenching New Orleans and wind damage would be extensive, but it was sidestepping the city, aiming its worst winds at the coast farther east. Thus was born the cliché that echoed through parts of the city Monday morning, "It could have been worse." At Brennan's, the revered restaurant on Royal Street in the French Quarter, the relief was palpable. "I hate to say it," said Jimmy Brennan, the part owner, "but it turned into a hurricane party. We had a great time."[40] The wind blew out a wall on the fourth floor of Antoine's, another landmark, but overall the intransigents in the French Quarter looked on Katrina that morning as more exciting than unnerving. Burlesque diva Chris Owens, for example, was surviving the hurricane just fine, drinking her new "Chris Owens" bottled water and cooking up chili over a sterno can. "I kept looking out around the French Quarter," Owens recalled. "Everything looked fairly all right. I was still worried, though, that Sheriff Harry Lee would be furious I hadn't left."[41]

VI

At the Superdome, where close to 10,000 people had taken shelter in the 69,703-seat steel-framework stadium, the situation was harrowing for those sitting in the stands or lying in the corridors. Families tried to carve out their own turf where troublemakers would leave them alone. After the hurricane had clawed at the building with 100-mph winds for more than an hour, a section of the roof, made of a combination of foam and rubber sheeting, gave way. If only the cheapskate state of Louisiana had spent the $14 million for a metal roof repair instead of a subpar $4 million foam-rubber renovation, the Superdome wouldn't have been compromised. Nineteen stories above the playing field, two holes burst open in the bubble and let the rain shower down. A few people got out their umbrellas. Others feared they were going to be crushed. The modern structure had been built to withstand winds of 200 mph, or so the press releases boasted when it opened in 1975. That was before the New Orleans skyline closed in around it with skyscrapers that intensified the winds, turning them into effective drills for boring. When the storm poked through the roof, many people inside panicked, assuming that the whole cap would peel off the building by the time the wind was through. "I was okay until that roof fell off," NBC producer Heather Allan told a reporter. "I was terrified then. . . . Otherwise it hasn't been too bad. People are so nice and the people staying here have been really cooperative. But the washrooms are terrible."[42]

Allan distinctly remembered the moment when the Superdome became an observatory for Katrina. She was standing on the floor, along with Brian Williams and cameraman Dwaine Scott. They had been doing a running series of interviews, basic "evacuee plight" segments, when rain started blowing inside. Holding a satellite telephone, Scott captured the terrifying hole. It was the first of many news breaks for NBC and for Williams as an outstanding anchorman during Katrina. "My first concern was that the huge lights would come crashing down," Allan recalled. "But the people were impressive. Three times the National Guard moved them as units, and there was no pushing or shoving. Considering we could all have been squished, you had to admire the collective calm."

For the time being, the Superdome was adequate for those sitting in the stands, trying to read or sleep or gossip. The Louisiana National Guard kept everyone under strict control, while the Superdome's regular managers passed the long hours in the office suites. Secretly they worried that the roof would cave down on the evacuees, killing thousands. Those who stayed in the Superdome overnight still had no real idea what was going on outside, except that rain was falling through the two holes in the roof. And the wind never let up, with its overcharged howl. "I could've stayed at home and watched my roof blow off," forty-three-year-old refugee Harold Johnson said. "Instead, I came down here and watched the Superdome roof blow off. It's no big deal; getting wet is not like dying."[43]

At City Hall, Councilman Oliver Thomas kept checking in at the Emergency Operations Center on the ninth floor, and then walked down to his fourth-floor office. Something about Katrina made him want to be alone. He was full of trepidation. With the winds at a feverish pitch, he kept wondering if the shotgun houses in the "fighting" Ninth Ward were already coming apart. (They were.) The screeching noise was unbearable, and he had flashbacks to that beast Betsy, recalling how, as a child, he had been petrified by the very same hurricane uproar. He lay down on his office sofa. The windows were jangling, shaking like tambourines, about to pop out of their frames. "I heard the glass shattering out on Poydras from the Hyatt, where the mayor was," Thomas recalled. "And the car windows just burst out in the street. Each popped windshield sounded like a little bomb. It was deafening. It was frightening. The lights in City Hall had gone out. All darkness. I kept being drawn to the window. I feared that the windows would break out. But it was calling me. It was like 'Come see the devastation.' And you could feel it. It felt like the end of the world."[44]

As Katrina approached, Councilwoman Jackie Clarkson was in the Hyatt Regency. She was only on the fourth floor, but even there, where Entergy had set up an emergency-response hub, you could hear windows popping out like mad. Clarkson was evacuated from her room down to a 25,000-square-foot ballroom and spent the storm alongside frightened city workers, the press, and a few hundred evacuees. "My deputy and I got a little spot on the floor," Clarkson recalled. "We were supposed to be in City Hall initially. There were bunks there, but City Hall began swaying at the top, so they moved the mayor and all of us to the Hyatt. First they put us in rooms, but that was too scary. Then they moved us to two different ballrooms."[45]

The director of Safety and Loss Prevention at the Hyatt Regency in New Orleans was forty-four-year-old Gralen Banks. According to Banks, the Hyatt became Mayor Nagin's "command center" as a matter of "convenience." Entergy had generators, telephones, and computers on the fourth floor. Nagin would use this as his base, while Terry Ebbert, a true hero of Katrina, ran City Hall. Because the Hyatt was an atrium hotel, Banks's fear of falling glass was acute. He was gripped with fear when "Flying Geese," a huge aluminum sculpture hanging from the roof, started to sway. "That thing was being moved by the wind," Banks said. "That meant the wind was moving the roof. Yeah, I had fear."[46]

Everybody in the Hyatt Ballroom and the French Market Exhibition Hall lamented that cell phones were down. They were isolated from the world outside. But Clarkson, a professional telephone busybody, discovered that if she stood in the atrium and pointed her phone in a certain direction, she could receive calls. "At one point I was standing out on the balcony and the windows in the atrium began blowing in and you could feel the glass," Clarkson recalled. "I shouldn't have been out there, but I couldn't help myself. Suddenly, my phone started ringing and I was able to say hello out there and it was my daughter in Italy."

The daughter—Patricia Clarkson, who had been nominated for an Oscar for her role in 2003's *Pieces of April*—was in Venice, watching the coverage of Katrina on CNN International. (She was in Venice promoting the movie *Good Night, and Good Luck*.) Her earlier efforts to get her mom to flee New Orleans had been futile, but she comforted herself with the thought that Mom would at least be at City Hall or the Hyatt—two ultrasafe buildings. So you can imagine her dismay as CNN showed baneful images of the Hyatt's windows popping out, and beds and desks being sucked into the morning sky. "She wanted me to leave for someplace safe," Councilwoman Clarkson later laughed. "Where was I to go? She put George Clooney on the phone. He tried to persuade me that the Hyatt wasn't safe. It was strange. I'm standing on the atrium balcony of the Hyatt, and I'm watching these windows blow in and this fierce wind and rain. And I can feel speckles of glass hitting me while I'm talking to Patty and George Clooney in Venice, of all cities, the city of canals. Then we lost our connection."

Among the guests staying in the Hyatt were the Black Men of Labor, a social club that was a legendary part of New Orleans's annual Labor Day Parade. Five members of the club—Fred Johnson Jr., David Sylvester Jr., Todd Higgens, Reynard Thomas, and Roland Doucette—had boarded up their homes, sent their families away, and evacuated to the Hyatt. They were worried about the poor and elderly who hadn't evacuated New Orleans. Together they decided to be *first* first responders. While almost everybody else was looking for cover, the Black Men of Labor stared out of the hotel lobby windows, anxious to help storm victims. "Divine inspiration brought them together," Banks recalled. "These weren't guys who waited around for FEMA. They didn't get a quarter. They were native sons of this city. What the Black Men of Labor understood was that it was our friends and family members stuck. Nothin' could hold them back. They weren't going to sit around with thumbs up their butts beggin' for federal help."[47]

VII

In the early morning hours of Monday, August 29, holed up about eight long city blocks away from Clarkson at the Hyatt, was Deputy Chief of Police Warren Riley, his face bitten by anguish. About five inches of water was already flowing down South White Street behind the NOPD headquarters and the storm had just started kicking in. The telephones and radios were still working; in the first twenty-three minutes after wind speeds exceeded 80 mph, there were more than six hundred 911 calls, mostly from New Orleans East, the rest of the Ninth Ward, and Lakeview. His dispatchers were overwhelmed: roofs were blowing off, levees were breaching, storm surges were topping flood walls, sewers were backing up, homes were being destroyed, and people were dying. The NOPD saved the SOS tapes—the most heart-wrenching historic artifact of the entire Katrina saga. Just imagine the agony of being a dispatcher, receiving a 911 call with a woman screaming that her son or daughter was going to die, and all you could do was say, "After the storm." The NOPD was going to need therapists and psychiatrists when the atmosphere settled—a trainload of them. An impatient Riley paced about like a caged tiger, tormented that he couldn't do anything to help these poor drowning souls. He especially worried about fellow officers stuck in New Orleans East and Lakeview. "The dispatchers were very upset," Riley recalled. "We can't send people out if the wind is above 50 miles an hour."[48]

One harrowing 911 call came in from fellow NOPD officer Chris Abbott, trapped in a curse of water. With the 17th Street levee break, his Lakeview house was swamped, Lake Pontchartrain gushing into his living room as if the Hoover Dam had broken in his front yard. As Officer Abbott pleaded for help, NOPD dispatchers huddled around the radio switchboard more embarrassed than scared. "I can't get out of my attic," he said. "The water is rising. It was up to my waist. I'm trying to get out. It's up to my neck." Police Captain Jimmy Scott, with Riley at his side, tried to talk Officer Abbott out of his ominous predicament. "Listen, I can't get out," Abbott said, trying to camouflage the obvious panic in his voice. "I apologize for asking for help."

Captain Scott tried to calm Abbott. "Chris, do you have your gun?" Scott asked.

"Yeah," Abbott said.

"How many magazines you got?" Scott asked. To which Abbott replied, "Three . . . there's water up to my chin!"

Scott said, "Listen, take your weapon, take your three rounds, and fire a circle into the roof."

Abbott, every fiber in his body tingling, reported back that his service revolver was wet. All the officers around the radio apprehensively glanced at one another. One despondent officer feigned a kiss and went out of the room. Every second was an eternity to all who listened. Refusing to give up on Abbott, a calm Captain Scott instructed him on how to fire a gun single-action in order to shoot a circle of holes in the roof. The bullets would structurally weaken the roof and then a fist would finish the job. "Punch your way out!" Scott shouted.

"The water's up to my mouth," Officer Abbott said. "I really don't know that I'm going to make it. I really apologize."

At that moment the radio went dead.

Silence fell over the Rampart Street headquarters. Everyone in the room was motionless. Rain was pelting the glass; the Grim Reaper was at their windowpane. Many of the officers, particularly those who knew Officer Abbott well, were in tears, ashamed that Katrina had rendered them impotent to save him. And they couldn't believe Officer Abbott had *apologized* for asking for help.

Since 2001 Officer Abbott had been a folk hero in the NOPD. That May, heading to court to testify in a criminal trial, he eyed a suspicious-looking man named Brandy Jefferson walking down Dumaine Street with what looked like a concealed weapon. When Officer Abbott approached Jefferson—a known felon with "No Mercy" tattooed on his cheek—the hoodlum opened fire. Officer Abbott took three bullets and lay in a pool of blood. Rushed into surgery at Charity Hospital, he had bullets removed from his neck, spine, and chest. It was a miracle that he wasn't paralyzed for life.[49] Before long, the ten-year NOPD veteran had fully recovered and was back at his job.

After Abbott survived that ordeal, it seemed the height of injustice that he would drown in his own home. A few of the officers closed their eyes in mourning. Other 911 calls came into the command center, but the last words of Chris Abbott—his apology—made the dispatchers and officers feel helpless. Although they wouldn't admit it, many of the officers had written Abbott off as dead, an early victim of Katrina. "We heard absolutely nothing for a good thirty to forty minutes," Riley recalled. "Then . . . we heard from him again. He's out of breath, saying, 'I'm on a roof. I need someone to come and get me.' Everybody wanted to help. It was one of the most dramatic moments for us that there was."

Officer Abbott was rescued by the NOPD, and even though he lost his bungalow, in a day the indefatigable cop was back on the job, rescuing stranded New Orleanians. "That shows the courage, the calmness, and the desperateness of the entire situation," Riley later explained. "He couldn't have been calmer."[50]

What Abbott couldn't communicate over the radio was that along the Orleans–Jefferson Parish line pure anarchy reigned. Lake Pontchartrain was unbelievably rough, with huge waves slapping away at the Bucktown marina, and the storm surge was overtopping jetties. Boats were being lifted in the air and dropped down on top of houses. The wind was otherworldly. A spate of 911 calls had come in before the levees even breached, as the wind pulled off the sides of hundreds of homes. "Windows begin shattering between 9 and 10 A.M.," wrote Jennifer Broome, describing her hotel in suburban Metairie. "Glass is falling into the lobby. . . . The wall of my room faces the elevators and it looks like it's breathing." She feared for her life. "I run down the hallway, praying that this is not the end," Broome continued. "I have never been as terrified as I was in that moment. I walk through darkness. With my flashlight, I see people huddled together in hallways, many saying prayers."[51]

VIII

Officer Abbott wasn't the only person ravaged by the 17th Street Canal and London Avenue Canal breaches in Lakeview, the first section of New Orleans to flood when the Lake Pontchartrain storm surge rushed in. He was one of thousands.

The breach in the 17th Street Canal levee opened up in stages. At the London Avenue Canal farther to the east, though, the levee simply burst in two separate explosions powered by nothing more than the water pouring in from the lake and pressing outward against the canal's inadequate walls. The levee walls along London Avenue were about fifteen feet high, consisting of dirt embankments at the base and concrete sheetpile walls farther up. Katrina pressed the storm surge into the lake on Monday morning, and then left the bulging water behind as it moved farther north. By that time, the swirling winds, no longer pulling the water north from the sea, were positioned to push it back south against the city of New Orleans. The water was forced into the London Avenue Canal at a terrific rate. It soon rose to within three feet of the top of the levee wall.[52]

More water kept pushing its way into the canal. In a "normal" flood situation, the water would just have kept rising, finding the least resistance from the air above. But the sheer weight of the angry waters, and the force that drove them into the canal, was more than the walls could withstand. For more than an hour, water seeped through the embankments holding the sheetpiles in place. Had the sheetpile walls been secured by deeper pilings, the seepage in the embankments might have been the only problem—a small one at that. But with the earthen embankments compromised, the inadequate pilings that supported the sheetpiles could not remain anchored, and they toppled from the force of the water. At approximately 9:30 A.M., the embankment on the east side of the canal's southern end crumbled, and the flood wall toppled with it. The Ninth Ward neighborhood near the Mirabeau Bridge was overwhelmed by what amounted to a flash flood. Even so, the London Avenue Canal was not relieved of the water pressure. In madly swirling waters, there is no consistency to the pressure brought to bear on a vessel: in this case, the levees of the canal.

Already weakened, the levee on the west side of the London Avenue Canal, farther north toward the lake, gave way at about 10:30 A.M. Like a river unleashed from its bounds, eight feet of water from the canal poured into the neighborhood near Robert E. Lee Boulevard. It was still gushing from the south end of the canal, too. In both places water would not stop pouring out of the London Avenue Canal for more than a full day. Even though senior citizens John and Annie Kelt of Warrington Drive had life preservers on, they drowned, the crude violence of the London Avenue breach flattening every object they had accumulated since they were married in 1949.[53] According to Ceci Connolly and Manuel Roig-Franzia of the *Washington Post*, between twenty and thirty corpses were eventually found on the streets of Lakeview, a neighborhood where houses sold for $1 million.[54]

Fifty-three-year-old Michael Prevost was just one of many residents of Lakeview who found themselves in dire straits. Prevost had been born in the Lower Ninth Ward, but when he was about five, his father worried that the neighborhood was becoming "too black," and moved the family just over the city line, to Metairie. "I was a white-flight kid," he recalled. "Back then Metairie was pretty rural, we used to get out our BB guns and pellet guns and then shoot them off into the bayou." Eventually Prevost took an undergraduate degree at LSU–Baton Rouge and an MA in social work at Tulane. Despite his upbringing, or maybe because of it, he became a champion of civil rights. At the time of Katrina, the six-foot-tall Prevost—who wore his hair parted down the middle, had slightly graying sideburns, and wore scholarly glasses—was in his eleventh year as the head of counseling at the prestigious Isidore Newman School on Jefferson Avenue in New Orleans. Considered the best private school in Louisiana, it boasted among its distinguished alumni former *Time* magazine editor Walter Isaacson, bestselling author Michael Lewis, and NFL star Peyton Manning. "It's like belonging to a community," Prevost said. "It's not a job."[55]

Prevost was in the city during the hurricane Sunday night, but there was nothing high-minded about the reason he stayed. Basically, he didn't want to be stuck in bumper-to-bumper traffic. Divorced, with a daughter in Virginia, Prevost owned a sturdy ranch house on Paris Avenue just off Robert E. Lee Boulevard, a major city thoroughfare five blocks from the London Avenue Canal. As an adult, he was still an outdoor enthusiast, the kind who subscribes to *Louisiana Conservationist.* Prevost had a seventeen-foot red Coleman canoe, which he pulled into his living room as Katrina approached. "I never expected to use it, but I did make sure it had paddles and life jackets," Prevost recalled. "You never know about Lakeview." Truth be told, Prevost wasn't even sure if he lived in Lakeview—a blanket term for a whole host of "lake" housing developments.

Lakeview had once been a marshy spillover area from the lake; its main structures were small fishing camps and dockside shacks. In the 1930s, when the edges of Lake Pontchartrain were dredged, the land was raised and a seawall was built. The area then quickly became a magnet for second-generation Irish Americans. As one of the newer neighborhoods in New Orleans, with postwar architecture and wide, logically planned streets,

Lakeview lacked the charm of distinctive neighborhoods like the French Quarter and Bywater. But the waterfront neighborhood, 90 percent white, developed an atmosphere and character like that of San Diego, with a residential-beach-area vibe. Its gems were West End and Lakeshore Park, where locals like Prevost could picnic, jog, and sunbathe. Sportsmen docked their boats in the Lakeshore Drive Marina, and when the weather was good, people sailed, fished, or water-skied in the lake. Spirited summer volleyball games played out daily at Coconut Beach, and bustling, no-frills crab shacks like Jaeger's and seafood joints like Bruning's dotted the shore. It was a family haven within the city limits, with a decidedly safe and suburban feel. Large manicured lawns in front of two-story homes, well-lit streets with sidewalks, and choices between decent public schools and excellent private ones made it a "safe place" to live, without being too far from Galatoire's or Commander's Palace. In the 2000 census, Lakeview was among the more wealthy neighborhoods in the city, with an average household income of $63,984.[56]

When Katrina made landfall, Michael Prevost was holed up in his house with his mixed-breed SPCA dog, Chelsea. He wasn't concerned as his front lawn filled with water. He thought it was rainwater. Chelsea, however, was petrified, glancing at her owner as if he were insane. Water had started bubbling through the front door. Within what seemed like minutes the water was thigh-deep. It didn't come in a mad rush; it came in an insistent creep. Chelsea was curled up on the sofa, looking scared when her furniture-of-last-resort started floating. Realizing that the London Avenue Canal must have breached or been overtopped, a methodical Prevost brought an ax and bottled water to his attic. By 1:30 P.M. he had an urgent choice to make: head for the attic or the canoe. "I chose the canoe," Prevost said. "I filled it up with a quilt, knife, parka, water, and my dog. Purposely I left my gun behind. I just didn't want to have to use it."

The trick was getting the canoe out of his house. With the water about five or six feet high inside, he opened the sliding door and pushed his canoe out into the rain-blown afternoon. The winds were still 80 mph. Defying nature, he paddled out to his neighbors' backyard. They had an awning still up, which provided cover. All the houses in the area were flooded over, with only second stories above the waterline. "It was beautiful in a strange way," Prevost recalled. "Everything had an odd glow to it. That's what I was thinking when I saw two black guys screaming for help. They weren't in acute danger, but one of them was clinging to his roof. When the storm died down a little more slightly, Chelsea and I were going to have to go get them."[57]

Chapter Five

WHAT WAS THE MISSISSIPPI GULF COAST

High noon I can't believe my eyes
Wind is ragin' there's a fire in the sky
Ground shakin' everything comin' loose
Run like a coward but it ain't no use
Edge of the river just an ugly scene
People getting pushed, and people gettin' mean
A change is comin
and it's gettin' kind of late
There ain't no survivin', there ain't no escape.
—John Fogerty, "Change in the Weather"

I

MAYOR EDDIE FAVRE HAD continued patrolling the streets of Bay St. Louis until about thirty minutes before Katrina made landfall. When the raging winds became too much for him, he made his way to the police station on the Old Spanish Trail. Begun in 1915, this trail was part of the "Good Roads" movement, a 2,743-mile route that ran the length of

America, all the way from San Diego, California, to St. Augustine, Florida. It was the oldest automobile road along the Gulf Coast, the first blacktop link between New Orleans and Mobile. The police station was located along the trail on the highest knoll in Bay St. Louis, just across the street from the railroad tracks. Virtually all of its policemen, a total of twenty-seven, stayed on duty, ready to start rescuing people when the weather permitted. 'We received emergency calls from about sixty residents before communication broke down," Favre recalled. "We took note of them all and were determined to check up on them the second the storm passed."[1]

The Mississippi Gulf Coast was getting hammered even harder than had Louisiana—if that was possible. At 10 A.M. Monday, when the worst of the storm was through with New Orleans, the easternmost towns of Mississippi came into the grip of Katrina's second landfall. By then Katrina was a Category 3 storm, bringing winds of 125 mph, along with a 20-foot storm surge and 10 inches of rainfall.[2] Edwina Craft was visiting her mother, a patient at the Biloxi Regional Medical Center, when Katrina reached the Mississippi Coast. The only illumination came from a couple of generator light systems that kept going *Zzz! Zzz!* "The room was shaking," she recalled, "the windows were shaking. They moved us from the sixth floor to the bottom floor, but that was not much better. It flooded down there, and then the lights went off."[3]

In Gulfport, to the west of Biloxi, the fire chief reported that buildings downtown were "imploding" from the force of the storm. Sand from the beach whipped around in squalls. The high branches of ancient oaks suddenly defined the waterlines. Electrical wires dangled downward in every direction, snapping back and forth on poles. Sparks jumped from wet wires along the Gulfport boat harbor.[4] Leslie Williams, a reporter for the New Orleans *Times-Picayune*, was at her brother's sturdy house in nearby Bay St. Louis. She and several other family members gathered there, looking out the window at the glade of pines in his backyard. "The trees snap like twigs in a child's hand," she wrote. "One breaks several feet from its base, then another, then dozens."[5] Finally, her brother suggested that they get away from the windows and take shelter in an interior hall.

"This hurricane was like God and the Devil fighting it out here with Godzilla as the referee," said a soldier at Keesler Air Force Base.[6]

Soon after the winds came the water. As an officer at Keesler put it, "The base looked like an ocean."[7]

"A boat floated up my runway," marveled the commanding officer of another military installation, on the grounds of Gulfport-Biloxi Airport.[8] Water covered the land for at least a half mile inland. Along Mississippi's coast, that half mile included charming old towns, new gambling districts, and thousands of stilted houses built to take advantage of beach living. Six feet of water swept in with the hurricane's initial surge and more kept coming. The water would pull backward like a slingshot and then let loose with a punishing force, with waves capable of literally punching down brick walls. Fifty-year-old Huong Tran and her fisherman fiancé ran out of their house when the storm surge suddenly threatened to sweep it, and them, away. The Vietnamese couple helped each other climb to the top of a tree, a vertical refuge of last resort. Before the storm was through, they would spend six hours there, clinging to the branches and hoping the trunk wouldn't break. "I thought I was going to die," Tran said. "The water was over the house." She prayed to a Buddhist goddess. "I called to her, 'Help me, help me. I think I'll die.' "[9]

First, the hurricane winds chopped everything in their path into pieces, and then the storm surge, twenty-five feet at its worst, swept it away. Julie Goodman, a reporter for the *Clarion-Ledger* in Jackson, Mississippi, described what it was like to realize that your house was disintegrating around you. "You see it tear through your walls, forming bubbles as it bolts between the paint and your family room sheetrock," she explained. "You hear the nails pop, one by one, off your shingles. You see water come through your light switches and drop down toward your bathroom sink."[10]

Hurricane survivors claim that sustained winds of over 140 mph create a ghastly howl similar in timbre to a freight train crashing through the room. This is too dramatic a metaphor. Think instead of the grating pitch a dentist's suction tube makes after it's extracted the last molecule of spittle from your mouth. It's a dry vacuum drone that over a two- or three-hour period starts chiseling away at morale. The wind makes a white noise, really, except that it's punctuated by startling thuds and explosions from blocks away. Your nerve-racked mind imagines that each distant crash represents a collapsing bridge or a crumbling building or a falling oak. All you can do is pray that each blast or blare doesn't signify human death.

A family named Taylor, terrified by the noise, panicked and tried to evacuate from Bay St. Louis on Monday morning. By then it was too late. Their white SUV was trapped on Highway 90 as the road turned into a raging river. Afraid of drowning, the Taylors climbed on top of the vehicle's roof. Volunteers from the Bay St. Louis Emergency Management Agency, including two police officers, suddenly appeared, seemingly out of nowhere, to rescue the family, quickly strapping life preservers on each of them and shepherding the four terrified people to safe ground.[11] An Associated Press photograph of the daring rescue became one of the most memorable images of Katrina. "Volunteer rescue teams all over the region saved more families like the Taylors than history will ever properly count," Bay St. Louis Police Chief Frank Griffith recalled. "Two of my officers, Tom Burleson and Jennifer Favalora, helped rescue the Taylors by holding the vehicle down so it wouldn't get swept away."[12]

II

Chief Frank Griffith had once been a homicide detective for the NOPD but decided he preferred the down-home friendliness of the Mississippi Gulf Coast to crime-plagued New Orleans. He opened a combination seafood restaurant and deli in Waveland for a while, but in 1991 he went back to law enforcement, becoming head of the Bay St. Louis Police Department (BSLPD). Griffith was proud to have his police force ready if a Camille-like hurricane ever again reared its ugly head. When Katrina hit, his BSLPD officers huddled together, drinking coffee, lying on cots, and peering out across the street at the Senior Citizens Center, which was starting to flood. "Thank God the storm happened during daylight," he said. "At night the death toll would have been higher."

With Chief Griffith throughout the brawl was fifty-four-year-old David Stepro. A stickler for detail and a computer whiz, Stepro understood that three-quarters of being a police officer was paperwork. But he didn't like being deskbound. Unlike most of the twenty-six other BSLPD officers holed up on the Old Spanish Trail, Stepro didn't hail from the Gulf South. His father had been with the Army's Criminal Investigation Detachment, so Stepro grew up in such disparate locales as Ohio, Okinawa, Michigan, Pennsylvania, New Jersey, Thailand, Indiana, and Kentucky. He worked for the corrections department in Louisville, Kentucky, for twelve years, but he fell in love with a Mississippi woman and moved to Hancock County.

That marriage didn't work out, but Stepro rebounded quickly. He developed a crush on Laura Lachin, a Waveland cop thirteen years younger than he. With her shoulder-length brown hair and green eyes, Lachin was tough and unassuming. After graduating from the University of South Florida with a degree in criminology-sociology, she joined the Waveland Police Department. Everybody in the county knew, liked, and respected her. Eventually David proposed. They got married at a Louisiana plantation and moved to a cottage house on Leopold Street in Bay St. Louis. You couldn't have found a happier couple. "We were both police and that helped our marriage," she recalled. "We've both worked Narcotics, understood what late nights meant. We were on the same wavelength."[13]

As Katrina took aim, both David and Laura Stepro determined that they were going to stay on duty. First, they had to drop their Great Dane off at the Arcadian Grill on Highway 90, where a friend would take care of him. Laura bought a blue heavy-duty raincoat at Wal-Mart. The couple would be separated during the storm, David at the BSLPD headquarters on the Old Spanish Trail and Laura at the Waveland headquarters on Highway 90. But they had cell phones, and if those went out, they could stay in touch by radio dispatch. It was a sensible game plan. The only problem was that Katrina wasn't a sensible storm.

On Monday morning, things had turned topsy-turvy at the Waveland police station. Because the brick headquarters was nearly two miles inland from the Gulf of Mexico, Laura and the other Waveland officers felt immune from Katrina's ravages. They had patrolled Waveland until 4 A.M., when the winds increased threefold. They then holed up at the station, ready to emerge again to help rescue folks when the storm passed. At least that was Chief Jim Varnell's plan. But as Laura told David on the radio around seven that morning, the station was flooding. They were going to evacuate to the Coast Inn and Suites farther up on Highway 90. The Waveland police had real doubts whether they would all survive the ordeal. "We thought the roof might blow off," twenty-five-year-old officer Michael Prendergast recalled, "but the building survived Camille and we didn't expect flooding. But the water started pouring in like mad. We were just flushed out of the building."[14]

Often the Waveland police, both on and off duty, would meet for po'-boys at Beningo's in Bay St. Louis, which was almost a mini-museum to Hurricane Camille. On the restaurant's brick wall, in fact, was a crevice with a sign posted next to it that read, "Camille Crack, August 17th 1969." Surrounding it were newspaper articles from the *New Orleans States-Item* and a poster that read, "This Property Is Condemned." But many of the Waveland police had been born after Camille, and they tended to admire the harpoon and redfish taxidermy at Beningo's more than the hurricane memorabilia. So they were all the more astonished on August 29 when the water kept rising, filling the police station in only an hour. All the buildings near them, like Daddy O's Café and Grill across the highway and the Cycle Barn Harley-Davidson dealership, were collapsing. When the Pepsi machine at the station started floating, the officers feared the tall blue water tower near the station might come tumbling down on them. By the time Chief Varnell decided to evacuate headquarters, all twenty-nine members of the Waveland police were in the ice-cold water of the fierce storm surge, struggling not to drown. Everybody scrambled for their lives.

Washed out of the headquarters, Laura Stepro, along with fourteen other Waveland officers, looked for something to anchor onto. Yards away from Highway 90 was an ugly, flimsy tree, a tall bush really, that looked like the leftover orphan shrub at a Kmart garden center—Officer Prendergast later referred to it as that "butt-ugly bush." Desperate not to be swept away with the current, the fifteen officers clung to the branches of this bush for three or four hours. "We wanted to make it across Highway 90, which had become a river," fifty-six-year-old Glen Volkman recalled. "But we settled for the tree."[15]

Keeping her cool, even though she thought she was about to drown, Laura put a foot in a crook of the bush, and then, after a few failed attempts, stuck her other foot in another. She clung on for dear life. What an odd sight it must have been: police officers clinging to a bush in the middle of a river, only one snap of a branch away from death. In Katrina lore these washed-out officers became known as "the Tree Cops of Waveland." Other police officers and staff climbed atop an elevated generator on the southeastern side of the building and made it to the roof. "The combination of the wind and water left us freezing," thirty-three-year-old Israel Neff recalled. "Many of us developed hypothermia."[16] When the winds died down, they noticed that both the water tower and the flagpole were still standing. Their headquarters, however, was gone, all the radios and computers and file cabinets washed away. "I didn't have time to worry too much about my husband," Stepro said. "We were trying to stay alive."[17]

III

Along the coast, twenty-nine-year-old Michael Veglia and four buddies—Alex Coomer, Ronnie Williams, Chris Stephens, and Joey Lee—were trapped in a prefabricated house on Edna Street, midway between the waterfront and Highway 90. The aluminum-sided ranch house was owned by Stephens; everything inside was both cheap and new. A gully ran in front and sometimes in the spring, when the rains came, it would fill up, perfect for catching frogs and crawfish. Veglia had spent Sunday at his father's popular bar, the Knock-Knock Lounge on Highway 90, moving equipment like pool tables, beer kegs, and the jukebox out of harm's way. As water filled the Edna Street house on Monday morning, Veglia and the other young men dove into the ten-foot-high floodwater. They started swimming down the 1400 block of Waveland Avenue, hoping to find dry land. Eventually, they came upon a small house where the porch was not yet flooded. They clambered onto it, even if it offered only temporary safety. All around them pine trees and TV antennas were snapping, causing what Veglia called "the scariest damn noises I ever heard in my life."[18]

As they were resting on the porch, shivering from the cold floodwater, fearful for their own lives, they noticed somebody peeking out behind the window shades inside the house. When Veglia and his friends turned to look, the shade dropped down. They heard a woman inside screaming,

"What they doin', what they doin' out there?" Within minutes the waters flooded the porch, blowing the front door off its hinges. Inside the house, a terrified African-American family screamed for help. A rattled grandfather started pleading with the five white strangers, "Save my babies, save my babies."[19] Veglia and his buddies waded inside the sinking house, trying not to be knocked down as water continued to gush in. Besides the old man there were his wife (Grandma) and his daughter and her two children, around eight or nine. If somebody didn't think quickly, they were all going to perish. Realizing that the porch would flood soon, they all climbed into a nearby tree. "We used this as our base until we could figure out what to do next," Veglia recalled. "We looked after the family but were nervous." That's when twenty-seven-year-old Coomer spotted a boat, which had slammed into a trailer across Waveland Avenue. Perhaps, he thought, they could get it started, or, at the very least, use it to stay afloat. Coomer and Williams plunged back into the dark water and swam toward the crashed twenty-foot boat. Quickly, they rummaged around the marooned vessel and grabbed ten life jackets and float cushions. They made their way back to the tree, where Veglia, Stephens, and Lee were trying to protect the family. They fastened the life jackets onto the women and children, none of whom could swim, and all of whom were trembling. As the young men surveyed the flooded street, they would see neighbors pop out of rooftop attics like jack-in-the-boxes, crying for help. An entire family right behind the house with the porch did, in fact, drown that very afternoon in Katrina's floodwaters. Waveland Avenue, a two-lane road the five men had known so well, was terra incognita.

Luckily for Veglia and his group of nine castaways, a pine tree had fallen nearby between two ancient oaks. It formed the only potential safe haven on the block, a tree-house-like bench twelve feet above the flood. As they let the swift current take them to the horizontal pine, the wind was still treacherous, but eventually they all clambered onto the log. It offered their first respite.

After a few hours, when the winds died down a little, they made their most daring move. Veglia and Coomer ventured across Waveland Avenue with ropes. They then carried the two children to the rope and instructed them to cling to it, pulling themselves along as if on a conveyor belt. In a similar manner, they all then advanced up Waveland Avenue using the rope for balance. They repeated this process a number of times, moving along telephone pole by telephone pole, before they finally came to a flooded apartment complex. The first floor was underwater, but there was a second-floor balcony, a patch of concrete above the raging waters. Slowly, but surely, they all made it safely to the outdoor stairs and then up to the balcony. There they huddled together for two hours, hoping to survive the storm's end, praying this wasn't the quietus.

All five of these young Mississippians, a couple of them great-great-grandchildren of Confederate soldiers, were true citizen-heroes. Although Mississippi had earned a reputation for outdated Jim Crow racism toward blacks, here in 2005 were five white men risking their lives for a black family unknown to them. "I would have done that for anybody," Williams later recalled. "I didn't dream water could come that fast. The whole damn Gulf Coast was underwater."[20]

Veglia, when telling their story, was extremely concerned that he wasn't portrayed as more heroic than the others. "Sure," he said. "We acted fast. Just make it clear that we acted as a team."[21]

IV

Like the Veglia gang, fifty-three-year-old Hardy Jackson of Biloxi suddenly found himself floating in the water along with fallen branches and refrigerators and garage doors. "Everything happened suddenly," Jackson recalled. "We were drinking coffee, I was pacing about scared to death, and I grabbed a hammer, stuck it in my pants just in case we were forced to the attic." Jackson and his wife, Tonette, had evacuated their two children who still lived at home, but had chosen to stay behind to defend their rented house. The powerful surge forced its way through the Herndenheim Street home, just a block from the Gulf, creating a thunderous roar. "It was a big house," Jackson later explained. "And water just kept filling her up. Then a wave came and almost cut the house in two. First we headed into my daughter's bedroom. Then we headed to the attic. I punched a hole in the roof. I looked out of an air vent and saw our house was flooding."[22]

The wind pounded Jackson's head, causing severe damage to his eardrums. "It felt like somebody had stuck a screwdriver in my ears," he recalled. "It was awful, awful pain." Desperately, he tried to keep the flood from dragging his wife out of his grasp. They had been an inseparable team for twenty-eight years, raising four children together. His "full glass," he used to call her. Life without her would be empty. "She told me, 'You can't hold me. Just take care of the kids and the grandkids,'" he said. Those were her last words. "I was going to save my wife. She came up out of the water and I grabbed her by the wrist." The water rose over her eyes and then swept her away like debris; her outstretched arms shot up convulsively, waving stiff-fingered hands in the air as if to say good-bye.[23]

Hardy started hollering. All he could think was that somehow, for some reason, this was part of some divine plan. As he watched his wife float away, all he could scream was "Oh Lord, please no!" Then like an incantation that never ended, he kept hollering, "Oh Lord, why? Why? Why?"[24]

Tonette Jackson's waterlogged body was found days later.

For most victims of Katrina, the wind was horrifying, but it was the storm surge that struck with cobralike suddenness. At Ralph and Joan Dagnell's house in Bay St. Louis, the water rose 15 feet in just a half hour. When Ralph Dagnell, an electrical engineer, designed this house in 1969, he specifically built it to be hurricane-proof, specifying among other things reinforced concrete for the walls. Ralph and Joan lived at the house, gardening and taking care of their thirty-four-year-old pony. In anticipation of Hurricane Katrina, the Dagnells' daughter and her husband joined them in the house, believing it to be safe from the storm. An hour after Katrina struck Bay St. Louis, however, the water ripped the house apart. Slashing waves instantly separated the four family members. The daughter and her husband were each caught in the high branches of trees. Somehow, the pony survived too. Ralph and Joan Dagnell drowned along the coastline they loved, on the worst day it has ever known.[25]

Flooding had biblical antecedents. Children learn the story of Noah's ark from Genesis: a kindly Old Testament bearded man with a curved cane brought two of every animal species on his wooden vessel and floated away in Dr. Dolittle fashion. Then there was Moses parting the Red Sea, made memorable by Charlton Heston, or the passing over the

River Jordan. Great floods sent by God as a divine act of retribution were part of the human mythology worldwide. There was Matsya in the *Puranas* of Hinduism, and Utnapishtim survives such a deluge in the *Epic of Gilgamesh*. In Greek mythology, Zeus sent a flood to end the Bronze Age, and in the Anglo-Saxon epic *Beowulf* there is also a flood. The Incas had Hurracan, who caused a great flood, while in ancient Chinese mythology a goddess, Nuwa, "fixed" a "broken" sky after widespread flooding. But there was nothing storybook or mythological about the flooded streets of Bay St. Louis and Waveland. These were rivers of death, the pale rider or the grinning skull of the Styx ferryman, translated into a mortuary roll where victims were slowly tortured by incremental rising water, gasping for air—like Tonette Jackson or the Dagnells. No Noah's ark or River Jordan or parting of the Red Sea.

Eventually the lungs of the drowning victims filled up like water balloons and then burst, which started the quick course of decomposition. What the mythic flood stories don't convey is the reality of decomposing flesh, a process that is accelerated when a body is waterlogged. Decomposing tissue emits a gas that causes the skin to blister and turn greenish-blue. The abdomen swells, the tongue protrudes, and blood from the lungs comes flushing out of the mouth, nostrils, and eyes. A hideous, repugnant, rotten-egg stench, caused by the release of the methanelike gas, permeates the dead skin, making it especially awful for those who found the corpses after Katrina.

V

At 5:30 A.M. on August 29, electrical power went out in the Fire Dog Saloon in Bay St. Louis. Nick Breazeale's cell phone, amazingly, was still operable. Then it, too, went silent. The bar manager was all alone. He couldn't sleep in the all-pervasive darkness. He knew from the sound of the wind that this hurricane was no Camille or Ivan. It was even more violent. Looking out he saw automobiles lifting up on two wheels with every major gust. The thought of flying cars was too much to handle. His spirits sank. No more machismo. "I suddenly knew I had bitten off more than I could chew," he told a Dallas reporter. "That's when all the big plate-glass windows began blowing out of the bar front. Now the wind started whipping through the place and ripping crap apart. The roof also started disappearing in sections. It wasn't very long, maybe fifteen minutes, when water started coming up."[26]

Determined not to panic, to stay optimistic in this increasingly frightening atmosphere, Breazeale pretended that the foot of water in the Fire Dog was the result of drainage problems. He headed to the rear of the bar, but there was no escaping the water, which had blown out the bottom part of the back metal door. "Suddenly the water washed me clear across the room and back, on a quick in-and-out surge," he went on. "The storm surge was now chest deep. Somehow I had to escape the building. I didn't want to drown in there. The door was now completely gone. So when the out-surge came, I went out with it."

Breazeale's chances of survival weren't much better outside. The entire town was swimming in seven or eight feet of water. He eyed his car floating down Main Street, spinning like a toy top. He himself floated and dog-paddled, surrounded by crushing debris. He kept pushing it away, worried that some heavy object would knock him out and cause him to drown. Light bits of flotsam must have already rammed his head, since whenever he wiped water off his face, there was blood on his hands. Debris had also smashed into his sides, breaking his ribs. "Then I watched the building next door just sort of fall," he recalled. "It was Big Daddy's, a popular bar. It just collapsed. This wall that was standing just fell over in one big piece. That's probably what saved my ass, because it captured lots of debris. That kept more debris from coming in on me, and it also kept the debris that I was in from washing out when the water went down."[27]

According to the National Hurricane Center, the eye of Katrina slammed its way through southeastern Mississippi between 8 and 10 A.M. Late in the morning, the winds finally died down and the floodwater was starting to recede into the Gulf of Mexico. There had been hours of misery for everyone in the region, but especially for those hunkered down, like the Waveland police and the Veglia gang. Nick Breazeale was caught under a pile of uprooted trees, shattered glass, mountains of mushy wood, and the orthopedic shoes of an old disabled man who lived across from the saloon. Miraculously a Disney-like Dalmatian roof ornament remained poised on top of the Fire Dog, oddly unscathed by the maelstrom. "I don't know whether I was unconscious or not, but all of a sudden I heard the voices of two old ladies that live behind the Fire Dog," Breazeale recalled. "I started yelling that I needed help. At first I didn't get any response. But pretty soon these guys came over and pried and pulled and everything else and got me out of the debris. They put me on a stretcher thing and carried me back to City Hall. From there I went to Hancock Medical Center. I was in *real* bad shape."[28] (As for Henderson Point, home to Breazeale's bosses, Greg Iverson and Sue Belchner, the condo complex was washed away.[29])

During that miserable time when Breazeale was adrift, the Gulf Coast, according to the NHC, had been battered by a furious storm surge that reached as high as 30 feet. The winds had blown steadily at 125 mph, gusting to over 160. More than 8 inches of horizontal rain pelted down in twelve hours, sounding like a heavy Wyoming hailstorm. Nine out of Biloxi's ten floating casinos were totally destroyed (the tenth was only partially intact).[30] As if the hurricane's rage wasn't enough, it created weaponry consisting of millions of pounds of airborne debris, cleaved from beachside buildings and out of unlucky houses. From garbage-can lids to Ford 150 truck doors to Maytag refrigerators to 120-foot yachts, Katrina made shrapnel of everything. Many people died from head injuries after being hit by flying or floating objects. "If they're dead, they're dead," Gary Hargrove, coroner of Harrison County, Mississippi, stated bluntly in what was essentially the prevailing ethos of the week. "We've got the living to take care of."[31]

The entire Gulf Coast region, from southeast Louisiana to Alabama, was declared a federal disaster area; an estimated 284,000 homes were destroyed.[32] People in Mississippi who didn't evacuate faced death or near-death experiences. From one disturbing instant to the next, over the course of eight or nine relentless hours, Katrina inflicted cruel punishment. No one could function at even half capacity after such an unnatural day, full of so many horrid ordeals. It was both exhausting and heartbreaking. Watching one's own community drown and crumble caused disorienting passion fatigue. A collective numbness came over the surviving population as if they had been asked to inhale chloroform. A human mind can only stand so much grief before it cracks.

VI

With great determination, Bay St. Louis Mayor Eddie Favre sprang into action. His biggest concern was Hancock County's Emergency Operations Center (EOC) located in a county building downtown on Ulman Avenue. All sorts of local officials and volunteers had stayed there and he didn't know how they had fared. The EOC building was less than a mile from the Old Spanish Trail police station, where Favre was stranded. He wondered whether the EOC was still standing. "Somehow I had to get to them," he recalled.[33]

All the roads in Bay St. Louis looked as though somebody had dumped parts from a difficult home-assembly toy on the streets, bludgeoned them with a sledgehammer, and then opened the fire hydrants and the sewers. Police cars were of no use; they were flood-damaged anyway. Suddenly, Favre had a flash of inspiration. The only vehicle in Bay St. Louis that could plow through water was a county Public Works Department dump truck parked behind the police station. Favre asked Ron Vanney, the head of the town's Public Works Department and a former Pascagoula shipyard foreman, to prepare the truck for action. The door on Favre's passenger side wouldn't close, but with Vanney at the wheel and the mayor riding shotgun, they rumbled toward downtown Bay St. Louis and the EOC. Vanney later called the gut-wrenching destruction he drove through as "shock and awe." Usually when Bay St. Louis flooded from heavy rains the ditches would back up. That's what happened during Camille. Katrina was, as he put it, "a whole different enchilada."[34] The Fire Dog was gone and so were all the rest of Favre's and Vanney's favorite haunts: Bookends Bookstore, Winn-Dixie, La Coffee Cafe, and KG's Cajun Seafood and Poboys. The Bay-Waveland Yacht Club, where Favre sometimes held court, was in splinters. A quick glance indicated that his office on Court Street had been flooded out. According to a stranger they met on the roadside, Waveland, Bay St. Louis's sister beachfront community, had been obliterated.

"We felt totally isolated," Favre recalled. "No roads were open in or out of Bay St. Louis." Bay St. Louis sits on the western side of St. Louis Bay, while historic Pass Christian was on the opposite, eastern shore. (Local lore pretends that during Camille the bay parted in biblical fashion.) The Highway 90 bridge across St. Louis Bay was toppled in the storm, as the water inside the bay became a drill, whittling out anything solid. Seeing that the bridge was gone made the full brutality of Katrina clear to Favre and Vanney. They just stared, slack-jawed, at the devastated bridge. Before long, people they encountered reported that three other nearby bridges were also destroyed. Even the NASA Stennis Space Center roads, usually offering a guaranteed and highly secure route out of harm's way, were washed away. "Our dump truck reconnaissance mission told us that we better open up a shelter fast," Favre recalled. "All those people who survived Katrina but lost a place were going to need a dry place to sleep. It was already noon. We had to think and act for ourselves. This was Bay St. Louis's problem. I understood that FEMA and the federal government would eventually come to our aid, that President Bush would help us out in a couple days. But Katrina wasn't his fault. Like in all natural disasters the first responders had to be the local and county officials. We went into action."

Everywhere they looked, people were in trouble. To stay alive, they had hung on to anything that offered hope of survival. Because the railroad tracks were relatively high, many Bay St. Louisians, flooded out of their homes, clung to the track ties for dear life. While Favre and Vanney surveyed the damage from their truck and wrote down locations where people were stranded, Officer David Stepro was also taking action. Stepro had waited out the storm in the Old Spanish Trail station with his colleagues.

No sooner had the eye passed than Officer Stepro and his chief, Frank Griffith, hopped into a Jeep Cherokee and raced toward Waveland—or so they thought. Less than a quarter mile from the station the water was six to eight feet high. "You just couldn't drive through the trees and power lines and water," Griffith recalled. "It was impassable." But Stepro wasn't going to be thwarted. With the permission of his chief, he headed toward the EOC. Perhaps somebody there knew an open road to Waveland. It was a wise move, for suddenly in front of him like a mirage, there was an empty yellow school bus, the keys in the ignition. Officer Stepro commandeered the bus. He tried to start it, but it was coughing water. No luck. He tried to start it again. This time the motor took hold, spitting water out of the exhaust pipe, gradually emitting a throaty roar. He shifted the bus into gear and drove out of the parking lot. He headed straight for Highway 90 and the butt-ugly bush that was saving the Waveland cops. At the EOC he hooked up with Richard Fayard of American Medical Response. By the time they approached Waveland, the water was receding and, in front of them in the distance, they saw about fifteen people just standing on the road. They had on gray uniforms that looked torn and tattered like those of the defeated Confederate soldiers heading home in the movie *Cold Mountain*. Stepro cocked his head over the steering wheel, his nose almost touching the windshield, searching for his wife. And there she was in her blue rain gear.

David Stepro stopped the bus. Out the door he went, picking Laura up off the ground and swirling her around, a modern-day equivalent of the famous Alfred Eisenstadt photograph of a nurse and sailor kissing, taken in Times Square on a V-J Day in August 1945. They still had their lives together. The specter of death had vanished, and their hearts flooded over in joy. It was only then that Officer Stepro learned that the "Tree Cops of Waveland" were actually the "Shrub Cops of Highway 90." It became a joke among them that the ugly shrub that they had intended to chop down as an eyesore just weeks earlier had become a sacred plant. "I am going to get a sampling from the bush's roots," Stepro said, "and plant one in our new front yard *away* from the Gulf of Mexico."[35]

Quickly the Waveland officers boarded the bus. Many were bruised and cut. David Stepro drove the bus to the Hancock Medical Center about one mile down Highway 90. Along the way, he picked up dazed citizens who had abandoned their flooded homes and were looking for water and medical attention. Many had broken limbs, long gashes, or other injuries requiring serious care. The bus quickly filled. When it turned into the flooded parking lot of the Hancock hospital, nurses Sydney Saucier and Angie Gambino gazed at the filthy vehicle rolling toward them. "At first we thought, great, it's a FEMA bus or something bringing us aid," Gambino recalled. "But as it got closer to our emergency entrance door, we realized that wasn't the case. People were longing for medical attention. That bus started a nonstop rush to our doors, as people walked toward us from all over, desperate for help."[36]

Meanwhile, Favre and Vanney soon reached the EOC, or what was left of it. Half of the roof had blown off, and the ten-foot storm surge had filled the building up like a tub. "Many of the EOC folks thought they were going to die," Favre recalled. "They wrote their names and Social Security numbers on their arms, so their bodies could be properly identified. I knew they were probably in bad shape."[37] Those inside eventually made their way through the water to safety in another building. Whatever destruction Katrina wrought on Mississippi Gulf Coast buildings, the psychological damage to the people was, from Mayor Favre's perspective, far greater. He saw many survivors wandering through the rubble unsure where to go or what to do. They were frayed zombies, Katrina survivors, in search of a hug.

On Monday afternoon, as Favre and Vanney drove about in the dump truck, thousands of Hancock County citizens were stranded on rooftops or in attics desperately screaming for help or sending out SOSs by waving sheets, homemade placards, towels, or pickaxes. They also lit fires and fired guns. The cell phone towers had all been knocked out, and the landlines were down, so communication was as primitive as a rag waved frantically by a person in trouble. The survivors whom Favre and Vanney picked up were the lucky ones. They had weathered Katrina and all of them had wild-eyed stories to tell. "My God," thirty-four-year-old Dara Adano of Bay St. Louis exclaimed describing the sweeping damage. "Katrina obliterated the landscape. Roads were torn up like they were sheets of paper and buildings looked like a huge hammer had pounded them into rubble. It was a wasteland."[38] Throughout Mississippi Gulf Coast towns, in parking lots once jammed with fast-food drive-through traffic, only an occasional local emergency responder showed up. For four long post-Katrina days not a single federal first responder came to devastated Bay St. Louis. By Friday, September 2, federal police protection *did* finally arrive, followed by four law-enforcement groups from Orange County, Florida, who helped out for the next five weeks. The word was out that FEMA, the Red Cross, and other relief agencies were probably days from being able to turn cavalry.

Mississippians—with the air help of the Coast Guard—were going to have to take care of themselves. "We knew we would have to pick ourselves up," Favre recalled. "And guess what? We did. Or at least we're trying to."[39]

Surviving the storm came with recriminations. Others would be less forgiving of the delay in state and federal help. Katrina fixed itself permanently in the thoughts of survivors, pushing some into deep neurosis. A woman in Bay St. Louis named Terry Lucas, who lived next door to her parents, Gloria and Luke Berigno, was speaking to her mother at eight in the morning. Suddenly the storm surge flooded over the houses on their street—not in a creeping way, but as a swirling, insistent torrent. "This is worse than Camille," were the last words that Lucas heard her mother say. Both of her parents drowned inside their house.[40]

Seventy-five-year-old Pete Fountain, the legendary clarinetist, tried to stay positive. He was safe, but his $1.5-million Bay St. Louis house was gutted, and his most treasured possessions, which documented his illustrious jazz career, were lost forever: correspondence with Frank Sinatra, gold albums, signed pictures of himself with presidents, Louis Armstrong memorabilia. Losing the gold record of his signature song—"Just a Closer Walk with Thee"—was a truly personal blow. Fountain's vintage gun collection was also gone. After Katrina, it all came under the heading of one word: debris. Fortunately, Fountain had evacuated with his family to Hammond, Louisiana, seventy-five miles west. Katrina "really got me," he told the Associated Press. "But I have two of my best clarinets, so I'm okay."[41]

That's what Fountain told the press after Katrina. He had that Mississippi Gulf Coast male mentality that no matter what your age, you don't complain. His post-Katrina sadness, however, was acute. For forty-six years, Fountain had participated in New Orleans's Mardi Gras parades. His krewe was called Half Fast Marching Club and was a huge tourist draw, parading down St. Charles Avenue while Fountain played his clarinet, a modern-day Pied Piper. When February 2006 rolled around, however, Fountain stayed at his rented home in Hammond, Louisiana, refusing to participate in Mardi Gras. His heart wasn't in it. "I think maybe it was just depression about all the stuff that happened," he said. "All the things we lost. All the disruption. And then you look around and see all the stuff messed up. It just sort of grinds you down."[42]

VII

During the long hours when Katrina passed over the Mississippi Gulf Coast, the doctors, nurses, and patients at Hancock Medical Center had watched in utter disbelief as whitecapped water kept rising through the parking lot and up to the hospital doors. Then, like an artillery barrage, in rapid succession they saw ambulance mirrors blow off and the *Sun-Herald* and *Sea Coast Echo* vending boxes tossed skyward in a mean gale. Water was slowly starting to creep into the first floor. The parking lot was badly flooded. Frantically, staff members pushed sandbags against the doors and jammed towels into the cracks hoping to stem the flood tide. CEO Hal Leftwich, staying with his team, realized that his home in the Cedar Point neighborhood, virtually facing the bay, was already washed away.[43]

By 10:30 A.M. on August 29 the water was up to the windowsills. The staff started moving the thirty-three sick patients to the unflooded second floor via the elevator. There were five ICU patients, but luckily nobody was on a ventilator. Outside, the hospital looked like it was an island surrounded by the Gulf of Mexico. Because the hospital was designed with lots of big picture windows on the first floor, those inside must have felt they were in the Aquarium of the Americas looking at the floor of the Gulf. The second floor was the only safe refuge because the third-floor roof was blowing off. The stress of the harrowing situation took a toll on everyone. A quivering woman who had given birth forty-eight hours earlier worried that her baby was going to die. There was a guy who had just had abdominal surgery, a woman with cancer, and a few emphysema patients. Patient Crockett West, who was nonambulatory because he weighed 500 pounds, was the last person transported in the hospital's elevator before the generators went out. "I swear they just got him out of the elevator and then everything went dark," Saucier recalled. "We were forced to carry the remaining patients on our backs up the stairwell. We had some patients in the ER. Unfortunately, the wind-driven rain prohibited use of several patient rooms on the second floor, forcing patients to be cared for in the hallway."[44]

West was growing increasingly uncomfortable after the electricity went out. His chronic obesity caused health problems galore. Unable to walk around because of back surgery, a bad heart, and painful knees, West was essentially a hostage in his oversize bed. He lived by himself on Longfellow Road in Bay St. Louis, not far from the Jordan River, with a regular home-care aide and housekeeper named Jo Ann Garcia visiting daily. Part of Garcia's job was to help him diet. "He was starting to show results," she recalled.[45]

As Katrina came around, no shelter would admit West because of his weight. No regular ambulance could fit him in back, so Garcia had to persuade the Red Cross to find a special ambulance, which took West to Hancock Medical Center. She was going to ride out Katrina with him, and whomever else she could help. The obese patient's worst fear was perspiring to death. He simply couldn't take the heat. As Hancock Medical Center turned into a sweatbox, and doctors and nurses peeled off clothes to stay cool, West was left in his own layers of fat. "It was miserable for Mr. West because it was so hot," Garcia said. "And I had other people to look after. But he survived."[46]

The staff waited and prayed for the hurricane's eye wall to pass. As the barometric pressure rose and the winds slowed, the seawater would roll back. Wouldn't it? With flashlights, nurses retrieved undamaged medical supplies, linen, and food from the first floor. A bucket brigade was formed to obtain water to flush toilets as water pumps failed.[47] "I was stunned at how the floodwater ruined everything," Angie Gambino recalled. "I never knew a refrigerator could float. We lost most of the stuff on the first floor. We thought if you put valuable stuff on top of the desk, it'd be safe. Boy, were we wrong. We had fish in the hospital swimming around the first floor. We felt sorry for one so we put it in a coffeepot. Kind of an orange-looking thing and we called it Katrina for days."[48]

Over the next few hours other animals emerged looking for safety in the hospital or just blown inside by the surge. The Gulf Coast was abundant with wildlife, and suddenly the staff at Hancock Medical Center saw catfish, crabs, and snakes in their corridors. A large snapping turtle even made it into the emergency room, and an armadillo raced about, clearly panic-stricken, desperately looking for an exit. "The armadillo became our mascot," Leftwich recalled. "Our engineer ad-libbed that we'd eat him if we had to. Fortunately, we didn't get that desperate."[49]

But Fredro Knight was starting to fear the worst. What if the water rose to the second floor, where all the patients were? "There would have been nowhere for them to go," he recalled. "All these patients in post-op had just come out of surgery with open wounds. With all that water and sewage, they could have died from massive sepsis. We had a lot of people panicking, employees panicking and passing out. We had one girl who worked in the lab. She had a bad valve, and heart palpitations started. She passed out."[50]

Fortunately just as things were starting to crack, the water started receding. It was only calf-deep in the streets when a black Chevy pickup won the distinction of bringing the first of many maimed survivors to the emergency entrance. The driver jumped out, opened the tailgate, and starting pulling a woman in her mideighties out of the truck bed. She was strapped down to a house door functioning as a makeshift stretcher, and she was shivering uncontrollably. The hospital was still in lockdown (no patients in or out). No generator was working, and the staff was in no shape to start admitting patients. Registered nurse Sean Graham told the people in the truck "No"—but that stubborn stance lasted just three or four seconds. Graham traded glances with Facilities Service Director Hank Wheeler; they both knew that lockdown was, in post-Katrina Mississippi, officially over. "Screw it," Graham said, waving them in. He helped carry the woman inside. She had a badly broken arm. "We all knew from that moment on, it was no sleep for us," Gambino recalled. Before long David Stepro's school bus, bringing in the Waveland police officers and the others he'd picked up, arrived. "Adrenaline, compassion and desire to heal fueled the next forty-eight hours," Janet McQueen, Hancock's PR person, wrote in her diary, "that is until the first disaster military assistance arrived"[51]—days later.

The survivors came wandering in from every direction, out of every alleyway and dirt road in the coastal county. As the *Houston Chronicle* noted, even though the "hospital's floors were slick with mud and an oatmeal-like mush had fallen from ceiling tiles . . . the emergency room was in business."[52] The medical facility suffered $20 million in damages, but it kept running on the heroics of its staff members.[53]

Many of the people who were admitted to Hancock were in shock.

Some just wanted to come into the facility for reassurance, to touch a cheek, to look for somebody. Others came seeking sustenance: water or juice, cheese and crackers, an apple or an orange. They were mired in grime, soaked in bilgewater. Some asked the hospital for shoes, gowns, or T-shirts. They offered to buy flashlights and wanted to pay rent for a spot on the hospital floor to rest on. One dazed family showed up with a shopping cart, children jammed in the basket, bedraggled beyond words. Many of the arrivals were, of course, in serious need of medical or hygienic attention. To handle them all, the hospital implemented a triage system. "We had to assess the ones who needed immediate attention," Sydney Saucier said. "You prioritized them. The ones that need immediate attention, you try to get them lying down. The ones who are partially hurt, you sit them in chairs. And the ones that are basically okay, we left them standing."

Fredro Knight became known as Mr. Stitch. "It went from three to four stitches to eighty-five stitches [each]," Knight recalled. "I had to repair thighs, legs, deep chest wounds, you name it. These people had horrible, horrible lacerations." Usually an ER doctor would x-ray a wound or make sure there was no foreign body in it to cause infection. Now, with the generators out and supplies rationed, it was prehistoric medicine—no painkillers or Novocain. The nurses were setting up suture trays in assembly-line fashion. "We were seeing eight to ten patients at a time," Knight said. "I stitched over ninety people in just a couple days."

The hospital treated about 800 people during the course of the first two or three days after the storm. Only one person died—a man in his late seventies who had a chronic upper respiratory problem. He had hoped to weather the storm in his son's house. When it filled up with eight feet of water, and everyone else headed for the attic, the old man kept saying, "I'm too old! I'm too tired! I'm not going to die here!" The son got his dad into the attic and kept poking him with a stick when his head drooped into the floodwater. "No! No! No!" the son pleaded. "Wake up!"

All members of the family survived the storm and then they rushed the father to Hancock at three that afternoon. "At 4:15 he basically died," Saucier recalled. "But we resuscitated him in the dark, with no electricity. He came back to life." In fact, the old man even cracked a joke. "Why'd you bring me back?" he asked Saucier. "I'm tired. I'm ready to go." Saucier had a ready answer: "Because it would ruin my day and it would ruin yours. I've already had a bad enough day." They continued to joke about it over the course of the next day: he had to stay alive for her. "Then at seven at night I went to bed," Saucier recalled. "We had got him stable, he was fine. I couldn't sleep so at around eleven I went over and I checked on him. He was fine. But then at 12:05 he had died. I was like, 'That little stinker, he waited until the next day to die!' Because he kept on saying, 'I'm ready' and I would say, 'Well, not today. You better wait until tomorrow.'"

A good way to describe Hancock Medical Center was as a MASH unit stuck in a flood zone. Out in the parking lot all of the vehicles—including Saucier's Durango and Knight's Land Rover—had become fishbowls. Inside, maintenance staff hung glow sticks in the stairwells and hallways to provide some illumination. By Monday night, supplies were running low. There was no food, oxygen bottles, baby formula, linen—basically nothing but these medical responders' own wits and determination. Everybody was practically naked, trying not to sweat to death. Clothes were taken off and washed. "Scrubs, jeans, shorts, whatever, off they went," Saucier later laughed. "It looked like *Petticoat Junction*. We had our bras hanging out of the third-floor windows. We were dealing with patients and had to be sanitary. We suddenly smelled good and it raised morale."

One of the most common injuries surprised the doctors and nurses. All over the Gulf Coast, in hospitals like Hancock Medical Center, patients wandered in with the skin ripped off their hands, the flesh lacerated almost to the bone. During the storm, those who were strong enough survived by hanging onto treetops for dear life. The result was that many people had hands swollen and raw like slabs of red meat.

In Biloxi, Mabel Walker spent more than seven hours in a tree with her eighty-year-old uncle, hoping they would avoid what poet Robinson Jeffers called "the seamouth of mortality." Her hands were cut but not badly. At one point, she spotted a man without legs floating past. The swirling current was too strong for her uncle to manage by himself. Helpless, all she could do was watch in stark horror as the stranger floated off, probably to a watery grave.[54]

VIII

At twilight on Monday, many coast survivors came outside for a look around. What they saw was like a scene from *The Night of the Living Dead*, as shocking Katrina stories started coming to light. More than fifty people had been "riding out the storm" in the Quiet Water Apartments in Biloxi when the three-story brick building was inundated by the angry waters of the storm surge. Fifty-five-year-old Joy Schovest barely escaped. "We grabbed a lady and pulled her out the window," she said, "and then we swam with the current. It was terrifying. You should have seen the cars floating around us. We had to push them away when we were trying to swim."[55] Landon Williams had also been able to dive clear of the Quiet Water Apartments, in company with his grandmother and uncle. Williams had the chance to look back, like a sailor forced to abandon his sinking ship. The walls collapsed and then the contents, including the frantic people, were sucked under the water. "We watched the building disintegrate," Williams said. As it crumbled, he knew that his neighbors inside were doomed. Thirty of them perished. All he could mutter was "God bless."[56] In the aftermath, neighbors who walked to the site of the Quiet Water Apartments—a complex that used to live up to its name—saw nothing more than a concrete foundation.

Death was everywhere along the Gulf Coast that Monday; it was difficult to avoid encountering floating or crushed corpses. Tonya Walker, a forty-one-year-old maid on one of the oil platforms in the Gulf, had recently moved into a house in Long Beach, Mississippi. She told her story to Renee Montaigne of National Public Radio's *Morning Edition:*

> Well, I had a bathroom collapse around me. I rode rooftops in a little dinghy and mowed over trees and telephone poles and wires like it was nothing. . . . I was on the eastern eye wall. My house was in direct—I only live, like right across [Route] 90 from the beach. And I never realized how furious Mother Nature can be and how unforgiving and—I mean, I watched cars and big screen TVs and buildings and everything just flying around me like it was a kid taking toys out

of a toy box. Fin[ally]—and we were in the storm for eleven hours before it calmed down enough.

We amazingly seen this little dog. We knew the water was gonna be going back out, and we had to find shelter or we would end up back out into the ocean with it going back out. And there was this little dog that was barking, and we made our way to where this little dog was. And the water had already started receding, and we stood behind a brick wall while the rest of the wind and everything was going on.

And I don't know if the dog was real or not.

I really don't because when it was all over with, I couldn't find the dog anymore. But I think there were a lot of miracles that happened that day. And I will say this, I've always heard, you know, that before you die, you see your life pass before your eyes. I believe that, because I'd seen it. And I didn't think I was gonna make it.[57]

But many did make it. After Mabel Walker climbed down from the oak tree, her legs were bruised ivory and purple. The stagnant water that remained late on Monday changed from muddy to oily, and from no odor to a pungent, vomitlike stench. Walker saw a stray dog approach the fetid water to drink, but it backed away apprehensively, as if it knew the water was undrinkable. Instinct was all that man and animal had left in Katrina's distorted wake.

The Veglia gang and the family they rescued eventually made it to the Knock-Knock Lounge, which suffered only minor structural damage. As night fell that Monday, they were exhausted. Dumbfounded, they watched as looters broke into the Hancock Bank ATM machine across the street and raided the liquor store next door. They wanted to attack the thieves but decided instead to remain undetected in the darkness. "There was nothing we could do," Veglia recalled. "We just stayed silent, glad we were safe."[58]

A distraught Hardy Jackson, still calling for his dead wife, was saved by a Biloxi neighbor he barely knew. "That man came to my rescue," Jackson later told the *Citizen News*. "He waded over to me in the water and yelled, 'Hardy, hold on, I'm coming.' He got to me and put his arm around me. Then he took me to his house and gave me dry clothes and tennis shoes and something to drink. That was a little after four o'clock in the afternoon." Eventually Jackson was evacuated to Palmetto, Georgia, where he was reunited with his children.[59]

Signs reading "Turn Back: Hazard Zone" should have been posted every quarter mile along the Mississippi Gulf Coast. Where there had been a neighborhood neatly filled with houses, there was attendent despair, only empty space, stripped even of greenery, telephone poles, and crooked water tanks. Salt-tolerant plants, however, were strewn all over, looking like wet spinach. A number of old-growth oaks—trees that had dangled moss for four hundred years, predating the Pilgrims at Plymouth Rock—had been toppled. Their cable-thick roots, as pointy as a witch's fingers, stuck up into the air. Everything smelled like rot, and a gritty film coated every surface exposed to the storm. Locals rubbed their eyes in disbelief. They were ill equipped to digest the unfathomable scope of the catastrophe. They felt deserted on some sort of twisted Gilligan's Island, unable to communicate with the outside world. On the west side of Gulfport, more than forty tons of chicken lay exposed, rotting in the dogged heat.[60] Nearby, hundreds of Mazola corn oil jugs had exploded, turning the street yellow.[61] Survivors swapped information (or misinformation) about whether their local McDonald's or Sav-a-Center or Kentucky Fried Chicken was still standing. Some, though, were too exhausted, too battered, or too stunned to see anything clearly.

Those on the scene at dusk on Monday had trouble describing the damage in terms of a hurricane. Reporters all coughed up essentially the same statement: Katrina's magnitude had surpassed anything they knew or could imagine. No Gulf Coast Cassandras or Nostradamuses had anticipated the magnitude of the destruction. It hardly seemed possible. *Harper's* editor Lewis Lapham wrote that the barren Gulf Coast beachfronts were pure "Book of Revelation," a dismal eighty-mile-long junkyard of block-by-block torments.[62] "Right now," said the director of the Harrison County Emergency Management Agency, Colonel Joe Spraggins, speaking of Gulfport, "downtown is Nagasaki."[63] Another man, Blake Beckham, in D'Iberville, north of Biloxi, used essentially the same metaphor: "It looks like Hiroshima," he said.[64] Governor Haley Barbour also made reference to Hiroshima, after he flew over the coast on Monday evening. A few days later, novelist Richard Ford, a native of Mississippi, wrote a *New York Times* op-ed piece, refuting the Fat Man and Little Boy comparisons. "'It's like Hiroshima,' a public official said," Ford wrote. "But no. It's not like anything. It's what it is. That's the hard part. He, with all of us, lacked the words."[65]

But people still looked for words. Some insisted that America had been hit with a tsunami, for no hurricane could have done so much damage. "Katrina wasn't a hurricane," Greg Iverson, owner of the Fire Dog Saloon, said. "It was a hungry Cyclops settling scores."[66] Most of the analogies were over the top. The number of Mississippi casualties—approximately 220 Katrina-related deaths—though high by any standard, was not comparable to the number wrought by the atomic attack or other natural disasters. In 1931, nearly four million Chinese died in a flood of the Yellow River. In December 2004, close to 300,000 perished in the South Asian tsunami. But the landscape painted a picture as grim as any disaster in history. Along the Gulf Coast, it looked like a product of the scorched-earth policy initiated by William Tecumseh Sherman during the Civil War. Not since Atlanta had been burned to the ground had a swath of Dixie looked so wretchedly barren.

Part of the eeriness of the immediate aftermath of Katrina was the incongruously clear weather along the battered coast. On Biloxi Bay, after the storm, the water was smooth and placid, under what one meteorologist with the National Weather Service called "the prettiest blue skies you've ever seen."[67] Jim Butler, a reporter for the Biloxi *Sun-Herald*, did not have to look the following morning, but just listened in order to hear the difference. "Most noticeable," he wrote, "for those living through the day before was the seemingly muffled sound. Gulls, mockingbirds, cardinals, passing vehicles—anything making noise seemed muted following the hours and hours of the storm's roar."[68] Perhaps the hardest chore along the Mississippi Gulf Coast was retrieving the bodies. Calmly walking the streets, men were picking up the dead and carrying them on stretchers, looking for a funeral home or a compound that would take the bodies. Bay St. Louis Police Chief Frank Griffith, for example, had to retrieve a body from the beach; it had just washed ashore. He was a former NOPD homicide detective, so death was nothing new. But bagging hurricane bodies was a different experience. When he saw those blank faces, he thought that but for the grace of God, it could have been him. "One of the hardest moments for me was seeing a mother that drowned," he recalled. "She was under a mattress with her three dogs lying next to her. They had all died together."[69]

The decision made by Charlie West, the Bay St. Louis vet, to stay at the Ramada Inn in Diamondhead proved wise. The motel was one of the few places in Hancock County not destroyed, suffering only roof damage from a tornado that spun off the hurricane. But Pethaven, Dr. West's hospital, didn't fare well—not at all. Water rose to six feet and drowned all the cats, including Dr. West's beloved gray tabby. Some of the dogs, including his springer spaniel, also drowned, trapped in their pens as water filled their lungs. "I felt so stupid," Dr. West recalled. "All my life I've tried to help animals and now I failed. I lost both Puss and Brandy. It was just awful finding them dead." One of his pets, however, the bloodhound named Flash, survived; apparently it dog-paddled for hours until the waters receded.[70] Tragically, the animal shelter in Hancock County, which was built in a flood zone, didn't evacuate, and the dogs and cats in its care all drowned.

The Marine Life Oceanarium in Gulfport was also demolished, but an adult sea lion survived. He sat in the rubble barking for buckets of slimy water to be dumped on his back. "You can't just walk away from him," his fifty-one-year-old savior, Jeanne Robinson, said. "Here's something trying to live."[71] The all-encompassing fear at the aquarium was that its eight bottlenose dolphins had been killed. Their saltwater holding tanks had been destroyed and none of the dolphins were inside. The dolphins' owner, Moby Solangi, launched a search for the star attractions, although he assumed they were dead. Solangi was pleasantly surprised when his helicopter mission succeeded. The dolphins were in Mississippi Sound, huddled together around Gulfport's main pier. "They were really glad to see their trainers," Solangi recalled. "They recognized the whistles [that were] used to train them. They were jumping out of the water when they heard the whistles."[72]

The fringe of Gulfport Bay was littered with the remnants of casinos that had once been moored there. Scattered slot machines had the surreal look of Dalí's melted clocks. Bar stools popped out of the mire like mushrooms. Some structures were ripped open and left hollow as caves. Harrah's Grand Casino was sitting in the middle of Route 90. The President Casino had smashed into a Holiday Inn on the highway.

Buses were in the water. Boats were on land. Some oil platforms in the Gulf had crumbled like Legolands, although the Navy immediately protected the ones still standing. This was considered a Homeland Security measure—at all costs, the oil supply had to be saved. In Gulfport, a fleet of eighteen-wheelers bearing the yellow-and-blue color scheme of the Dole company had rolled over like fallen dominoes, fruit spilling everywhere.

In a typical tropical storm or hurricane, buildings located between the beach and Highway 90 might expect some flooding. Katrina pulled ocean water far past 90—for instance, past the Hancock Medical Center—and in some places, water even washed all the way to Interstate 10. Buildings as far as 100 miles north from the Gulf—in Hattiesburg, Mississippi, for example—were ravaged by Katrina. Coyt Bailey, a helicopter pilot for WLBT-TV, an NBC affiliate in Jackson, reporting on the Mississippi Gulf Coast region, radioed a numbing message that "everything that was within a quarter mile of the beach has just either been leveled or destroyed or just consumed with water."[73] Houses that were built before the Civil War were severely damaged, many of them reduced to rubble. All the beachside dunes had vanished, along with the tall summer grasses, as if by an eraser wiping a blackboard clean. They had withstood strong winds before, but never a storm surge like Katrina. From the sky, as WLBT reported, it was obvious that town after town had been largely expunged from the map by some evil magic wand. That evening, all over America, aerial views of the devastation dominated the news broadcasts of every station. Local politicians were among the first to talk to the nation at large on radio and television. "Highway 90 is destroyed," said A. J. Holloway, the mayor of Biloxi. Sand and stones buried the road up to several feet deep in places.[74]

IX

In Pascagoula, at the eastern edge of the Mississippi Coast, Beach Boulevard had been the crowning glory of a pristine neighborhood that sat on a jut of land between Highway 90 and the sea. Pascagoula was the industrial heart of the Mississippi Gulf Coast, home to the state's largest employer, Ingalls Shipbuilding. Ever since explorer Hernando de Soto established a relationship with the Pascagoula tribe in the 1540s, the community prided itself on its nickname, Singing River City. The town had a literary history too, with Henry Wadsworth Longfellow penning "The Building of a Ship" during a visit and William Faulkner writing *Mosquitos* by the beautiful homes along the beach one summer during the 1940s. One of those homes belonged to a favorite son of Mississippi, Republican Senator Trent Lott, the Senate Majority Leader from 1996 to 2002. "It wasn't a fancy house," he lamented after the storm. "Just a Creole cottage, but it was built in 1854."[75] In the wake of Katrina, the Lott homestead was nothing but a yard strewn with water-soaked wreckage. "Absolutely nothing was left of our house," Lott explained. "It was built a good eleven feet above sea level and survived dozens of hurricanes. But Katrina wiped out everything, even the foundation."[76]

It had been a tough couple of years for Lott. When he made racially insensitive remarks at Senator Strom Thurmond's one-hundredth birthday party in 2002, he was forced to resign as Senate Majority Leader. What really hurt wasn't that John Kerry and Ted Kennedy attacked his comment—he expected that from his Democratic opposition—but that President Bush hadn't come to his defense. Even as the painful political misfortune lingered, Lott's ninety-one-year-old mother died in July 2005, and he was still in mourning as Katrina struck. He had been in Birmingham, promoting his book, *Herding Cats: A Life in Politics*, when the hurricane made landfall. He wanted desperately to be with his wife, Tricia, who had evacuated to Jackson. As soon as the winds subsided, Lott went to the Mississippi state capital, which—located 192 miles from the coast—suffered only minor damage. Lott had been dealing with the policy ramifications of hurricanes for thirty-seven years, and he immediately launched into action. The main lesson he had learned from Camille, in fact, was "not to panic or yell at people." With Tricia, he drove to Pascagoula to survey the damage, and then rolled up his shirtsleeves and started helping his friends and neighbors dig out. Pascagoula was one of the Jackson County towns that was decimated. Ocean Springs was yet another. In all, one-third of county residents lost their homes. In Pascagoula, from the beach to the railroad track, everything had been submerged under twenty feet of water: you could see the waterline.

When the Lotts arrived at their Beach Boulevard house, they were surprised to see their favorite old oak tree, plucked bare of Spanish moss, its thick trunk boldly standing; but everything else gone. Together they walked over to the tree, touched it, and sobbed. For a while they felt sorry for themselves. But their daughter, Tyler Armstrong, calmed them down. "Oh, Dad," she said, "it's just things."[77]

Like many along the Mississippi Coast, Lott quite naturally believed that his insurance policy would cover his Pascagoula home—after all, he had hurricane coverage. But his insurance company would not cover the damage from Katrina. State Farm contended that it shouldn't have to pay for the water damage caused by Katrina. Lott and other residents thought that State Farm was trying to weasel out of its obligation to home owners in order to save billions, splitting the difference between wind damage and storm surge and saying hurricane insurance didn't include the latter. "I have joined in a lawsuit against my longtime insurance company because it will not honor my policy," Lott said, "nor those of thousands of other South Mississippians, for coverage against wind damage due to Hurricane Katrina."[78]

But in Harrison County, in the center of the Mississippi coastline, the damage to the coast was even worse. For the week after Katrina, around 13,000 Mississippians remained stranded at 101 Red Cross shelters. Thousands more were gathered at makeshift shelters in schools and churches. The biggest city on the coast was Gulfport, just to the west of Biloxi. In Gulfport, as the fire chief explained, most downtown buildings seemed to have "imploded."[79]

A 40-foot fishing boat in Gulfport was carried hundreds of yards over a marsh to rest between a pair of trees. What was remarkable was that the trees were not scarred; this meant that the boat had been carried inland on a storm surge at least 35 feet high, to be dropped down between the two trees.[80] Another boat, a 22-footer, was neatly parked in the drive-through window at a Burger King restaurant in Bay St. Louis. But, perhaps because the citizens of the Mississippi Gulf Coast live with the Gulf as their front yard, there was little panic in the post-Katrina air. Just a lot of blank stares. With determination, some unfazed survivors went looking for boats that worked, knowing it would be days before FEMA or the National Guard would arrive in full force. Mississippians started rescuing victims themselves in fishing skiffs, motorboats, canoes, and even Jet Skis.

Recognizing the dearth of a FEMA presence by Monday evening, Eddie Favre and Ron Vanney drove their dump truck to Bay St. Louis High School. They were looking for a shelter for the many people made homeless by the storm. Favre had a key and went inside to inspect the hallways and corridors. To his surprise the high school not only hadn't flooded but had experienced only minor wind damage. Although it didn't have provisions, it did have a roof, and that was more than could be said about a lot of structures in Hancock County. Because all the bridges leading to Bay St. Louis and Waveland were down, isolating the two towns, the high school was their best bet. "We traveled everywhere we could," Favre said. "We opened the high school gym as a shelter. It was all spur-of-the-moment. FEMA was essentially nowhere to be found. But we didn't whine about that or complain. We acted. We commandeered what we could. Wal-Mart, for example, allowed us supplies. Needless to say, in my town there was no police looting. Eventually churches, like Calvary Independent Baptist Church, came in with food and water. Private sector help came pouring in in droves. The loss of human life was hard to take. We'll never know the exact count. We lost about twenty folks in Bay St. Louis alone."[81]

Bay St. Louis and Waveland and their neighbor across the inlet, Pass Christian, were largely destroyed, the sections of each town closest to the water suffering damage listed officially as catastrophic. Overall, one-fifth of all housing in Hancock County was made uninhabitable by Katrina.[82] "What really hurt," Favre recalled "was seeing Our Lady of the Gulf Catholic Church and Christ Episcopal gutted by God."[83] The amount of debris strewn over the flat, bare, frightened-looking Gulf Coast landscape was estimated at 50 million cubic yards. The garbage that Katrina made of people's homes, their businesses, their boats, their cars, and their lives would fill 400 football fields to a height of 50 feet.[84] Throughout the Gulf Coast states, approximately one million people were without power. Clean running water was almost nonexistent. That was, however, not all that was lost in Mississippi. A way of life went, too. "Generations of Mississippi Coast dwellers enjoyed their piece of paradise with a certain enthusiastic embrace of the good life that is a part of our heritage," wrote Stan Tiner, executive editor of the Biloxi *Sun-Herald*, within a day of the hurricane. "The good times have rolled through the decades with a party that never quite ends fueled in more recent times with the glitz of electric lit rows of casinos and a booming economy." Since 1969, hanging over the "good life" of the coast had been dread of the next big hurricane—"the next Camille," as Tiner put it. "Monday, August 29, 2005," he wrote, "our worst fears were realized."[85]

However terrific the good life of po'-boys and daiquiris and boiled shrimp had been along the Mississippi Gulf Coast, the destruction brought by Katrina made some people want to leave at once. They were ready to move away to Iowa, Wisconsin, or New Mexico—anywhere without a storm surge. They couldn't go for a while, though, with all the major roads closed. Others lost no time in declaring that the coast would rise again, "better than ever." They were stuck, too, in the wretched present that forestalled their hopeful future. Everyone along the coast was trapped as Monday night led gently into Tuesday, with nothing to do but to separate the living from the dead (and mend the injured). "We thought the shelters would be needed for a day, maybe two," said Joe Spraggins, head of Harrison County's relief effort. "No one knew the catastrophe we would be dealing with. It's a big problem."[86]

Joe Scarborough of MSNBC was keenly attuned to the devastation along the Gulf Coast. A former Republican congressman from Pensacola, Florida, Scarborough pulled no punches in describing the lack of federal aid reaching the region. Every evening of the coming week, with sledgehammer directness, he shamed FEMA and the Red Cross, among other relief agencies, for the snail's pace of their rescue and relief process. His emotions were raw, and his diagnosis of the failures was right on the mark. "And here in Biloxi," he erupted a few days after the storm, "a place where, when we traveled around, we couldn't find enough federal agents, enough state agents, enough emergency personnel around to even begin to take care of those young children and elderly adults that are still without food, still without water, still without the most basic of necessities. Friends, I have got to tell you, I have been involved in a lot of hurricane relief before, and what I have been seeing these past few days is nothing short of a national disgrace."[87]

Chapter Six

THE BUSTED LEVEE BLUES

I seek ye vainly, and see in your place
The shadowy tempest that sweeps through space
A whirling ocean that fills the wall
Of the crystal heaven, and buries all.
And I, cut off from the world, remain
Alone with the terrible hurricane.

—William Cullen Bryant, "The Hurricane"

I

STRANGE TO THINK THAT the editor of the New Orleans *Times-Picayune* started off early Monday morning, even with Katrina taking aim, in a relatively chipper mood. Hurricanes often forged a special camaraderie between Jim Amoss and his reporters. He could feel the electricity in the newsroom and smell the storm. It was exhilarating to lead a team of committed journalists on the front line, poised to offer blanket coverage from landfall to aftermath. Every summer, just when he was feeling deskbound, a touch too complacent, the possibility of the Big One reared its ugly head and Amoss got in combat mode. His hunch about Katrina, however, was that it would strike around Mobile. "I came back from this lunch in Chalmette [on Friday] and into the newsroom and I was just standing around and Mark Schleifstein, our hurricane reporter, happened to be standing next to me," Amoss said. "I said something about the unremarkable weekend ahead, and Mark looked at me with this strange pallor, with this vacant stare in his eyes, and said, 'Jim, I have to show you something.' He motioned me over to his terminal, and on his screen was displayed the latest cone of the hurricane aimed straight at New Orleans."[1]

As editor of the newspaper since 1990, Amoss had prepared his company's offices at 3800 Howard Avenue for hurricanes many times before. Diligently he had ordered extra food for the cafeteria, welcomed employees (and, in some cases, their extended families) to sleep in the newsroom hallways, and shooed everybody away from the storm-vulnerable windows. Regardless of whether Katrina was a Category 1 or 5, he planned on bringing out a newspaper Tuesday morning. As the cliché went, come hell or high water, there would be exclusive coverage of the breaking news in his pages. Inside his third-floor office hung framed front pages from the now defunct New Orleans *States-Item* with World War II headlines like "Invasion" (June 7, 1944) and "Germany Surrenders" (May 8, 1945). Who knew, maybe late Monday he might have to run a historic banner of his own for Tuesday that read "Bull's-eye," "Direct Hit," or "Big One Smashes Big Easy." But it was just as likely the bravado headline would read, as it usually did during hurricane season, "Dodged Bullet Again" or "Alabama Shore Takes Brunt." That's what was great about both the newspaper racket and the hurricane season—they were ultimately unpredictable.

The fifty-seven-year-old Amoss looked like a Hollywood casting agent's perfect big-city newspaper editor. The native New Orleanian was conscientious and fair-minded and had a deep feel for local history. Even though he had spent ten years of his childhood in Germany and Belgium, where his father worked for Lykes Brothers steamship company, his life orbited around his hometown. The trilingual Amoss graduated from Yale, was a Rhodes scholar, and was a conscientious objector during the Vietnam war.

Instead of doing military service, he had worked as an orderly, often pushing patients' gurneys from ward to morgue. In the Massachusetts hospital where he was assigned, he witnessed firsthand the plight of the poor and underinsured. Without universal health care, children died unnecessarily of everything from asthma to pneumonia. Prenatal care just didn't exist for the underclass. "Being in the gritty world of Boston City Hospital, which is kind of like Charity Hospital in New Orleans, cured me of any desire to be a professor of comparative literature or some other such marketable skill," he said. "I met reality head-on, saw how tough life was for average people. I wanted to write about the real pulse of life that I was experiencing. It gradually dawned on me that journalism might be something I was able to do."[2]

So Amoss came home to New Orleans, and got a job as a *States-Item* intern for the summer of 1974. He handled general assignments, covering auto fatalities, drug-related murders, suburban fires, city council squabbles—standard cub-reporter fare. "The paper was heavily dependent on street sales," he recalled. "So crime was played up." For six years, Amoss worked as a reporter and loved every minute of it. He broke some big stories with his investigative partner, Dean Baquet (now managing editor at the *Los Angeles Times*), on such topics as police corruption in the French Quarter and mafia boss Carlos Marcello's crime syndicate. By the time New Orleans became a one-paper town in 1980—the *Times-Picayune* acquired the *States-Item* and dropped the name—Amoss was St. Bernard Parish bureau chief. He was promoted to an editorship two years later and ascended the ladder during the 1980s. When he took the helm in 1990, he knew every aspect of the *Times-Picayune* inside and out. "I took pride in knowing my community," he said. "Everything that happened in the New Orleans area was of interest to me and my paper."

As Katrina hit New Orleans, a calm Amoss was stretched out in his sleeping bag, in front of the very office where he had been hired when Richard M. Nixon was about to resign. Things had changed in the ensuing thirty-one years, but the paper was still a place where young talent was nurtured. Amoss, in fact, was a talent scout. Mentoring came easy for him. He always saw a flicker of himself in those upwardly mobile reporters trying to succeed in the Newhouse empire, the chain that owned the *Times-Picayune*. His patience with guiding new blood straight out of schools of journalism—particularly Louisiana State University—was legendary. He always had around him a bevy of talented young journalists. And who knew? Perhaps the next Ida Tarbell, David Halberstam, or Maureen

Dowd had just joined his stable. Without exception these newcomers wholeheartedly admired Amoss—even if, behind his back, they sometimes ridiculed his pop culture doltness (for instance, he mistook Rambo for the French symbolist Rimbaud). But his employees truly respected his unflinching Brooks Brothers demeanor and the courtesy he gave every honest request for an assignment. You had to. Under Amoss's dedicated leadership, the paper won the 1997 Pulitzer Prizes for both public service and editorial cartooning; the *Times-Picayune*'s first since its inception in 1837.

What it all added up to was that Amoss—with his wife and two kids safely evacuated to Texas and his Esplanade Avenue home boarded up—was in for the Katrina long haul. Whatever it took, the *Times-Picayune* ink barrels would spill damage assessments and death tolls. "It was a community obligation," he later explained. "I wasn't thinking about getting out and I don't think anybody else was. We'd planned on not only weathering the storm here, but also weathering the aftermath here. We'd get up this system of computers powered by generators in the core of the building, just for the purpose of putting together the newspaper, designing pages electronically, and then transmitting them to some remote location, where the paper would theoretically be printed and trucked back to New Orleans. We were still in that mode as of Monday morning, when the storm was at its height. We thought, We're going to be here two or three days and then things will gradually return to normal and we'll be able to get power restored and run our presses."[3]

Theoretically this sounded noble. Such confidant musings, however, didn't last long. One of the big plate-glass windows on the business office floor exploded Monday morning. Katrina had arrived. Amoss wondered how his intrepid reporters deployed in suburban New Orleans were holding up in the colossal winds. Word was out, through Garland Robinette on WWL radio, that the Industrial Canal levee had breached and the Ninth Ward was flooding. *Times-Picayune* photographer Ted Jackson volunteered to go to the area around the St. Claude Avenue Bridge. Amoss was anxious to see those photographs; they would be early indicators of how his old stomping ground St. Bernard Parish was holding up. And he also knew the National Guard was holed up in the Ninth Ward, at the Jackson Barracks along the Mississippi. If the Industrial Canal levee had breached—as the rumor mill was asserting—400 Guardsmen wouldn't be able to be first responders. Instead they would have to save themselves from what the early-twentieth-century journalist Frank Harris called "the illimitable prospect of the waste of water."[4]

Phone service at the *Times-Picayune* that morning was sketchy, but occasionally Amoss tapped into an open line and was told horrific stories about the devastation in coastal parishes. All he could mutter was "Poor Grande Isle," "Poor Buras," or "Poor Venice." Since the building was operating on just a couple of generators, there was no air-conditioning. By noon, the staff was transported back to the days when Southerners used hand fans and ate pecan pie on porches. All the *Picayune* employees suddenly sympathized with what the heat-exhausted throngs were enduring at the Superdome; it was enough to turn you into Camus's Meursault, losing your equilibrium in the heat with a pistol in hand. "We still hadn't put two and two together," Amoss recalled. "What we didn't realize was what Katrina meant, personally, for us as a news operation. We knew that it was disastrous, but still pictured ourselves operating out of Howard Avenue and reporting on the disaster in the days ahead. It was only really that evening that we noticed that the water was creeping up on our front doorstep here. We kept looking at it, thinking, This is really deep for rainwater. When's it going to start draining out? We knew the pumps were incapacitated, but we thought it gradually had to subside. We were downtown; the London Avenue Canal and 17th Street Canal levee breaches, which Doug MacCash and James O. Byrne reported on from their bicycles, and the Industrial Canal in the Ninth Ward were all pretty far away from us."[5]

II

Around two on Monday afternoon, two of the *Times-Picayune*'s indefatigable team, features editor James O. Byrne and art critic Doug MacCash, volunteered to do a reconnaissance mission, to see what damage the 17th Street Canal and London Avenue Canal breaches had wrought. They were an unlikely pair, but it was an unlikely hour. Both had brought twenty-speed bicycles to work—Byrne a forest green Diamond Back and MacCash a blue-gray Giant model. Venturing outside the *Times-Picayune* building, MacCash later recalled, they saw "trees down in front and a good bit of flooding, but we didn't recognize it as being catastrophic flooding. The industrial park in which the *Times-Picayune* sat frequently flooded. But it was worse than that. Looting had begun, and we could stand in our back parking lot and watch people carrying stolen furniture on Earhart Boulevard. That was our first glimmer that this was not business as usual. Things were different."[6] According to the forty-five-year-old Byrne, looters had already ransacked Coleman's retail outlet near their headquarters. "We knew it was going to be dicey on the streets," Byrne said, "that no business was safe."[7] Born in San Jose and raised mostly in Denver, Byrne was a proud Irish American. As of Monday afternoon, he didn't know if his Lakeview home had survived: the verdict for the area along Lake Pontchartrain was still out.

The first rule of hurricane reporting was "See the damage for yourself," so the two men cycled out into the residual winds. MacCash was in shorts and waterproof sandals, Byrne in standard jeans and T-shirt ensemble. They were fortified with granola bars and bottled water. Byrne had also stashed two fine bottles of champagne in his office, so when they came back, if the mood struck, they could celebrate their surviving the twelfth named storm of the season. With helmets strapped on, they waded and cycled their bikes up I-10, headed toward the Lake Pontchartrain Causeway. There was no traffic, of course, and they certainly didn't see other bikers. Their first destination was Byrne's home on Louisville Avenue. Was it damaged? Did the street flood? Were people really stranded on rooftops? They would soon find out. They pedaled, wind in their faces, onto an overpass. They could see the flooding was extensive, kicking up whitecaps over miles of residential streets. Everywhere they gazed over the omnipresent water were triangles, rooftops peeking out of a glassy lake like shy survivors. They stopped at the Old Metairie Road exit, which was severely flooded, and passed a failed pumping station. St. Patrick Cemetery, where corpses were buried in mausoleums, was likewise inundated, and they kept a watchful eye out for caskets. What happened next was straight out of a Mad Max movie. A spiky-haired punker suddenly emerged from the floodwater, introducing himself as "Grizzly Bear." He had been stranded in Metairie and decided to "swim the pond" to make it to the French Quarter, where he had friends in a garret. He spoke in alarmed survivalist terms about how widespread the flooding was. "Okay," MacCash recalled thinking, "we're in for a strange run."[8]

Byrne and MacCash had their own obstacles to overcome. They lifted their bicycles over a highway fence and headed to the railroad tracks, which, they knew, stood on high ground. By following the train tracks, they eventually made it to the quaint Plantation Coffee House on Canal Boulevard—or saw it drowned from a distance—the road a torrent of raging water, like the Missouri River way up in the Rocky Mountains after spring thaw. "For James it was clear that he was wiped out, that his house had to be real bad," MacCash said. "The water was moving so quickly that I watched a watercooler flow under the railroad tracks, toward the city and then make a left. The path of water, at that point, was flowing from Lakeview up Canal Boulevard and making a left into City Park. It was very distinct. Roads were now rivers." Although Byrne didn't see his home with his own eyes, he knew it was lost. He used his cell phone to call his wife in Shreveport and, quite astonishingly, reached her. No punches were pulled. "Our house is gone," he said. You had to admire a man who could give bad news so cleanly.

Before long, Byrne and MacCash ran into Generation Z first responders who had saved a mildly retarded woman who thought she could walk across the riverlike Canal Boulevard. The current snagged her. If nineteen-year-old Joshua Bruce hadn't harnessed her in, MacCash recalled, she would have surely drowned.[9] The only organized rescue going on was one NOPD officer trying to throttle-start a clunker speedboat. "We asked the officer what the Coast Guard was doing," Byrne said. "He didn't know—they had no communication. I knew then and there the people in Lakeview were in deep trouble."

On their bicycles Byrne and MacCash followed the levee along the Marconi Canal and eventually got near to Lake Pontchartrain. People flushed out of their houses by Katrina were wandering about in chest-deep water, only a few screaming or yelling or even crying about a lost dog or heirlooms. Most seemed in shock. For a moment the *Picayune* team stood still and watched the Southern Yacht Club in Bucktown burn to ruins. The seventeen-hundred-member institution traced its heritage back to 1849, making it the second oldest yacht club in America. Over the decades it had survived good times, wars, bust years, yellow fever epidemics, and storms of varying magnitudes—but not Katrina.[10] The stately club had served as the home dock for some of the best-designed yachts and sailboats in the city (and as training ground for four Olympic medalists). Now nothing was left but embers and a billow of smoke.

As dusk approached that Monday, Coast Guard helicopters turned on their searchlights to survey the damage to Lake Pontchartrain. The Filmore Street Bridge over the Marconi Canal, which didn't breach, had become a refugee camp for flood escapees. Some of the misfortunates had been left there by an engine house, the first responders in crippled Lakeview. "What struck me about the encounter was that people were happy to see reporters," Byrne recalled. "They knew that somebody would tell their story. They were right. We did. But they also believed that people would rush to their assistance. This was America, after all. Boy, were they wrong in that regard!"[11]

Around Lake Terrace, a wealthy subdivision along Lake Pontchartrain, Byrne and MacCash stumbled upon a couple of carefree yahoos watching the water rise, sipping cocktails with ice made from their home generator. "Hey," one of the guys shouted territorially at the reporters. "If you see liquor bottles down there, don't take 'em! They're ours." Like Barataria Bay pirates, they had already claimed first dibs on the bounty of washed-out booze from a nearby bar. Whiskey bottles and Heineken cans were floating everywhere. Once the men got a little more inebriated, they were going to collect them, like kids hunting for seashells. "Just like New Orleans," MacCash quipped. "Alcohol always matters."[12]

Byrne and MacCash scribbled down notes on every flooded structure in distress, for example, the Robert E. Lee Shopping Center (7 feet), Hynes Elementary (8 feet), the Plantation Coffee House (7 feet), Walgreens (8 feet), Blockbuster (7 feet). All the businesses on Harrison Avenue had water to their rooftops. Battery-operated home alarms were going off every five or ten minutes, giving Lakeview a feeling not unlike London in the Blitz. Furiously, they recorded details in their spiral notebooks. "This was my neighborhood so it had special meaning," Byrne said.

"You know, 'That was *my* Baskin-Robbins with seven feet of water. That's where I took *my* kids.'"

MacCash also had children, two of them, along with his wife, Melanie Tennyson, whom he had evacuated to St. Louis. As an art critic, one with a long graying ponytail and a David Crosby moustache, MacCash was used to bizarre juxtapositions and abstract images. But Lakeview blew his mind like no canvas ever could. Manhole covers had been lifted off, warm water shooting out of them like geysers. Oddly, many flood-ravaged senior citizens they encountered refused aid or assistance. They had survived World War II, Korea, Vietnam, and disco; surely they could handle a flood. Eventually, most of the house alarms died out. Eerily, thousands of frogs croaked in unison like a ghastly swamp chorus. Dogs were stuck in trees. Cats were dead in the water. Scum was rising. Sewers were spewing. Trunks of memories had been lost. A woman in a green summer dress stood on top of her two-story Lakeview home, screaming out to MacCash and Byrne to telephone her dad that she was all right. This request struck Byrne as being ludicrous. "She wasn't all right," he said. "She was the furthest thing from it."

The strange, incongruous experiences didn't stop. An oddball couple they encountered were watching the brown water rise, saying that they had conducted a scientific experiment and discovered that a house brick was three inches tall (they measured one) and that the floodwater rose a brick every twenty minutes. The couple were clearly crazy as loons. Their primitive measuring system, however, was probably accurate. At one juncture Byrne nonchalantly mentioned to a few strangers that his camera needed a battery. A young boy volunteered his services. "Oh, yeah," he offered. "I've got some batteries." In typical reckless New Orleans fashion, he dove into the flooded street, swam fifty yards to his gutted house, entered through the front door (and oddly closed it behind him), and eventually brought out a package of AA batteries in a plastic Ziploc bag. "Thanks to him," Byrne recalled, "we were able to get four or five Lakeview pictures in the next edition."[13]

Unable to get to the 17th Street Canal, MacCash and Byrne started backtracking to the *Times-Picayune* building, feeling like war correspondents with a wild-eyed tale to tell. From a media perspective, they in fact discovered the Lakeview flooding. The return trip was arduous. Biking toward downtown, they waded in three feet of water at the Old Metairie trestle. Luckily, MacCash had a headlight on his bike, so they could see a few yards ahead. They chained their bikes on the first floor and then trudged up the marble *Times-Picayune* stairs. "You can imagine how ripped up we were," MacCash said. "Splendidly dirty. Eaten up by red ants. James had lost his house but wasn't complaining. He just didn't mourn stuff. I would have cried a good bit more. Actually a lot more. Actually come to think of it, he didn't cry, but I would have."[14]

Bubbling over with information about the Great Deluge, they found Amoss in an editorial meeting. The glean in their manic eyes said, "Hold the presses, boys." They had been out and about for six hours in 35 mph winds. By mingling with dozens of refugees on the Marconi Bridge, they learned that there were multiple levee breaches and tsunamilike flooding. "Forget everything you're planning," Byrne said. "This is news. This is the real catastrophe just beginning. The city is going to be inundated."[15] Amoss told them to write up the front-page story ASAP. "In order to write," Byrne said, "I need to have clean socks." A jangled MacCash, sitting down at a typewriter, suddenly got a pang of writer's block. He took a break. Soon, however, they coproduced the lead story for August 30, "Catastrophic: Storm Surge Swamps Ninth Ward, St. Bernard; Lakeview Levee Breach Threatens to Inundate City." (A month later, Amoss made sure a blown-up version of "Catastrophic" hung on the wall outside his office, a souvenir twice as big as the vintage *States-Item* ones he so admired.)

A taciturn Mark Schleifstein, the *Times-Picayune* reporter who had cowritten the alarming 2003 doomsday series about New Orleans, was at the editorial meeting when MacCash and Byrne burst in. Schleifstein also lived in Lakeview, and his home was destroyed as well. Science-minded and reticent, a longtime true believer that coastal erosion along the Gulf of Mexico would someday turn New Orleans into Atlantis, Schleifstein had suddenly become a hurricane prophet. For about two years he had endured being the Cassandra of the newsroom. Colleagues had called him Mr. Big One every time it rained. Obviously, circumstances had now changed. "Quite suddenly he had enhanced credibility," Amoss recalled.

"We treated him with a bit of reverence." All Schleifstein could say, armed with alarming telephone reports from the breach-beleaguered Army Corps of Engineers, was a matter-of-fact "The bowl is being filled."[16]

III

Indeed it was. As the *Times-Picayune* reported, at 2 A.M. Monday, August 29, the National Hurricane Center had confirmed that Katrina, within hours of reaching the Louisiana coast, turned east.[17] Moving quickly on its new northeastern track, it swept past New Orleans. In hooking to the east, Katrina swept almost due north over the southeastern tip of Louisiana, hauling its storm surge along. That near miss, seemingly fortunate for Greater New Orleans, only saved the city from some of the strongest winds. The storm remained in a perfect position to cause enormous water damage. In the first place, with winds moving in a counterclockwise direction, the water picked up from the Gulf was directed to the west, toward New Orleans. Second, the hurricane was moving over two Louisiana parishes—Plaquemines and St. Bernard—rife with waterways, both natural and man-made, that provided a deadly conduit for those waters. Between 6 and 7 A.M., Katrina had moved over the coastline, bringing with it a massive storm surge of a depth of 18 to 25 feet.[18] It overtopped the earthen levees along the Mississippi River in many places, like Chalmette and Mereaux, below New Orleans.

At 5 A.M., the *Times-Picayune* reported, even before the punishing storm was at its strongest, the human situation took an ominous turn. The power failed in the Superdome. The blackout was met with a sigh from those who were awake to notice it. Babies whimpered. A loud commotion ensued. Backup generators clicked on and off, but the end result was largely unrelieved darkness. Lighters flickered, and flashlight beams bounced aimlessly around the Dome. You could barely see the ghostly outlines of the scoreboards. More than ten thousand people were trapped in a haunted hall, fertile ground for the malevolent to flourish. The air-conditioning was off for good. General Ralph Lupin of the National Guard was in charge of security; he had about 450 troops stationed in the Superdome (the number varies, with some sources going as high as 550). Although the media later focused on the disgruntled, many inside were grateful just to have a dry, relatively safe place to sleep. "[If] they hadn't opened up and let us in here, there've been a lot of people floating down river tomorrow," sixty-four-year-old Merrill Rice said. "If it's as bad as they say, I know my old house won't stand it."[19]

Brian Williams of NBC had stayed in the Superdome until the wee morning hours. For security reasons, he eventually left for a room in the Ritz Carlton on Canal Street. He was sickened by the deplorable conditions in the Superdome, particularly the fact that nobody offered these evacuees any information. They were indeed being kept in the dark, herded into a white spaceshiplike corral, given food and water, denied the basics of personal hygiene, no generators for air conditioners. Colonel Terry Ebbert, Homeland Security Director for New Orleans, commenting to CNN about the Superdome before Katrina hit, minced no words: "It's going to be very unpleasant. We're not in here to feed people. We're in here to see that when Tuesday morning comes they're alive."

The Superdome symbolized New Orleans as a big-league city, the site of a 1987 visit from Pope John Paul II, the 1988 Republican National Convention, six Super Bowl games, and two NCAA Final Four basketball tournaments. Most famously, every January, the Sugar Bowl, an All-American gridiron tradition, was played there. Courtesy of Katrina, the Superdome's image changed overnight; *USA Today*, for instance, deemed it "the epicenter of human misery."[20] While Williams was still at the Superdome, a man dived off an upper deck, committing suicide. The artificial-turf playing field was getting soaked because of the hole in the roof; in fact, a large puddle had formed at the fifty-yard line. Concession stands and luxury suites were trashed. Grown men were defecating in front of little children. But, for the most part, the Louisiana National Guard did a good job of keeping the situation safe from crime. And most of the "shelter of last resort" citizens in the Dome were—at this juncture—pretty well behaved. "People for the most part were taking care of their own," newsman Brian Williams recalled. "But it was stifling hot inside. Some people were glad there was a hole in the roof because the rain cooled them off. Rumors of gangs were everywhere. Nothing was going very well. Nothing at all. The humidity was unbearable."[21]

Humidity was a benign term for the water vapor in the air that made it hard to breathe. Nobody liked its side effects: frizzy hair, sticky shirts, and plugged sinuses. But humidity was in actuality a greenhouse gas, and senior citizens with respiratory problems couldn't take it. Their lungs contracted and they gasped for air. Hyperventilation was very common. No doctor would recommend that any elderly person spend even one minute in an airless enclosed stadium during a Louisiana summer. In all, four people perished from "natural causes" in the Dome. There was the man who committed suicide and another who overdosed on drugs. Although six deaths were bad, in truth, it could have been much worse. According to Louisiana National Guard Colonel Thomas Beron, at one point he received a doctor's report that almost two hundred people had perished in the Dome. "Don't get me wrong, bad things happened, but I didn't see any killing and raping and cutting of throats or anything. . . . Ninety-nine percent of the people in the Dome were very well-behaved."[22]

As the eye of the hurricane progressed to within about 70 miles south-southeast of the city, the *Times-Picayune* reported that rain fell at the rate of over an inch an hour.[23] There was no rhythm to it, just chaotic lashing. Water, pulled by the surge or pouring down as rain, filled Lake Pontchartrain. It was as though the lake were being forced to hold the contents of the entire sky, as well as that of the Gulf of Mexico. According to the *Times-Picayune*, the water level rose by seventeen feet in about five hours. As the murky water churned in the wind, it pushed up against the earthen levees that protected the north sector of New Orleans. Constructed of packed clay and dirt, carefully laid beneath boulders set in a "rip-rap" pattern, earthen levees were typically ten times as wide at the base as they were at the top. The lakeside levees, about eighteen feet tall, were half as wide as a football field. Storm-maddened waves swept over them occasionally, but only a few gallons at a time. Overall, the berms bordering New Orleans held firm against the worst that Katrina could deliver, protecting the homes and businesses lying just a few yards away on the other side of the swirling water. "The old earthen levees along Lake Pontchartrain," James Byrne recalled. "They took the very best Katrina could offer and withstood the test. They simply fared better than more modern man-made contraptions."[24]

However, once the lake was riled by Hurricane Katrina, it had other ways of attacking New Orleans. Solid ground was disappearing. Three major drainage canals poked into the central city, which normally sent water flowing through them into Lake Pontchartrain. That water, runoff from rainstorms, was propelled by pumping stations located at the end of each canal. The 17th Street Canal, the largest of the three, was originally bordered by earthen berms, but in the 1930s, the walls were extended—not, however, by raising the height of the dirt embankment. That would have eaten up land, as the base of the berm would have to have been extended as well. Instead, a concrete flood wall was built over and above the berms. Over the years, improvements were made, patches introduced, and the need for repairs noted and sometimes neglected. Incredibly, no one was in charge: no one was fully responsible for overseeing just who was doing what to the levees. Various entities had a hand in the fortunes of the system—and *fortunes* are all that the levee system meant to a great many of the greedy scoundrels involved through the years. The Orleans Levee District was a state-chartered organization with two hundred employees and a peculiarly independent board of directors. For example, in the months just before Katrina, while a $427,000 repair to a crucial floodgate languished in inexcusable bureaucratic delay, the board went ahead with happier pursuits, building parks, overseeing docks that it had constructed, and investing in on-water gambling, leasing Bally's *Belle of New Orleans* casino boat on Lake Pontchartrain in Gentilly.[25]

In setting itself up as a second city hall, the Orleans Levee District Board seemed to do everything except oversee the levees. At the city level, there was a Sewerage and Water Board, which had charge of the pumping stations, and some parts of the levee system of which they were an integral unit. On the federal level, the Army Corps of Engineers oversaw the Lake Pontchartrain and Vicinity Hurricane Protection Project. The project was a joint federal, state, and local effort, authorized to enlarge and improve flood-protection structures. As fast as Congress could pry money out of the federal budget, the Corps of Engineers built levees and flood walls in New Orleans, but it was also easily distracted. Over the years, Congress habitually diverted funding from flood-protection projects to economically promising ones, notably dredging docksides and improving shipping facilities. "It was not always clear," concluded a National Science Foundation report on the levee failures, "which agency had responsibilities for what."[26]

The result was community confusion regarding the levee protection around New Orleans. The Corps of Engineers asserted that the system could withstand a fast-moving Category 3 hurricane. That was the assurance in place when Beth LeBlanc of Lakeview, a savvy, attractive middle-aged woman, saw water rising in her yard alongside the 17th Street Canal on Bellaire Drive in late November 2004. It soon became a pond, 75 feet long and 10 feet wide. The mystery water was taking over her well-manicured lawn and turning it into Swamp Hollow. An agitated LeBlanc appropriately reported her front-yard flood to the Sewerage and Water Board, which sent several investigators. One of them concluded that the water was coming from the canal. "They sent repair crews out," LeBlanc said. "They tore up sidewalks and driveways. Things got better, but it never got dry."[27] That ought to have shocked Sewerage and Water Board officials into fast-track action, but instead reports on the seepage disclosed by LeBlanc—and many others concerning the same vicinity—were filed away and forgotten. Out of sight, as the adage goes, out of mind.

Without question, the Sewerage and Water Board was lackadaisical in its response to the Bellaire Drive complaints. Twice—on December 7, 2004, and February 8, 2005—it sent work crews to Bellaire Drive but couldn't diagnose the problem. The crews just gave up, writing in one report, "Need environmental to find source of problem." "Environmental" referred to the geological experiments used to determine the root source of the mystery water, but they were never performed. Protocol instructed that the crew should have immediately reported the Bellaire Drive leak—what *Times-Picayune* reporter Bob Marshall called the "wading pond"—to the U.S. Army Corps of Engineers. The Lakeview residents were terribly ill-served. Obviously, the heavy seepage pointed to an engineering flaw in the 17th Street Canal, one that had to be fixed immediately. "If someone had told us there was lake water on the outside of that levee—or any levee," Jerry Coletti, a New Orleans operations manager of the Corps, told Bob Marshall, "it would have been a red flag to us, and we would have been out there, without question."[28]

Not calling the Corps was just one bungle in what post-Katrina locals call the "Nightmare on Bellaire Drive." University of California–Berkeley engineering professor Bob Bea, who later spearheaded an investigation into the 17th Street and London Avenue breaches, said that more than a dozen people had reported other leaks, or water-soaked yards or sand boils—all signs of underground leakage. Residents had given Sewerage and Water and the levee boards kicks to the solar plexus, but they were just ignored. "Some of them said they contacted the Sewerage and Water Board, most contacted the levee board, but in all cases, no one even came out to investigate," a disgusted Bea recalled. "These are all signs that something is wrong. . . . It means your system is stressed."[29]

At 5 A.M. on Monday, as Katrina swept along the Louisiana coast, a man telephoned the Army Corps of Engineers in New Orleans to say that he had heard from a state policeman that the 17th Street Canal had been breached.[30] Since the hurricane was still approaching, the Corps couldn't confirm the report. But in fact, it occurred just yards from the LeBlanc house, a crack in the flood wall.

The London Avenue Canal, which runs south from the lake, in between the 17th Street Canal to the west and the Industrial Canal to the east, also burst in two spots late on Monday morning.[31] By noon, Lake Pontchartrain was pouring into city streets through three breaches. Lakeview, as MacCash and Byrne reported, had become, along with New Orleans East and the Lower Ninth Ward, ground zero. "These three levee failures were likely caused by failures in the foundation soils underlying the levees," Raymond B. Seed, professor of civil and environmental engineering at the University of California–Berkeley, told a congressional hearing in November 2005.[32] The earth that was supposed to be holding the walls in place was too soft to offer the proper resistance. And the piles used to anchor the walls were too short to be effective braces under stress. To simplify the complicated expert testimony, the levees were poorly built, with too much poured concrete and not enough engineering integrity. The shoddy Army Corps engineering crippled the Greater New Orleans flood-control system. In addition, all of the water that leaked through or "topped" the levees further weakened the supporting earth. The walls, incapable of holding back the inordinate weight of the flood-level waters, toppled.

But the deluge began far less dramatically than that: in both the 17th Street Canal and London Avenue Canal, something in the wall shifted and water began to trickle through. The greatest violence to be perpetrated on New Orleans started with a noise no louder than that of a gurgling fountain.

IV

Across the Mississippi River from the French Quarter, a famous old business called Mardi Gras World had been whipped by the Katrina winds. Situated two long blocks from the Crescent City Connection bridge on the West Bank, Mardi Gras World was a local institution. Most of the festive and world-renowned fiberglass Mardi Gras double-deck floats were stored in the campus of warehouses situated along the river in Algiers Point. (The deepest part of the entire Mississippi, in fact, was between the French Quarter and Algiers Point at 196 feet deep.) Started in 1947, Mardi Gras World made and preserved most of the blinking floats used in the annual February parades, which brought around $1 billion annually into New Orleans's economy. It became a popular tourist attraction in the 1980s, complete with a souvenir shop and snack bar. In 2004, 150,000 people toured the facility to photograph the oversized storybook characters that bobbed and weaved and waved past adoring revelers during Mardi Gras.

The owner, seventy-nine-year-old Blaine Kern, was known locally as Mr. Mardi Gras, a poker-playing, gregarious entertainment wizard who often wore a seersucker suit during business hours. Drenched in jewelry and Grecian Formula, like Jerry Lee Lewis (a native of Ferriday, Louisiana), he used to paint murals for a living but, due to his fun-loving floats, had become a multimillionaire. Back in 1959 Kern's skills as a float sculptor were coveted by Walt Disney, who admired the eighteen-foot-tall gorilla Kern had created and arranged for the fake ape to make a guest appearance on *Walt Disney Presents*. In the coming years Disney often commissioned fiberglass creatures from New Orleans's float master.[33] To Kern they were his papier-mâché babies. When Katrina came, Kern, who ran the business with his son, Barry, was working especially hard preparing for the 2006 Mardi Gras, which would be the 150th anniversary of the pre-Lent bacchanal. In advance of the hurricane, his work crews, following his instructions, hammered plywood over the windows of the gift shop and tied down the floats so that they wouldn't drift in case of West Bank flooding. "Then I headed to Houston with my daughter and nine grandkids," Kern recalled. "I had been at the Algiers location for over fifty years and I believed we would survive Katrina."[34]

Not only did Katrina blow out the twenty-foot-high metal doors on several Mardi Gras World warehouses, or "dens" as Kern called them, it created a wind tunnel that screamed through the krewe floats. Among other things, the Mummy lost its bandages, the Werewolf lost half its face, Dracula lost his black cape, and the Devil had his due, having his plaster flames wiped out. Although the damage could have been worse, it was bad enough to put the idea of the 150th Mardi Gras in doubt. That is, for a week or so. Caught inside the warehouse, Katrina's winds turned the floats into so much trash, like the residue of yesterday's parade. About six feet of water filled part of Mardi Gras World, but this hardly tapered Kern's morale. "Heck," he said, "we might leave the water mark around one or two just for publicity."[35]

Nevertheless, watching CNN from Houston, Kern had real worries. Windows blowing out of the hotels in the French Quarter and the central business district filled him with dread. Before Katrina hit there were 36,000 rooms in New Orleans for Mardi Gras tourists to enjoy. After the storm there were only 25,000. Not too bad, all things considered, but it translated into lost revenue for New Orleans—and Mardi Gras World. Kern employed about one hundred people, most of whom had evacuated. Right up until the weekend before Katrina, they all were hurriedly trying to construct the krewe of Alla—a monster-mash-themed float. Kiss the progress made on that project good-bye. Kern, however, left behind two armed guards, tasked with protecting the property. According to Kern, they earned double-time. Still, they were helpless against Katrina's winds.

Kern later took float losses with a sense of irony, humor, and chin-up spirit. He still had his eighteen-foot Hercules at a Tokyo fast-food restaurant and a fifteen-foot Elvis in a Paris bar. No matter what, Kern insisted in November 2005 that Mardi Gras 2006 would proceed, so keep the trombones dry and the bead boxes above water. He had approximately 400 floats to showcase to the world. All the masked riders were ready. The krewes of Rex, Zulu, Endymion, Proteus, Muses, Shangri-La, and Pygmalion were trying on their costumes by winter. "Mothballing is not allowed," Kern said. "We took losses of floats, about fifty of them in Gretna, and our Rex den on South Claiborne Avenue is full of muck, but so what? You don't stop Mardi Gras. Nothing stops Mardi Gras. The parades must go on."

Besides the storm-related damage, Kern had faced the problem of looters. He had hired two security guards to protect Mardi Gras World on August 29—they did. "I gave these guys permission to shoot," Kern recalled. "I told them if we let looters mess with Mardi Gras World, then there won't be Mardi Gras." Sure enough, as he guessed, a few guys tried raiding his souvenir shop, stealing snacks and vandalizing property. One of the looters fired at Kern's guard, Lynn Pitre. And Pitre fired back. "I think my guy got the looter," Kern later said. "He heard a scream which echoed around the large hall. But the looters escaped the same way they came in."

Some media types were scornful of Kern, Mr. Mardi Gras, having guards fire at intruders. He, however, had no remorse. "They called us vigilantes," he later complained. "Forget it. If you come on my property, you're going to get shot."[36]

V

By midday Monday Jeff Goldblatt of Fox News was out and about. Unfortunately his rented SUV was locked inside the W Hotel parking lot—the gate, like everything else electrical in New Orleans, was out of commission. He only had one option: he stuck his thumb out on Poydras Street hoping he could bribe somebody into giving him a devastation tour. Suddenly, a pickup slammed on the brakes and told Goldblatt and his cameraman, Robert Lee, to hop in. The driver was Kevin "Fish" Williams, a jack-of-all-trades with longish curly hair and left-wing politics. Sometimes he barbecued for the NOPD to make a little cash and other times he dabbled in real estate. "We jumped into the back of his truck," Goldblatt remembered. "This freak was drunker than a skunk, but we were just happy to have a ride. He was our tour guide. He took us up Martin Luther King Boulevard. That's where we realized there was more damage. We saw all the shotgun shacks, they had just fallen apart, and we heard reports about bodies floating around."

Around Central City, New Orleans had become a war zone. Lee kept film running, bouncing around in the truck bed with Goldblatt, as Fish headed for the Superdome. Goldblatt slipped Williams forty dollars, gas money, and asked to be taken to Bourbon Street. Although the wind damage was extensive, there was no flooding around the French Quarter. It was getting late in the afternoon. Goldblatt found a spot to set up shop on the corner of Bourbon and Conti, and that's where he reported in to Bill O'Reilly and Greta van Susteren. He homed in on the Oceana Bar, where gumbo and beer were being served, French Quarter customers convinced the Katrina damage wasn't too bad. "It was eerie in the Quarter," he recalled. "All the lights were off. It was dark. People were joking, 'Hey, I can actually take a leak on Bourbon Street and I'm not going to get arrested or clubbed.'"

Because Fox News had field lights and water, a group of NOPD officers started hovering around their Bourbon Street base camp. All of the cops were complaining about no communications. "I had a couple officers take me aside and tell me, 'I'm in the dark out here, so screw it!'" Goldblatt said. "They kept saying, 'I'm going to get myself a good meal. I'm going to keep myself safe.' They could have cared less that bodies were floating all over the town."

Once the worst of the winds died down, the NBC cameraman Tony Zumbado and his soundman, Josh Holm, were instructed by Heather Allan to head back into the streets of New Orleans; the story had dramatically shifted from gale-force winds to massive flooding. After checking out Governor Nicholls Street Wharf along the Mississippi, they headed to the Morial Convention Center. Not a soul was around. All was as still as an empty church. "I got beat up in the winds of other hurricanes much more than Katrina because in downtown New Orleans the buildings provided protection," Zumbado recalled. "But the water. Never, ever, had I seen water like that."[37]

The hurricane veteran and the neophyte headed over to the Superdome and found—just as Brian Williams reported—utter chaos. The satellite was down, so they weren't able to do a live shot. Water was creeping all around the area. Brackish water. Menacing water. Black water. At one point, their van almost started to float. Heavy debris was swirling around the area. The NBC team headed down Canal Street, amazed at the various survival tactics being employed. At one point they saw a sea of shoes floating down a street—clearly, an athletic store had been ransacked. Footballs were bobbing around like tub toys. White tube socks drifted around in the water, rejected by looters in favor of Nike shirts and NFL hats. Suddenly Zumbado and Holm saw about twenty people looting a mom-and-pop store. "We parked the van about a half block away from them," Zumbado recalled. "I told Josh, 'Stay in the van and watch me.' All around I videotaped people coming out of their apartments, down staircases, going directly across the street, looting the place, grabbing whatever they could, and taking it back to their apartments. They were taking everything. I have video of all this."[38]

Zumbado was being cautious with his camera. In Miami looters would have shot him, or at the very least fired their weapons in the air to frighten him away. "But that didn't happen," Zumbado recalled. "They just ignored me." Some seemed to enjoy his video presence, offering up a V sign or a wave. Calmly and with no interference from the looters, Zumbado recorded their privateering. That simple mom-and-pop store was just the opening salvo for the looters. As they patrolled the Tremé neighborhood around the Lafitte Housing Projects, Zumbado filmed pillagers breaking into Publix and other stores. At one shop, people were hauling out about ten or fifteen cases of beer on a hand truck.[39]

Once back at the Ritz, Zumbado showed the footage to a New Orleans narcotics cop named Mike. Startled, the lieutenant tried to report the mass purloining to his dispatcher, but his walkie-talkie wasn't functioning. His car was broken. All he had left was instinct and it also seemed to have eluded him. He was a personal wreck. Fraught with indecision, unsure how a cop should act in a deluge, his nerves frayed, he looked around for guidance. "So I said to the officer, 'Do you want to get in my van? I'll show you,'" Zumbado said. "He gets in my van and we drive out, and he tries to make contact with some other cops. We eventually met up with them and guess what? None of them had rain gear. Can you believe that? That tells you how ill prepared they were. So they started chasing looters down, walking in two or three feet of fetid water, in regular police uniforms. Then it turned awful. There was a pregnant lady with a shopping cart full of baby supplies: Pampers, powdered milk, oil for the baby. She was pregnant and they chased her down, yelling, 'Stop, lady, stop.' And she dumps the basket and bolts. But the police stay after her, and go and arrest her and a few other guys for looting. And she kept crying out, 'It's for my baby, I'm just doing it for my baby.' And she fell to the ground. I'm going, 'Oh Lord,' because she's pregnant, [and] you know what can happen."

Zumbado felt guilty for leading the NOPD to the band of looters. In roughshod fashion the police handcuffed her and said they were taking her to jail. What jail? All the prison facilities were flooded. (They didn't know that fifty-four-year-old state prison director Richard L. Stalder was on his way to New Orleans, about to rescue prisoners and open up a makeshift jail at the Greyhound Bus Terminal. He eventually sent prisoners to thirty-seven safe locations in an amazing rescue operation that deserved high praise.[40]) About four or five police got into a heated argument about where to incarcerate her. Zumbado hadn't realized they were so poorly trained to handle a crisis. When in doubt, they turned fascistic. "They were like a Keystone Kops kind of thing," he recalled. "They were trying to do their jobs, but it was like they had no plan on how to do their jobs. They were clueless. So I left that particular street and went around town videotaping some more looting."[41]

In a city surrounded by water, the police had only a handful of operable boats. Their radio system, cellular communications, and landlines went down simultaneously. They were without satellite phones. Because of flooding they couldn't even send couriers from one part of the city to another. They had no strong leadership. Their police superintendent, Eddie Compass, at the urging of Terry Ebbert, assumed the role of media spokesperson for the city. But the gregarious Compass was elusive. Rumor spread that he had fled to Texas like other NOPD officers. (He insisted to *The New Yorker*'s Dan Baum that the rumor was false.[42]) In Houston local police officers, disgusted by the unprofessional behavior of the NOPD, started photographing the Louisiana police cars, putting the images on their cell phones as a daily chuckle. It became sport in Houston: see how many NOPD cars you could photograph in town today. High-tailing NOPD officers had lost track of rules and regulations; many just drove their patrol cars straight out of the bowl to Texas. A few of those who stayed in New Orleans were "outlaw" cops, not accountable to anybody—desperadoes looking for a quick score like a Panasonic plasma TV or a Sony CD player. Thugs with a silver badge. Pure and simple. Only now, standing in floodwater, their gunpowder wet, their toughness evaporated. They turned yellow. Dozens sank to looting and freebooting—and roughing up the occasional pregnant woman. Most were stranded, just like everybody else left in the flooded parts of the bowl. At least the stranded police, unlike the 15 percent of the force who fled, didn't face post-Katrina disciplinary problems for fleeing New Orleans.

As the storm raged, and Zumbado and Holm were filming the flooding on Canal Street, their shoes sucking in mud, and a cabal of NOPD officers unsure what to do, NBC's Allan slept soundly in her Ritz-Carlton room. When she got up around 7 A.M., she looked outside and saw water creeping down the street. Sensing that the flooding was not going to quit anytime soon, she quickly called a conference. The NBC team decided that once the winds slowed down to 40 or 50 mph, they would move their satellite trucks farther up Canal Street toward the Mississippi, which was the highest point in New Orleans. Allan herself waded in the floodwaters to find the right spot. Late on Monday, the NBC team was working out of a few cars and trailers on the Canal Street median between the U.S. Customs Building and Harrah's Casino. NBC had broken the stories of both the breach of the Superdome roof and the looting mania. But in the highly competitive world of TV news, it wasn't enough to cover a story, you had to *own it.* Allan said, "I just told Tony and the other cameramen, 'Just *go*. Shoot what you can—there is a story on every block.' I didn't corral Tony or pin him down."[43]

Eventually, NBC would show the world Zumbado's images of people emerging from shattered houses and wading through heaps of damaged goods and worthless debris. And then there was the looting. Certainly it was sad to see families walking through broken storefront windows, emerging with blenders and hair dryers. There was something wrenching about sitting comfortably in a living room watching such thievery. Commentators kept asking: Is this America? Analogies were made to Third World countries. Anderson Cooper of CNN eloquently captured the sense of outrage just a few days after Katrina, saying, "The truth is people aren't *frustrated* here. People are *dying* here. Walking through the rubble, it feels like Sri Lanka, Sarajevo, somewhere else, not here, not home, not America."[44] From Cooper's vantage point, however, many of the poor were "commandeering" provisions—not looting—for their families to stay alive.

Because virtually all the looters Zumbado and others captured on film were African American, the anarchistic New Orleans street scenes made the issue of race impossible to ignore. It was impossible not to notice that the vast majority of those stranded in New Orleans were poor blacks. (Whites did loot the Wal-Mart Super Center on Tchoupitoulas.) As CNN's Wolf Blitzer awkwardly quipped: "You simply get chills every time you see those poor individuals . . . so many of these people, almost all of them that we see, are so poor, and they're so black."[45] Predictably, blogs filled up with harsh assessments misinterpreting Blitzer. Aside from racist responses, there were ill-advised religious ones. Some Christian fundamentalists went so far as to claim God created the New Orleans deluge to rid the world of riffraff and sexual deviants. One of them quoted Hosea 8:7: "For they have sown the wind, and they shall reap the whirlwind." Mayor Ray Nagin later himself claimed that Katrina slammed New Orleans because "God is mad at America."[46]

Zumbado showed Allan the looting footage late Monday afternoon; she had never seen such widespread looting before in all of her globetrotting. She was concerned. "The vast majority were just poor people," she later recalled with a rueful reproach in her voice. "Don't get me started. Nobody . . . *nobody* would help all these people desperate for water and food." Her real ire, in the looming hours, would be directed at the NOPD. Virtually all of them were "cocky, arrogant, and cruel," she said. "Virtually none showed an act of human kindness." During the twenty-five years she had been traveling in the world's hot spots, she had seen such things as a police officer beating women in Dakar and a man executed

Mayor C. Ray Nagin spent most of the first few days of Katrina holed up in the Hyatt Regency. On Thursday, September 1, he emerged to speak on WWL radio, urging federal officials to save New Orleans. Kevork Djansezian/AP

FEMA Director Michael Brown *(left)* listened as Homeland Security Secretary Michael Chertoff announced on September 9 that Brown was being relieved of his on-site command of Hurricane Katrina relief efforts. Rob Carr/AP

Louisiana Governor Kathleen Blanco bore the brunt of the criticism from the Bush administration and the press for her purported lack of preparedness and inadequate response to the devastation wrought by Hurricane Katrina. In truth, it was her around-the-clock efforts that finally brought buses to evacuate New Orleanians. Shawn Thew/EPA/Sipa

Marty Bahamonde, regional director for External Affairs for FEMA, attempted to relay to federal officials firsthand information about the desperate conditions inside the Louisiana Superdome at the height of the crisis in New Orleans. Brendan Smialowksi/EPA/Sipa

New Orleans Police Superintendent Eddie Compass spent most of the crucial days of Hurricane Katrina and its aftermath as the city's public spokesperson for the deluge. He is seen here touring the Lower Ninth Ward. Ethan Miller/Getty Images

Displaced persons waded through high water toward the Superdome, the supposed refuge of last resort, on Tuesday, August 30. MARK WILSON/GETTY IMAGES

The Interstate 90 bridge over St. Louis Bay in Pass Christian, Mississippi, photographed on Tuesday, August 30, was destroyed by Hurricane Katrina. Not only was the Mississippi Gulf Coast isolated from the rest of the world due to severed electricity and telephone service, but the roads and bridges leading in and out of these communities were cut off. PAUL J. RICHARDS/AFP/GETTY IMAGES

A pickup truck rested in the Back Bay of Biloxi, Mississippi, near the collapsed Palace Casino more than a week after Hurricane Katrina slammed the Gulf Coast. EDWARD A. ORNELAS/AP

Reverend Willie Walker of Noah's Ark Missionary Baptist Church spent weeks rescuing hundreds of people from neighborhoods throughout Orleans Parish. His close friend Diane Johnson died of post-Katrina stress months after the storm. Willie Davis/Veras Images

New Orleans Deputy Superintendent Warren Riley talked with the press on Sunday, confirming that two New Orleans police officers, Paul Accardo and Lawrence Celestine, had committed suicide. Riley was sworn in to replace Eddie Compass as New Orleans police superintendent on September 28, 2005. Dave Martin/AP

Captain Chad Clark of the St. Bernard Parish Sheriff's Department was responsible for the survival of untold numbers of residents when the Wall of Water swept away much of Chalmette, Louisiana, in mere minutes. Lindsay Brice

Michael Veglia and four friends used ingenious methods to save a family, including small children, from certain death when the storm surge swept through Waveland, Mississippi. Lindsay Brice

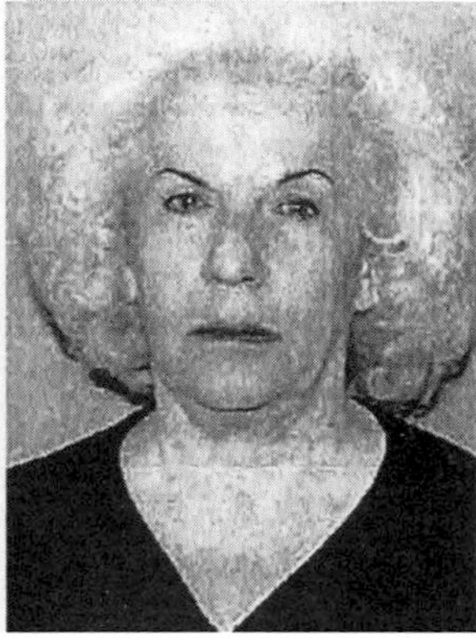

The booking photos of Salvatore Mangano Sr. and Mabel Mangano, owners of St. Rita's Nursing Home, who allegedly refused an offer of buses from St. Bernard Parish authorities to evacuate the residents of their facility. They were charged with multiple counts of negligent homicide. Louisiana Department of Justice/Reuters/Corbis

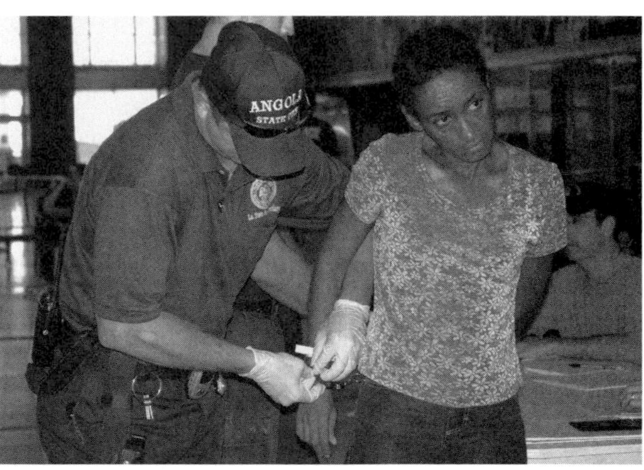

A female prisoner was processed by Angola State Prison authorities at the makeshift jail at the Greyhound bus terminal in New Orleans, as the local law enforcement facilities were flooded and the New Orleans Police Department struggled with disorder within its ranks. Lindsay Brice

A cyclist navigated floodwaters as the National Guard transported residents to the Superdome on Tuesday, August 30. Although officials called for a mandatory evacuation of New Orleans, many residents remained, some by choice, and many because they had no financial means by which to leave town. Eric Gay/AP

Houses slid off their foundations and collapsed during and after Hurricane Katrina. This home was on Frenchmen Street in the Seventh Ward of New Orleans. LINDSAY BRICE

The landmark Fire Dog Saloon in Bay St. Louis, Mississippi, was destroyed by Katrina's 127 mph winds and the hellacious Lake Borgne Surge. LINDSAY BRICE

Fats Domino was unaccounted for and presumed dead. He was found to have been rescued from his flooded home by Harbor Police, taken to the Superdome, and bused to Baton Rouge. He was later housed in Baton Rouge by star LSU quarterback JaMarcus Russell. Upon his return home Fats was delighted to find that several of his gold records had survived the deluge. LINDSAY BRICE

Lt. Col. Bernard H. McLaughlin of the Louisiana National Guard performed heroically throughout the Katrina ordeal. A decorated Iraq War veteran, he spent Monday, August 29, through Friday, September 2, at the Superdome, and then helped to open the food and bus lanes at the Convention Center. Courtesy of Bernard H. McLaughlin

Lt. Commander Jimmy Duckworth of the U.S. Coast Guard led first responders in boats and helicopters from his command post in Alexandria, Louisiana. Due to the foresight and prudent planning of Duckworth's commanding officer, Captain Frank Paskewich, the Coast Guard became the heroes of the great deluge. During the two-week period, the U.S. Coast Guard rescued thousands of displaced persons from the Gulf Coast floodwaters. Lindsay Brice

in Tangier, but never had she witnessed hundreds of police so callous in the face of human suffering, refusing to shift gears and place a priority on compassion and rescue work. The NOPD's mode was singular: self-preservation. "I saw a ninety-eight-year-old man paralyzed in a wheelchair ask for help and they just scoffed at him," she fumed. "They kicked three little huddled women with nowhere to go out of the Marriott because that was where the NOPD was sleeping . . . just tossed them on the street. We took them under our wing at NBC. They wouldn't even answer questions of people who asked which way the Superdome or Convention Center was. They basically mocked all the homeless. I have never, ever, seen such a cold, I-don't-give-a-shit attitude from cops in my life."[47]

Boiled down, Allan's central complaint echoed that of the sheriff in Cormac McCarthy's novel *No Country for Old Men:* "A crooked peace officer is just a damned abomination," McCarthy wrote. "That's all you can say about it. He's ten times worse than the criminal."[48]

VI

Commander Tim Bayard, the head of Vice and Narcotics, didn't give a hoot what Heather Allan or Cormac McCarthy or Anderson Cooper thought about the NOPD; he had a job to do—rescue folks. He tried to find fuel for his departments' five dime-store boats. It was embarrassing. Luckily, that wasn't their only "last resort" mode of transportation. They had some vans to ferry Katrina victims to the Superdome, I-10, and I-610. Once again, however, gas was a problem. His modus operandi was that he—Tim Bayard of the Ninth Ward, a true public-spirited cop—must give 110 percent during this crisis. His city needed him, just as it had needed his firefighter dad during Betsy. This was the time to prove that he was a chip off the old block. He would soon learn about the "Cadillac cops," who stole vehicles from Sewall Cadillac Chevrolet dealership in the central business district, and their thievery angered him, but he figured the judicial process would later sort out the "rank thieves" from the honest cops. He didn't have the time to think about the 200 NOPD officers who fled, even those who had lost their homes in New Orleans East or

Lakeview. "They shouldn't have left," Bayard said. "I have no idea why they left, but they shouldn't have left. Not when you're in the service. You're here to serve and protect. People expect you to be here. You had Friday and Saturday to take care of your wife and kids, to get your family out of town. And then your behind should have been at work. Bottom line." As for the NOPD not being properly equipped—for instance, stocked up on MRE emergency food or bottled water—he blamed his higher-ups. "That's the command staff of the police department or City Hall's fault," he said. "That's the mayor, the Office of Emergency Preparedness, the superintendent of police, the deputy chiefs. All those guys are supposed to give us equipment we need to work in the streets. It's their job to provide us with gear that we need to go do the job."[49]

Tim Bayard wasn't the only good leader in the NOPD trying to give his all. When the storm subsided, Deputy Police Chief Warren Riley left headquarters and with a few officers got in one of the NOPD's five boats and headed down Tulane Avenue. A little progress was being made. He was going to get a high-water military truck. There was about a foot of water covering Tulane, but Riley felt relieved. "We thought we survived," he said. "We thought everything was fine." Then they saw the busted-up Superdome and all the high-rises in the central business district with their windows blown out. Filled with fury, he knew immediately, from 911 calls which parts of New Orleans were the worst off. "We decided to go east," Riley recalled. In their truck, the team reached Chef Menteur Highway, a local road running into New Orleans East that *The Guardian* later described as "a dilapidated stretch of fast-food joints, strip clubs and hot-sheet motels."[50] They spotted a Hispanic couple in a parked car. A concerned Riley hopped out of his truck and went over to them. All around him were storm-torn buildings and random debris.

"What are you doing here?" a baffled Riley asked the man. His wife was eight and a half months pregnant.

"Well," the man said, "we were trying to leave during the storm and we couldn't. So we just parked here."

Riley couldn't believe *anybody* would park under the Chef Menteur overpass, arguably the most flood-prone spot in the entire city.

"So you've been here all night?" an incredulous Riley asked.

"Yeah."

Shaking his head in wonderment, bewildered at how a pregnant woman could put her unborn baby at such unnecessary risk, Riley had them get into his truck. They were the first two Katrina victims he rescued that Monday. Together with his NOPD driver, the four of them ventured farther east to survey New Orleans East, the true Land of No Return. They got as far as Morrison Road and stopped. The water was seven to eight feet high. Everything was underwater—houses, day-care centers, grocery stores, massage parlors. It was beyond the realm of words. Since Riley's military truck could handle only five and a half feet of water, they were stopped dead in their tracks. The Medium One had hit—all of New Orleans East had been obliterated. Heading back downtown on I-10, Riley saw storm refugees congregating on the concrete highway, looking lost and forlorn. His first instinct, however, was getting this poor pregnant woman to a shelter. He didn't want to play midwife in a military truck if she went into labor. Riley had her rushed to a shelter outside of the city. "That was incredible," he said. "We heard the woman had the baby a couple of days later."

Riley's cursory survey of New Orleans East made him understandably depressed, in no mood to twirl his nightstick. Would reconstruction even be possible, he wondered. His thoughts returned to manpower in his department. Besides the 200 NOPD officers who fled, there were others who were stranded in hotels. A few hundred more were trapped inside the LSU School of Dentistry on the third floor. The entire Second District force in New Orleans East was stuck in a hospital, which was inundated with ten feet of water. And now, adding insult to injury, hundreds of New Orleanians were looting, smashing plate-glass windows and burglarizing homes. Riley had long prided himself as the NOPD's expert in crowd control. Every year he dealt with barricades and drunkards and neighing horses at Mardi Gras with great success. Even critics of the NOPD had to admit that when it came to major events like Super Bowls or NCAA Final Four championships, they had crowd control down pat. "When LSU won the national championship," he recalled, "nobody tore down a goal post, nobody flipped a car like they do in New England. We believed that we had it under control."[51]

But a combination of bad luck, poor preparation, and corruption proved that the NOPD was not in control of anything. In the coming days, with his boss Superintendent Compass spinning exaggerations, Riley would try to put on a brave front. But he knew Katrina had gotten the best of the NOPD. He stiffened when he heard reporters lampooning his broken-down force. When a U.S. soldier (or Louisiana National Guardsman) fought in Afghanistan or Iraq, they knew back home in Lake Charles or Covington was safe, their parents were all right, that their spouses were living normal lives back home. Not the case with Katrina and the police. Approximately 900 of the homes of NOPD officers were destroyed. How do you help others when you're in such distress? Just when you thought you lost everything, you find out hundreds of fellow cops had bailed out on you. Welcome to the Land of Sinking Hearts. "As a commander, as a captain, you prepare for weapons of mass destruction, for a terrorist situation, for hostage situations, SWAT situations, things like that," Riley later explained. "We prepared for terrorists trying to take over the Superdome. We prepared for terrorists to come down on Bourbon Street during Mardi Gras. But this storm was the ultimate enemy. It cut off the food, the water, the transportation, the lights. It segregated your units and stranded them where they couldn't do their assignments. . . . This storm was absolutely beyond plausibility. How do you prepare for this?"[52]

Given the complete collapse of the NOPD, and with some of the Louisiana National Guard—the Jackson Barracks contingent, which was getting flooded due to the Industrial Canal breach—incapacitated, it was clear that New Orleans had a serious Homeland Security crisis on its hands. Because so many NOPD officers had fled their posts, while others were shell-shocked and homeless, unsure how to proceed, the question rose as to whether U.S. special forces (SEALs, Marshals, Rangers, FBI, etc.) would be needed to control New Orleans. Should the 82nd Airborne be brought in? History showed that widespread looting was often the first action in a full-fledged race riot. An executive decision was made to quell the civil unrest and essentially federalize the city. Somehow the stranded people at the Superdome would need to be bused out by FEMA to Houston, Atlanta, Baton Rouge, or elsewhere. But where were the supposed 500 FEMA rescue buses? Nobody seemed to know. As for the looters, they would be arrested and brought to the Greyhound Bus Terminal, the makeshift jail. If they resisted arrest, however, or tried to escape, they would be shot. New Orleans was inching toward a state of martial law. It didn't need to be declared. Silly lawyers could debate that later. Any fan of Louis L'Amour or Zane Grey knew that in a lawless town, everything was simply *understood.*

VII

At around 8 A.M. on Monday, just as the electricity went out in Baton Rouge and the winds were still ripping apart the Gulf South, Jimmy Duckworth placed a call on his government-issued push-to-talk Nextel phone to the Coast Guard's temporary headquarters in Alexandria. To his surprise he got right through to Captain Bob Mueller, the deputy sector commander. "Get your butt up here," Mueller barked over the receiver. "Now." Duckworth replied, "Yes, sir." That was the extent of the conversation. He knew the Medium One had hit and his expertise was needed in Alexandria. Lives were at stake. Tens of thousands of Louisianans and Mississippians were stranded. As Duckworth and Shelley Ford climbed into their Chevy truck, it was still windy out, blustery as hell, but they weren't in danger. Katrina had sheered off to the east and Alexandria was to the west. The work of the Coast Guard, however, was just beginning.[53]

Duckworth and Ford made it to Alexandria that day without incident. At 6 P.M. Duckworth stood watch at the incident command post (ICP) that he had helped create. The ICP was a large partitioned area with tables and metal folding chairs set up. The tables were divided into different groups with the search and rescue desk being the vortex. A few Coast Guard officers were sent to the Emergency Operations Center (EOC) in Baton Rouge, but Alexandria was the beehive. He was in charge of the Boat Deck, while Shelly Decker manned the Air Desk. Together they worked in close tandem. "We kept CNN on all the time in our command post," Duckworth recalled. "We had plenty of food and water. And willpower to save lives pronto. No waiting, no hesitation."[54]

The New Orleans Sector Coast Guard did not sleep that night. Fueled on adrenaline, they rolled. Having positioned forty aircraft and thirty boats and cutters in areas just off of Katrina's course, the Coast Guard was "making preparations to conduct immediate post-hurricane search, rescue, and humanitarian aid operations."[55] The Coast Guard was officially a component of the Department of Homeland Security, making it a sister organization to FEMA. Although the Coast Guard answered to Homeland Security Secretary Michael Chertoff, it had its own objective to save lives in the water—so it did not have to wait for either permission or an invitation to prepare for Katrina. Unlike out-of-towners, first responders from the Coast Guard—rescuers like Jimmy Duckworth—knew the waterways of southern Louisiana like the backs of their hands.

Duckworth, a World War II buff, immediately dubbed the New Orleans Area rescues "Dunkirk Two." But it was the New Orleans flood of 1927—he had also read John Barry's *Rising Tide*—that was at the forefront of his mind. "I felt that history was almost repeating itself with this," he said. "This is what the city fathers were worried about during the flood of '27. To save the city at all costs. And I felt like in all these years we'd done a good job of saving New Orleans from the flood, and now we've lost it. So I looked at the storm like the perfect enemy because the storm denied us logistics; it denied us communications, it denied us everything. If you look at Sun Tzu's *The Art of War* and if you look at the enemy called Katrina, it did us in perfectly."

Not so fast. Captain Frank Paskewich's prescient decision to set up headquarters in Alexandria, with plenty of food, dry accommodations, and proper equipment, gave the U.S. Coast Guard—unlike the National Guard, the Navy, the Army, or FEMA—at least a fighting chance to save the stranded.* Duckworth was convinced that his enlisted ranks would perform with vigor. Unlike the NOPD, they had plenty of fuel to keep both aircraft and boats going around the clock. Besides overall coordination, Duckworth's job in the Alexandria command post was to take care of the Coast Guard troops in the field. Duckworth was constantly on the telephone with the state EOC in Baton Rouge, which he believed carried out its mandates to the "best of their abilities."

* The Eight Coast Guard District headquarters had set up their EOC in St. Louis, Missouri. Duckworth reported to them.

The various communication breakdowns were another important part of the news from New Orleans. Duckworth was quick to point out one high-tech item that didn't fail him: satellite imagery. The standard search and rescue grid was approximately fifteen nautical miles square. "That's too damn big," he said. "You're talking about urban S and R." What Duckworth needed was something about a half mile long, images covering one-thirtieth the territory. He needed to see exact Lakeview addresses and Lower Ninth Ward homes. By zooming in on the worst flooded streets, Duckworth could "put our helicopters right in the game." Duckworth wanted to be able to see an SOS sign on a tenement or a family waving for help from a rooftop. With the assistance of a Coast Guard map expert, Duckworth took a big satellite image of the greater New Orleans area and then overlaid it onto a marine grid, thereby achieving finer granularity. "So we were carefully looking at the color of the snapshots from the satellite," he said. "We were using these pictures to determine the worst impacted area."

In this way, Duckworth determined which flooded neighborhoods should be given top priority for search and rescue missions. Nobody knew that Monday exactly where Jefferson Parish was flooded after the 17th Street Canal and London Avenue Canal breaches. Duckworth looked at the satellite map of the parish, which looked beaten up. Yet the close-up maps showed something different. Certainly the area between Airline Highway and Metairie Road was flooded due to the 17th Street Canal breach. But there were two tones apparent in the more detailed images: brown, which meant dry, and dark, which meant wet. "So for miles away, when you looked at the city from the air, you saw lighter colors or darker colors, and anything that was lighter was good to go and anything that was darker was flooded. Most of Jefferson Parish was the lighter color. It didn't go underwater. That meant our helicopters and boats would head first to the darker areas."

The darkest area of all was around the breached Industrial Canal, the 5½-mile-long waterway that connects the Mississippi River to Lake Pontchartrain. Duckworth knew that lock master Michael O'Dowd stayed in the canal's lock house during the storm and was worried about him. O'Dowd had brought his family to ride out Katrina in the lock house, which had hurricane-proof shutters. In its report on the Industrial Canal break, the National Weather Service said that three to ten feet of flooding was possible. "You've got no idea how bad the water got," O'Dowd, a forty-one-year-old U.S. Army Corps of Engineers veteran, recalled. "I just worried about the poor people who didn't evacuate."[56]

None of Duckworth's Coast Guard boat team would go AWOL during Katrina—nor did Lieutenant Shelly Decker's helicopter pilots—even though many of them lost their homes. Every night over the next week Duckworth would ask his exhausted lieutenants and enlisted men, "Are you in the game? Have you heard about your house?" One of the computers in the ICP was set up so the boatmen and helicopter pilots could find their houses on Yahoo! Earth to see whether they had flooded. It was an ironclad Coast Guard rule that when people needed saving no heirloom and brick-and-mortar mourning was allowed. But one lieutenant, Hector Clinton, got off watch and came to make a special appeal to Duckworth. "My house is in Slidell," Clinton said. "Everything we've got is in the house. My wife's jewelry and everything else. And I'd really like permission to go to my house and gather some stuff." Duckworth played firm. "Hector," he said, "we're in Alexandria. You've got to be back on watch. I need you at 1800 hours. You want me to give you permission to drive home to Slidell, more than 200 miles away, when I'm not sure you can get there and come back in one piece? And then I need you to stand watch again?"

Duckworth figured that would do the trick. Stellar lieutenants like Clinton never talked back to superior officers. But he persisted, looking Duckworth straight in the eyes. "Can I please, sir?" Taking a deep breath, pausing for ten or fifteen seconds, Duckworth bit his lip and said, "Just go." That green light was all Clinton needed. He was Slidell-bound. And sure enough, when it was his time to go back on watch, there he was back in Alexandria, all salutes. Pleased to see him safe, Duckworth walked up to Clinton. "Did you flood?" he asked.

"Yes, sir. About eight feet."

"Are you in the game?"

And Clinton replied, "I'm in the game, sir. Don't worry about me, I'm in the game."

Ever since Katrina had made landfall, Duckworth had been on autopilot, refusing to be emotional about the fact that Venice and Buras and Empire and Belle Chasse, all those marvelous Plaquemines Parish communities he knew so well, had been washed into the Gulf of Mexico. He concentrated on the task at hand and was so pumped up he could barely eat a ham sandwich. But something about Clinton, losing everything but staying on duty for a second shift, brought tears to his eyes. Not wanting Clinton to detect his sentimentality, Duckworth turned his back to him, regained his composure, and barked, "Carry on." So many of these Coast Guard youngsters had lost their homes. Lost *everything*. A one-hundred-dollar bill would have meant a lot to them. But they continued to perform, hovering over floodwater in helicopters and saving Katrina victims from roofs. "You know," Duckworth later said, echoing Warren Riley, "God bless our GIs working overseas. But when they go to sleep and no matter how bad it is, there's a home somewhere. There's a home and you've got a mental picture of your house and it exists. It's a reality. A focal point of your life. To watch these Coast Guard people work after Katrina, knowing that home is no more, was humbling. They never—not one of them—put themselves first. I'm proud. That's the best I can say."[57]

Others were also proud of the U.S. Coast Guard. Reporter Amanda Ripley of *Time* magazine dubbed them "the little agency that could."[58] Stephen Barr of the *Washington Post* wrote, "Let's have a round of cheers for the U.S. Coast Guard."[59] Although the Coast Guard had only 45,000 uniformed and civilian employees, they outshone the National Guard, FEMA, the Red Cross, and everybody else rolled into one. The TV images of them plucking stranded Katrina victims off of rooftops in HH-60 Jayhawk helicopters with drumbeat regularity became the most breathtaking moments of the Great Deluge. Over the ten-day period following Katrina they evacuated more than 33,500 people using orange helicopters and flat-bottom boats. According to *Time* this was six times more people than the Coast Guard rescued in the entire previous year. "The pace we kept up was amazing," Duckworth recalled. "When I say we were working around the clock, I mean it. Both boat and air. We were all go, go, go. Every minute of delay meant a possible loss of life."

Unlike the Marines, who are given macho monikers like "jarheads," the Coast Guard had long been denigrated in military circles as fey "puddle jumpers." But just as 9/11 brought a newfound respect to firemen, Katrina did the same for the reputation of the Coast Guard. At the peak of rescue operations they had 62 aircraft, 30 cutters, and 111 small boats stepping up in rescue and recovery operations. They did it all one person at a time. And with virtually no exceptions, they treated the suffering with respect. They didn't wear their humanity on their patches, it was in their hearts. When Sheriff Jack Stephens of St. Bernard Parish was asked later by the U.S. Accountability Office how to improve FEMA, he said, "I would abolish it. I'd blow up FEMA and ask the Coast Guard what it needs."[60]

VIII

Governor Blanco was in a disheartened mood on Monday afternoon. Every time the word "breach" was uttered, she visibly cringed. What she called "the double punch" of Katrina—hurricane winds and surging floodwaters—had left southern Louisiana in dire peril. Murphy's law was clearly in effect: everything that could go wrong did. Especially concerning the National Guard.

Unlike the Coast Guard, with its Alexandria ICP, the National Guard was in Jackson Barracks in New Orleans, and the 280-acre compound was now swamped by the Industrial Canal breach. The National Guard didn't have a "Jackson's Victory" fiddle moment or "The Eighth of January" War of 1812 jubilee, just a Katrina fist in the face. More than 400 Guardsmen were marooned at the barracks at the time of the breach. According to local lore Andrew Jackson had designed the barracks in the 1830s as a fortress against attacks from the British, French, Spanish, and Native Americans. Over the decades its brick walls had intimidated all of those intruders. But not Katrina. "It basically didn't stop until it got six feet inside the building, and the building was three feet off the ground," Master Sergeant Stephen Cockerham, of the Louisiana Air National Guard's 236th Combat Communications Squadron, recalled of the water. "I'd say within fifteen minutes it was six feet deep within the building. We were right in the midst of the strongest winds when the levee broke." Desperate, Cockerham and others waded into the water to save their truck batteries before their vehicles floated toward Arabi and Chalmette. They would need them for radio communications.[61]

The Guardsmen were instructed to congregate at the headquarters building, built on higher ground in the 1830s. It had eighteen-inch-thick walls, which withstood the storm's hammer just fine. Like soldiers looking for reprieve in a foxhole, the Guardsmen took stock of the situation in the safe haven. Many of them, spitting in the eye of Katrina, wanted to go save lives—that's why they joined the National Guard. Even though most of their equipment was destroyed, they had boats, and they knew their Lower Ninth Ward neighbors probably needed help. With bolt cutters in hand, they swam in the tempest and liberated four or five chained-down boats. All around them was debris. "I was not in favor of staging assets in the Lower Ninth Ward because I feared flooding. Luckily they had gotten the five trucks out of Jackson Barracks at midnight before the Monday storm," Terry Ebbert recalled. "They were pre-positioned at the Hilton next to the Convention Center. In the hours after we got hit, it's those Guard trucks that brought us around. I heard about the flooding of Jackson Barracks, but we got some boats out. But, yes, it was a mistake to leave so many assets there."[62]

The enemy had arrived at Jackson Barracks without a sword or tomahawk. There was no general calling "Charge!" or horse hoofs galloping to battle. At the Military Museum, housed in an 1837 powder magazine, the water level inside reached ten feet. Like the scene from *The Adventures of Huckleberry Finn* when the *Robert E. Lee* sinks in the Mississippi River, hundreds of Jackson Barracks historical artifacts were churned into debris. Confederate documents, Civil War knapsacks, and the saddle of General P. G. T. Beauregard were underwater, a muddy circle of wet wood and clayey yellow paper virtually impossible to reclaim. A buffalo soldier mannequin toppled into the water, its head slamming into a brick wall. There were families living at Jackson Barracks who wanted to rescue the heritage crumbling all around them, but there was nothing they could do. The muzzles of antique rifles that used to guard this Mississippi River fortress had become nothing more than scrap metal in the muck.

At Jackson Barracks, like everywhere else, it was the isolation that sent the mind racing. There was no way to know whether the French Quarter or City Park or Hollygrove was underwater. That was the consensus, the one fiasco everybody agreed on—whatever else Katrina did to New Orleans, it had clearly broken down all standard modes of communication. President Bush, in fact, later chose this communication breakdown factor as the one that delayed federal assistance.[63] Telephone lines, computer networks, and cable systems were knocked out of service. Many cell phone antennas were damaged or destroyed, making wireless communication in the region extremely patchy. Cellular service for phones registered in the 504 area code was vastly overloaded. Satellite telephones, which cost more than $2,500 to purchase, with commensurate charges for calls, worked perfectly, as long as they were charged up before the deluge. In the aftermath of Katrina, however, virtually no city government official was equipped with a working satellite telephone, even though in 2003 the federal government had provided New Orleans with a $7 million grant to connect all first responders like ambulance drivers, police, and firemen. What had happened to the money? Nagin, a former communications industry executive, was himself without communication after his cell phone battery died. "People that are too hard on Ray need to remember how difficult it was for him to operate," Boisie Bolinger, a Nagin advisor, said in his friend's defense. "He couldn't get in touch with anybody." But when pressed to explain why Nagin had never procured a satellite telephone and hand-crank recharger before Katrina, Bolinger turned quiet. "That's a different story," is all he said. "That's a fair question."[64]

For stranded individuals, the best chance at communication was a handheld computer device, like a BlackBerry or a Treo 650, which sent e-mails using relatively little bandwidth, meaning that the networks were not easily overloaded. The arduous process, however, of typing on a tiny keypad made for abbreviated messages, decreasing their effectiveness. Brief text messages like "I'm ok" or "We survived just fine" or "Any damage?" were common. Renee Marcus, a mother of five, spent Katrina in a high-rise New Orleans condominium with her family. Her mother, however, had stayed at her house across from the New Orleans Country Club. Marcus tried to reach her mom by phone, to no avail. Somehow, however, on something like the hundredth try, she got through, relieved to hear her mother's sweet, loving voice. "The water is rising, every minute," she told Marcus in a terrified tone. "I don't know what to do."

Then click—the phone died. Marcus was left in the lurch, fearing her mother had drowned.[65]

With the lines of communication shattered, the general feeling of relief emanating from New Orleans on Monday night was not based on anything like accurate fact gathering. Individual impressions replaced the consensus necessary to draw a true picture of a whole region's plight. With gusts continuing to blow debris around New Orleans and power lines down, few people ventured very far that afternoon. A falling street sign or tree branch to the head and you could die, becoming another Katrina statistic. Some, like the Zumbado-Holm and Byrne-MacCash duos, did so in the name of journalistic obligation.

One person not often seen on the streets, at the Superdome, or on a rescue boat of any kind was Mayor Ray Nagin. Occasionally he'd pop up inside the Superdome, clinging to the exit doors, then disappear. Since the storm had approached the Crescent City, Mayor Nagin had been cloistered in the Hyatt, lording over the Superdome. From the get-go he was terrified for his own personal safety. And for good reason. At the storm's peak, many of the windows of the Hyatt blew out. The high-rise was a jagged, ripped concrete-and-steel monstrosity, swaying in the feverish winds. Frightened, Nagin refused to make City Hall a command center. Terry Ebbert, the New Orleans director of Homeland Security, ostensibly ran the city. "I went over to the Superdome numerous times," Ebbert recalled. "I didn't carry a weapon. I walked all around without a real problem."[66]

Unlike Ebbert, Nagin was apparently repelled by the idea of speaking at the Superdome, to offer the evacuees both information and a morale boost. He refused to give a pep talk, blaming the city's communications breakdown for his decision. His primary post-storm initiative was to get a generator hooked up to the elevator so he wouldn't have to walk all those stairs. A timid Nagin had squandered a historic opportunity for a "bullhorn moment." With a touch of guts he could have walked over to the Superdome with Teddy Roosevelt exuberance and tried to calm the jittery crowd. When Martin Luther King Jr. was assassinated in April 1968, riots broke out in thirty-one American cities, but Robert F. Kennedy, shirtsleeves rolled up, fearlessly marched into the midst of an angry African-American mob in Indianapolis, easing their confusion and hurt with words of uplifting encouragement. RFK had seized the "golden moment" that Maureen Dowd wrote about.[67] At the Superdome in New Orleans, scared citizens needed Nagin. But he feared that if he mounted a soapbox at the Superdome, he'd get shot, lynched, or bloodied up. He made the costly mistake of viewing the displaced persons as malcontents. He had squandered the golden moment, putting his own personal safety ahead of those poor and elderly in trouble.

While Mississippi Coast mayors like Eddie Favre in Bay St. Louis, A. J. Holloway in Biloxi, Brent Warr in Gulfport, Tommy Longo in Waveland, and Matthew Avara in Pascagoula were out and about, putting their lives at risk on Monday afternoon, checking up on everything after Katrina's onslaught, Nagin found out what was happening by turning on his battery-operated, hand-cranked radio straight out of *The Waltons* and following WWL reports. He didn't realize, as the *Times-Picayune* reporters did, that it's best to see with your own eyes, boots on the ground. Go see the 17th Street Canal breach for yourself.

All that Nagin knew was basically what every land-bound reporter knew late that Monday morning; and that was only what he could survey from his Hyatt perch. The *Chicago Tribune*'s first story on the storm was typical. It led with the observation that, "Until nearly the last minute Monday, it looked like Hurricane Katrina might deal New Orleans the cataclysmic blow that scientists have long feared for the low-lying city. . . . What saved [it] from even worse damage was the storm's last-minute turn to the east."[68] But CNN and Fox News reporters warned that if a levee had broken, as claimed, then parts of New Orleans might soon be underwater. Meanwhile, at City Hall, Ebbert stayed cool, collected, and in charge. While others were cracking up all around him, he concentrated on problem solving. "My Vietnam experience probably helped," he said. "Once you've seen the Maker a couple times like I did after being shot, you don't get rattled too easy. The Marines teaches you that."[69]

IX

Much of the water pulled from the Gulf of Mexico by Katrina was trapped in Lake Borgne, which is not a lake at all but a lagoon on Louisiana's eastern shore. A protected sac-shaped body of water, Borgne was filled to bursting by the storm surge—and burst it did, down the path of least resistance: the Mississippi River–Gulf Outlet (MRGO). The MRGO—or Intracoastal, as it is also known—was completed in 1965 as a 500-foot-wide shortcut into New Orleans that bypassed most of the serpentine Mississippi River with a deep-channel, straight-shot, cement-sided waterway connecting the city and the Gulf. In the 1990s, federal funds that might have been used to strengthen the levees around New Orleans were diverted to widen the 75-mile-long MRGO, which watermen and captains called "Mr. Go." Running almost due east from the city, the ship channel was built to shepherd oceangoing vessels as expeditiously as possible to the Gulf. Boosters hoped it would be a good investment in industrial development. It was never fully utilized, though, and in trying to entice more traffic, officials forced through improvements that saw the MRGO grow in some places from 500 to 2000 feet wide. The result was the same as if a team of top-flight engineers had been assigned to build an instrument for the quick and effective flooding of New Orleans; they could not have come up with a better design than the MRGO. When the water came flashing out of the sea, via Lake Borgne, it was forced through the MRGO and then crashed into the perpendicular Industrial Canal, causing an 800-foot-long breach at 8:14 A.M. on Monday.[70] The MRGO channel also overflowed on the eastern side of the Industrial Canal, destroying New Orleans East communities like Huntington Point, Lake Barrington, and Spruce Lake. St. Bernard Parish was also laid waste by the drastic overtopping of the MRGO.

The water from the broken Industrial Canal, shooting out into the streets, added more water to the flood in St. Bernard Parish, bordering New Orleans East. Water also gushed into parts of the Ninth Ward, the New Orleans neighborhood that was not only closest to the canals but low enough in elevation to draw the overflow. By 9 A.M., up to six feet of water flooded the Ninth Ward, and it was still rising. On Japonica Street, the Louisiana SPCA took eight or nine feet of water. If Laura Maloney hadn't evacuated the animals from there to Houston, they all would have died.

St. Bernard Parish was even worse off. The water was deeper there, and it rose more quickly. Most people scrambled to the roofs and attics. Some couldn't. At St. Rita's Nursing Home in St. Bernard Parish, between the towns of Violet and Delacroix, the owners had decided *not* to evacuate residents on the assumption that Katrina would be less potent than originally predicted as so many New Orleans storms were, most recently Hurricane Ivan. Of the five nursing homes in the parish, St. Rita's was the only one that did not evacuate. According to state law, a nursing home had to have an evacuation plan, but the law did not specify under what conditions implementing the plan became mandatory. Approximately seventy residents were in St. Rita's when Katrina hit, along with about thirty staff and friends, who believed that the twenty-year-old, single-story brick building would be a safe place to weather the hurricane.

At about 9 A.M., the people inside heard a rumbling sound. It was the water, coming from the storm surge. Within twenty minutes, the rooms of the nursing home were flooded to the ceiling. Those twenty minutes were horrific. "People were screaming like somebody was murdering them," said Gene Alonzo, a retired fisherman who was at the home to be with his brother, a resident, during the storm.[71] The able-bodied frantically tried to put the residents onto mattresses and float them out the windows. When the windows filled with water, they chopped a hole in the roof and tried to pull people through, still on their mattresses. Many of the residents were in wheelchairs. Some couldn't get out of bed at all. Others were too much in shock to move. Trishka Stevens hadn't walked in five years when the surge broke the air-conditioning ducts and filled her room. Holding the sides of her bed, she used her strength to tip her head above the water. Steve Snyder, a twenty-nine-year-old oil worker, had just been flooded out of his own house nearby and was driving by on a boat when he heard the screams from inside the building and from the roof, where survivors were clinging to life. Snyder broke the window in Stevens's room and pulled her out. "I thought that man was my angel," Stevens said later. Snyder tried to help others, but time was running out. And even for a strong young man, it was almost impossible to move the inert weight of a debilitated person. Still, he kept going back inside the building as long as he could, to rescue more. "It was such hard work because they were all wet and heavy," he said. "After we pulled out as many as we could, I had to swim out of there. There were bodies floating around us."[72] In the end, thirty-five elderly or handicapped people died in the flood at St. Rita's; five more would die within a week, probably from stress of the ordeal. (Louisiana Attorney General Charles Foti launched an investigation into the deaths shortly after the revelations were made public, and later had Mabel and Sal Mangano Jr., the owners, indicted for negligent homicide.

With St. Bernard Parish under ten feet of water and New Orleans's Ninth Ward continuing to fill up, those managing the disaster still didn't know the extent of the destruction. Nita Hutter, the state representative from Chalmette, for example, had decided to weather the storm with her constituents. Along with other local leaders, she holed up in the parish government complex on Judge Perez Drive. At the height of Katrina, they clocked 175 mph winds just outside the front doors. "I thought the building would collapse," Hutter recalled. "The women were crying and the men were crying too. I called my sister in Kansas City and told her, 'I'm not going to survive this. Tell everybody I love them. Good-bye.' "[73]

At only fifty-six, the irrepressible Hutter wasn't ready to die. If this was her fate, however, she wanted to make sure everybody in the steadily collapsing building wrote their name on a list. It would help the FEMA body removers and cadaver dogs to properly identify them. "So I started a head count," she recalled. "Just then somebody shouted that the parking lot was filling with water. We were now in double jeopardy, wind and water. In just ten minutes the water was fourteen feet high and we were floating." Just down the road in the town of Arabi, Hutter's house for thirty years also filled up like a fishbowl, destroying everything she owned. "Overwhelming is one word that comes to mind," Hutter later explained, choking up. "But that would be trite. Maybe unreal, frightening, heartbreaking . . . there is no word. I just assumed I was dead."

But Hutter didn't perish. A couple of hours after the tempestuous winds died down, she changed her attitude, from passive resignation to glorious action. Putting fear at bay, she waded into the floodwater to save human lives—or at least to lend a helping hand. Hutter now saw herself as an eyewitness to history. She watched the most incredible heroic acts by everyday people—people she had thought of as clerks or bureaucrats. It was mind-boggling. She never suspected Chalmations had the gallantry of Superman or Wonder Woman in them. Hutter herself was worried that Chalmette had become *Waterworld*, and like Kevin Costner in that flop movie she would have to find food and water for desperate people. At one point over the next few days she actually got in a boat, a true plow-woman of troubled waters, and fished for food. She gave her catch away to people stranded along the slips and docks. "We had to help ourselves," she recalled. "And I thought about all those Americans out there, well removed from Chalmette, not able to imagine what was going on. I wondered if they realized that someday it could happen to them."[74]

As CNN, Fox News, and the other networks showed high-tech simulations of the New Orleans levees cracking, and the consequences thereof, TV viewers all over America did worry about their own neglected infrastructure. Citizens in northern California worried whether the Bay Area could survive a huge earthquake, and some fretted about the deteriorating levees along the Sacramento and San Joaquin rivers—a collapse could poison the state's water supply. Then there was also the possibility of a tsunami overwhelming Puget Sound or northern Oregon. People in Montana and Idaho had a volcanic eruption of the Old Faithful at Yellowstone National Park to fear, while those in the Great Plains wondered if there wasn't a more sophisticated way to prepare for tornadoes than merely to head for basements and cellars.[75] People in otherwise idyllic waterfront communities in Cape Cod, Massachusetts, and Suffolk County, Long Island, knew that their own areas were due for a Category 4 or 5 hurricane sometime in the future.

X

The Ninth Ward, the Hyatt-Superdome area, New Orleans East, Lakeview, and St. Bernard Parish weren't the only Greater New Orleans locales battling Katrina's wicked aftermath. Although it generally didn't really flood, something was *seriously* wrong in Uptown—at least that was what fifty-one-year-old Jimmy Deleray sensed late Monday. A cagey working-class New Orleanian, Deleray had weathered Katrina at his small shotgun house on Dante Street in the Riverbend section of Uptown. It was an unpretentious part of New Orleans filled with singles like himself (i.e., musicians, artists, and laborers who didn't want to raise children). On any given night from his front porch, he could see the Mississippi River levee and hear the streetcar clank down Carrollton Avenue. Just a few blocks away was his "living room," the Maple Leaf Bar, where New Orleans music greats like the Radiators, Snooks Eaglin, and the Dirty Dozen Brass Band would perform, sometimes until dawn.

Because Deleray was a Hurricane Betsy survivor and full of frank bluster, he wasn't going to let Katrina drive him from his comfortable Riverbend home. Always thinking ahead, he had purchased a generator a while ago. If New Orleans experienced a prolonged blackout, he would have light. Fortified by his trusty generator, he certainly wasn't going to flee down I-10 to sterile places like Houston, Dallas, or Beaumont. All three Texas cities, he believed, lacked culture and old-time soulfulness. So he was staying put. Taking a page from New Hampshire, he adopted a "N'Awlins or Die" philosophy as his rough-and-ready motto. He actually made posters of the Louisiana state flag, a real "Pelicanhead," as they joked at the Maple Leaf Bar. Besides, he owned an ax. He wasn't afraid to chop a hole in his attic to get out. He'd rather be stuck on a New Orleans rooftop flagging down the Coast Guard for help than be ensconced in some high-rise condo with the Lone Star flag flapping on top.

For a while, his strategy made perfect sense. Then things went awry. Staring out his attic window, he watched as a gale blew away his fence. Then he saw a thirty-foot pine tree break in half. That made him queasy. His house wasn't particularly well built. Whenever the water truck drove past, in fact, his wooden floor used to rattle, not to mention the dishes in the cupboard. By contrast, Katrina was like a fleet of water trucks racing down Dante Street in madcap procession. Unsure how to cope with the relentless caterwauling, he put his pillow over his head and went to sleep. Dealing with Katrina was like dealing with a drunk gone bad; it was best to pass out and wait to attack the hangover with coffee in the morning.

When he woke up Monday evening, it was silent outside, although there were still residual winds. He strained his ear, in fact, determined to hear a peep, but heard nothing. The stillness was eerie. Immediately, he turned on WWL to find out the score. A public service announcement was being made for citizens with boats to make their way to Kenner. Parts of New Orleans had flooded and fellow Louisianans needed help. To Deleray's surprise, Dante Street was bone-dry. It appeared, however, that some crazy King Kong creature with an enormous buzz saw had whacked every tree and bush on his street. The cleanup was going to be a pain in the ass. But he didn't mind laboring for a worthy cause, and nothing was more noble than helping out his Riverbend neighbors. He was a chivalrous community guy with a sailor's taste for liquor. The yeoman cleanup would come soon enough with no complaints. Perhaps they would buy a keg and do the dirty work inebriated. First, however, he had to drive to Kenner and volunteer to rescue stranded folks. When he arrived in Kenner—not far from the airport—he was struck by a disconcerting sight. There was New Orleans Police Superintendent Eddie Compass in Jefferson Parish, all out of sorts. It wasn't his jurisdiction. "I went up to him, but he was crazed," Deleray recalled. "He looked stoned or drunk. All he could manage to mumble was 'It was bad, man, real bad.' People had arrived with boats, but nobody knew what to do, so I decided to head home."[76]

Back in the central business district, in his high-rise office on Poydras, blogger Michael Barnett of directnic.com was firing off communications to the outside world. He felt New Orleans survived Katrina fairly well, but he mentioned debris being everywhere. His biggest concerns were how to use his toilet, looters, "idiots" driving around, and whether the NOPD had its act together. "Last night there were three flags flying across the square from us—the IRS building—and one on the building on the eastern side of the square—the Federal Court of Appeals building I believe it is," he posted. "Last night the flags were all flying strong! Visibility was low all night and into the morning but when I walked out onto our balcony on the 11th floor overlooking the square, two of the three flags were still flying. Well I guess you'd have had to have been there, but man, it reminded me of [Francis Scott] Key and the SSB [Star-Spangled Banner]. When I got out there, the flags were flapping and you can just imagine the feeling of not being sure if anything was gonna be there when all was said and done and there she is."[77]

XI

Around nightfall Governor Blanco was in a helicopter flying over the Louisiana devastation. She refused to let a couple of pool reporters come with her. Her lack of media savvy proved to be a serious Achilles heel. Only "lifesaving people can go out," she scolded the press. That was her prerogative. It was also the media's prerogative to stick a knife so deep in her back that even Karl Rove winced. A perception grew nationally that the Louisiana governor was a hack. She had never explained to the press properly that her 200 Wildlife and Fisheries boats were in high-rescue mode. Another 200 were on the way to the flood zone. And likewise Louisiana sheriffs and firemen were operating at breakneck speed to save the stranded. Refusing to throw in the towel, she was encouraging every available boat from St. Augustine to Corpus Christi to help out. She launched unprecedented state rescue initiatives to help Plaquemines and St. Bernard parishes, both of which had been nearly wiped out. "We didn't have room in boats for media people," she said. "We needed every spot on the boat to put survivors in the boat. Media kept begging us to go out. I think it cost us."

Meanwhile, her aides were nervous that the breaches were growing in size. They began studying the big three breaches: Industrial Canal, 17th Street Canal, and London Street Canal. "Monday afternoon, Ray Nagin called me and told me that he was going to get the water in downtown from the 17th Street Canal breach," Blanco said. "I hung up the phone and called General Landreneau [of the Louisiana National Guard] and I asked him if he had any engineers who could start dropping sandbags to try and stop some of the action already. He couldn't get anything in the air on Monday. Not all helicopters can fly at night. You had to have helicopters that were able to do those big lifts. They started dropping three-thousand-pound sandbags into the breach on Tuesday. The general tracked me down after they had done some of it and told me, 'Governor, the hole is so deep we drop the sandbags and it's like nothing.'"[78]

Every second in the helicopter, Governor Blanco, surveying the devastation, grew more and more bitter at President Bush. She felt that too many Louisiana National Guard members were in Iraq. It was a specious, liberal, Democratic argument. "In the early hours of the disaster you need as many bodies as you can get for lifesaving missions," she said. "We supplemented the lack of our own people in the early hours with Guard members from some of our neighboring states. But it's always easier if you have your own people and you don't have to deploy all the men from somewhere else. That's what they're designed for. Their first responsibility is to local disaster recovery efforts. We didn't have our Louisiana National Guard stateside."[79]

Yes, she did. Instead of setting up outside the New Orleans bowl, north of Hurricane Alley, like the U.S. Coast Guard did in Alexandria, the Louisiana National Guard chose to set up at Jackson Barracks in the below-sea-level Lower Ninth Ward. The rest, as they say, was history.

Although Jackson Barracks got socked the worst, it wasn't the only nerve center taking water. The FBI regional headquarters—built along Lake Ponchartrain near the University of New Orleans—got swamped. FBI officers, however, without missing a beat, embedded with the NOPD, doing an amazing job of stabilizing the discombobulated department in the coming days. The parking lot of the *Times-Picayune* was flooding an inch every seven minutes. A surprised Jim Amoss realized on Tuesday morning that if the water level on the first floor reached three feet, he would have to evacuate all one-hundred-plus employees and head across the Mississippi River Bridge, over by Mardi Gras World, to the *Picayune*'s West Bank bureau, which he knew was still dry. But as midnight approached Monday evening, with a stillness in headquarters, and everybody crashed on the floor, a party was under way.

James O. Byrne and a clique of fellow *Times-Picayune* employees who had lost their Lakeview homes commiserated on the loading dock, watching the water rise. "I uncorked my bottles of champagne," he said. "It was bittersweet. But the combination of Katrina winds and no electricity led to a sky full of bright stars. Just gorgeous you know, it made it all feel all right. Or so, for a few fleeting minutes, we pretended."[80]

Chapter Seven

"I'VE BEEN FEMA-ED"

Broken cutters, broken saws
Broken buckles, broken laws
Broken bodies, broken bones
Broken voices on broken phones.
Take a deep breath, feel like you're chokin'
Everything is broken.

—Bob Dylan, "Everything Is Broken"

I

HOUSTON WAS HOT ON Sunday, August 28—a scorching 97 degrees—the morning sun beating on the backs of joggers in Memorial Park, all trying to sneak in a few miles of exercise before the afternoon humidity became unbearable. Ralph Blumenthal, the sixty-three-year-old Houston bureau chief for the *New York Times,* stayed inside reading his paper's massive Sunday edition. With Katrina fuming in the Gulf of Mexico, he was also monitoring the Weather Channel; the farther east the storm went, the less likely it was the *Times* would have him chase it. By midafternoon, however, he started getting calls from his editors in New York.

They were sending down a photographer, Pulitzer Prize winner Vince Laforet, and they wanted Blumenthal to lease a helicopter and fly to New

Orleans. None was available. Blumenthal called the local Coast Guard to see whether it had any choppers bound for New Orleans soon, so they could hitch a ride. "A helicopter going in?" an incredulous officer asked. "We're trying to get everything out. No one flies into a hurricane! We're trying to get our planes off the ground so that nothing's damaged!" Undeterred, Blumenthal called Houston Mayor Bill White's office. "Are any Texas relief groups headed for New Orleans?" he asked. "I quickly found out that really nothing was going in, in the face of the imminent hurricane," Blumenthal recalled. "But the New York office had this idea that Vince and I could go in the air. Within an hour of calling around, I realized that was complete lunacy and I told them that and they agreed."

Plan B was put into action. Laforet would drive a rented SUV with GPS while Blumenthal followed him in his silver Pontiac Grand Prix. Together they started driving east on I-10, but got only as far as the suburb of Baytown by 1 A.M. Monday. Almost everybody in the Gulf South, even in Texas, had battened down the hatches and was out of harm's way. The motels in Baytown were full of Louisiana evacuees. Eventually, Blumenthal and Laforet found a flophouse tucked next to the Humble Oil refineries, and got a few hours of itchy rest. They woke up before dawn, splashed some water on their faces, and headed to Baton Rouge. "At six or seven in the morning, it was starting to rain and the weather was getting worse," Blumenthal said. "We just kept keeping tabs on the radio and we just kept driving east."[1]

As they crossed the Louisiana line, thinking ahead, they purchased five or six ten-gallon gas tanks at an auto supply store, filled them up at a Chevron, and lined them all up neatly in the back of Laforet's SUV. That was a mistake. "Fumes just overwhelmed him," Blumenthal recalled. "Even with the windows open everything stunk of gasoline."

Situated along the Mississippi, 225 miles upriver from its mouth at the Gulf of Mexico, Baton Rouge was not incapacitated by Katrina and therefore became a natural choice as a staging area for groups intent on rescue and recovery. When the *Times* duo arrived, it was like entering a ghostland only sodden with rainwater. Clearly a storm had whipped through, for there were lots of small leaves and pine limbs and oak branches littering the empty streets. Puddles had formed in parking lots and over two hundred power lines were down. Nearly two thousand people were still housed in Baton Rouge area shelters. It wasn't, however, a place without tragedy. Three senior citizens had died while being evacuated from a nursing home, and a tree had fallen on a man in nearby Livingston Parish, killing him.[2] Jeremy Alford, a local twenty-seven-year-old stringer, joined forces with Blumenthal and Laforet at the Baton Rouge Sheraton. A huge, beefy-necked man, Alford had never worked with either of them before. Constantly working the cell phone, which offered only intermittent service, Blumenthal found out that the Texas Task Force, a volunteer team of emergency workers, firemen, medics, and road removal specialists had linked up with FEMA and was going to rendezvous in Baton Rouge in four or five hours. Because all roads into New Orleans were closed, this would be the *Times*'s most realistic ticket into the bowl from the west—riding FEMA's coattails.

But waiting hours seemed like a lifetime, so they abandoned Blumenthal's car at the hotel and clambered into Laforet's SUV. Their destination was the northern shore of Lake Pontchartrain, particularly the towns of Madisonville, Ponchatoula, Hammond, and Slidell. There they would capture stories of Katrina's vengeance, in both words and images—typical posthurricane fare. Realizing that the state police weren't going to cotton to reporters cruising down closed roads, Laforet pulled out a yellow light strip, a long rectangular emergency flasher with suction cups that went inside the windshield, and plugged it into the cigarette lighter.

"If you ever find yourself in a situation like this again," Laforet told Alford, "make sure you have one of these lights. It gets you right through roadblocks." Racing up to eighty miles an hour, they made it to the north shore around two Monday afternoon; Hammond looked like hell. Blumenthal interviewed one family seemingly straight out of *Deliverance,* living on the porch of their ripped-up house, trying to cook a huge chicken on a charcoal grill. Because of the moisture it would just broil. "It was really kind of pathetic," Blumenthal recalled. "They were drinking beers. It was still raining, or raining again, so we took a lot of devastation pictures and tried to make our way to Slidell. But Slidell was completely cut off from every angle. We couldn't get all the way there. We had no idea how bad things were in New Orleans. We were listening to the radio station. There was no cell phone reception, or if there was, it was spotty."[3]

At one point the *New York Times* team became ambulance chasers, but even the ambulance couldn't find a clear path to Slidell. Huge oaks had fallen in the middle of the road, making it impossible to get around. A few locals were outside, yanking chain-saw cords, determined to cut through obstacles to the traffic flow. National Guard troops were also out and about in Tangipahoa Parish, which suffered mainly wind damage. Whatever flooding Hammond experienced was minor, affecting only low-lying areas. This was clearly not a bull's-eye locale. Dutifully, Blumenthal poked around truck stops, timber yards, and dairy farms. Southeastern Louisiana University looked generally unscathed. Little did the *Times*'s team realize that Hammond was about to double in size, and the Wal-Mart distribution center in nearby Robert would soon be the number one store in America. With New Orleans 80 percent underwater, the Hammond area was coveted high ground.

As they rode down a back road, Alford tapped Laforet on the shoulder and pointed to a weary old man pushing a wheelbarrow full of his belongings, a little boy, no more than five or six, at his side, trying to help. "Hey Vince, you gotta get that shot," Alford said. "That's your Pulitzer." The laconic response from Laforet was, "Yeah, I've already got two of those."[4]

It was hard for Blumenthal, Laforet, and Alford to focus on Hammond while rumors abounded that levees had breached and New Orleans was underwater. The fact that Tangipahoa Parish had a 100 percent power outage was hardly front-page news in New York. Just when their frustration peaked, they received a cell phone message from the press secretary for team leader Jim Strickland, FEMA's emergency director for Katrina. Because FEMA didn't have enough money, it relied on emergency management teams from around the country to be first responders. Strickland, a Virginian, was one of America's twenty team leaders, who rotated every four months. He was in charge of *all* FEMA task forces coming into Louisiana, not just those from Texas. "Change of plans!" the FEMA message said. "We're going into N.O. tonight. Be at the EOC in next forty-five minutes." Bingo. Here was their optimal chance for a free lift into the bowl.

There was, however, a problem. There was no way to make the run from Hammond to Baton Rouge—forty-five miles—in less than an hour, even in ordinary times. But they had to try. The flashing lights were turned on and off as they sped west. "I thought if we missed the Strickland task force," Blumenthal said, "we'd miss our chance to go in with them. We dropped everything and rushed over there to the Emergency Operations Center, and as it turned out, they didn't leave for hours."

Blumenthal and Laforet arrived at the EOC, hooked up with Jim Strickland, and arranged to be part of the FEMA convoy. Alford would stay behind to report from the EOC for the *Times.* Located on Independence Boulevard, this EOC was mission control for the National Guard, the Red Cross, Homeland Security, and Louisiana state officials. Although the atmosphere was frenzied, there was a sense of doom in the air Monday. On the big room's wall, a map highlighted all the parishes without power (ironically, by lighting them up). Looking at the map you couldn't help but conjure up Randy Newman's piano ballad "Louisiana 1927" with its sad refrain "Louisiana, Louisiana / They're tryin' to wash us away." Meanwhile, Governor Blanco and other state officials were there intermittently making sure, among other things, that all the regional nuclear facilities had survived without leakage (they had).

At 9 P.M. Jim Strickland announced that they were headed to the Baton Rouge firefighting training camp, where the Texas Task Force (among others) was waiting for them. "It sounded good but the official FEMA convoy couldn't find the fire camp," recalled *New York Times* reporter Ralph Blumenthal. "We were in our SUV lined up in the EOC parking lot as part of the convoy. We just joined this line of vehicles. We went in one direction and then the whole thing stopped and turned around. Somebody shouted, 'Doesn't anybody know where the fire station is?' It was all very disorganized. But eventually they found it, just a few miles away from the EOC where we started from."

Once at the training field, Strickland ordered everybody to gather for a frank briefing. Volunteer task forces from Tennessee and Missouri were also waiting there. This was the FEMA rendezvous spot. Around the training lot were huge tractor-trailers with relief supplies—tents, kitchen stoves, food, chemical suits, first-aid kits. A lanky, mournful-looking man in his early fifties, Strickland was a rather unfortunate choice to lead the FEMA brigade into the New Orleans bowl. A retired fire chief from Fairfax, Virginia, he was a civilian clearly incapable of running a large-scale operation.

Constantly putting his fingers through his gray hair, he gave the impression that he was already exhausted, before "the game" had even begun. There was an old-school decency about him, however, and a can-do attitude that counted for something. No doubt he had been an expemplary fireman. He didn't mean to seem hapless; it's just that the magnitude of Katrina caught him by surprise. "We're going to go in tonight," he said to the various task force contingents. "We'll travel in convoy and go stay at the Convention Center."[5] The Texas Task Force was going to stay behind and get some sleep. It would follow on Tuesday to relieve the Missouri and Tennessee contingents.

At 10:30 P.M. the convoy finally left the fire training camp and got on I-10, heading east toward New Orleans. Sitting in the backseat of Strickland's command vehicle was Ralph Blumenthal. Full-time FEMA employee Dave Webb was driving. Following behind was Laforet in the rented SUV, with a FEMA photographer in the passenger seat. "There were no restrictions placed on me," Blumenthal recalled. "They let me listen in on everything. It was not off the record or anything like that. Everybody was focused on getting into New Orleans and no one was thinking about holding anything back. It was all very open."

As the Strickland convoy drew closer to New Orleans, you could sense an eeriness in the night. A spell had been cast on the city; there was a feeling of brutal displacement in the air. Because road signs were down, the convoy, which had been split in two, accidentally went separate ways. Such transportation blunders became part of life with FEMA-sponsored emergency volunteers. "The feeling really was one of disorganization," Blumenthal said. "They didn't know where they were going. We didn't have a driver who was familiar with the road and that, frankly, did strike me as off. To be going in an emergency convoy with FEMA and getting lost?"

As they approached downtown New Orleans, Strickland learned by radio that the areas around the Convention Center were unreachable. This report was untrue. With a GPS they could have glided into the French Quarter practically with their eyes closed. Why didn't Strickland have a scout who could have pointed out to FEMA that all you had to do was exit Causeway Boulevard to River Road and drive to Tchoupitoulas Street, a direct shot to the Convention Center? The whole area around the Convention

Center was bone-dry. The National Guard trucks—those five vehicles not left to drown at Jackson Barracks—were already being driven all over the French Quarter–CBD–Superdome area quite easily, not even debris in their way. The flooding misinformation caused Strickland to alter his New Orleans relief plan. The source? The New Orleans Fire Department had a representative at the EOC in Baton Rouge who radioed Strickland with orders to *abandon* Mission Convention Center, to "vector" to Sam's Club in Metairie. It was a fateful, poor decision, an abandonment of the largely African-American population left in New Orleans. "So we headed to a Sam's Club," Blumenthal recalled. "The fire department made that decision to change the destination. About midnight, as we rolled into Sam's Club, which was in Metairie—we had bypassed downtown. We had to change our route to avoid the supposed water around the Convention Center. We got into the Sam's Club parking lot on Airline Drive okay. FEMA decided on the store because there were bathrooms and water and whatever. But the manager from Sam's Club was there and said no—there was too much damage inside the discount store. It wasn't safe. The roof could fall down on them; it was leaking."

Sam's Club was a members-only warehouse association owned by Wal-Mart, and named after its founder, Sam Walton. Launched in 1983 to serve small-business customers, the warehouse sold merchandise in bulk, a debit-card charge at a time. At the time of Katrina there were 581 Sam's Clubs in America, each averaging about three acres of merchandise. Even if FEMA personnel couldn't sleep inside Sam's Club, they could commandeer provisions from the enormous store. The night hadn't been a complete washout, and as an added bonus there was this consolation: Jefferson Parish was far safer than Orleans Parish. Sheriff Harry Lee was not the kind of law enforcement officer to tolerate looting. Strickland, whose intentions were good, ordered the convoy to make a base camp in the parking lot, even though it was blanketed with broken glass from the windows of the nearby Galleria. A few boxes of snack packs were handed out to the emergency personnel, while Blumenthal and Laforet shared peanut butter and jelly sandwiches that Blumenthal had made back in Houston. Up until now Blumenthal and Laforet had been tolerating FEMA's fecklessness, figuring it was just getting off to a clumsy start.

That was until they learned that FEMA didn't have a single portable toilet for the hundred men and women in the convoy. "That struck me as beyond the pale," Blumenthal recalled. "They're going into an area like that and they don't have any emergency toilet facilities, so you just had to use the bushes?"

After sleeping for a few hours in the SUV, Blumenthal and Laforet arose on Tuesday, August 30, to discover that a primitive FEMA command center had been erected in the corner of the parking lot. Briefings were being held and rescue equipment unpacked. Meanwhile, all night long, private boats and Louisiana Wildlife and Fisheries vessels had been helping flooded Lakeview and Ninth Ward citizens. And, come sunrise, the "incident situation" teams from Missouri and Tennessee were eager to get going. Emergency help was needed in every direction. "It's worse than I thought," Strickland matter-of-factly told his team.[6] Reports came in of natural gas breaks, house fires, and breached levees. And the death toll, as conveyed by the *Times-Picayune,* was probably going to be catastrophic. Sergeant Chris Fischer of the Jefferson Parish sheriff's office asked a legitimate question: What to do if we come across a floating body? Eyes darted around, unsure of the appropriate answer, the thought of corpses making everyone understandably uneasy. Eventually Major Bobby Woods fielded it: "Let it go. Let's first go for life."[7]

While most of the Strickland team was anxious to save lives, Laforet was determined to get in a helicopter and survey the damage. His editor, in fact, wanted him to head for high-and-dry Lafayette, where private choppers might soon be allowed airspace. "We had a strong difference of priorities," Blumenthal recalled. "I wanted him with me, in case I went out on the water and participated in the rescues." Before their bickering commenced in earnest, the problem solved itself. Like a mirage, a Jefferson Parish helicopter touched down near Sam's Club. A determined Blumenthal started begging FEMA officials to let Laforet bum a ride; the *New York Times* needed a photo of the breached levees and flooded districts. It turned out that FEMA was only going to allow their official photographer in the air. The insistent Laforet hadn't won a Pulitzer Prize for nothing. He absolutely begged the photographer for the helicopter seat. "If I go instead of you, we can pool pictures together and then FEMA can still have them," he pleaded. "And my pictures, with all due respect, will be better than yours. So could I please, please, please, please have your spot?" His groveling worked.

Laforet hopped into the chopper and off it went. It didn't take him long to get chills: he cursed the miserable dividends of Katrina under his breath. Homes reduced to scraps. Homes on fire. Homes on top of homes. Homes no more. He couldn't believe his eyes. Water was up to the eaves in half of New Orleans's homes! Lakeview and New Orleans East were islets in Lake Pontchartrain. Out came the Leica. The Coast Guard was pulling women and children off of rooftops in drop baskets; for each one they could get, twenty were left behind. Never, in all of his thirty years, had Laforet seen such overwhelming destruction. "It's really hell," he told *Photo District News* a few days later. "This is by far the most logistically difficult assignment I've ever covered and I've done some wars. . . . Nothing comes close to this. I'm not sure the people outside of New Orleans understood just how dire the situation is out here. It's really a total disaster zone where people are starving to death, dying of dehydration and getting absolutely desperate."[8]

It was only a forty-minute copter ride, but that was long enough. Laforet emerged from the helicopter crestfallen. The breached levees were worse than expected. He had snapped shots of desperate people being pulled off rooftops by the Coast Guard, scenes that seemed straight from the *Inferno,* people in wheelchairs and baby carriages stuck in floodwaters. Upon landing, he relayed the horror he had seen to Blumenthal. Together the *Times* team strategized about how to get the news to New York. Cell phones weren't working, but rumor had it that a few pay phones in Metairie, for some odd reason, were still functioning. They pulled into a Shell station on Airline Highway; it had a pay phone where a frazzled refugee was talking away. Blumenthal waited and waited. He waited some more. Then he grew impatient. It was imperative that his editor learn what Laforet had witnessed from the sky. "Look," Blumenthal said, "could I use the phone?" All the guy said was "Sure," and hung up. That was oddly easy. It triggered a double-take by Blumenthal. Apparently this defeated man and his buddy were living out of a broken-down old car in the Shell station lot across the street from where Gram Parsons's ashes were buried, and a few blocks away from the Travel Inn where Reverend Jimmy Swaggart got nailed with a hooker. Besides them, there were four or five cats in the car. They had been living on handout food, like Snickers and Lays potato chips, from people in the area. "It was really pathetic," Blumenthal recalled. "I realized that now that I got to the phone, I didn't have any quarters. So the guy says, 'I'll help you out. Do you need quarters?' And he started giving me coins."

Blumenthal and Laforet were touched that a guy with all his earthly possessions jammed into a clunker was willing to give the reporter money, no questions asked, even after he had been rude. Thinking he was doing them a favor, Laforet snapped a picture of the two guys. That act started a ruckus. "Wait a minute!" one of them said. "You can't take our pictures! This is *very* embarrassing. We're upstanding citizens and we're just sort of down on our luck. I refuse to let myself be photographed. I'm going to take your film out of your camera! You can't do that!" A diplomatic Blumenthal tried to fix the misunderstanding. "Look, I'm sorry to upset you guys," he said. "If you don't want your pictures in the paper, we won't do it. But what's your problem here? Why are you guys stranded here?" The refugee said they had run out of gas. He was embarrassed by this fact. "We'll give you gas," Blumenthal reassured him.

With that, the *New York Times* reporter and photographer handed over five gallons of fuel they had purchased near Lake Charles. "We felt like a million dollars when we left those guys," Blumenthal said. "Because without us they would have been stranded forever."

Over the pay phone, Blumenthal dictated a description of what they had encountered to his assignment editor in New York. He was contributing to a monster *Times* overview story along with Joseph B. Treaster, who was at New Orleans's EOC on the ninth floor of City Hall. But there remained the dilemma of getting Laforet's photographs filed. They needed electricity. It was 4 P.M. Central Time. They would have to race to Baton Rouge at breakneck speed to get wireless Internet service. "That's where the light strip came in handy," Blumenthal said. "It made us look like an emergency vehicle." And they had another ploy up their resourceful sleeves: Blumenthal was wearing a dark blue Ralph Lauren polo shirt, which just happened to be very similar to the FEMA ones. Blumenthal recalled, "As a bonus, I had a red hat that I had gotten in Colorado that had a cross on it from the Colorado Rescue, so it looked somehow that I was officially involved in the Katrina mess. Authorities had set up roadblocks on I-10 and we got right through them. We drove like crazy. . . . We logged the story from the Sheraton in Baton Rouge."

Laforet's emblematic aerial photo of a Ninth Ward woman, floodwaters beneath her, being pulled out of the sludge by an orange Coast Guard helicopter, dangling in a basket, not only appeared on the front page of Wednesday's *Times;* it was one of the most memorable photographs of 2005.

So ended FEMA's first attempt at relief along the Gulf Coast: lost and confused, it ended up at a Sam's Club, which didn't need help, miles and miles from those thousands who did. And while the *Times* team didn't make it into the bowl, they published the haunting photographs that spoke to the pathos of the post-Katrina moment.

II

While Jim Strickland was in Shreveport, Baton Rouge, and Metairie, searching for an auto route into New Orleans, Marty Bahamonde was the only FEMA employee in the Crescent City. After arriving on Saturday, the indefatigable forty-one-year-old weathered Katrina at City Hall, keeping a watchful eye on about 13,000 people in the Superdome (around 12,000 evacuees, more than 400 National Guard personnel, and an assortment of facility managers and medical personnel). Bahamonde, the public affairs officer in FEMA's New England regional bureau, had worked at the agency for twelve years, often as an advance man on the scene of an expected natural disaster. Bahamonde considered that his primary assignment under such circumstances was "providing accurate and important information to FEMA's front office and Under Secretary Michael Brown."[9] As for natural disasters team leader Jimmy Strickland, Bahamonde had never heard of him.

Born in North Dakota and raised in Illinois, Bahamonde had majored in radio and television at Southern Illinois University. After college he held TV sports anchor gigs in Illinois, Texas, and Guam. It was in Guam, in fact, on August 28, 1992, that he covered the deadly typhoon Omar, which savaged the island, an American territory at the forlorn southern end of the Mariana Islands chain. According to the *New York Times,* it destroyed about 2,000 homes and seriously damaged 2,300 others.[10] "Typhoons were regular events in Guam," Bahamonde recalled. "That August we were hit by seven of them. But Omar was the worst. The people needed help and FEMA had arrived to give it. It had a profound effect on me. You might call it an epiphany. I knew there was more to life than reading sports scores. I had to do something that had a direct, positive influence on people's lives. I really believed in what FEMA was doing in Guam and I wanted to be a part of their team effort."[11] Just like that, Bahamonde quit television and joined FEMA. He was in the trenches on the relief efforts either before or following practically every natural disaster to hit the United States between 1992 and 2005. A partial list includes the Midwest floods of 1993 (Missouri), the Northridge earthquake of 1994 (Pasadena, California), Hurricane Floyd in 1999 (North Carolina), Hurricane Isabel in 2003 (Virginia), and the 2001 Red River flood (North Dakota), among many others. He worked in Florida during the devastating 2004 hurricane season. "Nothing," Bahamonde recalled, "compared to Katrina."

After meeting so many bedraggled refugees at the Superdome, Bahamonde worried about New Orleans's future. He was surrounded by jittery talk of the Big One, as if the community were a metronome ticking away the seconds until the hurricane made landfall. "I was alarmed by what I saw at the Superdome," Bahamonde said. "I've always tried to stay focused on disaster victims. And I knew at once that if we got a Cat 4 or 5, like was being forecasted, the city of New Orleans had a big problem on their hands. Instead of evacuating people out, the city had packed them in a very unsafe place."[12]

All day Sunday, Marty Bahamonde's focus had been on the Superdome, and he sent repeated messages about it to FEMA officers in Washington and Denton, Texas, insisting that food, water, and basic medical supplies were already running out—and the hurricane had yet to strike. "Issues developing at the Superdome," Bahamonde e-mailed David Passey of FEMA. "2,000 already in and more standing in line. . . . The medical staff at the dome says they will run out of oxygen in about 2 hours and are looking for alternative oxygen."[13] If the supplies and expert personnel didn't arrive on Sunday night, Bahamonde explained, "it won't be until Tuesday and by then it could be too late for a few."[14] As the blitz of e-mails flew back and forth to FEMA officers around the country, Bahamonde was assured that supplies were just hours away. In the Superdome, Dr. Kevin Stephens was in charge of medical services. "We've got sick babies," he said in desperation, echoing Bahamonde, "sick old people and everything in between."[15]

Bahamonde and Stephens were right to worry. As Katrina approached, forty to forty-five Superdome refugees were in life-threatening situations. There were also hollow-eyed junkies jonesing in the shadows, hungry for crack or heroin, afraid that their next fix was days away. Prescription medicines were needed for the seriously ill—diabetics in need of insulin, for example, or schizophrenics needing Zyprexa. Babies needed milk. Garbage was piling up in heaps, dirty diapers and empty beer bottles tossed down aisles with abandon. Many people had evacuated to the Superdome expecting food, water, and medicine—the basic provisions. They just barely received the food and water. But what they mainly got was a human tragedy. By Monday afternoon, conditions were awful, with the electrical and pumping systems out of order. The people who had come for refuge had become a community of beggars, crowding the sports arena as the Hebrew cripples did the pool of Bethesda. "I was and am still most haunted by what the Superdome became," Bahamonde later explained. "It was a shelter of last resort that cascaded into a cesspool of human waste and filth. Hallways and corridors were used as toilets, trash was everywhere, and among it all, children, thousands of them. It was sad, it was inhumane, and it was wrong!"

Doug Thornton, one of the Superdome managers, stayed on the scene all the while, trying to placate the throngs. He could see the problems, even before they arose. "We can make things very nice for 75,000 people for four hours," he said. "But we aren't up to really accommodate 8,000 for four days."[16] Mayor Nagin's evacuation plan had turned into a debacle. But early Monday morning Thornton believed the Dome, the world's largest steel-constructed room unobstructed by posts or pillars, would survive. The Superdome had weathered 87,500 Rolling Stones fans in 1981 and 80,000 Catholic schoolchildren anxious to see the pope in 1987. (It would, indeed, survive with about $150 million in repairs.) Three months after Katrina, NBC's Brian Williams had returned to the Superdome to compare notes. "I guess what I want to remember most is that [frightening] sound," Thornton told the newsman. "It's a sound I'll never forget and my concern for the safety of these people in the building."[17]

Waiting out the storm in Baton Rouge, Judy Benitez, executive director of the Louisiana Foundation Against Sexual Assault, was terrified by images of the Superdome on television. A rape therapy specialist, she knew that the Superdome evacuation scenario was setting up conditions for unimpeded assaults or rapes. Ugly male behavioral patterns were predictable. In a situation like the Superdome, feeling trapped and abandoned, without electricity or supplies, powerless men may feel compelled to victimize women even more. "Everything at the Superdome was a worst-case scenario for rape," Benitez later explained. "There was no authority or supervision of any kind. About 200 New Orleans policemen had fled the city. When you have squalor and filth like in the Superdome, there will be anger and frustration. Some people in the sports arena were criminals, some were mentally ill, others had violent tendencies. So rape becomes a method of gaining power or control, even for a brief period of time."[18] Even under normal circumstances, it's estimated that only one in every eight rapes is reported, so it was unlikely that accurate numbers would ever be determined. As of December 2005 the National Sexual Violence Resource Center had received forty-two reports of sexual assaults in New Orleans (or host homes) following Katrina.[19]

Despite the promises from FEMA officials in response to Bahamonde's e-mails, no supplies and no medical team arrived on Sunday night. Instead, mass confusion reigned in the Superdome. Families huddled together, hoping not to be harassed. A number of women were raped and a couple of men were stabbed. Bathroom stalls became drug dens. With gun-toting punks claiming certain sections as turf and with the roof panels peeling off the famed arena, three police officers quit on the spot. History will remember the Superdome debacle—caused by the dearth of evacuation buses—as "Nagin's Folly," mayoral incompetence of the first order. "I feel like I am living in the middle of a horror movie," FEMA's Public Affairs Division director, Natalie Rule, wrote in an e-mail message to Marty Bahamonde, just before she left her Washington office and went home for the night.[20]

Sunday night Bahamonde had settled onto a cot at the EOC in City Hall. There was a feeling of gloom in the air, and the rattling of the shuttered windows added to the tension. Bahamonde had spent much of the day setting up a helicopter ride with the Coast Guard over the post-Katrina devastation. What baffled him that night was that Mayor Nagin had chosen not to be with his emergency assistance team in City Hall, opting instead for the comforts of the Hyatt. "It was very tense," Bahamonde said. "I didn't go to sleep until 1 A.M. There was no light, just that constant rattling. It got really bad at 7 A.M." When Katrina was at its zenith, Bahamonde got a telephone call from FEMA headquarters in Washington, saying that Brian Williams had shown a photograph on NBC of a leak in the Superdome roof. His bosses wanted Bahamonde to investigate whether the leak was actually there. "You couldn't see out of the windows so I went downstairs," Bahamonde recalled. "I went out to the parking garage and looked at the Superdome. It looked structurally sound but you could see foam or whatever blowing off. And I could certainly see that the windows of the Hyatt had blown out. I reported back to headquarters an affirmative response."[21]

With the mayor at the Hyatt, Colonel Terry Ebbert of municipal Homeland Security was running EOC, directing panicked first responders on what to do. Sporting black hair and a perpetual grin, an American flag pin usually on his lapel, Ebbert was a twice-wounded Vietnam vet. After Vietnam, he became a security officer for the Pacific Fleet, responsible for policing all naval base and vessel security in California, Hawaii, Guam, the Philippines, and Japan. As the Homeland Security representative of New Orleans, he turned native, adopting localisms and cheering on the Saints. In Nagin's absence, someone had to become essentially the proxy mayor, answering questions like how the Superdome could get toilet paper or why in fact the mayor was AWOL from City Hall. To his credit, Ebbert handled the minutiae the best he could. The ugly truth was that New Orleans under Nagin had no serious plan to manage a catastrophe. New York developed one *after* 9/11, and in Bahamonde's hometown of Boston, they constantly had emergency drills and training sessions, but very few other American cities took disaster response seriously. New Orleans wasn't alone. Disaster planning was an inconvenience. Ebbert, unsure what the next steps were with the hole in the Superdome's roof, out-of-commission buses, and breached levees, pulled Bahamonde aside. "Hey, Marty," he said. "You've done this before. What do we do now?" It was a little late to be asking this question. But without Ebbert *everything* in New Orleans would have collapsed; he was almost a one-man Command Center, an unsung hero.

At 5 P.M., Bahamonde got into a Coast Guard helicopter and, as planned, surveyed the Greater New Orleans area. Forget Guam after Omar or North Dakota after the Red River flood of 2001—Bahamonde had never seen, or imagined he would ever see, such widespread devastation. "We had the worst-case scenario," he said. "I had never seen anything like it. The sheer magnitude of it all. I hardly saw anyplace not flooded, except around the Superdome and French Quarter. People were on rooftops and balconies waving for help." Staring blankly in the white glare at the breached levees, a demoralized Bahamonde understood that a deluge had occurred. At this juncture only two other Coast Guard helicopters were surveying the damage.

After the fifteen-minute trip, Bahamonde filed a report. With just enough daylight left, he convinced the Coast Guard to take him up for a second, longer trip. Upon returning to City Hall, he called his boss, FEMA Director Michael Brown, at the Baton Rouge EOC and filled him in for a good ten minutes. "I told him everything I knew and saw," Bahamonde said. "I felt very good. I let the senior guy at FEMA know just how desperate New Orleans and Slidell were."[22]

For Bahamonde the horror movie continued well into Monday evening. Finally, Mayor Nagin decided to show up at EOC for an 8 P.M. briefing. He was stoic, his downcast eyes fixed on a wall. He didn't yet understand the flooding situation in Lakeview. About twenty other emergency responders were in the conference room. Astonishingly, the mayor had yet to helicopter over his own city—the only way to get a real assessment of the damage. It was embarrassing. He asked questions like, "What did it look like in the Lower Ninth?" or "What's exactly a breach?"[23]

Times-Picayune reporter Gordon Russell, who had been tossed out of the evening briefing by Nagin's advisors, was able to grab Bahamonde in the corridor for about five minutes for details about his overflights. "At that point I had only a worm's-eye view of the City Hall and Superdome area," Russell said. "Marty had actually *seen* the extent of the destruction. He actually made a drawing for me of the breached 17th Street Canal. That was the tipping point for me. I now knew that with Lakeview gone, New Orleans was going to flood."[24]

III

Late on Monday morning, Michael Brown, the director of FEMA, arrived in Baton Rouge. He used the state EOC as his makeshift office. He spoke with Blanco at length. Almost immediately, he briefed President Bush, in a videoconference, on the impact of Katrina. Unfortunately, Brown had not been on the scene in Louisiana long enough to understand the impact. Indeed, with communications hampered in the affected areas, the situation demanded more investigation. At just about the same time, a rare telephone call reached the emergency team in New Orleans. "At approximately 11 A.M.," recalled Marty Bahamonde, "the worst possible news came into the [Nagin] EOC at the Hyatt in New Orleans. I stood there and listened to the first report of the levee break at the 17th Street Canal. I don't know who made the report, but they were very specific about the location of the break and the size."[25] That was not the first report of the breach, in actuality, but it was from an authoritative source. At the time, the break was small, just a few feet wide, but the force of the water was bound to change that, if it wasn't fixed immediately—and with a growing flood surrounding it, engineers from the Army Corps could not even get near enough to make a proper assessment, let alone a plan for immediate repairs. Even as that early report was filtering through the EOC, the breach in the 17th Street Canal was bursting open; the water won, and more of the concrete wall crumbled.

Tens of thousands of other people had meanwhile joined Michael Brown in descending on Baton Rouge, which had mushroomed overnight into the largest city in the state. By late afternoon Monday, Louisiana's capital was teeming with people emerging from hiding. Katrina was gone. They had escaped its stranglehold. There was reason to celebrate. The college town's asphalt parking lots were suddenly brimming with SUVs and minivans as frenzied refugees hunted for somewhere to perch. Utility vehicles flashing yellow lights—legitimate ones—crawled up and down Main Street, unsure where to go first in order to help. Gasoline lines were long, and with fuel at over three dollars a gallon, nobody filling up was polite. Yet Baton Rouge had become Luckyville, a place to plug in laptops or cell phone chargers. Electricity, more than hard cash, was the hot currency of the frenetic, taxing moment. The toxic zone was somewhere else, back in the blackout area that President Bush neglected while visiting San Diego, where on Tuesday morning he barely mentioned Katrina in a speech on Iraq.* And for a surreal moment, even though your house may have vanished and your cherished belongings may have been ripped away with one giant sucking sound, you felt like an endurance champion. Your number wasn't up. Life was still yours to live.

However, as FEMA and others tried to coordinate the cleanup of the Gulf South, it soon became obvious to those in the capital that the National Guard was not there in full force.

New York Times stringer Jeremy Alford headed over to the Pete Maravich Assembly Center, where the LSU basketball team played during the season. A triage center had been set up by Monday evening, with old people in need of medical attention packing the arena. A few had Tupperware out, trying to eat a normal meal amid the bedlam. All over Baton Rouge, there were reports of fistfights, gunfire, and auto theft. Nearly all were false, but the evacuees at the Maravich Center had no way of knowing. "You could feel the fear," Alford said. "Two women, seeing I was a reporter, begged me to help them find their lost children. Just as I was taking notes, two police officers threw me out of the facility."[26]

As in most situations, dark humor was on display at the filling stations and motel parking lots of Baton Rouge. "At least we don't have to watch the goddamn 'Aints' lose another goddamn game," one evacuee driver chuckled. Another guy walked out of a convenience store holding a copy of *People* with Jennifer Aniston on the cover. "Next week she'll be on

* On Monday, August 29, President Bush was in Arizona to participate in a birthday-cake photo-op with Senator John McCain and visited a resort to promote Medicare drug benefits. At 4:30 P.M. (CDT) he flew to California.

again," he laughed, "only wearing rubber boots." The story of Strickland's FEMA convoy getting lost had spread like wildfire. No matter how large the population in Baton Rouge had grown, it was still a small town. People began to joke that they would name their next baby "FEMA Aid" or "Federal Government," because it took nine months to arrive. David Letterman later devoted one of his nightly top-ten lists to "Questions for the FEMA Director Application," asking "Are you able to convey a false sense of security?" and "Can you screw up bad enough to take the heat off the President?"[27] All over the Gulf South, a new verb had been created: "I've been FEMA-ed."

But there was nothing funny about being in Marty Bahamonde's shoes. Using his trusty BlackBerry, a flabbergasted Bahamonde continued to warn of the dire conditions in New Orleans. He had seen the flooding with his own eyes. In all his years with FEMA, he had never seen such disorganization as there was at City Hall EOC and the Superdome. His urgent e-mails to Brown, asking for food, water, and medical aid seemed to fall on deaf ears. "I got the word to him as strongly as I could," Bahamonde would later tell the Senate Homeland Security and Governmental Affairs Committee. "I don't know where that information went."[28]

IV

By all accounts, Michael Brown was an unlikely head of an emergency organization like FEMA. It's been said that under his leadership, FEMA lost the "M" for Management. His official curriculum vitae was larded with troubling discrepancies. It was certain that he was a native of Oklahoma, earned a bachelor's degree in public administration and political science from Central State University in Ohio, and went on to get a law degree at Oklahoma City University. But from there his biography got murky. In the wake of Katrina, *Time* published a scathing story titled "How Reliable Is Brown's Résumé?" Brown claimed, for example, that from 1975 to 1978 he had been "overseeing the emergency services division of Edmond, Oklahoma." In fact, he had been "more like an intern," according to the city's public relations head. Worse, he claimed on his résumé to have been an "outstanding political science professor, Central State University."

False—he was only a student there. Anxious to showcase evangelical credentials for the Republican right, he claimed he had been director of the Oklahoma Christian Home from 1983 onward—a stretch, since he had never been affiliated with it in any capacity. There were many other falsehoods on his padded résumé, which the White House swallowed when he was appointed head of FEMA in 2003.[29] "It's the kind of thing you'd think somebody would have caught by now," *Time* reporter Carolina Miranda told CBS's *The Early Show* following Katrina, "but, clearly, nobody has."[30] Without being specific, Brown claimed that *Time* distorted his record. But as of this writing, Brown has not shown any proof to refute *Time*'s story, nor did he file a libel suit.

Nothing in Brown's résumé, whether exaggerated or not, recommended him to lead America's disaster relief efforts. Self-centered and comically suave, Brown was a cuff-link-shooting Republican dandy. Unflaggingly loyal to bosses, he was a kind, capable administrator, known for his dry, deadpan humor. Bill Dashner, the former city manager of Edmond, Oklahoma, recalled that the dapper Brown "always had on a suit and a starched white shirt."[31] At the time President Bush tapped Brown, he was second in command at FEMA; he'd been a staff attorney there, brought in with his longtime friend Joseph Allbaugh. Before that, Brown was the failed head of horse-show judging at the International Arabian Horse Association. His tenure there was fraught with lawsuits filed against the organization over disciplinary actions.[32] Eventually, he was forced to resign. "There was a feeling," said Stephen Jones, a prominent Oklahoma lawyer who was Brown's boss in the early 1980s, "that he was not serious and somewhat shallow."[33]

To understand why President Bush would appoint Brown to lead FEMA one must go back to the relief organization's founding. The Federal Emergency Management Agency was the brainchild of President Jimmy Carter, who created it in 1979 as a direct result of a National Governors' Conference petition. "At that time there were sixteen different federal agencies that dealt with disasters like Katrina and we put it together with three specific commitments that I hoped at the time were permanent," Carter explained in a speech on September 19, 2005, at American University. "One was that it would be headed by highly qualified professionals in dealing with disaster. Secondly, that they would be completely independent and not under another agency that would submerge it. And third, that it would be adequately funded. Well, I thought that that'd be a permanent commitment. But, as you know, all three of those provisions have been violated in recent years and obviously there were deep problems at the local and the state level and at the federal level, so I think now is the best time not to look at blaming about Katrina, but to try to correct the defects that have evolved in recent years and make sure they're not repeated."[34]

President Carter was barely able to get FEMA off the ground before leaving the White House. The incoming Reagan administration saw the outfit as a feel-good liberal money drain, a cousin to HUD and HEW. And, although Carter wasn't complicit, by 1981, the agency already smacked of patronage. For a president, appointing a person as director was akin to giving a donor or friend the ambassadorship to Luxembourg—a cushy, largely honorary post. When Hurricane Hugo slammed into the Carolinas during President George H. W. Bush's watch in 1989, FEMA was roundly criticized for being a bureaucratic joke, a disaster relief organization with no clout. Senator Fritz Hollings, a South Carolina Democrat, lambasted FEMA employees as "the sorriest bunch of jackasses I've ever known."[35] Under President Bill Clinton's leadership, FEMA was run by Arkansan James Lee Witt, a true emergency management expert.* The agency professionalized its response times and fine-tuned its preparation models. Then came George W. Bush.

As Texas governor, Bush had learned to rely on his chief of staff, Joseph Allbaugh, who would also be his campaign manager during the 2000 presidential election. Allbaugh knew how to raise funds and troubleshoot, but he knew absolutely nothing about disaster relief. He was rewarded for his election effort by being named director of FEMA. When Allbaugh retired from the post in 2002, owing to FEMA's absorption into the new Department of Homeland Security, his job went to Michael Brown, his old college friend. To give Brown credit, he performed ably in Florida during the series of hurricanes there in 2004. From August 13 to September 26, two Category 3 storms (Charley and Frances) and two Category 4 storms (Ivan and Jeanne) walloped the state. They were a relentless quartet, leaving 124 people dead

* Governor Blanco hired James Lee Witt as a post-Katrina consultant on September 3, 2005.

and a fifth of all homes with at least some damage.[36] The aid to ravaged areas was delivered in above-average time. It was a good show. However, the South Florida *Sun-Sentinel* later went on the attack: while Brown had turned a blind eye, FEMA had given millions of dollars in aid to counties that had virtually no hurricane damage. It was an embarrassing scene, as Brown was at the center of a costly controversy. In blistering language, the *Sun-Sentinel*—along with Democratic Congressman Robert Wexler—demanded that Brown resign at once.[37]

Whatever the circumstances the year before, Brown was still very much on the job on August 29, 2005. He was head of the U.S. government response to Katrina. The White House took its cues from him. Brown sent e-mails that would become permanent monuments to a detached Washington mind-set, out of touch with the needs of everyday Americans. One politician who was onto Brown's ineffectiveness early on was Mississippi Governor Haley Barbour. After Katrina decimated the Gulf Coast, Brown assured Barbour that the disaster would soon be handily addressed. "FEMA," Brown boasted, "had lots of hurricane practice in Florida." A gruff, blunt Barbour wasn't impressed. "I don't think you've seen anything like this," he snapped. "We're talking nuclear devastation."[38]

During 2004, Brown's first full year as director, FEMA had responded to the four big hurricanes in Florida. The state of Florida had been well prepared, though. And with the 2004 presidential election approaching, so was President George W. Bush's financial response. FEMA was conspicuously generous in its open-checkbook response, taking criticism afterward only for handing out too much money. Over $100 million went to people and businesses that hadn't sustained any damage at all.

Brown, whom the President nicknamed "Brownie," served the administration fairly well. If a rash of deadly disasters can ever be said to have "gone well," the 2004 Florida hurricane season had met the criteria. State officials in Tallahassee, in fact, knew days before each hurricane made landfall exactly what recovery funding and rescue operations they might need, and Brown was given a blank check with which to satisfy their robust requests. With the help of state officials and relief workers, FEMA moved displaced victims into hotels and produced fleets of trailers for the homeless. A political lesson had been learned, one that unfortunately wouldn't help the Gulf South in 2005: it's best to have a natural disaster in the heat of campaign season, when your state is up for grabs during a presidential election year. "FEMA's often invisible and incompetent reaction to the devastation in New Orleans stands in sharp contrast to the way the relief agency and the entire Bush administration sprang into action last summer as a series of deadly hurricanes—Charley, Frances, Ivan, and Jeanne—battered the crucial swing state of Florida just weeks before election day," editorialized Eric Boehlert of Salon.com. "Partisan politics were certainly in the air during the busy [2004] hurricane season."[39]

On a less cynical note, Florida had a distinct advantage over Louisiana. Once the Florida hurricanes were over, they were over. Damage assessment teams could file reports and begin the arduous cleanup processes, including blue-tarping houses with damaged roofs and offering trailers to the victims. Before insurance companies could deliver on their contractual obligations, they needed to enter the vast hurricane devastation zone and make damage assessments. In Greater New Orleans, however, the hurricane disaster was followed by the flood disaster, which was followed by human disasters—the Great Deluge broke down relief efforts on every level. It was one thing for FEMA to help 20,000 seriously distressed people in Florida, and quite another to provide relief for 400,000 in Louisiana, Mississippi, and Alabama.

By Monday afternoon, with New Orleans streets full of water and the entire Gulf South region full of victims in need of rescue, FEMA was being tugged on by thousands of disparate entities. At the Baton Rouge EOC out of which FEMA was working, Louisiana Lieutenant Governor Mitch Landrieu, for example, was handed a list of dire necessities by A. J. "Junior" Rodriguez, the president of St. Bernard Parish. A distraught Rodriguez, recognizing already that FEMA was MIA, lamented that his parish had "no body bags, no ice." There was an immediate demand for medical equipment, chlorine bleach, cleaning supplies, inoculations, tents, food, and boats. Rodriguez also asked Landrieu for generators, lots of them—to be dropped off at the Chalmette slip.[40] This list was indicative of the first wave of materials needed from FEMA by all the parishes in Louisiana and the counties in Alabama and Mississippi.

Inundated with dozens, if not hundreds of such lists, Brown needed the multifaceted imagination of General George C. Marshall, Paul Hoffman, and Jean Monnet to meet the plethora of Louisiana and Mississippi relief requests. All that he possessed, however, were the routine skills of a quartermaster—a linear bureaucratic breed for whom creativity or improvisation was decidedly a detriment. For individuals of Brown's mind-set, there was only the "recovery plan," which was to be executed as fastidiously as possible. Any deviations from this plan could lead, so the mind-set believes, only to prolonged lawsuits, distribution glitches, and chronic heartburn. Most seasoned rescue and recovery experts, however, understood that improvisation was the fundamental modus operandi in a disaster, flexibility being the true guiding principle. At headquarters in Washington, D.C., Leo Bosner, a watch officer in FEMA's National Response Coordination Center and the person who wrote the "national situation updates," seethed at Brown's thumb-twirling slowness, later telling PBS's *Frontline* that his boss was "out of his depth" during the debacle.[41]

Within hours of the hurricane, an astonishing array of offers of help had been made to FEMA, from Amtrak, faith-based groups, small businesses, and large companies like Marriott, Home Depot, and Continental Airlines. Over the ensuing days, those would be followed by offers from foreign governments, such as Cuba (Cuban medical brigade), Germany (military planes with fifteen tons of military rations), and Kuwait ($100 million in cash plus $400 million in oil products). Far-flung American communities galloped to give all they could, volunteer firefighters from Houston, doctors from the University of North Carolina at Chapel Hill, and five hundred airboats from Florida. What angered Bosner and other relief professionals at FEMA was how Brown delayed the deployment of all such offers of aid, insisting that they wait until a chain of command could be established. FEMA "will not authorize the airboats to enter New Orleans," a FEMA report stated. "Without that permission, they would be subject to arrest and would not receive security and supply services."[42] Then there were the inquiries from standard FEMA contractors, who couldn't understand why they were still awaiting orders. For example, Cool Express, an ice company in Blue River, Wisconsin, had a standing contract for ice deliveries in disaster situations. Yet the company didn't receive permission to send trucks to the region until 4 P.M. on Monday. By that time, the ice wouldn't even reach the staging area in Dallas until late Tuesday.[43] After that, it would require an eight-hour drive to southern Louisiana.

Most FEMA employees were mortified by such failures, for they prided themselves on helping those in distress, not turning a cold shoulder or fumbling as if the football was greased. According to Bosner, top FEMA officials, furious at Brown, were saying, "My God, why aren't we doing more? Why aren't we getting the orders? Why isn't this being treated like a real emergency? People were just lost."[44]

For the insurance companies, it was not as if New Orleans were dead, but there had been a massive stroke. The recovery process would be hard and slow, and the victim never quite the same again. After hurricanes, insurance companies usually rush to set up "catastrophe centers," to help with the losses, but in New Orleans, due to Mayor Nagin's decision not to evacuate hotels before the storm, State Farm had no place to lodge its team of damage assessors. "Katrina was odd," David A. Ross, claims manager for Louisiana, recalled. "They wouldn't even let us into New Orleans to help out. But at various blockades, we'd flash our State Farm badge and the police would let us through." Eventually State Farm took over a building in the Warehouse District and began to assess the widespread damage. "We Indian-traded our way around," Ross said. "I swapped a bunch of generators to a hotel in exchange for eighty rooms."[45] In Louisiana alone, State Farm would eventually pay out more than $3 billion in claims to homeowners and over $300 million in automobile claims.

V

Wal-Mart, the world's largest retailer and America's biggest private employer, stepped up to the plate by offering vast warehouses full of essential supplies to those stricken by the Great Deluge. Under the lightning-quick leadership of CEO Lee Scott, Wal-Mart used its muscle to meet the needs of the victims in the three ravaged Gulf Coast states, donating emergency supplies ranging from Strawberry Pop-Tarts to Hanes underwear. The company also opened its stores to emergency workers, who were allowed to take the supplies they needed. "Wal-Mart was our

FEMA," said Warren Riley of the New Orleans Police Department.[46] For a company often criticized as a money-grubbing monolith hell-bent on destroying small Main Street businesses, the relief effort was a masterstroke of public relations. As FEMA sputtered, Wal-Mart filled the void, offering provisions to officials, giving cash advances to employees forced to relocate from the Gulf Coast, and even guaranteeing employees jobs. As the Pittsburgh *Tribune-Review* noted, Wal-Mart "stepped over or around the confused, floundering and sluggish bodies of federal, state and local government relief agencies and sprang into action."[47] *Fortune* magazine praised Wal-Mart for having its own in-house meteorologist who recommended, six days before Katrina made landfall, that the company prepare itself to rush into the projected destruction zone. "The Red Cross and FEMA," Bay St. Louis Mayor Eddie Favre suggested, "should take classes on logistics, mobilization and compassion from Wal-Mart. They opened up their Waveland store and gave us whatever they needed. It was awesome."[48]

At that Waveland Wal-Mart, in fact, CEO Scott directed that a Wal-Mart Express be opened, a smaller version of the discount department store, specifically created to cater to the post-Katrina needs of the Mississippi residents. And the philanthropy kept coming: the chain donated $15 million to the Red Cross and other Katrina-related funds. The Walton Family Foundation, the charitable legacy of founder Sam Walton, gave an additional $8 million. Critics of the massive company charged that even those sums were piffling—chump change for a corporate giant that in 2004 earned $285 billion in revenues and $10.3 billion in profits—but the fact remains that all across the Gulf Coast, Wal-Mart brought relief during the first hours and days of the recovery effort.

Within a day of the hurricane, Coca-Cola had assigned an incident management team at its headquarters in Atlanta to coordinate a relief effort on behalf of the company and its hundreds of independent bottlers. The next day, the company gave $5 million to the Red Cross, the Salvation Army, and other charities. A portion was dedicated to assistance for Coca-Cola employees in the affected region. The Coca-Cola example was a reminder that it was one thing to donate food or supplies, and quite another to distribute them. In many cases, relief agencies look on donations of materials as a burden, since they cannot move or track them. Coca-Cola had the means to deliver its goods—with its own trucks—and the foresight to do so.

American Airlines also stepped up. CEO Gerard Arpey opened a command center to locate and evacuate all of his company's employees in the Gulf South. With that mission accomplished, Arpey ordered relief into New Orleans's Louis Armstrong International Airport. American Airlines planes were the first to land, just twenty-four hours after Katrina made landfall. The company flew in boxes of food, hygiene supplies, pallets of water, and other emergency supplies.[49] Then company planes took victims away. Over 1,000 evacuees were flown out of the New Orleans bowl after Katrina before FEMA even figured out how to move a Porta Potti down I-10.

Forbes kept a tally of the big donors post-Katrina. Oprah Winfrey, a Mississippi native, gave $1 million and FedExed a half million bottles of water. Microsoft's Bill Gates distributed $1.5 million among the Red Cross, the Baton Rouge Area Foundation, America's Second Harvest, and the NAACP. The principal owner of the Houston Texans, Robert McNair, gave $1 million to help Gulf South evacuees feel more comfortable in their new Lone Star State environment. It seemed that every musician who ever sang the lyric "New Orleans" held a benefit concert. Oil magnate T. Boone Pickens gave money and chartered planes in support of animal rescues.

Ben and Sarah Jaffe of Preservation Hall, after escaping New Orleans on August 30, eventually made their way to New York. Glad they had gotten banjoist Narvin Kimball, the old-time Preservation Hall veteran, evacuated to Baton Rouge, they decided to create the New Orleans Musicians Relief Fund. The idea behind the nonprofit organization was straightforward: purchase instruments for flooded-out New Orleans players so they could book gigs around the world. They ended up assisting more than 775 musicians to purchase keyboards, guitars, drums, brass instruments—and, in some cases, to pay rent. Among those they assisted were Bennie Pete, leader of the Hot 8 Brass Band; Benny Jones, leader of the Tremé Brass Band; and Doc Watson, leader of the Olympia Brass Band. "We did what we could," Sarah Jaffe recalled. "We just felt so bad for all those musicians. We tried to help Katrina victims, those who played, the best we could. Every day though, while in New York, we cried. We kept hearing bad story after bad story."[50] The worst news of all came on March 17, 2006, when they learned that Kimball had died in South Carolina, heartbroken he couldn't be in New Orleans.

VI

At midday on Monday, Brown began to realize that Katrina was not going according to the rules of the Florida hurricanes of 2004. It was on a scale never seen before: eventually, an astonishing 1.5 million people would register for FEMA assistance. Luminous clouds, a bright sun, blue skies with hundreds of skimmers flying about, and an invigorating breeze were supposed to follow hurricanes—not waves of scummy water full of heavy debris and six-year-olds carrying belongings in plastic bags over their heads, hoping not to drown. Along the Gulf Coast, the humidity, which had plagued the region more than usual that summer, continued to hang heavy.

At noon on Monday, Brown made an incredible announcement: he directed all emergency responders from outside the region to stay home, until specifically requested by local authorities. "The response to Hurricane Katrina must be well coordinated between federal, state, and local officials to most effectively protect life and property," he stated. "It is critical that fire and emergency departments across the country remain in their jurisdictions until such time as the affected states request assistance."[51] Unfortunately, many outfits that could have helped in the scramble to save lives listened to this overly cautious FEMA order. Others, who rushed to help, were stopped at gunpoint before reaching the disaster areas and told to go back home. National Guardsmen deployed to guard the approaches to New Orleans were in fact given specific orders to "keep emergency responders out." On *Meet the Press*, Jefferson Parish President Aaron Broussard said, "We had a thousand gallons of diesel fuel on a Coast Guard vessel docked in my parish. The Coast Guard said, 'Come and get the fuel right away.' When we got there with our trucks, they got the word. 'FEMA says don't give you the fuel.'" Broussard also charged FEMA with severing his parish's "emergency communication" lines.[52]

VII

While officials in Washington, D.C., and the Gulf South were grappling with myriad problems—or avoiding them altogether—dung-colored water from Lake Pontchartrain and the Gulf had been draining into New Orleans's Ninth Ward. By nighttime, the foul water was at least four feet deep. The Ninth Ward, scene of so much misery as the Katrina disaster unfolded, was divided into three sections, the names of which would become entwined with the emerging stories. If the overall area could be pictured as an E, then the backbone would be the Industrial Canal. The top prong would be the lake. The middle would be the Mississippi River–Gulf Outlet. The bottom prong would be the river itself. In that schematic, the area commonly called the Lower Ninth Ward was in the lower section formed by the prongs, south of the MRGO and north of the river. The upper section formed by the prongs of the E was New Orleans East. It was north of the MRGO and south of the lake. The area known as the Upper Ninth was to the west of the Industrial Canal (the backbone of the E). As of Monday morning, all were filling with water.

Both the Ninth Ward and the Lower Ninth Ward had once been the proud domains of Irish, Italians, and Germans. By the time Katrina hit, they were predominantly African American. Many of the clapboard houses, dating from the early or middle twentieth century, were in the "shotgun" style popular in the South, with two adjoining homes laid out front (living room) to back (kitchen and bedrooms). The Ninth Ward developed much of its defiant personality during the late era of segregation. But even with white oppression, the African-American residents built a close-knit community. It was a musical neighborhood where on any given night you could hear Delta blues, Dixieland, bebop, swing, or hiphop emanating from the simple, colorfully painted houses. People there respected musicians. The spirit of jazz cast a distinctly democratic, freewheeling aura over the Ninth Ward. Each section embraced a variety of working-class people—black and white—storekeepers, seamstresses, carpenters, Sheetrock finishers, dockworkers, and day laborers, as well as a broad range of professionals and artists.

Keith Calhoun had lived in the Ninth Ward all his life. He had been born at Charity Hospital and had grown up on Deleray Street. Back in 1955, the Ninth Ward was a middle-class black neighborhood. Segregation was the social order of the day, as white New Orleanians didn't want to live on the same street as blacks. So they had long since given them the Ninth Ward, in part a low-lying, marshy scrapland. Money was to be had in the Ninth Ward, however, by working the docks, joining the Longshoreman's Union, and earning good wages. At the Alabo Street Wharf, Calhoun's father made a living wage working for shipping companies like Cooper T. Smith and Lykes Brothers. In the eyes of young boys like Keith during the 1950s, dockworkers were heroic figures; one of those who loomed large was Dave Dennis, the gutsy president of the local Longshoreman's Union. "Guys like my dad lifted 200-pound sacks of coffee and cotton all day long," Calhoun recalled. "We all hung out at the wharfs, fishing for speckled trout, redfish, and catfish, sometimes crabbing, while our daddies brought home the bacon. A real community had been created."[53]

Calhoun and his wife, Chandra McCormick, became the premier social realist photographers of the Ninth Ward. Their neighborhood was a visual smorgasbord. Calhoun's vision of it was like John Dos Passos's "camera eye" in his famous U.S.A. trilogy. If you had a swooping camera panning across daily life in the Ninth Ward circa 1963, you would have seen a culturally vibrant neighborhood obsessed with wild music, hot food, external merriment, and a Caribbean flair for the exotic. Creole dishes were served at Miss Ross's Eatery, but the best meals were bought for seventy-five cents a plate directly out of neighbors' houses. There were no Popeye's or Church's fast food to contend with, just homemade fried chicken, shrimp étouffée, and okra gumbo. Live music abounded. The local favorite was Boogie Bill Webb, an irrepressible R&B and country blues prodigy. One family, the Barnets, had gospel tents set up in their backyard where, like some scene out of Eminem's *Eight Mile,* contestants would try to outdo one another in their singing. Because many of the African-American residents were middle class, and therefore better educated and better organized, it was not surprising that the Freedom Movement took root in the neighborhood as far back as 1896, when Homer Plessy refused to sit in a blacks-only train compartment headed to Covington, Louisiana. The Supreme Court verdict went against Plessy, and "separate but equal" remained the law of the land until 1954's famous *Brown v. Board of Education* reversal. But residents of the Ninth Ward were proud that they had a robust history of standing up against oppression.

This defiance became even more evident in 1960, when six-year-old Ruby Bridges became the first black pupil to integrate a school in the city. Overnight she became a symbol not only of the civil rights movement, but also of the Ninth Ward. John Steinbeck featured her in *Travels with Charley*, his classic road memoir, and Harvard psychologist Robert Coles analyzed her plight in *The Story of Ruby Bridges*. On her first day at William Frantz Elementary School, so many bigots were throwing tomatoes and eggs at her that she thought it was a Zulu parade. Four armed U.S. marshals had to accompany Ruby to class; she was in grave danger of being killed. Four years later Norman Rockwell painted *The Problem We All Live With*, a heart-wrenching portrait of six-year-old Ruby trying to enter school, accompanied by bodyguards.

The twenty-first-century Ninth Ward had more than its share of poverty; nevertheless, the residents didn't *feel* poor. Everybody knew everybody, and engaged in the neighborhood rituals: sharing big pots of red beans and rice or playing dominoes on old-time porches or walking to the "snowball" stand at dusk. They privately visited with ninety-eight-year-old Reverend Pappa Brown at New St. Matthew Baptist Church for wise counsel; attended Mother Washington Church (Holy Family) on Lamanche Street, where tambourines were rattled for salvation; and drank quarts of Budweiser at Junior's Bar, sitting in cozy shag-carpeted booths. The jukebox at this watering hole, just across the street from where the 17th Street levee broke, often played Little Willie John and Professor Longhair. As former resident Patricia J. Williams, a professor at Columbia University, pointed out, the Ninth Ward was an area "that has perhaps more African-American property owners than anyplace in Louisiana." Williams noted that in the midst of the flood following Katrina, she "heard a black woman on the radio describe how jarring it was to see the media describe her neighborhood as one driven by poverty and desperation. She was about to get her MBA, her brother already had his MBA, their extended family owned nine homes there, had insurance and owned cars in which they had fled for their lives."[54]

But there was that other side to the Ninth Ward. Nearly 40 percent of its residents did live below the poverty line, twice the statewide rate. Per capita income was only $10,300, less than half the U.S. average, according to the 2000 census. Many of the houses were raised two feet on cinder blocks, a precaution against flooding. The area of the Lower Ninth was 97 percent black. Crime in that part of the ward was off the charts. If you wanted to save anything, especially a car, in the Lower Ninth Ward, you practically needed to drape it in coiled barbed wire. But then there were the indelible Afro-Caribbean charms: many of the elegant street names were the same as those in the French Quarter, like Dauphine, Burgundy, and Royal.

After Katrina, Deborah Sontag wrote a heartbreaking series of *New York Times* articles on the Lower Ninth, all related to Delery Street. She reported on how the Calhoun-McCormick team lost much of their art in the flood. "They did not expect their living history of the Lower Ninth Ward to become actual history in their lifetime," Sontag wrote. "And they did not prepare for disaster. They did not digitize their negatives or create a secure storage system for their photographs. . . . When the hurricane destroyed their house at the corner of Chartres and Flood streets, they lost two thirds of their life's work."

Although much of the Lower Ninth Ward was seven feet below sea level, over a dozen blocks were actually *above* sea level. According to *USA Today*, most of the homes in that area were worth somewhere between $50,000 and $75,000 in the pre-Katrina real estate market.[55] These "high ground" homes suffered terrible wind damage, but they were standing and they were dry. The media reports that claimed that the Lower Ninth Ward was gone were incorrect. Weeks after the storm, the Bush administration had it wrong as well: Housing Secretary Alphonso Jackson rather offhandedly suggested not rebuilding the Lower Ninth on the grounds that it had washed away with the flood. But on that Monday, as Coast Guard helicopters flew above the neighborhood, Jackson seemed on point. It certainly looked gone forever, and FEMA was treating the Ninth Ward that way.

The problem the U.S. Coast Guard had with FEMA was that FEMA didn't know the local geography or place names or wards of New Orleans. They couldn't pronounce Tchoupitoulas (choppa-too-liss) Street, let alone spell it. They couldn't cross over the Crescent City Connection bridge because they thought it was a shuttle service to Houston. Without maps, it was hard to understand the 300 miles of levees around the coastal parishes. "The biggest challenge I saw for FEMA was having an appreciation for the nuances of New Orleans and the surrounding area," Jimmy Duckworth of the Coast Guard recalled. "The fact that the river doesn't necessarily run north-south was a big deal. The Port of Plaquemine, St. Bernard–New Orleans, Port of South Louisiana, they were often confused. I found myself and some of my fellow operators from New Orleans acting as translators for a lot of FEMA people. We helped them understand what it's going to take to do something, what are the logistics involved in getting from this point to that point."

VIII

The most famous Lower Ninth Ward resident at the time of Katrina was undoubtedly the seventy-seven-year-old Fats Domino. He had grown wealthy with hit songs such as "Blueberry Hill" and "Walking to New Orleans," but after gigs around the world he returned to his house on Marais Street, located just ten blocks from where he had grown up. Back then, in the Depression, Antoine Domino used to play piano for money, with a jovial style all his own. His very first record had been about cocaine and heroin in the Ninth Ward, an antidrug anthem titled "The Fat Man." The single brought Domino his initial success and his nickname. With a husky voice, Creole accent, pounding piano, and heavy backbeat, Fats Domino became a pioneer rock 'n' roll sensation.

As his legend grew, the self-reliant Domino stayed wedded to the Lower Ninth Ward. With his colorful neckties, bright pink Cadillac, and brood of children, Fats was the unofficial mayor of the Lower Ninth Ward. He lived in a modest yellow house with "FD" in front; inside the decor was all pink, lavender, and yellow. His favorite object, in fact, wasn't his piano but a "Cadillac couch," designed after the rear of a 1950s car. He appeared at the yearly New Orleans Jazz and Heritage Festival and occasionally at Mississippi Gulf Coast casinos, but stayed largely out of the limelight. At the time of Katrina, his international popularity was such that he could have sold out arenas from Dublin to Moscow; to Europeans he was rock 'n' roll royalty.

But Fats Domino didn't like traveling. He was a homebody. As Katrina approached Domino refused to evacuate because his wife, Rosemary, was sick, and he had five of his children, plus their families, with him at his compound. "He said to me, in that wonderful Southern accent of his, that no, he was staying on," said Charles Amann, a friend who called Domino on Sunday, August 28. "He had gone through the last one and he could go through this one."[56]

When the booming gusts ripped through the Lower Ninth Ward early on Monday, Domino got his family up to the attic in the nick of time. A headlong rush of dark water pushed into his house, and within minutes it rose to eight feet, slapping the bottoms of the first floor's chandeliers. Fats watched helplessly as his rock 'n' roll trophies wallowed in the muddy waters of the storm. A virtual shrine to New Orleans's music was destroyed on Marais Street; his famous "pink-walled room" was splattered with mud, his famous white piano churned into debris. Domino had sold more than 110 million records during his long career but the gold records that had lined his wall were plucked down and sucked into the muck. The yellow mansion, once the delight of the Lower Ninth, was gutted, although the basic structure survived. Someone, assuming that the family had been killed, spray-painted "R.I.P. Fats" on the side.

Meanwhile, just about a quarter mile away, Domino's Jourdan Avenue birthplace, located across from the Industrial Canal, was, as biographer Rick Coleman explained in *Blue Monday,* "torn in half, split in splinters . . . in jumbles."[57]

But reports of Domino's death were premature. Harbor Police, hearing that the legend was stuck, sent rescuers. Officer Earl Brown and Sergeant Steven Dorsey made sure all of Domino's extended family was hauled to safety. A *Times-Picayune* photographer was on the scene Monday afternoon to snap a shot of Domino, dressed in blue jeans and a blue striped silk shirt, being dropped off at the St. Claude Avenue Bridge. From there Domino and his family were taken to the Superdome, where, by then, close to 19,000 people were already stranded. Few recognized Fats Domino, lustrous in sweat, standing with his wife. Things were too crazy for the perks of celebrity to mean anything. All that Domino could do was wait for a bus to Baton Rouge. His high blood pressure was causing him grave problems. Around the country, news reports listed Domino among the missing. He was lost in plain sight along with everyone else at the Superdome.

IX

Thousands of Lower Ninth residents found themselves stuck in the floodwaters on Monday afternoon as a horrific stench filled everyone's nostrils. Those who had cell phones and managed to call for help only learned that none was coming. The Coast Guard was preoccupied with helicopter airlifts, although dozens of their boats were on the way to the region. The police and fire departments were as helpless as the residents. The only hope lay with the small fleet of Louisiana Wildlife and Fisheries boats, along with a growing flotilla of flat-bottomed boats brought to the scene by local sportsmen. There were fifty to seventy-five boats working the flooded regions of the Ninth Ward and St. Bernard Parish on Monday evening—and about fifty thousand people in need of rescue there.

One of those who needed saving was Diane Johnson, the sickle-cell anemia sufferer whom Reverend Willie Walker had tried to persuade to evacuate after church services at Noah's Ark on Sunday. Stuck at Tricou Street, Johnson and her husband, Daryl, thought they could weather Katrina's brutal force. Never did they expect mountains of floodwater to break into their single-story home. "The water was rising very fast," Johnson later recalled. "We made our way to the attic. All of our furniture was floating. We had nothin' to grab on to. We prayed for God to help us, we were in need. And we prayed that Reverend Walker and the members of our church were safe."[58]

Fortunately, Johnson, even in the panic of that violent moment, retrieved her medication from the bathroom cabinet. She was on Folit for her sickle-cell anemia, Coumadin for her bad heart. As the water in the house rose to five feet, and strange water-soaked floating objects started to rattle about in the living room, the family huddled together in the attic, holding hands and saying prayers. They brought a couple of paintings, and boxes of snack food with them. Within a couple of hours all their family heirlooms—except for those in the attic—were scuffled away. Three generations of precious scrapbooks and souvenirs became a ruined jumble. "She lost her motorized chair," Daryl later lamented. "And her asthma inhalers."

With temperatures in the nineties, and the humidity stifling, the Johnsons' attic felt more like a Hopi sweat lodge than a refuge from fetid water. Afraid they were going to die of heat exhaustion, they occasionally walked down the attic stairs into the toxic sludge, just to get a breath of the comparatively fresher air outside. The entire neighborhood, however, reeked of gasoline and sewage. Dead dogs floated by and all the porches were underwater. The Johnsons were scared, exhausted, beaten down, rotting in a mud hole. Chronically ill, Diane was dehydrated and in need of insulin for diabetes. With the local pharmacy gone with the waters, and hospitals much too far away, she was in dire straits. But a Wildlife and Fisheries boat came Tuesday and saved them. They were dropped off at the St. Claude Avenue Bridge. Diane had trouble walking without her Pronto chair. That machine was her legs, her baby. It broke her heart to leave it behind. "Last thing I did as we got rescued was I blessed myself," Johnson said. "And I waved good-bye to my favorite oak tree, across the street. It was still standing, praise the Lord. Eventually, we made it to the Superdome. Daryl got me a wheelchair, but I missed my chair. . . . But they tried to help me at the Superdome. They really did."[59]

X

The part of the Ninth Ward northeast of the French Quarter was known as Bywater, named in 1947 by a group of businessmen. World War II had stimulated some growth in New Orleans's economy, including the naval facility adjacent to the Industrial Canal commonly called the Port of Embarkation. This facility served as a Navy base and was home to the Panama Canal Commission. At night, the best Bywater jazz or funk in New Orleans could be found at Vaughn's Lounge, home turf of Kermit Ruffins and his Barbecue Swingers. A comical sign behind the bar read "No Whites Allowed" and the walls were lined with photographs of the Mardi Gras Indians who lived in the area. Nearby were B.J.'s and Sugar Park Tavern, which taken together with Vaughn's were known as the "Bar-muda Triangle." It was a poor, mostly black neighborhood that maintained a funky, bohemian air. Everybody in New Orleans knew Charmaine Neville, whose father, Charles, was the saxophonist for the Neville Brothers. She was a fixture around the neighborhood and its music lounges. She was a real homegirl, and *Time* magazine claimed she had the "best pair of lungs in New Orleans." The forty-nine-year-old Neville sometimes sang with her family's band but essentially she was a solo artist, singing "Sometimes I Feel Like a Motherless Child" and "St. James Infirmary Blues" at Snug Harbor or Tipitina's. She also had the distinction of being a mother of seven. She was living in Bywater when Katrina came bursting into town. Her son Damien, who lived in California, tried to persuade her to evacuate the Big Easy, to no avail. "I didn't want to leave my dog and cat," she said. "Plus my neighbor, Railroad Bill, was around to keep an eye on me."

Like many who stayed on, Charmaine Neville underestimated what high winds do to one's nerves. The rain beat at a slant on the windows, eventually popping them out, and then the water came pouring in. All of Bywater had a good four or five feet of water. Gagged by Katrina, she had to get out. Railroad Bill tied a gray rope around her waist and the two of them entered the flood zone, essentially swimming their way to Drew Elementary School on St. Claude Avenue, their wet clothes as heavy as lead. About 500 people had congregated at the school, huddled together in misery and discomfort. "It was hot, funky," Neville recalled. "People were sick, hungry. There were screaming babies, old people with parched lips, handicapped people in such agony. Never could I have imagined such a scene in America."

Neville became proactive. She and Railroad Bill waded back to her house, salvaged some food, and brought it back to the school. She urged others to go out as well and collect food from homes and stores before it spoiled. They would have a feast. A huge makeshift pot of beans was cooked up with every kind of meat you could imagine. They used a broken upside-down desk as the pot. Before long Railroad Bill was cooking up chicken and shrimp on a little grill. Neville felt that they had to feed the refugees, at least for Monday night; she hoped that come daylight FEMA or the Red Cross would arrive at Drew Elementary to take care of their needs. She kept a special eye on a sick baby and a dying old man, wondering if she would have to send them by raft, Huck Finn–style, to the Superdome. That evening a group of the Bywater evacuees with flashlights started walking in the dark, through the chest-deep water, to the Superdome. They would let authorities know that hundreds were stranded at Drew. "It was terribly hot in the school, suffocating, so after making sure everybody was fed, at around 2 A.M., I went on the roof," Neville recalled. "I was tired and exhausted. It was cooler up there so I just lay down and looked at the sky. I fell asleep."

That's when Charmaine Neville's life changed forever. A drowsiness came over her. She dozed off in a purplish glow. Suddenly her body must have sensed danger because she jerked awake. Standing over her was a large man brandishing a knife. "I was scared to death," Neville recalled. "I can't really describe anything about him but his white teeth in his face. It was too dark and he had the knife at my throat. He threatened to kill me, to slit my throat and toss me into the floodwaters if I didn't cooperate. So I was violated, raped. I just did what I needed to do to survive."

From the broken-bottle rooftop of Drew, where Neville was raped, you got a panoramic view of the blasted Ninth Ward and, that night, you could see the stars. Still, the foamy water was impossible for boats to navigate under such conditions, except for the rubber Zodiac rafts. Stray dogs were now howling like hoarse coyotes in the fullness of the moment. Occasionally a helicopter flew over the floodwater, shining a searchlight downward. The vastness of the flood was hard to fathom; the scale of the ruin and destruction was mindboggling. The situation was a prowler's delight. In Mississippi the analogy heard over and over again was that of a nuclear bomb. You felt naked and exposed and vulnerable in the barren beachfront towns of Mississippi. But the New Orleans night was suffocating and cruel, as if some sci-fi monster had poured an ocean of black slime down all the streets, gutting architectural gems, transforming the sturdy gilded-age buildings into something akin to a rotting Hollywood movie set, flimsy and ready to topple. A common sight in the Ninth Ward was staircases that led to nowhere, the buildings around them long gone. A few red emergency lights could be seen from atop buildings downtown, but not even a dull lamp radiated in the Bywater. There was no jack-o'-lantern effect (one generator-lit house every block or so). Just darkness. Every few minutes a scream or siren echoed in the night from somewhere around the St. Claude Avenue Bridge. But these were the exceptions. It was the deadly silence, the lack of life, modernity tossed to the wind, that hovered over New Orleans, as the citizens who stayed just flailed about as if trapped in a huge spiderweb with nowhere to go. As Jean-Paul Sartre said, "No exit."

The rapist fled into the night. Charmaine was left sitting on the roof to cry. All she could think about was getting out of Drew Elementary, wading back to her house, away from the outlaw aggressor, and hiding in the recesses of her closet, away from the world. But then there was reality, such as it was on the day Katrina struck. Being alone on the empty, water-filled streets was terrifying. Traumatized in the darkness, she claimed she saw an alligator thrashing about, pulling an old man into the murky water. She said, in fact, she saw many alligators. It's a doubtful claim. Traumatized, her mind was playing tricks on her. Evil shadows were lurking all over Bywater. You'd see alligators too if you had been raped on a dark roof, then waded through three feet of sewage-infested water with dogs stuck on roofs and babies wailing. Eventually she made it home. When it was too hot inside her house, she sat on her front porch, crying and dozing intermittently. "The next day when I came to Drew everybody in the school was mad," Neville recalled. "They had heard what had happened to me. Other women were attacked. They told me. The police are lying when they try to downplay the rapes."[60]

XI

At 6 P.M. on Monday, George Bush telephoned Kathleen Blanco in Baton Rouge. Shell-shocked, disconcerted, and running on no sleep, the governor was insistent, telling the President that Katrina had devastated much of Louisiana. She was near tears. "We need your help," she pleaded. "We need everything you've got."[61] That was a leading phrase for a notably incurious president to absorb: "everything you've got." The open-endedness of Blanco's request must have told the President that there was a tremendous leadership problem at the governor's mansion in Louisiana. Emphatic generalities didn't tell the President *what* supplies and services were needed. Governor Blanco should have rattled off specifics like water, nonperishable food, medical teams, buses, boats, and helicopters. Perhaps Blanco's own vague knowledge about what actual assets the federal government possessed may have prevented her from being specific. Perhaps she was right to be wary of President Bush: behind her back, he was trying to federalize her Louisiana National Guard. "You know, I asked for help, whatever help you can give me," she later snapped about her inability to be specific. "If somebody asks me for help, and—I'll say, 'Okay,' well, I can do this, this, and this. What do you need?' But nobody ever told me the kinds of things they could give me!" Eventually, in an interview with CNN's Anderson Cooper, Blanco admitted that she should have "screamed louder."[62]

With no detailed request from either Blanco or Brown, Bush didn't pursue the matter actively enough. Louisiana was a notorious black hole for pork-barrel funds. He wasn't going to write a blank check. He also wouldn't be inclined to make up for Blanco's inexperience; if she was floundering, he didn't leap to save her reputation. As for "Brownie," Bush trusted him wholeheartedly. While the President's hesitation may have been understandable from a bottom-line, CEO perspective, it was a mistake on his part not to take action, even if action would entail letting Blanco take credit for positive results. Great presidents in a time of crisis govern by instinct, bypassing the limitations of novice governors. But given Governor Blanco's vagueness, President Bush, somewhat understandably, demanded specifics. What he failed to understand was that FEMA should have been providing them.

President Bush only made a mental note to look into the financial obligation of the federal response. But he was dragging his feet. Certainly he discussed the matter with top White House advisors, like Chief of Staff Andrew Card, Deputy Chief of Staff Joe Hagin, and Counselor Dan

Bartlett. FEMA's Michael Brown told the White House he had everything under control. Why not just trust Brownie? A classic federal-versus-state showdown was occurring, and President Bush, one of the more stubborn American presidents, wasn't going to take orders from a rookie like Kathleen Blanco. After all, he had been governor of Texas, and he knew that Blanco had allowed Louisiana to be terribly ill-prepared. During his time in the governor's mansion, Austin had geared up for the Big One hitting Galveston or Corpus Christi; he had made sure Texas had enough food and water to take care of hurricane survivors for months. As a tough, resolute president, Bush didn't respect Blanco's uninformed emotionalism. A Bush-versus-Blanco square-off—"a dance," as Nagin called it—ensued. It was childish and unhelpful. Petty squabbles and deliberate miscommunication needed to be shelved for the good of the people of the Gulf South. FEMA saw the squabbling through a different lens, as Blanco versus Nagin. "My biggest regret," Brown later said, "is not getting the governor [of Louisiana] and the mayor of New Orleans to sit down and iron out their differences."[63]

As fate would have it, however, Brown was in no position to point fingers that Monday. His foolish e-mail exchange that morning certainly belongs in Bill O'Reilly's feature "The Most Ridiculous Item of the Day." Brown, busy doing TV interviews Monday morning in Washington, before leaving for Baton Rouge, received an e-mail compliment from a female FEMA colleague saying, "You look fabulous and I'm not talking about makeup." Flattered, Brown fired back, "I got it at Nordstrom's. E-mail [FEMA spokeswoman LeaAnne McBride] and make sure she knows! Are you proud of me? Can I quit now? Can I go home?" This fashion chitchat in the midst of the mayhem was reported by the *Times-Picayune* under the headline, "FEMA Chief Dawdled, E-mails Show; As N.O. Suffered, He Made Small Talk."[64]

At 7 P.M. Monday, Michael Brown received an urgent telephone call from Marty Bahamonde, his FEMA representative on the scene, who had just toured the New Orleans region in a helicopter to assess the damage. "I explained what I saw," Bahamonde recalled. "And then provided my analysis of what I believed to be the most critical issues we were facing." He listed his dominant concerns:

* Ground transportation into the city was virtually nonexistent because of the massive flooding.
* Search and rescue missions were critical as thousands of people stood on roofs or balconies in flooded neighborhoods.
* Supplying commodities would be a challenge as more and more people were headed to the Superdome to escape the floodwaters and food and water supplies were already very short at the Superdome.
* Medical care at the Superdome was critical because the staff there had run out of oxygen for special-needs patients and more and more people needed medical attention.
* Housing an entire city of people would be a major issue as approximately 80 percent of the city was underwater to varying degrees and many areas were completely destroyed.[65]

Bahamonde's telephone call had a sobering affect on Brown. He was jarred, genuinely concerned, almost at a loss for words. For the first time he fully understood that the entire future of New Orleans was at stake. "I was beginning to realize," Brown recounted, "that things were going to hell in a handbasket."[66] Under the crisis circumstances, and realizing that he was outmatched by Katrina's devastation, Brown did exactly the right thing. Within the hour, he called his boss, Michael Chertoff, pleading for help. Drowning in his own inabilities and convinced that Governor Blanco was "dysfunctional," Brown had the good sense to seek guidance from a superior.[67] It was the first time he and Chertoff had spoken together all day. Once again, the communications breakdown within FEMA was proving to be extremely detrimental. Dotting i's and crossing t's has a seminal place in our highly legalistic society, but not when fellow citizens are in desperate peril. At this juncture, as Dwight Eisenhower used to say, "People come before paper." In point of fact, the ultimate responsibility for the lackluster federal response to Katrina lay entirely with Chertoff, the secretary of Homeland Security.

Under rules instituted in January 2005, Homeland Secretary was in charge of *all* major disasters, whether from international terrorism, Mother Nature, or infrastructure collapse. Until Chertoff designated it "an incident of national significance," and appointed someone (presumably the FEMA director in the case of hurricanes) as the "principal federal official," relief would be halting at best. Without that designation, Brown could not legally take charge, giving orders to local and state officials and overseeing deployment of National Guard and other U.S. military personnel. "I am having a horrible time," Brown admitted to Chertoff in a telephone conversation on Monday. "I can't get a unified command established."[68]

A stronger personality than Michael Brown might have seized command anyway. But even Brown's GOP allies knew he was weak-kneed. The question that still haunts the events of Monday, August 29, was not, however, *why* Michael Brown needed post-Katrina direction and so much instruction from his boss. The important question was *why* Chertoff was so callous, both to Brown's specific relief needs and to the apocalyptic needs of the entire Gulf Coast region. Brown tried to maneuver around Chertoff, to appeal directly to President Bush, but it was hard to get through to the White House.

With his nonplussed countenance, sunken cheeks, oppressive quietude, and closely cropped beard, Chertoff exuded the contained anger of a haggard academic denied tenure. Although only fifty-one, he had accomplished a lot since his childhood in the blue-collar city of Elizabeth, New Jersey. His sterling résumé, in fact, qualified him to be a Supreme Court justice or U.S. Attorney General. His trajectory was classic Eastern Establishment fare: Harvard University; Harvard Law; admittance to the bar in the District of Columbia, New York, and New Jersey. He clerked for U.S. Supreme Court Justice William J. Brennan Jr. By the time Bill Clinton was in the White House, Chertoff had developed a reputation as a first-rate federal appeals judge, a former gutsy prosecutor who didn't flinch when going after terrorists, mobsters, or political bosses. During any of the big U.S. political moments from 1992 to 2005, Chertoff was somewhere in the mix. He was, for example, both Senate Republican Whitewater Counsel and assistant to Attorney General John Ashcroft before the 9/11 terrorist attacks. "My style was always very straightforward and simple," he told the *Washington Post*. "Here's what I know. Here's what I don't know. If I've made a mistake, I'll admit it."

Clearly Chertoff didn't just make a mistake during the first days of Katrina—he did virtually nothing at all, which was by far the greater sin. With the hurricane approaching Louisiana and Mississippi, Chertoff never even went to his office, staying at home for the crucial forty-eight hours before landfall. Most astonishing of all, as Katrina ravaged nearly 29,000 square miles of America on Monday, Chertoff didn't even speak to Brown until 8 P.M. When CNN, Fox News, ABC News, and the rest started reporting the horrific flooding in New Orleans due to the levee breaks, Chertoff scoffed, dismissing media reports of human suffering as melodrama. With a cavalier wave of the hand, according to the *Washington Post,* Chertoff downplayed the bleak reports as "rumored or exaggerated." Worse yet, Chertoff insisted that Brown and FEMA as a whole were doing an "excellent" job.[69] Evan Thomas of *Newsweek* was closer to the mark when in his seminal article "How Bush Blew It," he declared that FEMA was "not up to the job."[70]

Chertoff's inaction cost lives. FEMA had been brought into the gargantuan Department of Homeland Security after 9/11; now it was clear somebody needed to pull it out again. It was a huge black eye for Homeland Security. The Harvard prosecutor performed just as poorly as the Oklahoman—even worse. Brown, to his credit, kept trying to get the Bush administration's full attention. Chertoff had assumed his important cabinet position with big talk about keeping Americans safe from man-made and natural disasters. He was a principal engineer of the USA Patriot Act and wrote an article in the neoconservative publication *The Weekly Standard* full of bravado about fighting the war on terror "beyond case-by-case." He fancied himself an intellectual, but one who understood trench warfare. President Bush, in selecting Chertoff to replace Tom Ridge, said that "Mike has shown a deep commitment to the cause of justice and unwavering determination to protect the American people." His determination to protect the American people did not seem to extend to those who lived in Gulf towns like Grand Isle, Louisiana; Ocean Springs, Mississippi; or Dauphin Island, Alabama. The one quality, in fact, not evident in Chertoff's handling of Katrina was caring about what the storm inflicted. While fellow citizens were dying, screaming for help, clutching chunks of floating wood and palm fronds trying to stay alive, Chertoff, the one

American who could have helped the most, turned a casual, cold, indifferent eye to their plight.

When Brown put through his 8 P.M. telephone calls on that Monday, Chertoff was at his home resting. Chertoff's spokesman later claimed that the Homeland Security secretary "was hobbled by a lack of specific information" regarding Katrina on Monday night.[71] That clumsy contrivance presumed that Chertoff was discounting or ignoring the reports from Brown, who was then in the EOC in Baton Rouge, or those reports streaming in from the affected area that were all over various FEMA offices. Air Force aerial images of the swamped Gulf Coast were arriving with increasing frequency at EOC, each showing an obliterated landscape, with water towers and refineries among the only recognizable landmarks in St. Bernard and Plaquemines parishes. As Homeland Security chief, Chertoff had the most effective communications network of any cabinet office at his disposal, including the resources of the top brass in the Pentagon. He didn't use it. If nothing else, there were a growing number of images on television. But he seemed oblivious to Barbour's "nuclear devastation" metaphor, and allowed the Great Deluge to run its course willy-nilly. "What happened was Homeland Security was geared toward terrorism," Louisiana Attorney General Charles Foti said. "They knew that FEMA could cope with a hurricane. Okay. Maybe. But the Bush administration refused to come to grips with the flood. Wind damage was not water. They just didn't get that. In New Orleans, house after house, block after block, mile after mile was disappearing."[72]

On Monday evening, Bill O'Reilly opened his Fox News program with a stunning revelation: "At least forty thousand homes just east of New Orleans—*forty thousand*—have been destroyed."[73] He was referring to the flooding of St. Bernard Parish. On CNN, Paula Zahn spoke live to a woman who reported that on the Mississippi coast, "there are like eighteen wheelers on top of cars and homes in the middle of the streets. And there's people wandering down the streets with nowhere to go, homeless. They've got maybe a bag over their shoulder, and they're all in the middle of the streets, with nowhere to go. And the homes, houses and boats and cars are just . . . debris is just everywhere. It's just . . . it's very catastrophic down here."[74] Although many media reports on Monday morning had been tinged with relief that "it could've been worse," by nighttime the real situation was becoming apparent. People like Tony Zumbado of NBC News were on hand, supplying videotape evidence that the flooding in metropolitan New Orleans was of unprecedented proportions. It was inconceivable that with all the warnings raised before Katrina struck and the reports that filtered in (even through the general media) within hours afterward, Michael Chertoff could be unaware that "an incident of national significance" had indeed occurred that morning along the Gulf Coast. He instead waited for thirty-six hours.[75]

Regardless of the telephone call between Brown and Chertoff, the only productive action taken by FEMA that Monday was in the form of a straightforward memo from Brown. It formally requested Chertoff to make 1,000 Department of Homeland Security employees available, allowing them two days to report. Two thousand more were requested within seven days. In any other business, that might seem reasonable, but in response to what even Brown was calling in the memo a "near catastrophic event"[76] employees should have been at the ready within hours, not days. Wal-Mart and American Airlines were there in twenty-four hours. For those who needed them, every minute counted.

Meanwhile, Brown continued to send goofy e-mails all week long, exposing his vanity and embarrassing FEMA while the trapped were dying. On his "rescue and relief" watch, Brown found time to muse about life's frivolities:

- August 29 to FEMA Deputy Director of Public Affairs Cindy Taylor: "If you'll look at my lovely FEMA attire you'll really vomit. I am a fashion god."
- August 29 to Taylor: "Can I quit now? Can I come home?"
- August 30 to assistant Tillie James: "Do you know of anyone who dog-sits? If you know of any responsible kids, let me know."
- August 31 to Marty Bahamonde, still FEMA's only staffer in New Orleans: "Thanks for the update. Anything specific I need to do or tweak?"
- September 2 to a friend: "I'm trapped now, please rescue me."
- September 6 to [Sharon] Worthy (after her e-mail about fast-food choices during a trip to Florida): "Order a #2, tater tots, large diet cherry limeade."[77]

XII

FEMA wasn't the only outfit missing from New Orleans. Katrina had cut a path of destruction almost as long as the distance from Chicago to New York. Yet oddly missing from the zone was the American Red Cross. Certainly it opened shelters in Houston and Baton Rouge, cities away from the damage, and these shelters were well run. In Louisiana alone, it started 76 shelters to house 18,000 displaced people. But when word of the levee breaches hit, the sympathetic American Red Cross President Marsha Evans, who was in San Diego preparing to fly to China, claimed she would have let volunteers into New Orleans if Homeland Security hadn't stopped her. "Once the levees broke," she said, "they didn't want us in the bowl, they didn't want us to set up Red Cross stands at the Superdome or later, the Convention Center. They were trying to get people out of New Orleans." Throughout the week Evans grew frustrated by the bureaucratic way the state of Louisiana and Homeland Security were dealing with the biggest natural disaster in recent American history. But on Monday, the Red Cross was already hosting approximately 45,000 Katrina victims in shelters, 250 of which were in devastated areas. Throughout America 700 Red Cross shelters were eventually established. "This is our largest mobilization in the history of the organization," a spokesperson said. "We are focused on providing the most elemental essentials . . . food, shelter and water."[78] An astonishing 63,000 Red Cross volunteers and staff were mobilized to help the decommissioned Gulf South, a virtual army of Clara Bartons serving 7 million meals in two weeks. The Red Cross used mobile kitchens like *The Spirit of America*, a fifty-three-foot trailer with the capacity to serve 30,000 hot meals daily, and it collaborated well with faith-based groups like the Southern Baptist Convention, the Church of Scientology, and Catholic Charities. With FEMA getting hammered in the media, the Red Cross, sensing it might also be vulnerable to criticism, quickly launched a PR blitz, posting photographs by Gene Dailey on the Web showing Slidell children smiling after receiving a box lunch and smiling Baton Rouge girls handing out cheese sandwiches.[79]

While FEMA was certainly a flop, and the Red Cross was neutered by its questionable policy of not offering any direct assistance to New Orleans, ordinary Americans filled the void. For instance, seven-year-old Dan Noonan Day of Cedar Falls, Iowa, opened a lemonade stand and gave his profits to the Red Cross disaster relief fund. Californian David Perez, disgusted by FEMA's lackadaisical response, chartered his own Boeing 737 and airlifted in supplies to the Gulf South. A magazine called *The Razor's Edge* had women shave their heads like Sinead O'Connor and raffled off their tresses, fetching up to five hundred dollars per scalp.[80] The Aerolite Meteorites Society, a group of meteorite hunters and collectors, auctioned off some of their specimens, raising $12,437 for the American Red Cross.[81] Even celebrities like cyclist Lance Armstrong got in on the action. He donated $500,000 for cancer patients whose treatment might be dangerously delayed by Katrina's aftereffects. "It just seems like help was too late to come there," he said. "If you've started treatment and you miss a week or two weeks, it's potentially fatal."[82] Meanwhile, Marsha Evans defended the Red Cross's decision not to go into New Orleans. "The last thing we should have done was send out Red Cross volunteers into New Orleans to be victims," Evans said. "With that said, I wish we could have helped the people of New Orleans better." (Under extreme pressure from her board of governors, she would resign from the Red Cross on December 13, 2005.)

XIII

Many of those who were rescued from the flooded Ninth Ward were dropped off at the rusty St. Claude Avenue Bridge, which spans the Industrial Canal between the Upper Ninth Ward and the Lower Ninth Ward. The Red Cross was, of course, nowhere to be found. Part of the bridge was submerged in floodwater, making it an excellent dock for flat-bottom boats. Peering up at rescue workers and evacuees was a large white statue of the Virgin Mary, the water level just below her carved eyes. Jim Sohr was waiting on the bridge, looking for a friend in the boatloads of people who were dropped off, frail and shaking and clutching a few belongings. "When you watch the people get off the boats," Sohr said, "their faces have an unforgettable expression—they've been saved, but now what?"[83] The only thing to do was walk to the Superdome, as best they could, in a mad scramble. They had heard that the city was offering refuge there. As noted, all the RTA buses and yellow school buses were submerged in water.

Those who did make it to the Superdome were in for a rude discovery. The power had failed in the midst of the hurricane, and the sports arena had only dim light provided by auxiliary generators. Temperatures in the Dome heated up to more than ninety degrees, and the structure didn't cool with the setting sun. The air inside was moist, hot, and fetid. People did their best to maintain their composure. Water supplies were running out, as was food. You could smell revenge. The National Guard was getting edgy, as were the desperate evacuees, gesticulating wildly and pointing at those to blame. New Orleans's Homeland Security representative, Terry Ebbert, startled by the sheer mass of looting, dictated that the "cockroaches" needed to be dealt with.[84]

Between the many people who somehow walked to the Superdome, and those who were dropped off there from boats, the population in the arena grew from about 10,000 during the hurricane to 25,000 on Monday night. Managers had no choice but to lock the doors and stop admitting newcomers. Without having heard from Mayor Nagin—even though he was just next door in the Hyatt—or from anyone at FEMA, the managers couldn't offer any advice or support for those who continued to arrive from the Ninth Ward, Central City, and other neighborhoods. The new arrivals were desperate for shelter and water and medical attention. But they were greeted by National Guardsmen, rifles in hand, enforcing the lockout. "The first wave of folks that came to the Superdome were more docile," Gordon Russell of the *Times-Picayune* recalled. "The second wave that had arrived after the storm were more angry. While they were waiting for buses, some of them had looted liquor stores. There was a good deal of drunkenness. I knew it was going south."[85]

Turned away, survivors who were suddenly homeless looked for that most basic of requirements: dry shelters and scanty rations. Someone announced that the Morial Convention Center, about nine blocks away, was open as a refuge. It wasn't. Still, the bogus rumor spread like wildfire. A skeleton crew of about forty managers and maintenance people were inside the complex—it was almost a mile long, consisting of rooms and main halls that had grown along with New Orleans's robust convention business. The workmen were only on hand to see the facility through the storm. They had no way to provide for the hurricane victims who were streaming toward the building. With the doors locked, they tried to turn people away, but there was no controlling those who were frightened—or perhaps unleashed—by what the hurricane had done to their city. Someone found a chair and used it to break through the glass door at the front entrance. With that, the mob took over the convention center. There were no weapons searches or National Guardsmen, as at the Superdome, and so it was a cauldron from the first. For the time being, though, with the media trucks staged at the Superdome, the volatile Convention Center was hidden from the rest of the world. Looters and the Coast Guard rescuers were the stories of the moment.

Unknown to the media that week, there was a scatological phenomenon among some looters called the "big dump," defecating on property as a mark of empowerment. With the Convention Center quickly devolving into a squatters' chaos zone, many looters decided to commandeer businesses. Besides stealing, groups of angry looters would defecate in cash registers, on bartops, and around aisles. In Mulate's, a world-famous Cajun restaurant across from the Convention Center, looters "big dumped" in the deep fryers. At a Coast Guard station that was broken into, hanging uniforms were used as toilet paper. Lamar Montgomery, owner of a Shell station at Lee Circle for thirty years, had evacuated, going to northern Mississippi. When he came back to New Orleans, not only were his mini-market and auto-supply shelves looted, but the intruders had "big dumped" in his refrigerator numerous times. "They behaved like animals," he said. "They were making a statement, shit on you."[86]

The rest of the city was not much better. Shots rang out, which was not in itself a surprise. Gun ownership was high in Louisiana, and New Orleans in particular was a place where gun violence was common. On the first night after the hurricane, rumors spread that some looters were taking shots at Coast Guard rescuers, whether in boats or helicopters. According to the Coast Guard's Jimmy Duckworth, "Safety of our crews was the overriding imperative. I don't want to call somebody's mom or dad and say, 'He was killed while saving people off of the rooftops and he was killed by a sniper.' There was one point where we needed to know about the St. Claude Bridge and I called the St. Claude Industrial Canal lock master, Michael O'Dowd—I've known him for years. O'Dowd stayed in the lock house during the storm and I had one of my lieutenants call him because the captain wanted to know if the St. Claude Bridge would work. The bridge is at the Industrial Canal—it's a major choke point—and we needed to know whether or not we could get the interharbor navigation canal or Industrial Canal open. So we asked O'Dowd to open up the bridge. He called us right back and said, 'I shut the bridge. As soon as I started opening the bridge, they started shooting at me.' I know Breaux Brothers, contractors on my watch, were working on the Florida Avenue Bridge and they were getting sniped at. They're contractors trying to fix the bridge and they're getting shot at and their private security that they had contracted with to take care of them wound up shooting at least two of the perpetrators. So with this in mind, I've got crews in the field and I've got air crews in the field and just at any minute you were waiting for the phone call that said, 'Someone went down' or 'Our boat crews got shot.' It never happened. Nobody got shot, but there was shooting going on all over."[87]

Officials couldn't be sure whether the shots they heard were the work of snipers or stranded residents trying to attract the attention of rescuers. That night, officials couldn't be sure of anything in southeastern Louisiana. No one could. Storm victims were sobbing for breath. Whatever was certain on Sunday was all just an ugly blur on Monday. They were reduced to human luggage looking for a cargo bay somewhere outside of the bowl. And Chertoff sat idly in Washington, D.C. At least the Navy was sending four ships to the Gulf South carrying water and food and medical aid to the stricken region.

For a public official who liked to boast of candor, Secretary Chertoff didn't inform the American public until January 24, 2006, that on August 28—the day before Katrina hit—the Department of Homeland Security had compiled a forty-page "fast analysis report." The report, which was e-mailed to the White House at 1:47 A.M. on August 29, essentially predicted the Hurricane Pam doomsday scenario. The report even predicted

60,000 deaths. "The potential for severe storm surge to overwhelm Lake Ponchartrain levees is the greatest concern for New Orleans," the report said. "Any storm rated Category 4 on the Saffir-Simpson (hurricane) scale will likely lead to severe flooding and/or levee breaching. This could leave the New Orleans metro area submerged for weeks or months." Yet, as Bill Walsh of the *Times-Picayune* pointed out when he broke the story of the "fast analysis report," President Bush on ABC's *Good Morning America* just a few days after Katrina hit, said, "I don't think anyone anticipated the breach of levees."[88]

Holed up at the Grand Palace Hotel, the Ivory Clark clan had survived Katrina, peeking out their window as cars were blown off the top level of a nearby parking garage. Everybody on their floor was listening to WWL as people in Louisiana's crippled parishes telephoned in their litany of Katrina woes. Food was in short supply after the storm, so Ivory volunteered to go to a Winn-Dixie and "commandeer" water and food. He was trying to help suffering people, and the idea that he was *looting* never crossed his mind. What he saw on the streets of New Orleans was awful: rising floodwaters everywhere. When he got back to his room, distributing food to the nearly two hundred people in the hotel, his wife told him that a band of thugs had entered the Grand Palace Hotel, demanding jewelry and kicking people out of rooms. With his children now in danger, Ivory decided that come Tuesday morning he would first take his aunt to Charity Hospital, where her asthma could be treated. Then they would head to the Superdome, which he assumed would be a safe haven, with plenty of food and friends.

"We heard FEMA had five hundred buses at the Superdome," Clark explained in defense of his planned evacuation of the Grand Palace Hotel. "And I figured they had plenty of medical assistance teams over there to help folks like us out."[89]

Chapter Eight

WATER RISING

Fill my days with circulating rhythm
Where they spill, I will spill with them
Dip my bucket in the running stream
Try to go with it, whatever that means.

—Robert Hunter, "Circulate the Rhythm"

I

A SENSE OF DREAD emanated from Governor Kathleen Blanco early Tuesday morning, as she climbed into a Blackhawk helicopter at the state police helipad in Baton Rouge. The sounds of the hurricane, the detonations of thunder and the unnerving whirr of rattling windows were over. There was only fatigue in the air, and little puddles of water on the runway. A touch of grime coated the helicopter. A deflated attitude could even be seen on the face of the Louisiana National Guard pilot. It was not the time to crack a smile. Not in Louisiana, or at least nowhere south of Alexandria. With Blanco were three suddenly high-profile national figures: FEMA Director Mike Brown and Louisiana's two U.S. senators, Mary Landrieu and David Vitter. All over America the morning TV shows were whipping up passions about the dispossessed clinging to rooftops, the unhinged looters lurking down empty boulevards, and the jerry-built levees. If only the U.S. Army

Corps of Engineers had constructed the levees with the same exacting standards it implemented for dams, New Orleans might have escaped the deluge. With the whole world watching, Governor Blanco, along with her guests, was now going to get an aerial view of the Great Deluge for the first time. "By the time, the winds died down enough for all of us to fly on Monday, it was too dark," Blanco recalled. "We couldn't safely go up in a helicopter. But we sent the state police out and were in touch with the Coast Guard and National Guard. Tuesday morning was my first time up."[1]

Just two Saturdays before Katrina, Governor Blanco and her husband, Coach, had taken a helicopter ride from Lafayette to Venice and crossed Camanada Bay. "Look, Kathleen, I want to show you something," Coach had said. "You see those marks in the water? Look down. Those are canals. You don't understand, since we were down here last, look down to the south and look to the north. If a storm were to come up here, to Grand Isle, there's nothing to stop it. Look, there's no wetlands left!"[2]

The Blancos circled the area a few times in utter disbelief, stupefied by the extent of the coastal erosion. It was one thing to read about the phenomenon in *America's Wetland: Louisiana's Vanishing Coast* or talk with the director of the Coalition to Restore Coastal Louisiana, but to see so much marshland—marshland once so familiar to locals—just disappear in a year was alarming. Due to coastal erosion, the seawaters of the Gulf of Mexico would soon be slapping up against New Orleans's hurricane protection levees. Grabbing a BlackBerry from her assistant Sidney Coffee, Governor Blanco fired off an e-mail to Anne Williamson, secretary of the Department of Social Services: "I'm looking where the oil field canals were built and there's water on both sides. Remind me. I think it would be very important to get the President to come see this, so he can understand for himself what our challenges are."[3]

On Tuesday, August 30, Blanco was back in a helicopter, headed for the Hyatt-Superdome area in New Orleans. It was too late for President Bush to fly over Louisiana's wetlands—the wetlands were underwater. Peering out the window with Brown, Landrieu, and Vitter, all Blanco could do was sigh, and shake her head and cry. Humble tract houses, strip malls, drive-through banks—suburban structures of all kinds—were inundated with water. Senator Landrieu grew ill when she saw the pervasive damage, calling her trip "a helicopter ride from hell." She had spent the storm at the

Embassy Suites Hotel in Baton Rouge, shuttling to and from the EOC. Landrieu, a mother of two adopted children, and her husband, Frank, had boarded up their family fishing camp in Slidell the Sunday before Katrina hit. Gazing out the helicopter window, she knew it was probably gone. "Still, you want to see it with your own eyes," she said. "Only then do you truly believe."[4]

The first stop on the helicopter tour was Jefferson Parish, to meet Aaron Broussard and discuss the abandoned pump stations. Governor Blanco had agreed to rendezvous with him on Williams Boulevard in Kenner. "We landed on I-10," Landrieu recalled. "I was under the impression that we were going straight to the Superdome. But the governor insisted on meeting Broussard in Kenner. When we arrived at city hall, we were told Broussard was on the west bank."[5]

Somehow the communication between Blanco and Broussard (or their staffs) had gotten cross-wired. Philip L. Capitano, the angry, cantankerous mayor of Kenner, did meet with the delegation, demanding National Guardsmen, pallets of water, and debris removal equipment. Capitano fumed that the sheriff was useless and that Louis Armstrong International Airport, a prime economic engine in his city, had actually contributed to the flooding, because it had been constructed on a slope, allowing water to roll down into neighborhoods. "It was obvious to me that they weren't prepared to handle a storm of this magnitude," Capitano recalled. "They were in awe of it, totally confused, unsure what direction to go in."[6]

From Michael Brown's perspective, something wasn't right in Louisiana. Everything was off kilter. Mayor Capitano, for example, was operating with raw emotion, instead of cool pragmatism. Concerned, Brown caught Landrieu's attention and called her over for a private word. "Do you think everything is going to be like this?" he asked. "Because if it's like this everywhere, we're going to be in a lot of trouble." The PR-wise senator knew that having the head of FEMA bad-mouth Louisiana wasn't going to be helpful. She reassured him that Kenner was a sui generis situation, that most mayors *did* work in tandem with their sheriffs. "I hope so," Brown said. "I hope so."

The delegation climbed back into the Blackhawk. Ten minutes later, they landed on the helicopter pad at the Superdome and headed to a briefing with Mayor Ray Nagin, who was full of stories about how when Katrina hit, the Hyatt windows had shattered and curtains had blown out into the wind. The immediate future of New Orleans was at stake. Everybody was extremely cordial, Nagin more poised and collected than the rumor mill had led Blanco to believe. He seemed in control. The police superintendent, Eddie Compass, however, was in a bad way, trembling, his eyes glazed, mumbling to himself. Based at City Hall with Terry Ebbert, Compass had been working harder than any other NOPD officer since Katrina formed in the Gulf. But he was not working well. For a trained investigator, Compass was off his game, believing too many false stories that were floating around town. Instead of discussing specifics, he would say things like "All my men have left town" or "Rapes are happening in the hospitals." He kept whispering to the delegation that "the levees broke"—a fact they already knew. The shocked gleam in his eye told Landrieu, Brown, and Vitter to take his reckless statements with a few grains of salt. "I was working around the clock with no sleep whatsoever," Compass recalled. "I was on the front line. I was only human. People started saying I was drunk or on drugs. I wasn't. The only medication I was on was perscription stuff for my glaucoma. And I had recently had a hand operation. I had twenty-some stitches. I was trying to keep it clean but no painkillers."[7]

One of Nagin's most controversial appointments had been promoting NOPD Captain Eddie Compass to police superintendent in May 2002. The boyhood friend of the mayor had been a good street cop. It was always fun to be around Eddie—he was a hoot, wearing Mardi Gras beads every February and swapping stories by the watercooler. Raised in a housing project, Compass had worked his way to the top post by charming everyone he met; among his friends were the Neville Brothers and CBS's Ed Bradley. Early in his law enforcement career, before he was commander of the First Police District (1997–2002), he had provided security for Mayor Sidney Barthelemy (1987–1991).[8] A truly wonderful, honest human being, Superintendent Compass unfortunately didn't know how to run the New Orleans Police Department with an iron fist. He was quirky. Just months before Katrina, for example, he had suggested that the Nation of Islam be tapped for neighborhood watches. Instead of tough, no-nonsense punishment for rogue cops, he launched a "service first" campaign to change the mind-set of corrupt and rude officers. In most U.S. cities, the emergency response time to a 911 call was around five minutes; in New Orleans, it was over double that. On August 20, 2005—just nine days before Katrina—the *Times-Picayune* reported that the murder rate had risen 7 percent since the previous August. Incredibly, Compass tried to put a smiley face on this news by saying that "the vast majority" of New Orleanians killed were "related through blood or close friends or associates"—as if that were comforting.[9]

Like Nagin, Compass enjoyed being a celebrity, smiling for the cameras, getting recognized on the street, and schmoozing with the NOPD cops that he'd befriended during more than twenty-five years on the force. "I didn't want to do so much media," he later said of the impression left by his high-profile way of life. "But I was damned if I did or damned if I didn't. We had to give the impression that we weren't hiding anything."[10]

Nobody paid Compass much mind that Tuesday morning. Everyone thought he needed a time-out, a break. "We all sat down at a table," Landrieu said. "There wasn't any visible animosity between the governor and the mayor. We were all working as a team."[11] Nagin briefed them on the extent of the flooding, telling them there was only one way into and out of the city. Maps were unfurled and it was explained to Brown how FEMA supply trucks and rescue buses could arrive at the Superdome with ease. There were routes open into the city. "Michael Brown had to go, because he was going on to Mississippi and, as I recall, Alabama," Blanco recalled. "But I thought Ray did a good briefing and that we had our marching orders. Michael Brown had a schedule, so I couldn't stop that little procession."[12]

Governor Blanco and Senator Landrieu were surprised that the Superdome had received so much damage. Back in the early 1970s, when it was built for $156 million, a great debate went on as to whether the structure was hurricane-proof, whether its foundation designs were sound. To answer such concerns, architect Arthur O. Davis shipped each panel to St. Louis for a wind tunnel test. Every one easily passed muster. A "tension ring" was also installed, twelve feet deep, around the entire circumference of the Dome. "The roof was erected on a series of temporary towers, twenty-six in all, and once the structure was in place, the towers were lowered hydraulically, all at the same time," Davis wrote in his unpublished memoir, *Design for Life.* "That was the moment of truth as to whether the building structure would be safe. Since this had never been done, there was quite a bit of controversy about the process and there were even some predictions that the roof would fall straight to the floor like a pancake."[13] Of course, it never did.

The construction of the Superdome went smoothly in an engineering sense. Only one unexpected problem occurred: pigeons. Before the Superdome roof was to be placed, nobody had thought to chase out the thousands of pigeons that, during building, had made the open-air structure their nest. Nobody knew how to get the pigeons out. Should buses come with cages? Should the city just let the pigeons expire? Find them a new shelter? Poison them? "The story was widely covered in the media and amused citizens all over Louisiana submitted their solutions to the problem," Davis recalled. "This went on for some weeks, but fortunately this issue was not within our area of responsibility and we proceeded with matters of design and construction, while others shooed pigeons around the Louisiana Superdome."[14]

Thirty years later, the pigeons were long gone; the Superdome was now filled with human beings. The dilemma of how to empty the Superdome had become the foremost concern of Governor Blanco. You couldn't shoo them away or just ignore them. You couldn't evacuate them with RTA buses that were underwater. At all costs, though, buses were needed pronto. Nothing would start to heal in New Orleans unless buses were found for the Superdome and sandbags were dropped into the breached levees. As the Blackhawk left the Superdome, Governor Blanco was hopeful but still disturbed. Her people—Louisianans—were trapped. "I'd already started finding buses from all over the state to evacuate these people," Blanco said. "With FEMA's help we were going to prevail."[15]

Folk-rocker Lucinda Williams, a fine lyricist, wrote a song called "Joy," which appeared on her 1998 album *Car Wheels on a Gravel Road.* In an anguished, snarling chant, a determined Williams declared, "I'm going to Slidell and look for my joy." Unfortunately, as Governor Blanco's Blackhawk flew over that waterfront city of 26,845, no joy could be found. All of the fishing camps were submerged. So much lumber was floating around that it looked like Paul Bunyan Days in northern Minnesota. Senator

Landrieu elbowed the governor and asked if they could fly over her homestead, just a mile away. "Sure," Blanco said, directing the pilot to follow the devastation up the lake a little more. "There was nothing left of our little place," Landrieu recalled. "Nor our neighbors. A way of life was wasted away. I pointed down and said, 'Governor, that's where it was.'"[16]

Upon returning to Baton Rouge, the three politicians went their separate ways. They were scheduled to reconvene at a press conference at 3:00 P.M. All of them were concerned with the situation at the Superdome. Television commentators were saying ad nauseam that they couldn't believe the arena, in its current state, was "part of America." Governor Blanco was equally shocked. Deep down, she blamed Mayor Nagin for the debacle and wondered why he wouldn't meet with the Superdome evacuees. What was he afraid of? What kind of man was he? When she saw her husband at the EOC, she spoke about all the poor stranded people she had seen outside the Dome. "I need to find out what the tempo is," she said, "and what's really going on."[17]

With Coach at her side, and no media in tow, Blanco headed right back into the bowl via helicopter. She wasn't going to be like Nagin, indifferent toward and scared of Louisiana's most vulnerable citizens. She directed the National Guard pilot back to the Superdome. Within half an hour, there she was, standing in the largely African-American throng, trying to offer words of hope. She was confronting the forgotten. "Now there were a lot of upset people," Blanco said. "I vividly remember this lady who was very upset because in the rescue efforts she had been [dropped off at] the Dome. She was one who had been evacuated after the storm to the Dome and she was just fretting and fretting because the children in her family had left in a different helicopter and she didn't know where they were. She and her family had been separated and she was scared. She didn't know if they had been rescued or if they had perished."

Everybody, it seemed, wanted to touch (or shove) Governor Blanco, to tell her their Katrina story. Never once did she worry about her own safety. "The crowd rushed her and she immediately knew the misery," Coach Blanco recalled. "Believe me, she knew the misery. They were desperate, they had just run out of water."[18] A visibly frustrated man held a baby in front of her. "You know we don't have water here," he said, staring the governor in the face. "My baby needs to be bathed. My baby's going to get sick." She did not view these people as ruffians; she said they were simply "scared" and "traumatized."[19]

II

Given the tense, hurly-burly atmosphere, Governor Blanco knew that time was of the essence. As she continued her tour, she made an effort to inventory the problems. More MREs, bottled water, medics, etc., were needed. But what the people at the Superdome really needed were buses—fleets of buses to drive the evacuees out of the anarchy. She was concerned that the Orleans Parish water system had gone down, as had the sewage system. The lack of electricity she had expected, but not the other two factors. "The water was coming into the city all day Tuesday, and it was building up, so that was the fright," she said. "The water continued to rise up and people were still being rescued. I asked them if there was food and water in the Dome. I had been told by the Guard that there was and the people told me that yes, there was. It was irritating to have to stand there in those long lines to receive their food."[20] It was a struggle to maintain a supply line, but officials were managing, just barely. Blanco promised to help with that, but the main concern was buses.

It was on Monday that the governor had learned that all the New Orleans RTA buses were flooded and out of commission. She stepped into the void to procure others. By early Tuesday morning, even before the helicopter rides, she had tasked Leonard Kleinpeter, her special assistant, and Ty Bromell, executive director of the governor's Office of Rural Development, with becoming bus wranglers. A hunting enthusiast, Bromell, six feet tall with cropped hair, looked like a physical education teacher. That morning he had been jarred awake by a telephone call from Kleinpeter at his Christian Avenue home in Baton Rouge, informing him about the breached levees. "New Orleans needed buses," he said. "It wasn't just the Superdome, but the people flooded out in Lakeview, the Lower Ninth, New Orleans East, everywhere."[21]

Bromell started working the telephones. He called just about every school superintendent in northern and western Louisiana, in search of school buses for New Orleans evacuations. Diesel-powered, these standardized yellow buses could hold from forty-five to seventy passengers. He directed all of the volunteered buses to the Tanger Outlet Center in Gonzales, right next to the mall's Cracker Barrel restaurant, which would become the bus dispatch center. By early afternoon, Bromell had found one hundred buses from places like Shreveport, Monroe, and Natchitoches. The game plan was that once the bus drivers picked up directions at the Tanger Outlet Center, they would drive to New Orleans and bring hundreds of evacuees back to a new Hurricane Katrina shelter created at the Maravich Center on the LSU campus in Baton Rouge. "Our efforts were focused just on school buses," Bromell recalled, "and without an executive order, we couldn't force parishes to give us buses."[22]

Governor Blanco was wise to have Bromell search for yellow school buses. But she was making mistakes. Why *not* issue an executive order, which would have *forced* parishes to donate their buses to the evacuation effort? Instead of Bromell's one hundred yellow buses, she would have had thousands at her disposal. The same was true of city buses. For example, there were eighty buses in the fleet of Baton Rouge's Capital Area Transit System (CATS). Dwight Brasher, the CATS director, recalled, "At tops, I was running twenty buses after Katrina, giving people in shelters access to hospitals. We'd take people with kidney problems or dialysis to hospitals. But we had sixty buses which we weren't using." When asked why Governor Blanco didn't commandeer his buses, all Brasher could say was "Well, I don't really know."[23]

Again, all Blanco had to do was issue an executive order and all sixty of those CATS buses could have headed to the Superdome or the I-10 and Causeway intersection, where thousands were congregated, many being dropped off there by the Coast Guard and fire departments. "It wasn't that simple," Bromell explained in defense of Blanco. "Each bus had to come with a driver and a security guard. The story was bad on TV about the looters. So, many school bus drivers from upstate were afraid to come to New Orleans."[24]

In Louisiana, political feuds are quick to flare up and slow to heal. Blanco was taking the cautious approach, and not using up all her political capital quickly. She didn't want to start the response effort on a confrontational note. Knowing she'd need even more cooperation as the days wore on, she tried diplomacy first. Governor Blanco, however, *did* reach out to others for buses. On Tuesday, Blanco put Angele Davis, head of the state's Department of Culture, Recreation and Tourism, to work. Determined, chatty, and goal-oriented, Davis was the kind of state employee who actually enjoyed ribbon cuttings and awards ceremonies. Cynicism was not in her makeup. She was of the Old South school that believed that a smile gets you a thousand miles farther than a scowl. Her job was to find additional tour buses to complement a convoy of five hundred FEMA buses supposedly coming down from Arkansas. Davis worked the phones like a pro. Her first success was with the Travel Industry Association of America, headquartered in Washington, D.C., where she was looking for tour buses—ones that usually took vacationers on Cajun country or Mississippi Delta treks. "They pitched in," Davis recalled. "There wasn't any hesitation. I called another guy who owned two buses, woke him in the middle of the night. He said, 'You've got it. Take them; they're yours.' Such acts of generosity were repeated time and again."[25]

On Tuesday night, the first major convoy of yellow buses left the Tanger Outlet Center headed for the Superdome. The mere sight of these rescue vehicles would bring cheers from the people stranded there. According to the governor's own data that day, nearly 70,000 people were stuck in the flooded sections of greater New Orleans. Major General Bennett Landreneau of the Louisiana National Guard was in charge of the operation. Wearing U.S. Army fatigues, General Landreneau answered tough questions during daily press briefings in Baton Rouge about Louisiana's slow response. An unanticipated curveball, however, was thrown General Landreneau's way when his bus convoy neared the Orleans Parish line. At the intersection of I-10 and the Lake Pontchartrain Causeway, thousands of stranded evacuees, desperate for help, blocked the road. They were exhausted; with the sun beating down on them and with no water, they were in pathetic shape. "What I did not know that Tuesday night was that people were gathering on the interstate," Blanco said. "So when the buses went in, they ran into people. The general called me and said, 'We've got a situation developing. Do you still want buses to go on to the Dome? Or should we pick up the people on the Interstate?' "[26] Murphy's Law was still alive and well in Louisiana. "Obviously, they've been in the sun, they have no water, they have no food," Blanco told Landreneau. "The people in the Superdome have a nasty place and are miserable. But they have a roof over their heads and they have water and food. This is like triage. The ones coming off of their rooftops in ninety-eight-degree weather with children and elderly people had to come first. We had to get them out of the sun with food and water."[27] So the buses started picking people up on the highway before they could get to the Dome.

Exacerbating the situation was the shortage of bus drivers. Bromell and Kleinpeter were finding it easy to find buses but difficult to find drivers. "All day Tuesday, Leonard and Ty had been gathering up buses," Blanco recalled, "but the images on TV were scaring the bus drivers and they were going, 'Look, here's my bus, here's the keys, it's full of gas. You just get the drivers.' "[28]

It's now clear that Blanco should have better prepared Louisiana for the Big One, taking more seriously than she did the New Orleans doomsday scenarios like Hurricane Pam. The day *after* the hurricane wasn't the time to look for emergency bus drivers. A system should have been in place—like the one in Texas—for fleets of buses to gather at assigned spots near the population centers, ready to go after the storm. It also must be said that Blanco was god-awful on television, discouraged and sad. She needed to appear strong and tough, spitting in the eye of Katrina and reassuring her constituents, as did Mississippi's defiant Haley Barbour. Press Secretary Denise Bottcher, recognizing this deficiency, instructed her boss later that week to act more Barbouresque. "I'm now a bit concerned that we're doing too many 'first lady' things and not enough John Wayne," Bottcher e-mailed the governor's assistant. "Women are easily portrayed as weak, which KBB [Blanco] has had a hard time overcoming."[29]

But Blanco deserves credit for her ability to improvise in the days after Katrina. She barely ate or rested. She did absolutely everything she could to help victims. Mistakes were made, but she also dealt with each crisis that occurred in a thoughtful, rational manner. Not once did she crack under the strain. She didn't swear. She didn't denounce Bush, Chertoff, or Brown. "Never would I criticize my president in the middle of a crisis," she said. "That's not what I'm about." When she learned that Kleinpeter and Bromell couldn't find drivers, she made adjustments. "Who are the school bus drivers?" she later asked. "They're just little old ladies and old men from in town. They all saw the crime on the screen and they got scared. So we had a couple hundred buses parked at Tanger Outlet Center, full of gas with the keys and no bus drivers. So I grabbed [an officer] of the National Guard and said, 'Can you get me 165 bus drivers?' We had 40 courageous drivers who were willing to go in and drive. I had to get members of the National Guard to drive the school buses to go pick up the ones who needed rescuing. Wednesday evening, [still] no FEMA buses."[30]

Just picking up stranded Louisianans wasn't enough. The other dilemma was deciding where to take them. Under Governor Blanco's leadership three shelters were established in Baton Rouge. The predominant one was still the Maravich Center. Nobody would claim that the shelters were run with Swiss efficiency, but given the magnitude of the crisis, the Red Cross and others working the venues earned an A for effort and a C for providing services. "They did fine under the circumstances," Blanco said. "Nothing's easy about this. We were creating shelters all during the day, into the night. We were evacuating people continually all through Wednesday and Thursday. Now, I really needed the FEMA buses, and Social Services began to realize that Louisiana was backing up. There was no more room."[31]

You might say all of Blanco's bus wranglers were performing their jobs too well. Her logistical thinking was skewed, but to her credit, she continued improvising well. Clearly, the Tanger Outlet Center in Gonzales was simply too far away from New Orleans to be a huge dispatch center. While it was great being next to a Cracker Barrel restaurant, her team needed to relocate. Ultimately, they moved operations twenty-eight miles east on I-10 to mile marker 209 in La Place, where there was a huge Texaco truck stop. This was the closest place to New Orleans that still had electricity and phone service. "From 209 right into New Orleans, it was a communication dead zone," Bromell explained. "We were on the edge, so we'd found the best staging area we could."[32]

On Wednesday morning, FEMA telephoned the governor's office with the news that its phantom buses would be arriving at mile marker 209 that afternoon. Bromell, Kleinpeter, and Davis exhaled. From 10:30 A.M. to

2:30 P.M., they stopped making calls for school and tour buses because FEMA assets, after a long delay, were on the way. The only problem was that, once again, the FEMA buses never arrived. They wouldn't come until Thursday. Meanwhile, a number of fellow governors were stepping in and offering help. Arnold Schwarzenegger of California sent a team of underwater divers, Bill Richardson of New Mexico sent medics, Mike Huckabee of Arkansas sent Blackhawk medevac helicopters, and George Pataki of New York sent disaster relief specialists. Reaching out to Louisiana became the rule, not the exception. There was a three- to four-hour window, in fact, on Wednesday afternoon, when Blanco felt that she had the post-Katrina beast in a box. Up out of the mire—that was Blanco's attitude—Louisiana was rising like a dirty phoenix. At least she was trying to turn that corner and embrace hope. And Blanco was going to rise with the tide. "We thought everything was under control," Davis said. "We were getting the job done, slowly but surely."[33]

Well . . . not so fast. FEMA had essentially double-crossed Blanco again. There were no buses thundering over the Arkansas line on Tuesday or Wednesday, fully equipped with food, water, and medical supplies. It was a ruse, a lie. "When we realized the FEMA buses weren't coming, I got back on the phone, realizing we would have to do it all ourselves," Davis recalled. "Now we started getting help from out of state. I'd take anything. Trains, planes, buses. Lieutenant Governor Mitch Landrieu was just awesome, determined to get everybody out of the Superdome fast. The response I got was so impressive. Texas and Utah were trying to help us in every way imaginable. It just made you feel good about our country."[34]

On Wednesday evening, August 31, feeling duped by FEMA, Governor Blanco issued Executive Order 31. Bromell and Kleinpeter didn't have to politely ask for yellow school buses anymore—providing buses to help rescue Katrina's victims was now mandatory. Louisianans, with the help of the Coast Guard and their own National Guard, were going to have to save themselves. "Our own people saved thousands of our own people," Leonard Kleinpeter later wrote, angry at the way his state was portrayed by both the Bush administration and the national media. "This is our chance to show the world who Louisiana people are. The 'verbal carpetbaggers' need to be answered. Our own talk-show hosts are lying about each of us. We stood up for each other and saved our friends' lives."[35]

III

On that Tuesday morning, as Blanco, Brown, Landrieu, and Vitter were touring the New Orleans area, the *Times-Picayune* office, located only a quarter mile from the Superdome, was in crisis. The staff had awoken to find water at their doorstep, rising. Calmly, the editor, Jim Amoss, had the whole team convene in his office. Outside they were surrounded by three feet of water—a manageable amount. But with the 17th Street Canal levee breach growing bigger by the minute, it wasn't too far-fetched to imagine fat porpoises in the parking lot before long. Everyone agreed it was time to evacuate. "We had to make our move quickly," Amoss recalled, "before it became impossible and we were trapped in this building and couldn't report, couldn't function. The publisher at that moment stuck his head in my door and he had reached the same conclusion. So we started running through the building, shouting to everybody to head for the loading dock, rousing everybody. Just getting out was all we thought about. Just getting to the interstate, which was visibly dry."[36]

Timing was crucial at this juncture. Shortly after 10 A.M., a dozen delivery trucks pulled up and all the *Times-Picayune* employees started piling in, in place of the newspaper bundles that the drivers had expected. Photographs of the escape show these delivery trucks lumbering down the service road, coughing up diesel, with floodwater almost over the headlights. A dubious Amoss, worried that the twelve-truck caravan might only make it for a block or so, watched the electronic dashboard message flash: "Water in fuel. Water in fuel." The likelihood of the trucks' conking out, or an even greater calamity, was very real. He was nearly faint with dread. He knew there weren't just *Times-Picayune* reporters, but family members who had weathered Katrina with them, including an eighty-seven-year-old man, a three-month-old baby, and children of various ages. "The thought that we might stall in the middle of this deluge," he said, "and have no option but to drag these people into the water, had me on edge."

The plan was to get across the Mississippi River Bridge to Algiers, where the *Times-Picayune*'s West Bank bureau was reportedly dry, although without power. Somehow they made it off the flooded service road and then purposefully went the wrong way up the Howard Avenue exit ramp, made a U-turn off the interstate, and escaped the bowl. "I still remember our truck made it onto the dry road, and everybody inside cheered and everybody behind it thought, All right, we can make it," Amoss recalled. "Once on the West Bank we all pulled over and said, 'Okay, now how are we going to make a paper tonight?'"

That was a good question. After a rushed meeting, a varied group of reporters and editors, including the editorial-page editor, music critic, and sports editor, volunteered to return to the dry area of Uptown. They had to get the news out. A paper could be printed somewhere else, but reporters had to be in New Orleans. They couldn't abandon their readers in a time of peril. Unfortunately, the West Bank Bureau was small, without a printing press. With Amoss's approval, a group of daring volunteers took one of the delivery trucks and recrossed the Mississippi River Bridge, deciding they could all sleep at the home of one of the editors, a house that fortunately hadn't flooded. While heading back into New Orleans, the *Picayune* contingent stumbled upon the Wal-Mart being looted. Meanwhile, Amoss led the rest of the caravan on Highway 90 to Houma, where the New York Times Company owned a small paper called the *Courier*. One of Amoss's staffers was an alumnus of the *Houma Courier* and she had contacted it in advance to say that they were coming. The most populated town in Terrebonne Parish, Houma sat 57 miles southwest of New Orleans. This small city of 32,000 was heavily Cajun and filled with shrimpers and fishermen.

When Amoss's staff, packed into delivery trucks like sardines, the doors partially open for air, pulled up to the *Courier* building, they were greeted with warmhearted Cajun embraces. It was as if they were saying, "Please, just take over our offices. Whatever you need, computers, meeting rooms." After the delicate health of a few caravan members was addressed, Amoss acted fast, planting his production people and designers there to use the *Courier* presses to produce the *Times-Picayune*. "But we needed a real newsroom, in a bigger city, to make the newspaper," Amoss recalled. "So we left a core group there and the rest of us drove along Bayou Lafourche to Baton Rouge, still not knowing exactly where to go."

One possibility was calling up the *Times-Picayune*'s rival, the *Baton Rouge*

Advocate, to ask for help, but then it occurred to Amoss that he was on the Board of Visitors at the Manship School of Mass Communication at LSU. He could ask Dean Jack Hamilton if they could commandeer the journalism building. The dean readily agreed to provide space for 140 *Times-Picayune* employees.

For the first time in two days, Amoss felt relief. At 8 P.M. the caravan pulled up to the journalism school in utter darkness, with scarcely a streetlight for guidance. Hamilton and a few faculty members stood in front, waiting for them. "He just turned the whole place over to our use," Amoss recalled. "The students hadn't arrived yet for the school year and there were these wonderful, state-of-the-art computers. We instantly made ourselves at home. Hamilton had even fixed up a room for us with dozens of cots for us to sleep on that night. That was the beginning of our inventing our Baton Rouge future. We stayed for two weeks."

Obviously they were unable to get out a *Times-Picayune* newsstand issue that Tuesday. But, even through all of the commotion, they managed to produce a full paper, posted online in PDF format that night. It was comprehensive coverage. On Wednesday and Thursday there were further electronic versions of the *Times-Picayune,* breaking numerous stories about looting, flooding, rescues, and the Army Corps of Engineers. And then, on Friday, September 2, after not officially missing a single issue, they had a printed version issued from Houma. "It was a weird-looking paper because their format is different from ours," Amoss said. "It stayed that way for two weeks." Amoss had a new set of collectors' items for his wall.

IV

Just like the *Times-Picayune* caravan, Garland Robinette and his WWL team, who had broadcast during the hurricane from their downtown high-rise, now found themselves in a real dilemma. As of Tuesday morning, water was starting to flood around the Superdome, a couple of blocks away. Robinette didn't want to give up the microphone. This was one of the few times he felt good about being on the air. There was nothing narcissistic or egotistical about trying to bring families back together or report escape routes out of the bowl. But broadcasting from a closet on the fifth floor of a blasted-out building, shattered glass everywhere, made no sense. "At first I thought a broken pipeline or a manhole cover blew," Robinette recalled. "But people shouted at me that the water outside our building was rapidly rising. By the time we all fled down the stairwell, water was knee-deep. We had to get to the parking garage *fast*, real fast."

Because most of Robinette's colleagues had regular cars, they failed to make it through the high floodwaters. Robinette, however, had a hybrid, and his car cruised out of the garage with relative ease. With five WWL employees Robinette headed for thc West Bank. He dropped his passengers off in New Iberia and Lafayette, and then headed up to Natchez, Mississippi, where his family was. "When I saw the rising water on Tuesday I knew exactly what it was," Robinette said. "The very first thing that hit the brain was, I cannot believe you lived long enough to see it. Remember, I had done all of these doomsday documentaries about New Orleans flooding and coastal erosion. Some months I worked on these projects seven days a week nonstop. They were creative endeavors and when I was done I'd go eat and drink and say to myself, Not in our lifetime. Good story. But not in our lifetime."

Nothing about the levee breaches really surprised Robinette; he knew the Corps of Engineers had constructed them on the cheap. However, as he drove northward, following the Mississippi, he did marvel that his city was being destroyed by a tortuously slow trickle instead of a booming loud crash of twenty feet of water. Robinette seethed with anger at all the frivolous Louisiana greed-heads who had refused to properly maintain the levees. "We've known the storm was in the Gulf for thirty years," he fumed. "I used to go to government meetings on civil defense on West End Boulevard every year. These FEMA and Corps guys would sit in the back of the room while parish officials explained their obviously inadequate preparations and plans. We've all known about this for thirty years and the city, state, and feds did nothing about it. It had been predicted. When I saw those floodwaters on Tuesday it was like for thirty years telling your child, 'Don't walk in front of the train.' And now I had to stand there and watch him walk in front of the train. The levee breaches could have been easily avoided. In 1970 it would have probably cost them a million bucks to change the tide on wetlands. In 1980 it would have cost them ten million to fix the levees. They all knew about it, the crooks. And guess what? In New Orleans we also knew our education system was the pits. We knew we were corrupt. We knew we were going down the crapper but it was okay, because it was just New Orleans. Screw 'em. I get mad at people who say that New Orleans's 'do nothing' attitude is our culture—isn't it quaint? I'm an honest Cajun and let me tell you that's just malarkey."[37] The "do nothing" attitude may not have been universal in New Orleans, but for decades it prevailed, both in government and in levee management—and for that, every complacent citizen had some share of the blame.

Despite the anger Robinette felt after Katrina, he remained helpful on the air. When he resumed broadcasting on Thursday—from Baton Rouge, where Clear Channel helped him broadcast all over the world—he was the voice of reason in Louisiana. He constantly gave sound survival advice to screaming mothers, terrified children, and lost souls. He was emotional but not the least bit over the edge. He seemed to thrive on the Hunter S. Thompson adage "When the going gets weird, the weird turn pro." When asked how he stayed so professional, Robinette had a one-word answer: Vietnam.

When pressed to elaborate on how his Swift boat days there were related to Katrina he grew quiet, then spoke. "The water didn't remind me of Vietnam," he said. "The dying did. Knowing people were dying and hearing stories and talking to people who were in the process of dying, who were going to die as soon as we hung up. That reminded me a lot of Vietnam. When the people came out with me from our studio to the flooded cars, they were hysterical or shocked. I felt extremely calm. Vietnam. When the windows were blowing out (and I'm afraid of heights if you put me on a ladder), I wasn't afraid. I was calm. Vietnam. I think it was thirteen months of combat, and if you lose control, if you get emotional or you get afraid or too brave or just don't stay calm and think and be relaxed, you're going to die. So Vietnam was a gigantic plus. It comes back to the fact that I have been to a bad place. I had that bad place to go back to, that helped me remain calm enough to get the job done. . . . I had seen the birds disappear before."[38]

V

As of Tuesday morning, the intrepid Louisiana Department of Wildlife and Fisheries (LDWF) had been on the scene for a full twenty-four hours, with a force of two hundred rescuers with boats. By Saturday, September 3, they would have rescued more than 10,000 people in southeastern Louisiana—an astonishing number for an outfit usually concerned with issuing hunting licenses, preserving game, protecting habitats, and maintaining fish hatcheries. They truly were first responders and their tireless work immediately became the stuff of Louisiana legend—even if the media didn't cover their exploits properly. Using a site on the St. Claude Avenue Bridge, LDWF's people were unique in that they came with everything they needed, including their own food, water, and tents. Nonetheless, their launch site was more akin to a keelboat dock in Muskrat Flats than a city wharf. According to Sam Jones, an aide to Governor Blanco, they had become "fishers of men."[39]

Instead of wandering a fragile strip of beach in Port Fourchon, they were carrying patients out of Lakeview Medical Center. Instead of banding roseate spoonbills at Lake Martin, they were pulling children out of the rubble in the Lower Ninth Ward. And instead of patrolling the Freshwater Bayou in Southwest Louisiana, W&F, as they were called, were now wading in oily, chemical-tinged muck. Even when Katrina was still kicking with 45 mph winds on Monday, and other first responders were waiting to enter the bowl, W&F boats were already on the job. "Governor Kathleen Blanco has given LDWF and all state agencies the green light to do everything necessary to get our jobs done," LDWF Secretary Dwight Landreneau said. "And we're doing it."[40]

The overwhelming concern of health officials during the flooding was stopping the spread of disease. Reports of rashes, dizziness, dehydration, headaches, chemical burns, along with hundreds of other maladies were becoming frequent. Luckily, many of the rats in New Orleans drowned, flushed out of the sewers by the nestful. But rats being rats, many more managed to survive. Gnawing on rotted wood and electrical cables, scouring through mounds of debris for water-soaked food, the surviving rats stayed in packs, particularly along the wharf front. Fleas, trying to escape the floodwaters, burrowed in the backs of rats, making the rodents a double health menace. At night, those humans stuck in the bowl heard the hungry rats scampering about, brazenly prowling for food. "And dem big-ass rats dat live under da wharf," Rick Ray, who waited out the storm at New Orleans Pumping Station 7, told John Burnett of National Public Radio, holding his hands two feet apart, "dey got really bad, dem sumbitches was all over da place. We had to sleep in cars."[41]

An even bigger worry than rats were mosquitoes. Cans of Off! and DEET were passed around rescue boats and shelters like tanning oil at a beach resort. Nobody had much experience with floodwaters, so every mosquito caused a pang of worry about diseases like malaria or St. Louis encephalitis. Tulane University may have boasted the finest school of tropical medicine in the hemisphere, but when you're standing in stagnant water and see blood on your skin, that doesn't help much. You think of worst-case scenarios. You ponder flukes. Just because you escaped the Big One didn't mean that the Little One—the mosquito—wasn't going to knock you off on the viral back end. In such desperate conditions, every headache became dengue fever, every fever blister seemed to be some new strain of leprosy. Life in the deluge on Tuesday wasn't just a breeding ground for mosquitoes, it had become a nursery for newborn hypochondriacs, rightly worried that a paper cut equaled tetanus.

Although the New Orleans Tourist Board liked to minimize the statistic, Louisiana was a per capita leader in West Nile virus, with 177 citizens infected in 2005 alone. Mosquitoes were the carriers of West Nile, which caused an inflammation of the brain. If you were bitten in the Katrina floodwaters by a West Nile mosquito, within days you could expect swollen lymph glands, high fever, and local paralysis. The World Health Organization had little to recommend as a precaution. "To keep from being bitten by infected mosquitoes, health officials recommend staying inside around dusk and dawn, when the insects swarm," John Pope of the *Times-Picayune* wrote, "covering arms and legs, getting rid of standing water, and using repellent with DEET."[42]

Unbeknownst to most of the first responders was the fact that New Orleans had an ally in its fight against mosquito-borne disease. Salt Lake City has a monument to the seagulls, which in 1848 swooped down from the sky to devour a swarm of locusts, thereby saving Utah crops. They were known affectionately as the "Mormon Air Force." Someday New Orleans should likewise honor the dragonfly. With their large multifaceted eyes, two pairs of strong transparent wings, and outstretched bodies, dragonflies frighten most people. On Tuesday dragonflies blanketed New Orleans, hovering just inches above the smelly floodwater, eating every mosquito in sight.

Epidemiologists considered dragonflies—called in backwoods areas "the Devil's darning needles"—an insatiable predator, one usually found in ponds and bayous. Contrary to the popular misperception, they don't bite or sting humans; in fact, they are as harmless as ladybugs. Female dragonflies lay their eggs (nymphs) in water or on floating plants. After Katrina their eggs were deposited in the floodwater. Then aquatic larvae hatched. From then on, they started devouring mosquitoes, dive-bombing them with aerial acrobatics that made the Coast Guard helicopters look clumsy by comparison. With a nearly 360-degree field of vision, dragonflies can both stay stationary and soar to speeds of 60 mph (they are the fastest insects in the world).

Sometimes when a corpse was found floating around the streets of New Orleans, washed up against a chain-link fence or concrete wall, dragonflies hovered around the victim. Never did the dragonflies touch the flesh, hunting instead for maggots and fleas and mosquitoes, drawn to protect Katrina's victims just as they did its survivors.

VI

Because the water was being described as "toxic gumbo," in the annoying cliché du jour, the NOPD avoided the dragonfly zone. The officers seemed angry with those citizens who had stayed behind. Not many of the displaced were dangerous or high on drugs, they were only confused. A few were parolees, craving dowel-like instruments with which to smoke the resin in their crack pipes. The majority were only thirsty, maybe a little restless. But by that Tuesday, they had become objects of derision to the police. Many members of the NOPD, including black officers, were fixated on the fact that the "fire ants," as they called the stranded, had blown their chance to evacuate—in fact, they had been ordered to leave and had simply refused. All of a sudden, now, after defying common sense, they wanted help. Well . . . *tough.* That was the bitter NOPD attitude. They had no obligation to play Shaft or Zorro.

Many of the NOPD officers had lost their own homes in the flood. They were put out. Jeff Goldblatt of Fox News even saw police officers on Poydras Street pushing shopping carts filled with the only belongings they owned. You had to feel sorry for them. "On Tuesday morning, around one o'clock, we were getting ready to turn in for the night, getting my gear, just kind of cleaning up," Goldblatt recalled. "All of a sudden I see a line, a single-file line of about a dozen or so lights, flashlights kind of searching for something, coming down Poydras toward the Convention Center—toward us. I'm like, 'What the hell is that?' All of a sudden as it got closer I saw men and women in the blue uniform, New Orleans police, some of them holding the hands of little kids, all of them with supermarket carts with their worldly possessions, going into the W Hotel. In return for I guess some tacit guarding, some tacit security, they had a room and a place to stay. It was pretty sad."[43]

Not all of the NOPD officers were shirking their duty or copping cocky attitudes. For fourteen straight days Captain Tim Bayard and some of his troops joined the NOLA Homeboys efforts. Furious at officials who were failing to properly organize rescues, Bayard simply joined the homegrown effort. He launched boats from places in virtually every district. He put his good citizenship ahead of the red tape that may have bound his badge. Whether he was using a hammer to knock off roof vents, sticking his face in a hole to smell for death, borrowing a boat, or climbing wobbly apartment stairs to a stranded child, Bayard was never idle and he shirked nothing. "We were launching boats off I-10, off up ramps and down rescue ramps," he said. "We launched them off of bridges. We couldn't get to Claiborne—there was too much water—so we launched off of the St. Claude Bridge."

Because the Wildlife and Fisheries boatmen had heard about the NOPD looting Cadillacs and fleeing New Orleans, good cops like Captain Bayard paid the price. "Wildlife and Fisheries were a pain in the ass," Bayard said. "They did not coordinate with us. Absolutely nothing. They turned us down on the interstate when we tried to put fuel in our boat. They said, 'We can't put fuel in your boat. We need to put gas in our boats first. Then if we got extra fuel, we'll put it in your boat.' That's the kind of treatment we had. Everybody was giving us the cold shoulder. They got boats lined up on the parking lot of Memorial Avenue, waiting for Wildlife and Fisheries, so I grabbed a mayor from Texas. I said, 'I need your boats. We've got to go deploy.' He said, 'We don't work that way in Texas. We are working now in conjunction with the W&F here.' I went to a captain with the W&F and I introduced myself. 'We need those boats, brother. We've got to put them in the water—we've got to go rescue people.'"

There was something tragic about Bayard's plea. It was as if he were trying to redeem his departed force by outsaving everybody else.

"I can't give you the boats," the W&F captain replied. "We've got other things we're doing with them."

"Jesus" is all Bayard could say, walking away in disgust.

Later, Captain Bayard would complain that W&F was "locked in" with FEMA and that's why they were so rude to the NOPD. But nobody did a better job than W&F when it came to first responding—perhaps because it did stay focused on its own mission. For his part, however, Bayard stayed in the murky waters, borrowing boats wherever he could. Months later, Captain Bayard answered questions about Katrina in the clipped fashion of a latter-day Joe Friday—just the facts, folks. Did he see corpses? "Yeah," he said. "There were bodies all over the place. I think there were a lot of bodies that were never found. We were working with the missing-persons list and we found additional bodies and they're still finding more bodies." Did he experience gunfire? "Yeah," he said. "That ain't nothin'. That happens every day in the city of New Orleans. We didn't have long rifles and we were running out of ammo and stuff, and really had no boats, but we were still going out there. As far as the gunfire, yeah, it was out there and some of it was directed at law enforcement and some of it was directed at emergency workers. Which goes to show you the mentality of the element you're dealing with."

By Tuesday morning, as Bayard tried to rescue his fellow New Orleanians, lake water was flowing into the city through four breaches in three canals, and the flooding in New Orleans rose to an average level of eight feet. Dead bodies were a common sight in the stagnant water: discolored and bloated, the skin had taken on a marble pattern, with psoriasislike blotches on the arms from sunburn or chemical irritation. The elderly and ill remaining in the neighborhoods were dying for lack of emergency care. Because headquarters on South Broad Street was flooded, the NOPD command was moved to Harrah's Casino. "But a lot of the group stayed," Compass recalled. "History shouldn't forget that."[44]

VII

Stationed in Alexandria, Lieutenant Commander Jimmy Duckworth of the Coast Guard was receiving telephone calls from people with recreational boats who wanted to help. "The phone was ringing off the hook," he recalled. "We had one person who was doing nothing but fielding phone calls for private logistical support and I was taking anything that could float."[45]

Duckworth's proactive "anything that could float" attitude was in contrast to that of Governor Blanco, who wanted the Louisiana Department of Wildlife and Fisheries to run the water rescue operations. She wanted everything streamlined, even though it was a helter-skelter environment. With looting going on and rumors of chemical spills, she rightfully worried about people's safety. If every Louisianan with a canoe or washboard came paddling into New Orleans, the situation, she believed, would become unmanageable. "At first, yes, I was worried about too many civilians meaning well but making things more difficult," she said. "But, I must say, Louisianans with boats rose up and did a great job."[46]

Duckworth also worried about what vessels he green-lighted. For instance, he received a telephone call from two Texans with a World War II Duckboat. As a military history buff who had built a replica of a Higgins boat, the Louisiana-manufactured craft used on D-Day, Duckworth knew those old Duckboats had a terrible habit of sinking and could be dangerous. Crowd too many evacuees on one and it might just go under. Glad that they were at least *asking* for approval, Duckworth cut a deal. "Look," he said, "you come to Alexandria and let me inspect it and I'll put you in the game." This was essentially a ploy. The chance that two guys would drive from Houston to Alexandria towing a vintage vessel to be inspected by a Coast Guard commander was slim. But sure enough, the Texans reported for duty about six hours later. "God bless those guys," Duckworth said. "They show up with a huge custom trailer and a beautiful World War II Army Duck. These two great guys. They came into the command post. I walked outside. I inspected their Duck. It was in tip-top shape. I told them the Coast Guard could use them. I wrote down a bunch of notes on where they needed to go and get people. They followed orders, took off, and the last I heard had saved people in New Orleans."

The very first responders will never be known by name. When the levees broke, and the bowl started filling, hundreds of residents with recreational vessels went into high gear. The city had people conducting rescues with yachts, dinghies, ferries, canoes, rafts, sailboats, scows, skiffs, sloops, tubs, catamarans, dories, draggers, baiters, and ketches—and even a floating wagon. Who were these navigators? What was their story? They were known at Johnny White's Bar as the NOLA Homeboys, the oddballs who refused to evacuate, who were saving New Orleanians from the ravages of Katrina. They knew fellow citizens were suffering and even dying and they weren't going to sit on the sidelines, waiting for our-of-state help. These were their brothers and their sisters. These were the neighbors they never knew. These were the neighbors they were determined to save. They had no official insignia or training. They were bartenders and insurance salespersons and clerks—regular folks. It never dawned on them to wait for the FEMA trucks or the Oregon National Guard.

The *American Heritage Dictionary of the English Language* defines "homeboy" a few different ways. The common meaning of "a fellow male gang member" certainly didn't apply here. But another meaning was a friend or acquaintance (usually male) from "one's neighborhood or hometown." Without worrying about mosquitoes, chemical burns, or tetanus, with no surgical masks on their faces, the NOLA Homeboys plunged forward to help the terrified. There was no hesitation, just raw human instinct to risk your life to save another person. There was a part-time dishwasher at Cannon's Restaurant, a checkout clerk at the A&P, or a tacomaker from Juan's Flying Burrito. Some of them were considered misfits in town. Without search and rescue training, they were fueled by a combination of courage, decency, instinct, and adrenaline. Some had never heard of FEMA, but they intuitively knew that by the time "those guys" showed up, people would have drowned. "Law enforcement is never comfortable with common citizens coming in," Bob Mann said. "We didn't want to be rescuing the rescuer. We had widespread violence. So we were concerned about jeopardizing their lives. In the end, thank God they did."[47]

The NOLA Homeboys, in many cases, were the outcasts in the community. One of the NOLA Homeboys, for example, was Michael Knight, a thirty-three-year-old African-American reggae crooner and father of seven who risked his life to help others. With long dreadlocks and a carefree Caribbean demeanor, Knight was high just living on Jah's cool earth. Everything will be all right, man, was his general outlook on life. He was all about the second-line at Vaughn's Lounge, steaks from Winn-Dixie, and to-go cups at Lucky Charms. His entire life centered on the eighty-three blocks of the Seventh Ward. Occasionally he dreamed of moving away, and he had once spent a brief spell in Virginia, but the cold weather always brought him back to the bayous. "You can't go nowhere else in the world and get up at two in the morning and go to the bar to have you a drink and a po'-boy," Knight explained of New Orleans. "The simplest things is the things you miss the most."[48]

On the eve of Katrina, Knight and his wife, Deonne, lived in a blighted block of St. Philip Street, only a short drive from the waters of Lake Pontchartrain. Considering that Knight was living in a small shotgun house, it was odd that he kept his banana-yellow, twenty-foot boat in his driveway—or it seemed odd to those who didn't know Michael Knight. Although he worked for a framing company, and had seven kids to raise, he treasured his time alone on Pontchartrain, spool reeling in redfish, speckled trout, and flounder. Sometimes the kids would come along for a family outing, and he would tie ropes around them all to make sure they didn't dive in and drown. Instinctively, he was a lifeguard, a cautionary fisherman. Only a true lover of Pontchartrain knew how dangerous its riptides and mood swings could be.

Although Knight knew that Katrina would probably flood St. Philip Street, he decided to stay at home, with his family, including his mother-in-law, Rose, and his sister Theresa. Truth be told, he had nowhere to go—without a credit card and living from paycheck to paycheck, he didn't have the means to head to, say, Memphis. He did stockpile water and a little food and he had a couple of generators. Most important, he hitched his Sea Ray boat to the pickup parked in the street, hoping to anchor it down. He had gone to bed Sunday evening with his wife, deciding not to stay up to monitor the weather; they didn't want to make the kids nervous. Although Knight seldom went to church, he had developed a spiritual outlook, derived from listening to Bob Marley, Peter Tosh, Sly and Robbie, and Bunny Livingstone. Every evening, he played Marley CDs and caught the groove. You name the hypnotic Marley song—"Satisfy My Soul," "Buffalo Soldiers," "So Much Trouble in the World"—Knight knew the lyrics to it by heart.

When Knight woke up on Monday morning, St. Philip Street was badly flooded, and his truck was a total loss. But his Sea Ray boat was afloat, bouncing on the water unscathed, a cork in the maelstrom. "I said, 'I'm gonna get in the boat and just take a ride up the street to assess the damage,'" Knight recalled. "But when I got out there, everyone was hollering from every which way, 'Help me!' and 'Get me outta here! And there were elderly people, and young people, just walking around lootin'. It was sad. And no damn police. I refused to pick looters up, but you know, somebody'd say, 'Hey, my momma's ninety years old. She's on So-and-So Street.' They'd give me the address and I'd go ride and find it and knock on the door. Some of them wanted to leave, some of them didn't."

Infused with a missionary zeal, Knight started using his Sea Ray as a rescue vessel. Deonne, with curly bleached blond hair and a love for everything about St. Philip, was his deckhand. Wearing a colorful Hawaiian shirt, denim shorts, and no shoes, with Marley blaring out of his boat stereo, Knight became a one-man 911 operation. Whenever looters would shout at him for a ride, he shook his head no and said, "Fire's burning, man. Pull your own weight." To Knight, this meant that the water had risen, find your own way out; it was every able man for himself, but take care of the weak. And he drew inspiration from the Israeli Vibrations song "Ambush," written by Albert Craig, particularly the verse "I'm just a buffalo soldier / Survival is my game."

Realizing that many poor folks hadn't evacuated the Lafitte Housing Project in Tremé, he started shuttles to the 896-unit complex, parking his boat on the corner of Broad and St. Philip. "So I three stories up and there's an elderly white man and white woman," he said. "They might have been eighty or ninety years old. I grabbed the man out of his wheelchair and I bring him down to the boat. His wife could walk. We put her in the boat and then we get all set and ready to pull off. The old man said, 'Where's my wife?' and I said, 'She's right here in front of you.' And I took his hand and I took her hand and I put them together. When I did that, he smiled. It was just unbelievable."[49]

Over the course of the week Knight rescued around 250 people from the Seventh Ward, many of them infants and elderly. Locals brought gallons of gasoline to his house so he could keep up the work. Coast Guard helicopters, flying over Knight's boat, gladly pointed the reggae skipper in the direction of people stranded on balconies or rooftops. "Once the Coast Guard seen I was rescuing people, once they started seeing I was bringing people to them, they'd give the thumbs-up, 'You're doing a great job, guy,'" Knight recalled. "I'd look up and they'd be over my head and I'd go over there to the bayou and the helicopter would drop down the baskets and pick people up."

Refusing to stop even after dark, Knight commandeered a canoe from City Park and paddled around the Seventh Ward in the moonlight, looking for victims whom he could snatch up in the morning with his Sea Ray, which could fit in twenty people a run. "At night, people would shout at me, 'Hey, can you come rescue me in the morning?'" Knight said. "They'd say, 'I'll pay ya!' And I'd say back, 'No, don't worry! I'm comin' to get you first thing when the sun rise,' and I did. One woman was stuck in an apartment with fifteen kids. I got all dem to dry, safe land." Often, after a puff of sacrament, paddling alone in the flooded streets under the stars, Knight would think about Katrina, whether it was the result of the greenhouse effect or a hole in the stratosphere. Then he turned his mind to the Righteous One and all of his troubles evaporated.

On some of these daring rescues, Knight's wife, Deonne, was his co-captain, guarding the vessel from thieves when he went into the Lafitte Projects to carry out the infirm. She was constantly chasing away "poverty pimps" looking to hustle cash, and "urban foresters" gathering valuable lumber for later resale. She also made sure that there was plenty of Corona in the cooler, and that the pipe was full. Because her husband was a big eater, weighing 220 pounds, he needed to be fed well, preferably home-cooked food. His favorites were T-bones, rib eyes, mustard greens, wild game, and Lake Pontchartrain redfish. A neighbor even brought them a whole beef brisket. Every evening Knight himself cooked steaks and chickens on a grill and with the sky full of stars he launched into "Redemption Song," "Belly Full," "Natural Mystic," and "Kaya." Always, he said, he "kept the charcoal dry." And somehow, when nobody else in the neighborhood had cell phone service, Knight was able to get a signal out. He checked in with his father in Atlanta, who was worried about his seven grandchildren. "Get them the hell out of New Orleans," he instructed his son. "Now, Papa," Knight said, "I'm rescuing people. I just fine."

VIII

Operating with the NOLA Homeboys, but not in tandem, was a fiery African-American woman with a time-honored sense of human justice, reminiscent of such civil rights leaders as Fannie Lou Hamer, Daisy Bates, and Ella Baker. With an angelic smile and a bellyful of gripes about authoritarianism, Dyan "Mama D" French Cole was the Florence Nightingale of the Seventh Ward of New Orleans. Scorning handshakes for hugs, and often dressed in flamboyant West African garb, her Dorgenois Street home became a grass-roots post-Katrina nerve center. "I was sittin' on the porch watchin' Katrina when she came," Mama D recalled. "It was crying, howling, like she lost all of her children. But I didn't mind the storm per se. I've been livin' in the swamp for sixty-one years. We do hurricanes."[50]

A dedicated activist, Mama D had trained with Bayard Rustin back in the 1960s on direct-action campaigns for social change. She became the first woman to head the New Orleans chapter of the NAACP. When the 17th Street and London Avenue canals breached, she knew how to find boats. She always knew how to find everything.

She dispatched a group of "brothers" to pull the poor and sick out of attics in canoes and johnboats. "No person left behind" was her motto. Her heart was especially open to the unhinged underdog, the mentally disturbed, the barewall junkies shaking alone in their flooded quarters. She was the Guardian of the Outcast, her long dreadlocked hair usually pulled back, wrapped in a scarf, her deep hazel eyes always looking to assist the untouchables of the Seventh Ward. "You need something to eat?" Mama D would ask some confused, lost soul on Esplanade Avenue. "You need some food, baby?"[51] Mama D called everyone "baby."

Many of the unevacuated New Orleanians were afraid of the NOPD and the National Guard. Instead of fleeing the bowl, they were trying to avoid public notice, each bullhorn shouting from an armored vehicle or aid truck merely intimidated, pushing them farther back into their Seventh Ward garrets. Starting on Katrina Day—August 29—Mama D would parade down streets like Tremé, St. Philip, and St. Ann, carrying herself with down-home dignity, self-confidence apparent in her every stride. She pushed a cartful of baby formula, chicken noodle soup, toothpaste, deodorant, tampons, prenatal vitamins, and ham sandwiches, and handed them out. Mama D was a walking, singing mini-market, a one-woman mercy machine. She had no agenda but to help the afflicted. "So many of them are scared to come out of their homes," she told *Times-Picayune* reporter Trymaine D. Lee. "But they're hungry, I know they are. So, I just come by everyday and let them get used to my voice and hope they come out." The provisions she passed out had been, in her words, "liberated."

The posse of African-American men around her house was made up of Christians and Rastafarians. They were named the Soul Patrol by Rick Matthieu, after the Jimi Hendrix song "Power of Soul" from the album *Band of Gypsys (Live at Fillmore East)*. Matthieu was the leader of the Soul Patrol, the navigator-operator of their fleet of five boats: a pontoon, a 12-foot flatboat, a 21-foot fiberglass boat with a 200-horsepower engine, a cabin cruiser, and a 16-foot flatboat. What enabled Matthieu to act so efficiently was his childhood experience with Hurricane Betsy. He had been piloting boats since he was seven years old. "With my dad during Betsy, we saved lots of folks," Matthieu recalled. "I prepared on Friday night before Katrina because my dad had prepared in the same way for Betsy."[52] During that first post-storm week they saved at least five hundred New Orleanians. They were the very antithesis of looters. The Soul Patrol, like Michael Knight, just kept putting their boats in the sludge, saving babies, priests, and the disabled. Among the members of the Soul Patrol was Earl Barthé, his red, yellow, and green Rasta cap always on. He was the swimmer extraordinaire who dived right into the flooded houses. Some of the other members were Ortegas Coleman, a Black Cross medic on assignment from Mama D, who knew no fear; Ricky Thompson, another Black Cross member, who looked like the all-American athlete next door; and Ronald "Fat" Davis, a man never without a parlor joke. They all shared a deep-seated belief in direct community action, and were suspicious of the National Guard's rescue orthodoxy. "They were real good," Mama D said. "Unlike Mayor Nagin who was a poor excuse for a brother."[53]

Word was out in the Seventh Ward that the Superdome was a horrific place for evacuees, as was I-10. The Soul Patrol decided that many of the Seventh Warders were better off in the unflooded sections of their own shotguns or apartments than in the cesspool of those so-called staging areas. The Mama D gang had their own way of doing things: herbal medicines, folk cures, nonviolence, vegetarian food, no guns, One Love. It was a holistic haven that didn't care for the U.S. government or City Hall. In the post-Katrina days, Mama D built a cinder-block fire ring, a barbecue pit that served as a neighborhood powwow center. It was located outside her grandmother's house, where she had lived since the 1950s. With the Red Cross and FEMA nowhere to be found, Mama D and the Soul Patrol took it upon themselves to dole out compassion. In the Seventh Ward, they didn't just respect Mama D; they *loved* her. And for all her maternal bear hugs, when Mama D caught a whiff of inequity of any kind, she confronted it with the old Black Panther, in-your-face determination, which was stylistically anachronistic everywhere except in Oakland, Detroit, and Newark. She was a provocateur, a woman of the left, the breathing epitome of Sojourner Truth's "Ain't I a Woman" speech. Testifying before Congress on December 5, 2005, about the Seventh Ward, Mama D stole the show. "You know what we need right now?" Mama D asked the legislators. "We need y'all to take FEMA and the Red Cross . . . just get 'em outta my neighborhood."[54]

IX

Once Michael Prevost, the private-school administrator, left his Lakeview home in a canoe on Monday afternoon, he too started paddling toward NOLA Homeboy status. Every house in his upper-middle-class neighborhood, between the 17th Street Canal and the Marconi Canal and City Park Avenue, was swamped. All of the boats and yachts in Lakeview were piled up like firewood, were dumped on top of houses, or had become surreal decorative items on front lawns. Popular Lakeside businesses like the Robert E. Lee Shopping Center, Plantation Coffeehouse, and Basin Bar were under six to eight feet of water. Helicopters were hovering over the breached levee, shining huge searchlights on the gaping holes. As the winds continued to drop, Prevost was hailed by neighbors who shouted for lifts. Without a motor and with the floodwater still choppy, he cautiously started plucking down the "roof people of Lakeview," one by one. Early on, near the University of New Orleans (UNO) campus, he helped two Chinese men who had been in America for only two weeks. "Is the water going to get higher?" one of them asked, panic-stricken. "We've never seen this much water." All Prevost could say was "Neither have I."

That evening Prevost commandeered an apartment at UNO along with two African Americans he had rescued. Frogs croaked all night long like a choir—something never heard before in Lakeview. The sky was so gorgeous it was like camping in the Rockies on a perfect summer night. With his mutt, Chelsea, at his side, Prevost slept on the balcony. He enjoyed being out under the stars, an unexpected aspect of Katrina.

At dawn Tuesday NOLA Homeboys were operating all around Lakeview, rescuing people off of roofs. Prevost joined the action, leaving Chelsea on the apartment balcony. He had two simple missions: "Paddle around and save people, and survive." Although he later hated admitting it, there was something soothing about paddling around the flooded streets, all modern communications systems down, and picking shrunken oranges off trees for sustenance. He was like a pirate craving lemons and limes. He was an urban Thoreau who thought saving dogs was nearly as important as saving people. His greatest failure, to his mind, was his inability to lure to safety a scared dog that was stuck in some tree branches. "It wouldn't trust me," Prevost lamented. "It kept snarling at me and I was in no position to get bit. I had no first aid and was in the water. So, sadly, I had to leave it behind." Prevost had another bad experience. A friend had sent a text message about saving his two dogs left behind in a house. Prevost paddled to the appropriate address, only to find that they had drowned.

At dusk on Tuesday, Prevost thought about canoeing right into the Superdome. But he heard that dogs were banned there. He didn't want to go back to Lakeview, because looting was becoming a huge problem. At one point, Prevost spied two looters, each with a pack on both front and back. "I tried to stay out of their view," Prevost said. "They were almost drowning with the weight of their booty packs trying to cross Robert E. Lee Boulevard. I could hear one of them cursing at the other, saying, 'I told you to get rid of that big pack,'" Prevost said. "I knew they couldn't cross Robert E. Lee—they would have to swim, the water was eight feet deep."

To understand fully what NOLA Homeboys like Knight, Mama D, and Prevost accomplished, it helped to appreciate how dangerous New Orleans was when they were in the water. At dusk on Tuesday evening, Jimmy Deleray, walking in his Riverbend neighborhood, witnessed felonies being committed not far from his house. Riverbend businesses were looted that week. Deleray saw Jet Skis being stolen from the Aqua Marine Inc. at the corner of Oak and Monroe. "The looters had pickup trucks," he explained. "They would tie a chain to the door and pull. Suddenly, they raided the shops like a pack of rats. At this point I knew I had to protect the neighborhood."[55]

Boldly, Deleray decided to confront a group of looters stealing drugs from the Castellon Pharmacy. Packing the .22 rifle his father had given him as a boy, Deleray pointed the gun at a looter in his late fifties from the window of his Ford pickup. The guy dropped the drugs. Deleray drove his Ford up on the curb and started running over the plastic pharmaceutical bottles. "I established that they weren't in control of the neighborhood," Deleray recalled. "They threatened to come back with AKs. Screw them. Other neighbors joined me in controlling our neighborhood. We started boarding up places wherever we could. Then, as we patrolled around, we saw them looting. They'd stolen a forklift and rammed it into the front door. Now they were running out like rats with arms full of makeup and stuff. I yelled at them to stop. They completely ignored me. I then fired twice into the ground. That did the trick. They ran. It was sad—grandmothers were encouraging their grandchildren to loot. What kind of people would do this? What did they think? They acted like it was a shopping spree for free."[56]

And so the widespread looting continued. Houses were burglarized, clothing stores were trashed, and cash registers were stolen. It was the brazen nonchalance of the thievery that infuriated Deleray most. He reported that at one bar, somebody had defecated on a tabletop in "some kind of bizarre animallike toilet dominance trip." It was the big dump, and it was becoming commonplace. The looters had come from a poor neighborhood between Oak and Claiborne known as Hollygrove, back by an old waterworks compound. "I started filming them," Deleray recalled. "Once it got around that I wasn't going to shoot them, they ignored me. One guy said, 'Hey, man, don't be filming me, I's got to get stuff. You understand?' " All a ticked-off Deleray could shout back was a defiant "No, I don't."[57]

The people from Hollygrove who preyed on the Riverbend shops had long been angry about racism in New Orleans. It often showed up in their music. While Deleray listened to Cajun fiddle and hillbilly jazz at the Maple Leaf Bar, he had no ear for hiphop, not even the work of Lil' Wayne, who grew up nearby. All Deleray had to do was listen to Lil' Wayne's song, "Oh No," in which he rapped, "Get that look off ya face and recognize you got a crook in tha place." Or perhaps B.G.'s anthem "From tha 13th to tha 17th," where he and Lil' Wayne told the Deleray white-breads of Riverside exactly what was on their minds. Like cheerleaders they spelled out H-O-L-L-Y-G-R-O-V-E in the cut, where they warned, "Niggas be tryin' to creep / So you better watch your back." Deleray had no tolerance for the kind of inner-city hiphop that Lil' Wayne and B.G. slammed. The very thought of such violent lyrics made him angry. Songs like "Danny Boy" and "Whiskey in the Jar" were more to his taste. He had never heard such phrases as "bling bling," "drop it like it's hot," "H to the Izzo," or "Holla Back, Young'n," and if he had, he would have dismissed them as stupid, infantile, gutter babble. When the lights went out August 29, however, like it or not, Jimmy Deleray was living in a gangsta's paradise.

X

One leading Louisiana politician, a New Orleans native, also became an honorary Homeboy rescuer, a flood-dog, putting his life on the line to help others. Lieutenant Governor Mitch Landrieu had spent Monday night at the Baton Rouge EOC, trying to assess the damage and wondering how he could be most helpful. Deeply worried about towns in St. Bernard Parish, where he heard that a "Wall of Water" had knocked communities right out of the grid of civilization, he was desperate for details. State Representative Nita Hutter of St. Bernard Parish miraculously got through to him on his cell. "She was frantic," Landrieu recalled. "She was stuck along with Junior Rodriguez, Joey DiFatta, and other St. Bernard officials in the government building. She kept me updated throughout Monday night. I heard that the Ninth Ward in New Orleans had flooded. I decided that there was really no role for me at the EOC, where press conferences were advising the public about what was and what wasn't. I told my staff early Tuesday morning that we were going to New Orleans and Chalmette."[58]

Landrieu—accompanied by state troopers Troy McConnell and Sheldon Perkins and Lieutenant Jay Diez of Wildlife and Fisheries—headed down I-10 to New Orleans. Coming from the north, they never got as far as the city, or St. Bernard Parish. Arriving in Jefferson Parish, they could see it was a complete mess. With the pump houses still abandoned, a major portion of the parish was flooded. Parish President Aaron Broussard, who was doing a lot of national media and had been commended for his evacuation strategy, was starting to find himself the scourge of Metairie and other eastern towns in the parish for his Lake Pontchartrain pump policies. Taxpayers in these districts had always imagined that they were protected from flooding because of the pumping stations. But, as mentioned earlier, Broussard hadn't wanted his operators to drown and ordered them to evacuate their posts.

On the upside, Parish Sheriff Harry Lee had launched a massive boat rescue operation at I-10 and the Lake Pontchartrain Causeway. "We went to the mat," Lee said. "We put every asset we had into play."[59] With shirtsleeves rolled up and ball cap in place, Landrieu reached the Louisiana Wildlife and Fisheries boat launch. He was anxious to get to St. Bernard Parish or the Lower Ninth Ward. "At this site everybody was trying to figure out logistics," Landrieu recalled. "There were boats, but you can't put a boat on Lakeview and Clearview and go directly to St. Bernard. It took forever." Landrieu decided to weave his way east along any roads that weren't flooded. He drove his car down Causeway, all the way to the Jefferson Highway, then to River Road, near the Rivershack Tavern. From there it was straight into the Audubon Park area of Uptown. This was his neighborhood: he lived on Octavia Street, and his father, the former mayor Moon Landrieu, lived nearby with his mother.

Everywhere the lieutenant governor looked there were downed trees, utility poles, and telephone wires. Magazine Street, around Whole Foods, Scriptura Stationery, and St. Joe's Bar, was covered with debris, making it nearly impassable with anything less than a Hummer. Landrieu's instinct told him to cling to the Mississippi levee, which hadn't breached. To that end, he got on Tchoupitoulas Street and took it all the way down to the French Quarter, around Faubourg Marigny. Then to the partially submerged St. Claude Avenue Bridge, where he joined members of LDWF who were launching boats into the Ninth Ward and St. Bernard Parish. "When you went into the water in St. Bernard and Lower Ninth, there was always a whole story of a house underwater," Landrieu recalled. "The only way you could get down the street was to figure out where the big, tall streetlights were, so that you could identify the fact that you were indeed on a street. You were worried about whether or not you were going to hit a fence underneath you or whatever. People were up on roofs, it was one after another. We had W&F agents, who had never been in New Orleans, actually saving folks, and there were people on the ground saving each other. The citizens of the Lower Ninth Ward were saving the Lower Ninth Ward. There was really no federal presence at this point. Just W&F and local heroes."[60]

A lot of stranded New Orleanians clutching the rails of the St. Claude Avenue Bridge recognized the lieutenant governor. Just seeing such a high-profile politician working with Wildlife and Fisheries was a morale booster. Stunned by the damage in the Lower Ninth Ward, Landrieu realized that Katrina had wreaked at least ten times more havoc than Betsy. Immediately, he became a rescue ringleader, riding with firemen and police officers, extracting the helpless from flooded buildings. Most folks he saved were grateful beyond words, he recalled. It was sad watching people leave pets. "You've got to understand," Landrieu explained. "You go to one house and you put the victims in a boat, and the boat starts to fill up. You've only got so much room and all of a sudden, you've got fifteen people in a boat. And that's all it can take."[61]

What Landrieu gathered, talking to dozens of Lower Ninth Warders, was that they had truly believed that Katrina was going to hit Mississippi or Florida, and that living sixty miles inland from the Gulf of Mexico would protect them anyway. "Some of those people recognized me and we talked," Landrieu recalled. "And I'd ask them questions:

" 'You know who I am?'

" 'Yeah.'

" 'You heard us tell you to leave?'

" 'Yeah.'

" 'Why didn't you?'

" 'Well, we didn't think it was going to hit us. We didn't think the water was going to come up that fast. We thought we'd be okay. We didn't think it would be serious.' "[62]

Like the NOLA Homeboys and Soul Rebels, Landrieu later had lighthearted stories about some of the stubborn Ninth Warders he encountered. At one home, with floodwaters inundating the first floor, Landrieu saw a middle-aged guy stuck on the roof. "Come on down," Landrieu shouted up at him from the boat. "I don't want to go," the man shouted back. A perplexed Landrieu, cupping his hands around his mouth, asked "Why not?" The response was idiotic, yet somehow very New Orleanian: "Because," he yelled, "I want to get rescued by a helicopter." A helicopter? "You've got to be kidding me!" Landrieu shouted back. "You better get your ass in this boat and you better get your ass in this boat now!" The disappointed man murmured, "All right, all right, give me a few minutes."[63]

At another house Landrieu anchored his W&F boat in front of a Katrina victim who didn't want to evacuate because the stagnant water in the shade of his house was too frigid. "I don't want to come down," he pleaded. "The water's too cold in my house." All Landrieu could do was scoff and shake his head in disbelief. He tried to persuade the man and then gave up. There were other folks who needed saving. There was only so much you could do. "People were just funny," he said. "Many just didn't want to be bothered, or didn't want to leave behind whatever possessions they salvaged from the flood."[64]

After dark on Tuesday, when all LDWF rescues were halted for the day, Landrieu drove back to Baton Rouge to deliver the grim news. "I told the folks at EOC," Landrieu said, "that it's worse than we originally thought." Because of his Catholic training and penchant for Jesuitical thinking, Landrieu asked himself one question on Tuesday evening in Baton Rouge: Where am I most needed? The St. Claude Avenue Bridge had become a well-established rescue center. The LDWF colony (with some NOPD officers and fire department first responders, and a group of NOLA Homeboys) was working out of this staging area in overdrive. And the media had descended on the Lower Ninth Ward because it wasn't far from the French Quarter. It was ground zero for rescue efforts. A national spotlight was on the area, even though the federal government, with the notable exception of the Coast Guard and a few Louisiana National Guardsmen who escaped the Jackson Barracks flood, was still a no-show. The answer became self-evident to Landrieu: St. Bernard Parish. That's where he headed first thing Wednesday morning.

XI

Right after Katrina had moved northward on Monday evening, Heather Allan of NBC News asked Tony Zumbado and Josh Holm to find a route out of New Orleans with their Ford 350 Econoline van and bring back the Hallmark motor coach, which they'd parked in Gonzales. Her team gladly followed the order. Quite mistakenly, however, they first ventured to the Ninth Ward and encountered massive flooding. They circled back and made it across the Mississippi River Bridge to the West Bank. At an abandoned 7-Eleven store they saw a lone police officer. He was cordial but clearly frightened, acting like an escaped prisoner wondering which way to run, frozen in his own heels. Holm asked if he knew a way to Baton Rouge. "I know a way to get out of here," the officer said, "but I don't know how bad the roads are."

Zumbado and Holm decided to risk it. Anyway, they didn't want to face the wrath of Heather Allan if their mission failed. To their surprise the roads weren't too bad. There was a couple feet of water on the pavement, but lots of trees and power lines were down. "Around midnight, we just parked the van and slept in the middle of nowhere," Holm recalled. "There was too much water in the road, plus darkness." At sunrise they made it to Gonzales, retrieved the motor coach, and headed back to New Orleans; in daylight, the devastation was stunning. When they arrived back on Canal Street, they told Allan the good news that there was a way in and out of the city, so provisions and generators could easily be brought into the bowl. Their producer got ahold of the CEO of General Electric, NBC's owner, and Shazam generators were flown into Baton Rouge. "We were kicking ass," Zumbado recalled. "Heather Allan was on the phone saying, 'We need this' and 'We need that' and 'We need it now.' So General Electric sent us everything: motor homes, trucks full of food, rations, all kinds of stuff, prepackaged food. They even sent us a water tank, a gasoline tank, and a diesel tank." The supplies were for NBC employees, but half of them ended up going to displaced persons they encountered.

What NBC understood was that the levee breaks had turned Katrina into a long-term story. A decision was made that the company would eventually create a "Camp NBC" next to the headquarters of local affiliate WDSU just off of Lee Circle. Allan, realizing that Zumbado's skills as a cameraman were needed in the field, pulled him off of logistics. In the rented canoe, and with Holm at his side as soundman, he started videotaping in the Central City neighborhood, filming people stuck in homes, animals left behind, and octogenarians crawling out of windows and on top of roofs "wailing down the street," as Zumbado put it, for help. He saw people coming out of nowhere, water up to their chests, or even to their chins, jumping on their toes to keep the water out of their mouths. They were carrying personal items in travel bags and pillowcases full of photographs. One group had a floating table with two grandmothers sitting on top of it. Others were using coolers as rafts to float children or other loved ones down the street.

All the first floors of houses were underwater. The streets were full of everyday items bobbing in the water: tea kettles, wine racks, paint cans, paintings, gasoline cans, shower curtains, bed mattresses, juicers, and so on. "So we were in our little boat and people were yelling, 'Help us! Help us! Our grandparents are here!'" Zumbado recalled. "We went up to a house, we videotaped, and we helped the elderly into our boat. We were like, 'Oh my God, now we're caught up in rescuing.' And I'm sitting in the boat thinking What am I going to do? I have to videotape or I have to rescue? I was in a dilemma and I told Josh, 'Let's get out the camera here on this porch, you stay and film what you can. I'm going to go help rescue these people. We're going to get the people the hell outta here. But once I get them out, I'm picking you back up and getting out of this neighborhood. I was overwhelmed. There weren't many Coast Guard and [LDWF] boats at this point. Just me and a couple of locals in boats."

During the next hour, Zumbado ferried twelve or fifteen people out of Central City, along heavily flooded Claiborne Avenue, dropping them off on a dry corner of St. Charles Avenue. He then swung back to Canal Street to hand Allan a stack of his videotapes, which were soon running on NBC. Then Zumbado raced along Canal Street to tell the NOPD officers he encountered to join the rescue operations. They merely shrugged and said, "Yeah, there's people all over the place." The cameraman was incensed at how lethargic they were. He had heard that hundreds of people had been left behind at Memorial Medical Center on Napoleon Avenue. He enlisted an NBC correspondent and, along with Holm, started canoeing.

Meanwhile, NBC News anchor Brian Williams, after breaking the Superdome roof story, fell terribly ill from dysentery on Tuesday. He possibly contracted the disease by ingesting water containing bacteria, while doing a *Today* show appearance. He was standing next to flood water, sipping distilled Kentwood Water, when he noticed a trickle of brown on the side of the plastic bottle. A few drops of the sewage water had accidentally gotten into his mouth. "I was overcome with dysentery late Tuesday," Williams recalled. "I was fading in and out. Somebody left me on the stairway of the Ritz-Carlton in the dark on a mattress."[65]

The amoebas had flattened Williams, who was virtually unable to move. He was burning up with fever, experiencing sharp cramps. Just forty-eight hours earlier, the Ritz-Carlton had seemed like a smart place to be. It was a cavalier joke among NBC employees, in fact, that they were weathering Katrina ensconced in the spalike luxury of the Ritz. Only Tony Zumbado really felt the bad vibe. Without air-conditioning, by Tuesday the building felt like a stifling tomb. Armed gangs had broken into the 527-room hotel, brandishing guns and terrorizing guests. Williams, in fact, had seen his first corpse floating down Canal Street from his eighth-floor window earlier that day. Then the fever consumed him. Delirium. He couldn't eat or drink. Jean Harper, the producer of *NBC Nightly News with Brian Williams,* found her anchor in a very alarming state. "Jefferson Parish [Reserve Sergeant] Matt Pincus [of the sheriff's office] was in the Ritz and Jean brought him to me," Williams recalled. "I was about eight or ten steps from the exit door. They were going to lock in or down the Ritz, shut it to keep the gangs out. Nobody was allowed out. No exceptions. It wasn't the kind of situation where anybody recognized me or anything like that."[66]

Williams asked Pincus to help evacuate him out of the Ritz. A Newark police officer on the scene also intervened on Williams's behalf. They were both from New Jersey—an automatic bond. A triage center had been set up in the Ritz lobby, and Williams was desperately in need of an IV, but he declined. "I had to refuse," Williams said. "There were so many ill people in line who needed it more than me. My conscience wouldn't have felt right if I had tried to pull rank. But I was in pure hell. I had no medicine, nothing."[67]

In order to get picked up by the Louisiana state police, Williams had to wade outside, in two feet of floodwater, barely able to stand. A gang was waiting on the streetcar tracks in front of the Ritz, ready to "smash and grab," as Williams put it, to take the vehicle. Eventually, a huge Toyota Land Cruiser pulled up, thirty-six inches off the ground, manufactured for the wilds of Africa. It was one of the few commercially made vehicles which just might be able to persevere through such high water. At Williams's side were Harper and Jack Bennett, an NBC technical manager. A group of Louisiana National Guardsmen pointed guns at the gang, making sure the NBC trio didn't get their escape vehicle hijacked. "They aimed weapons at the men on the street," Williams recalled. "Then we were on our own."[68]

That Wednesday evening Williams, Bennett, and Harper slept in a trailer along Canal Street, trying to avoid looters and floodwater. "I couldn't keep anything down," Williams said. "That whole night was hazy. I couldn't get clarity of mind." The next morning, Thursday, they made it to Metairie. Even though he was consumed with fever, he continued broadcasting the Katrina story. Blanket coverage. When off-air he spent most of the day groaning in the NBC trailer. "Somehow," Harper recalled, "he marshaled his resources and kept on reporting." They were toughing it out. "That evening we slept in a Metairie car dealership parking lot," Williams recalled. "I don't remember much."[69] He got evacuated out of New Orleans on Friday night after his news broadcast and recuperated at his home in New Jersey. With proper pills, uninterrupted sleep, and plenty of liquid, he was back in Louisiana in a few days. "With the levee breaks," said Harper, "Brian realized that Katrina was one of the biggest news stories of his life. Even with dysentery he never missed a broadcast. He was an on-the-spot reporter, really, not an anchor in New Orleans. He was at his best."[70]

XII

Even with three of the nine canals leading south into the city from Lake Pontchartrain pouring water into neighborhoods, a few New Orleans businesses refused to close, like Johnny White's Sports Bar, located on Bourbon Street. The owner J. D. Landrum had gone to Baton Rouge for provisions, including lots of ice-cold beer. He was determined not to let Katrina upset his sterling record of staying open nonstop (just about) for sixteen years. With the electricity out, votive candles were used for illumination. It felt like a Revolutionary War–era tavern, with a bohemian cast, as customers' silhouettes flickered on the memorabilia-cluttered walls. Eventually somebody donated a generator, which allowed the ceiling fans to run. The old wooden sign outside hadn't been knocked off by Katrina, so advertising wasn't a problem. Huge boxes of Fritos, saltine crackers, and toilet paper were also brought in from out of town. A battery-operated boom box played Johnny Cash and Tom Waits. Quickly, the bar took on the role of a community center, a gathering spot for artists and other denizens of the French Quarter, a NOLA Homeboy crossroads. Generation Z twentysomethings with tattoos and body piercings also congregated there. Shell-shocked senior citizens likewise emerged to socialize while complaining about FEMA, President Bush, and the Red Cross. It felt good to carp, even if it didn't help the grim reality of post-Katrina New Orleans.

What they understood at Johnny White's, however, what most Americans had yet to accept, was that New Orleans was no longer a city. It was a smattering of islands, rising out of the 80 percent of the city's land that was submerged. Anywhere from 50,000 to 70,000 people were left in the city, two-thirds were in dire trouble on Tuesday—on the verge of drowning, or trapped on rooftops or in attics. The 17th Street Canal was running into Mid-City, and a 500-foot section of its concrete wall had already crumbled. Another twenty blocks to the east, the London Avenue Canal was pouring water into the Ninth Ward, which was already flooded from the overflow from St. Bernard Parish. That eastern suburb was almost entirely underwater, both from the storm surge pushing over the walls of two hurricane levees, and from the later breach in the Industrial Canal, which ran south from the lake to the Mississippi River.

In sum, the city lay in ruins, its sorrow palpable. Tuesday was a day when individuals recognized with a jolt that they were entirely on their own. New Orleans didn't make sense anymore. Protection from danger was gone. Water was undrinkable. Emergency care was nonexistent. Seniors were trembling all alone, left behind by careless family members on the absurd assumption that they'd be okay. Many of them died in terrified solitude—no white-suited nurse with a clipboard holding their trembling hands, no nun telling them God was near. Just a mad rush of sewage water, filling up their lungs, snuffing them out with cruel vengeance. The oxygen tanks emptied and, after that, in some cases there wasn't even a soul around to witness their last breath. One can only pray that they were in a coma or unconscious at the grim time of reckoning. "Last night when I went to bed, I didn't have no water in front of my door," one New Orleans senior, Albertine Arseneau, told National Public Radio on Tuesday. "This morning when I woke up, my house has water in it. I'll have to spend the night on the second floor if I stay the night. I don't know how to swim."[71]

Josephine Johnson, a neighbor of Arseneau, didn't have even that much of a chance. She lost her life when the water rose through the house she shared with her son, brother, and other family members. As fifteen feet of water swept into the living room, the rest made a run up the stairs to the attic. Josephine, well into her eighties, couldn't make a run toward anything. She drowned in the house and the family, crouched in the attic, knew it. After flagging down a passing LDWF boat, they escaped the house and were dropped, along with many others, at the rusty St. Claude Avenue Bridge, still jutting out of the floodwaters. Josephine's eighty-seven-year-old brother walked up the incline with a walker. For the moment, the family was safe. It was then that the moment tore open for them. Josephine's son broke down, overcome with grief. "My mamma drowned," he wailed, "my mamma drowned!" He couldn't seem to stop, and no one on the bridge—all knowing what he had been through—expected him to.

Not far away on the bridge, Daniel Weber was also crying uncontrollably. He and his wife had been trapped by the water in their home. They escaped. But they weren't out of danger. "My hands were all cut up from breaking through the window," the fifty-two-year-old man said, through his sobs, "and I was standing on the fence. I said, 'I'll get on the roof and pull you up.' And then we just went under."[72] Weber's wife didn't surface. After floating in the dirty water for hours, he was rescued. He was dropped on the bridge, where he huddled with hundreds of others, newly homeless or widowed, like him. "I'm not going to make it," Weber mumbled. "I know I'm not."

A reporter from the *Times-Picayune* described the scene at the drawbridge, a collection of highly traumatized people: "As they emerged from rescue boats, at times wobbling and speaking incoherently, many of the rescued seem stunned they had not died. Jonell Johnson of Marais Street said she had been trapped on her roof 'with a handicapped man with one damn leg.' Gerald Wimberly wept as he recounted his unsuccessful effort to help a young girl, whom rescuers ultimately saved. Dupree said he had seen a young man he knew drown. 'I just couldn't get to him,' he said, 'I had to tell his people.' "[73]

XIII

The sky on Tuesday was dotted with helicopters, including sixteen from the Coast Guard, another sixteen from the Louisiana National Guard's I-244th Aviation Battalion and 812th Med-Evac wings; five from the Navy's warship *Bataan,* five from the Wisconsin National Guard, and three from the Georgia National Guard. All together, National Guard units had thirty-two helicopters operating along the Gulf Coast starting late Monday.[74] The number steadily increased. The region's primary private ambulance service, Acadian Ambulance, also had its fleet of helicopter and ground ambulances continually at work. All of the rescue teams were staffed by people who didn't wait for orders from on high. Their mandate was from the people in need, of whom there were many.

The Coast Guard continued to gallantly rise to the dire occasion. While the section headquartered in Alexandria rescued people in boats and baskets, they welcomed the help from Coast Guard outfits all over America. "We have air crews that have been brought in from Cape Cod to Miami, west to North Bend, Oregon," said Coast Guard Captain Pete Simons. "In the New Orleans area alone, we have twenty-two aircraft, helicopters that have been conducting rescue operations."[75] One of the people taken from the roof of a flooded home was himself a helicopter pilot with a National Guard unit. He was no sooner dry than, putting thoughts of his own loss behind him, he took a shift rescuing others. Of the rescue process, Captain Simons explained, "It looks more precarious really than it is. The first step is generally that we lower a rescue swimmer down to brief individuals on what's going to happen. And then they lower the rescue basket down again and one by one retrieve them from the rooftops or wherever they might happen to be."[76] Viewed on TVs all over the country, however, it looked very precarious indeed and left searing images of Louisianans needing military-style rescues.

The efforts of the NOLA Homeboys, Mama D and the Soul Patrol, the Coast Guard, the Louisiana Department of Wildlife and Fisheries, and the various out-of-towners who had arrived to help rescue New Orleanians resembled the heroic evacuation of British troops from Dunkirk, France, in 1940. In the same way and with the same homespun navy, tens of thousands of victims of Katrina were rescued, one, two, or a half dozen at a time. As of Tuesday, the Coast Guard reported that it had saved 1,200 people.[77] "Nobody in the press knew," Jimmy Duckworth recalled, "but the name of our effort was Operation Dunkirk."[78]

Not all of the helicopters buzzing around New Orleans on Tuesday were emergency aircraft, however. Corporations had hired copters to help save documents from their buildings or to evacuate their employees. The broadcast media used helicopters to shoot the footage that showed Americans that the situation in New Orleans was not under control. By midday on Tuesday, the truth about New Orleans was revealed as CNN viewers saw a middle-aged woman climb from a rooftop into a basket that was then winched into a helicopter hovering a hundred feet in the air. All around, the streets were rivers and the houses barely protruded from them, but the drama was contained in the sight of a civilian so vulnerable that the only means of escape was through that terrifying trip in a basket. Virtually all those rescued praised the Coast Guard for showing tenderness under duress. In his blog, Raphael Obermann told of how his airman gave him "a big hug" for surviving.[79]

Because the first responders on the water weren't being filmed live, the Coast Guard helicopters received the most instantaneous acclaim. The most dramatic early film footage and photography came from the sky. "We photograph the rescues from 700 feet up," *New York Times* staff photographer Vince Laforet wrote in his blog later that week after getting to do aerial photos. "We can't hover so we make constant loops—hovering is too dangerous in a single engine chopper. It's dangerous, period. This isn't time to take extra risks anyway. I'm shooting with a 500 mm and a 28-300 zoom mostly. The door to the helicopter is removed. The rescues are amazing. These 'frogmen' that lower themselves out of the choppers from a hundred feet or less are incredibly brave and lucky. It looks likes a well-rehearsed ballet. People are so tired, dehydrated, hungry and/or scared that you don't see much emotion on their faces as they are being brought up into the chopper. Many of these people have never flown before. Some of these neighborhoods were so poor that they had few phones, power or

TV—and they never even knew Katrina was coming. I was mad at people who stayed behind at first—didn't they realize others would have to risk their lives to save them? Now I realize, many were too frail, too poor, or didn't know what the hell was on its way. Some were mothers who had handicapped children they couldn't get out—an impossible situation. All of this is crazy."[80]

XIV

As Jimmy Duckworth told the story, two of his petty officers came back to Alexandria exhausted. They truly needed some R & R. They had black circles under their eyes when Duckworth sat them down for a debriefing. "They hadn't bathed in days and they were just grungy and dead-ass beat," Duckworth said. "They sat down and told me that at the start of their watch they had shots fired at them in near proximity. They continued on with their mission. They wound up evacuating some people, including a nurse, from a hospital. The nurse had been severely beaten and told our crew that she had been gang-raped. The crew took them to safety and then, later on their watch, wound up doing a second-floor entry into a flooded building. Upon entering, they realized it was either an old folks' home or a type of hospital. Every room they went into, there were dead people."[81]

According to Duckworth, his two petty officers suddenly heard a noise and their hearts jumped, scared out of their wits. "They found an old black gentleman who had a cross on his neck and he said that he was religious," Duckworth recalled. "There were two other old people in the room who were foaming at the mouth and gasping and they were dying. And the crew said, 'Do you want us to call a helicopter? Get these people out of here?' The old man said, 'They're too far gone; they'll never make it.' A moment of silence ensued and one of the petty officers asked, 'Would *you* like to be evacuated?' and he replied, 'Come back tomorrow. I'll stay with them until they're gone.' "[82]

Sixty-year-old Mary Fortune lived on Frenchman Wharf on Downmann Road in New Orleans East. A disabled mother of four, she decided to stay home during Katrina, not wanting to be inconvenienced by a ride out of town. At her side was her feisty twenty-one-year-old daughter, Brandi Idris, who was seven months pregnant. Watching the bumper-to-bumper traffic on the WDSU news, Idris had reasoned that she was in greater risk of having a miscarriage via a fender bender than through the impact of Katrina itself. Anyway, *somebody* had to stay and take care of Mama. But when the Industrial Canal levee was breached on Monday, she knew they'd made a terrible mistake. "Water started racing into our apartment," Idris said. "Mama and I went to the second floor of our apartment complex. All around us there were bricks toppling out of buildings and pipes busting and roofs kicking off. I thought we were all gone dead for sure. I cried for my baby that I'd never see."

Mary Fortune and Brandi Idris spent Monday night stranded on a balcony without food or water. On Tuesday morning, however, they heard Coast Guard choppers in the air. As luck would have it, the Guard had just been at a rooftop in New Orleans East, rescuing a teenager in a wheelchair who had known of Mary and Brandi's plight. Coast Guard rescuers got the message to their headquarters in Alexandria and Lieutenant Commander Duckworth directed a helicopter to make the rescue. "They dropped a basket down for Mama and me to climb in," Idris recalled. "I was scared silly. I thought we were going to fall in the water. Because of the blades, water blew in our face."

Words can't describe how grateful they were to the Coast Guard, which then took them to Louis Armstrong International Airport. From there they were driven to an evacuee staging area in Metairie. Their only complaint was that they were given nothing more in the way of sustenance than a warm bottle of water and no food. Eventually, they were taken on a bus to a shelter in Houma. "We got treated well in Houma," Idris said. "They brought doctors and nurses to see me quickly. And they told me my baby was going to be okay. And guess what? She was born in Alabama on Halloween. And she came out fine. I named her Savannah."[83]

What perplexed many Americans was the absence of a U.S. Navy presence in New Orleans. Shouldn't they have been the first responders? Why didn't Navy ships just come up the Mississippi, dock near the French Quarter, and start rescuing folks like the Coast Guard? These were fair questions. However, the Navy had precious few resources to spare. Just like the National Guard, the Navy had a lot of people and ships in the Middle East. Second, the looting had changed the mood and tenor of New Orleans. Secretary of the Navy Gordon England had concerns about bringing warships into an urban riot, where the police force had disintegrated and Coast Guard helicopters were being shot at. "We can't blame the Navy for being cautious," Jimmy Duckworth recalled. "We had numerous reports of rounds being fired in close proximity. If their ships had come into the Port of New Orleans, without proper security, there is no telling what would happen."[84]

The U.S. Navy would, after a critical delay, make a contribution to the recovery of the post-Katrina Gulf South. On Tuesday, the Navy sent four ships to the region to aid hurricane victims: the amphibious assault ship *Iwo Jima,* the dock-landing ship *Tortuga,* the amphibious transport dock *Shreveport,* and the rescue and salvage ship *Grapple,* all based at either the Norfolk Naval Station or Little Creek Naval Amphibious Base in Virginia Beach.[85] Unfortunately, the trip took two long days so they wouldn't be on the scene until Thursday or Friday. The 844-foot-long *Iwo Jima* first stopped in Biloxi to offer hovercraft and supplies for Mississippi, and then made its way to New Orleans. "We've got a lot of work to do," Rear Admiral Reubin Bookert, commander of Amphibious Group II and the *Iwo Jima,* said when he arrived on September 2. "Totally devastated situation. My first reaction was shock."[86]

XV

Not all the local heroes were water rescuers. At East Jefferson General Hospital in Metairie, the staff was relieved that they and their 3,000 dependents had safely come through the face of the storm and the subsequent flooding. But looking out the window, they could see that the residents of a nursing home across the street were still in their one-story building. The residents were attended by caretakers, but all of them were trapped by the rising water—and by temperatures that hung in the low nineties. With the water chest-high, the hospital staff made their way across the street to help. "They had one fan," said a nurse who went with the search party. "They had a dog over there. They had that one fan on the dog." While that may have surprised the nurse, the deep importance of pets was a common theme during the hurricane. Many of those caught in the storm believed that their pets were no less valuable than their family. For many, a pet was the only family they had, in fact. As the hospital staff prepared to escort the nursing-home residents across the street, where they were promised a bus ride out of the flood zone, one of the elderly residents was too proper to allow her standards to relax, even in the face of disaster. "If I'm going on a bus," she said, "I have to get my good dress on." The hospital staff took the time to find a dress in her closet.[87]

"As many bad stories as you hear about looting," noted Coast Guard Lieutenant Chris Huberty, the pilot of a night-flying copter, "there were plenty of people sacrificing for others, regardless of their demographic. I can't tell you how many times a man would stay behind an extra day or two on the roof and let his wife and kids go first. It broke my heart. We'd go to an apartment building and you'd see that someone was in charge, organizing the survivors. We'd tell him, 'We can only take five,' and they'd sort out the worst cases. It happened many times that the guy in charge was the last to leave."[88] The dependence on society was replaced in such cases by instinct: the human instinct to care first for the weaker, the dispossessed. In the worst situations, the strong fend for themselves—they rush for the exits, for the food, for the only vestige of safety. In some subliminal way, that is what happened in the days preceding the storm—the strong (those with cars, money, and good health) rushed for the exit. In what appeared to be a civilized city, the human instinct of caring for the weak was nowhere to be found, least of all at City Hall. In the chaos that followed, however, there were many individual Louisianans who cared more about others than they did about themselves.

But if someday there is a Great Deluge Museum in New Orleans complete with a Hall of Heroes, there should also be a Hall of Shame depicting the human-created situations that exacerbated the damage of the hurricane. In the first place, the weak and impoverished should not have been left in the threatened areas. Second, once they were rescued, there was no practical place to put them. The general plan followed by the many disparate groups performing rescues was that the sick and ailing were taken to the Superdome. The rest were deposited on any dry land. In no way were these solutions adequate. People with no food or water were hard-pressed in Tuesday's stifling heat to survive, let alone to walk great distances. The situation at the Superdome was even more hopeless. On Monday, during the storm, the Superdome offered shelter to approximately 9,000 people. By Tuesday, that number had exploded to 24,000. Since that population had a heavy concentration of the elderly and ill, the staff was unable to provide even the most basic care. The toilets were not functional. Food was in scant supply. Medical help was rudimentary.

Scanning the sweep of the Superdome on Tuesday evening, watching the suffering and confusion, one became acutely aware that "the blues" was not all boodlie-bum-bum on Bourbon Street. The real House of Blues wasn't the franchised club on Decatur Street, but sections 230 and 237 of the Superdome. Musically, blues music reflects the full range of emotions, but it is best known for describing sadder, darker moods. Suffering was the cornerstone of the blues, and misdirection its sidekick. Traditionally, whether it was Trixie Smith moaning about her "aching head" or Blind Willie McTell cursing about a "hot-shot liar," the blues was about weariness, about the end of the line, the land of broken shoes and a face full of frowns. The crowd in and outside the Superdome was almost entirely African American. It was a reflection of the racial makeup of the city, skewed by the fact that Katrina's displaced tended to be poor, isolated, and out of touch with mainstream news. To some, the crowd stranded at the Superdome conjured up images of both slavery and slave insurrection. Of course, such over-the-top comments were irresponsible. Reverend Jesse Jackson milked the hollow analogy when he spoke with CNN's Anderson Cooper later in the week about African Americans congregated on a freeway interpass: "Today I saw 5,000 African Americans on the I-10 causeway desperate, perishing, dehydrated, babies dying," he said. "It looked like Africans in the hull of a slave ship. It was so ugly and so obvious. Have we missed this catastrophe because of indifference and ineptitude or is it a combination of the both?"[89]

Perhaps what Jackson was trying to communicate was that the faces of African Americans in slow, starving despair was not a new visual or historical novelty to black people; it triggered memories of Jim Crow injustice, of Virginia lynchings, Georgia beatings, and Mississippi murders. The screams for help at the Superdome were an echo of slavery's whip, a tortured plea not to be forsaken yet again. No matter the murder or rape statistics, the resentment in the eyes of those affected was undeniable. The good news was that, watching their plight, most Americans had a simple, merciful thought: Can't we do more to help? Where are the mercy brigades?

"These are people who fell through the safety net," said Sheriff Paul Valtreau of Orleans Parish, surveying the condition of the New Orleans residents he was pledged to serve. "They're hard-working, tax-paying citizens and they're being treated like trash."[90] Watching the Superdome scenes on television, Aaron Neville, who had lost his New Orleans East home after the Industrial Canal breach, shook his head in calm disgust. "These were my people," the singer later explained to Fox's Neil Cavuto. "Even if I didn't know them, they were my friends."[91]

Chapter Nine

CITY WITHOUT ANSWERS

The air is getting hotter
There's a rumbling in the skies
I've been wading through the high muddy water
With the heat rising in my eyes
Every day your memory grows dimmer
It doesn't haunt me like it did before
I've been walking through the middle of nowhere
Tryin' to get to heaven before they close the door.
—Bob Dylan, "Tryin' to Get to Heaven"

I

WHILE THE GULF COAST was sinking ever lower into the grip of the worst natural disaster in modern American history, Michael Chertoff, the man in charge of the federal rescue, was on his way from Washington, D.C., to Atlanta to attend a daylong seminar on the flu at the Centers for Disease Control and Prevention. He didn't yet think it necessary to visit the Gulf states or even to devote a part of his day to the disaster in those states. Influenza was a prime health issue, to be sure, and the avian strain (or "bird flu"), which threatened to be a worldwide pandemic, was dangerous. The Bush administration was rightly concerned. But the flood devastation in

New Orleans was killing people that very morning, on Chertoff's watch, and partly because of the lack of immediate federal help. "I remember on Tuesday morning," Chertoff said later to Tim Russert on *Meet the Press*, "picking up newspapers and I saw headlines, 'New Orleans Dodged the Bullet,' because if you recall the storm moved to the east and then continued on and appeared to pass with considerable damage but nothing worse. It was on Tuesday that the levee—it may have been overnight Monday to Tuesday—that the levee started to break. And it was midday Tuesday that I became aware of the fact that there was no possibility of plugging the gap and that essentially the lake was going to start to drain into the city. I think that second catastrophe really caught everybody by surprise."[1]*

The levee breaches did not catch *everybody* by surprise. Actually, Katrina had hit New Orleans as a Category 3 hurricane (nearly Category 4), the maximum strength that the 110-mile levee system could supposedly endure. But, as experts had long predicted, the levees were failing left and right, springing leaks that were already growing bigger. Yet the most damning aspect of Chertoff's remark was that despite all the reports from FEMA personnel—and the availability of descriptions from the Coast Guard rescuers who were even then entering a second day of their own version of Operation Dunkirk in New Orleans—he based his actions on what he read in the newspapers.

If Chertoff had "picked up" the *New York Times* on Tuesday, he would have seen a whole spread dedicated to Hurricane Katrina. The lead article *did* allude to a sense of relief in New Orleans, but it also stressed that the storm had been no spring sprinkle: "Hurricane Katrina pounded the Gulf Coast with devastating force at daybreak on Monday, sparing New Orleans the catastrophic hit that had been feared but inundating parts of the city and heaping damage on neighboring Mississippi, where it killed dozens, ripped away roofs, and left coastal roads impassable."[2] Chertoff may well have read the *Washington Post* that morning, too. Under the headline "Amid the Devastation, Some Feel Relief," the paper led with the fact that New Orleans had long lived in dread of the Big One. It immediately quoted a man named Demetrius Ralph who was walking his dog in the French Quar-

* The U.S. Coast Guard operated as part of the Department of Homeland Security—like FEMA—making Chertoff's claim that he didn't know about the levee break until Tuesday even more perplexing.

ter. "This wasn't it," Ralph said bluntly. The article, however, then went on to describe the many neighborhoods in the eastern part of the city that were flooded up to the rooftops.[3] The *Post* depicted the widespread devastation of the Mississippi coastline, in a second article called "Storm Thrashes Gulf Coast." It explained that "along the Mississippi coast, the storm pushed water up to the second floor of homes, flooded floating casinos near Biloxi, uprooted hundreds of trees, and flung sailboats across a highway."[4] And if Chertoff had read the *Times-Picayune* online later on Tuesday (the only way to read it, under the circumstances), he would have seen the "Catastrophic" banner, as well as James O. Byrne and Doug MacCash's story on the breached Lakeview levees.[5]

Secretary Chertoff apparently based his lackadaisical response on one report: that of dogwalker Demetrius Ralph. This wasn't the Big One, and with that, the only man—except for the President—who had the authority to rise above bureaucratic tangles and enforce an urgent, immediate response to the Katrina disaster boarded a plane on Tuesday headed for Atlanta. The Bush adminstration had found out about the levee break at 7:30 A.M. on Monday. Louisiana Congressman Charlie Melancon got it right when he complained that during the hours after Katrina, Chertoff was behaving like a "complacent, uninformed, detached bureaucrat."[6] Apparently Chertoff—with a war on terror to fight—was unfamiliar with President William McKinley's wisdom regarding "homeland security," offered back in 1898 during the Spanish-American War: "I am more afraid of the West Indian Hurricane than I am of the entire Spanish Navy."[7]

After the Russert interview, Democrats piled on, some demanding Chertoff's resignation. But in a rare sign of bipartisan unity, the GOP right, to its credit, also concluded that Chertoff had failed in his duty. Syndicated columnist Robert Novak, not known for attacking the Bush administration, led the onslaught. Livid at Chertoff's haughty, above-the-fray attitude displayed on *Meet the Press*, Novak went for the jugular. "Chertoff's miserable performance on the air reflected a fiasco at all levels of government," Novak wrote a week later. " 'There'll be plenty of time,' Chertoff told Russert, to 'do the after-action analysis.' That bloodless dismissal made the human tragedy and physical mayhem on the Gulf Coast seem like a bureaucratic mistake."[8]

Perhaps better than any other columnist, Novak explained exactly why Brown and Chertoff were so numbly ineffective: they were lawyers trying to "cover President Bush's behind."[9] Katrina was a gut-wrenching, emotional drama being played out on TVs all over America. Like Ronald Reagan after the *Challenger* disaster or Bill Clinton after the Oklahoma City bombing, the White House needed to touch the nation's heartstrings, to add epic oratory to the catastrophic moment as Reagan or Clinton would have done. Instead, George W. Bush gave cold, terse, lawyerlike speeches void of human pathos or deep regret or full of grief. A scene Joseph Conrad described in *Heart of Darkness* as "passivity, paralysis, immobilization"[10] described perfectly the White House's modus operandi. Bush's front man was Chertoff, who Novak claimed was the "quintessential lawyer" who had "surrounded himself at Homeland Security with more lawyers." As Novak knew well, you just can't do *Meet the Press*, with Russert as the inquisitor, and get away with legalese and gibberish, particularly while Americans were dying in the Gulf South, reaching out for a federal hand. Chertoff's performance, as Novak pointed out, deserved an F; it was an unforgivable combination of "political deafness" with "lawyerly evasion."

The second half of Novak's column called for the immediate beheading of FEMA's Michael Brown—"heads must roll," as Novak put it. "I didn't call for any resignations," Novak later backpedaled in an interview. "I got into a lot of trouble years ago when I suggested a Federal Reserve official resign. But I will say that Bush, Brown, and Chertoff all looked terrible."[11] In his column, Novak brought to light Representative Mark Foley's complaint that his community, West Palm Beach, Florida, had airplanes ready to evacuate stranded Gulf Coast residents on Tuesday, August 30, but FEMA refused the offer. Examples of FEMA's flat-out rejection of help were unfortunately common. It was a form of proactive dereliction that was downright baffling to most observers. When the Senate Homeland Security and Governmental Affairs Committee held hearings on January 30, 2006, it was revealed that FEMA even neglected a "red-high" priority plea on Sunday, August 28, for 300 rubber boats from Louisiana Wildlife and Fisheries. A FEMA executive in Denton, Texas, simply scrawled at the top of the document, "Request denied." As Lieutenant Colonel Keith LaCaze of Wildlife and Fisheries said, "We could have used them to tow additional evacuees, and in lower water the rescuers could have used them to load people who were sick and handicapped."[12]

At the same time, the U.S. Department of the Interior offered FEMA 300 dump trucks and vans, 300 boats, 11 aircraft, and 400 law enforcement officers to help in the search and rescue effort. FEMA turned it all down. This flat-out rejection of Interior Department assets was brought to light in late January 2006, in a cache of documents released by the Senate Homeland Security Committee—over 800,000 pages of e-mails, memos, strategy plans, and intradepartmental correspondence. Among other things, the Senate documents revealed that FEMA didn't take up an offer of help from the U.S. Fish and Wildlife Service to save people stranded in Orleans, St. Bernard, and St. Tammany parishes. "Here we have another federal department offering skilled personnel and the exact kinds of assets that were so desperately needed in the wake of Hurricane Katrina, and there was no response we can discern from FEMA," Senator Susan Collins (R-Maine), who chaired the committee, said. "That is incredible to me."

At the FEMA Emergency Center in Washington, D.C., seventy people who reported to Michael Chertoff were supposedly working to obtain food, water, buses, and other critical supplies for New Orleans.[13] Although Michael Brown had made repeated assurances that all such supplies had been pre-positioned in states surrounding the vulnerable Gulf Coast area, the documentary evidence suggests otherwise. These Washington-based FEMA workers frantically hunted around the country for supplies, as if running a Jerry Lewis telethon, knowing that, at best, it would be two days before anything reached the devastated regions. And they were constantly getting trapped in bureaucratic red tape, which a little preparation and a lot of leadership would have cut through. The department was unquestionably in a tizzy. "Katrina pushed our capabilities and resources to the limit," Homeland Security spokesperson Russ Knocke said, "and then some."[14]

The problem was worse, however, than simply pushing "resources to the limit." Once looting started in New Orleans, Homeland Security pulled the plug on helping Katrina's victims. Like the Red Cross, FEMA was not about to put its employees in harm's way.

Senator Joe Lieberman was particularly furious over a FEMA e-mail message pertaining to the cancellation of relief and rescue efforts just three days after Katrina hit the Gulf South. FEMA's reason for throwing in the towel was security. "All assets," the e-mail read, "have ceased operation until National Guard can assist TFs [task forces] with security." Lieberman criticized FEMA for having "left early," noting the stark contrast with such outfits as the Coast Guard, Wildlife and Fisheries, National Guard contingents from dozens of states, and faith-based organizations. They all had security concerns, but kept helping fellow Americans in need. "This is shocking," Lieberman said, "and without explanation."[15] And the trend continued.

II

As of Tuesday, August 30, 2005, President Bush had four days left of his five-week vacation, most of which he had spent at his ranch in Crawford, Texas. Tuesday, however, wasn't much of a day off. In fact, the whole vacation had been a little trying for the President. Just down the road from the ranch, hundreds of antiwar protesters had been making noise all month, capped with Joan Baez singing "Where Have All the Flowers Gone?," "Swing Low, Sweet Chariot," and "Song of Peace" on August 21, 2005.[16] It was almost enough to make a dry conservative reach for a bottle. The agitation was led by Cindy Sheehan, the mother of a soldier killed in Iraq; she wanted a meeting with Bush, though she had already had one with the President and other parents of the fallen on June 10, 2004. Bush held firm to his position that he had thought "long and hard" about Sheehan's antiwar feelings and had nothing new to say to her.[17] But liberals in the media, by and large, embraced Sheehan, whom they dubbed the "peace mom," and they were turning her into the number-one celebrity of the anti-war movement, a latter-day version of the Vietnam war's outspoken Dr. Benjamin Spock or Reverend William Sloane Coffin. "I don't want him," she said, referring to the President, "to use my son's name to justify any more killing."[18]

A private man by nature, Bush was never a particularly accessible president, nor an outwardly sensitive one. The average U.S. president looms as a patriarchal figure, but Bush conducted himself like the last of the Victorians: jocular, yet remote and undemonstrative, divulging drips of information only when absolutely necessary. Under the right circumstances, Bush's rough-and-ready Texas demeanor gave the impression of a tough, self-contained cattle rancher blessed with a decisive Harry Truman bent. Many Americans were attracted to his central Texas swagger and enjoyed seeing him drive a Ford F-250 or clear brush with his Scottish terrier, Barney, at his side. But he often waited before being decisive. He was a practitioner of the prolonged pause. It was as if his certitude was learned—not inherent in his makeup.

After the attacks on the United States on September 11, 2001, Bush was unseen during the first eleven hours, making only brief statements and effectively ceding the public leadership role in the crisis to Mayor Rudolph Giuliani of New York. When President Bush emerged from his isolation, it was in a role in which he was comfortable: the Crawford aggressor, intent on revenge. With local New York and Washington officials handling the immediate emergencies of their cities (and FEMA doing a good job assisting in relief), President Bush seized the leadership role that interested him most: commander in chief. A "war on terrorism" was waged, first against Afghanistan and then against Iraq. It was a bold offensive move. A Bush doctrine was even formulated, which stated that the United States had the option of presumptive military strikes in its superpower arsenal. If the White House thought the United States would be attacked, it would attack first. This took deterrence to a whole new level. You had to admire President Bush for being a man of steely conviction, even if you thought his policies were wrong. Munich-like appeasement was not in Bush's nature, and diplomacy was not his forte. But you had to give him this: if you even threatened to punch, he punched back harder. As he told White House aides on the eve of his 2006 State of the Union address, 9/11 was "tattooed" on his mind—always.[19]

A hurricane, though, offers no target for revenge. You could bomb terrorist havens outside Kabul or hunt for weapons stockpiles in Baghdad, but all you could do after a hurricane was shake your clenched fist at the sky. The damage was done and what remained were the unglamorous details of putting daily American lives back together, without a martial snare drum or a Sousa band. After Katrina, the Gulf South region—and the United States as a whole—needed compassion. What it got instead was the incompetence of George W. Bush, who acted as though he were disinterested in a natural disaster in which there was no enemy to be found. More than any other event, Bush's slow response to the Great Deluge made Americans ask if he was a "bunker" commander in chief, unfastened from the suffering of the Gulf South, relying too much on cautious paper pushers such as Brown and Chertoff for advice. The gutsy president, it seemed, couldn't find his gut, much less his heart, in his reaction to Katrina. He was starting to be called the "bubble president."

III

On Tuesday, however, President Bush's primary consideration shifted to the military—and, by extension, to the war in Iraq. With appearances scheduled in Southern California, he had spent Monday night at the historic Hotel Del Coronado, a seaside resort just outside San Diego. On his agenda were several photo ops, including one to honor Navy medics who had helped victims of the December 2004 tsunami in Southeast Asia. The keynote event, though, was to be a speech honoring World War II veterans who fought on flyspeck islands like Midway, Guadalcanal, and Tarawa. In going through with the appearances, Bush seemed to be sadly out of touch with the plight of approximately two million of his Gulf South constituents. First thing on Tuesday, his staff at the White House met about the impact of Hurricane Katrina. They concluded that the President would have to take action on the crisis, cutting short his vacation, flying to Washington after one last night in Crawford, and directing the response of the White House, or at least appearing to do so.[20] At 5 A.M. Pacific time, Joe Hagin, the deputy chief of staff who was traveling with the presidential entourage, ordered that the President be woken up. The scene, a red dawn on the Pacific, was less dramatic than it might have been: Bush calmly told Hagin that he had already decided that he would forgo the rest of his vacation to focus on Katrina. It was only three days, after all. A public-relations-savvy Bush may have remembered the flak his friend Vladimir Putin, president of Russia, had taken for remaining on vacation on the Black Sea four years before, when the submarine *Kursk* had sunk with the loss of all 118 on board.

Bush wasn't going to pull a Putin. He knew he couldn't stay away from the White House on a prolonged vacation in Texas or California while three states—Louisiana, Mississippi, and Alabama—were desperate for federal help. A worried Bush scheduled a meeting of White House domestic advisors for the following day, Wednesday. That would give them all a chance to return from their own vacations.[21] But there was one high-profile holdout: Vice President Dick Cheney stayed in Wyoming, fly-fishing for trout in gurgling streams. Professor Stephen Hess of George Washington University, picking up on the vice president's absence from Washington, quipped, "Maybe Cheney's been ducking this one because he wants to keep his shirt clean."[22] The quick-witted Paul Begala, a Democratic pundit, saw a PR-minded rationale for Cheney's nonappearance. "Nobody's going to confuse Cheney with a warm and fuzzy guy," he said. "You're not going to send *him* to be commander in chief. He's the type of guy who would look at them and growl, 'Life's tough. Get back to work.' "[23]

In Atlanta on Tuesday, Chertoff, listening to presentations about the flu, was apparently unconcerned with the massive flooding in New Orleans—even though he had been told by Brown the night before about the 17th Street Canal breach and even though reports on television on Tuesday morning were showing live footage of a city and a region falling into the very pits of inhumanity. During the first thirty-six hours after Katrina hit, Chertoff seemed, as Jimmy Duckworth would say, "out of the game." The same was true of Cheney. As for Secretary of Defense Donald Rumsfield, he had spent Monday evening enjoying a Padres baseball game in San Diego at Petco Park. Later, in October, when Chertoff testified on Capitol Hill about his lackluster leadership during the early days after the hurricane, he was lambasted by both Democrats *and* Republicans. "I have a feeling that the Department of Homeland Security is dysfunctional," Congressman Christopher Shays (R-Connecticut) scolded Chertoff. "I get the feeling that you were a little detached from this. It's kind of like Pontius Pilate washing his hands."[24]

Out in Southern California President Bush began his morning speech at the North Island Naval Air Station. Most of his words were about V-J Day and the United States postwar occupation of Japan, but he did make a reference to the hurricane damage along the Gulf Coast, saying, "Right now, our priority is on saving lives, and we are still in the midst of search-and-rescue operations. I urge everyone in the affected areas to continue to follow instructions from state and local authorities. The federal, state and local governments are working side-by-side to do all we can to help people get back on their feet."[25]

In San Diego around noon, Mark Wills, a country music singer known for the hits "19 Something" and "Don't Laugh at Me," presented Bush with a guitar bearing the presidential seal. The President took it and strummed a fake chord as cameras clicked away. It was an odd juxtaposition: the photograph of Bush playing air guitar, while Americans were seeing inexplicable agony in pictures from the Gulf Coast. Even President Bush's most loyal supporters were wondering about his disconnect from reality, and the media hammered him. Commenting on MSNBC, Howard Fineman of *Newsweek* recalled another presidential picture—one taken in 1965, when, within hours of Hurricane Betsy's destruction along the Louisiana coast, Lyndon Johnson had taken the initiative to fly immediately into New Orleans. "With no electricity in the darkness there," Fineman wrote, "Lyndon Baines Johnson held a flashlight to his face and proclaimed, 'This is the President of the United States and I'm here to help you!' "[26]

One person who believed President Bush's promises was Mayor Ray Nagin, who in the coming days would communicate regularly with White House Deputy Chief of Staff Karl Rove, bragging to Louisiana Congressman William Jefferson that he was a new buddy of *the man*. They were going to rebuild together, baby. "We have the highest levels of government in the United States, including the President of the United States," Nagin said on Tuesday morning, "focused on this issue and ready to send resources. They have told us to put together our wish lists."[27] The mayor was still holed up in the Hyatt Hotel, trying to turn it into a command center for New Orleans's recovery. Even so, he didn't know what he was up against, either in terms of the extent of the disaster or the degree of disunity and disorganization in the ranks of the various agencies charged with recovery operations. His telephones were dead and he had little contact with state or federal officials. "The madness of C. Ray is killing us," Clancy Dubos of *Gambit Weekly* wrote months after Katrina, evaluating Nagin's performance in office. "He should seriously consider resigning—but I doubt that he could muster either the intellectual honesty or courage to take such a step. His ego is too monumental, his myopia too mind-numbing, his detachment from reality too complete."[28]

Joseph B. Treaster of the *New York Times*, reporting from City Hall, noted that Nagin just never seemed to be around. Terry Ebbert was in charge. When asked why he thought that was the case, Treaster, careful in his choice of words, said, "Well, it's more comfortable in the Hyatt than City Hall."[29]

What Nagin spent the day doing was awaiting the federal cavalry, pointing fingers at everybody but himself, and swearing. Swearing a lot. Swearing all the time. Everything was Blanco's fault. He peppered his language with "asses" and "shit" and "damn" and "man." It was "Ray Speak" gone wild. "Analyze my ass," he told Gordon Russell of the *Times-Picayune*, "analyze everyone's ass, man." As for his failures, it boiled down to communications. "I got cell phones from as high up as the White House that didn't work," he said. "My BlackBerry pin-to-pin was the only thing that worked. I saw the military struggle with this, too. No one had communications worth a damn."[30]

Being in New Orleans at midday on Tuesday was dispiriting. The break in the 17th Street Canal had widened to 500 feet by Tuesday afternoon. It was a widemouthed spigot pouring water from the lake into the city, rendering the city's thirteen fully manned pumping stations worthless. Even if the pumps did their job and pumped water out of the city (at a rate of 29 billion gallons a day), the stations sent water into the lake—and then it came right back. The only rational decision was to shut down the pumping stations, several of which then flooded themselves, making them unusable. The lake won, and it wasn't even finished throwing its punches. The canals, meant to drain New Orleans, became, as the *Times-Picayune* put it, "major vectors for filling it with water, wrecking neighborhoods and killing many of the 1,000 people who perished" after Katrina hit.[31]

The entire city was seething in despair. The whole state was, as well: an estimated 22 million tons of debris lay scattered all over Louisiana, not including the 350,000 automobiles and 35,000 boats destroyed by Katrina. Pick a building, block, or coastal parish and something bad had happened there. Dillard College and Xavier University were severely flooded. Outside of the D-Day Museum a woman was raped. Tom Planchet of WWL-TV reported that on Tuesday 11:30 A.M. "Canal Street was literally a canal." He elaborated: "Water lapped at the edge of the French Quarter. Clumps of red ants floated in the gasoline-fouled waters downtown."[32] The French Quarter survived without much flooding because of its location: it was on slightly higher ground, and it was alongside the Mississippi River, where the levees did not break.

Because the breached levees of Lake Pontchartrain caused so much destruction, the fact that the river levees held firm failed to make headlines. Down in Plaquemines Parish the levees withstood masochistic winds of 115 mph with unremitting waves topping over ten feet. Also unreported at the time were runaway barges slamming into the levees over and over again. A number of "I" walls built on top of the Mississippi River levees for added security were damaged, but the main levee itself absorbed all of the storm's punishing blows. "While Katrina was making short work of many hurricane protection levees in the region, the levee system designed to guard the area from the ravages of the country's mightiest river survived the hurricane remarkably well," Bob Marshall wrote in the *Times-Picayune*. "And engineers familiar with the two different systems said there are lessons to be learned from the comparison that could help protect New Orleans in future storms, lessons that reach all the way to the will of Congress."[33]

Because the Mississippi River levees didn't breach, the fine old residential sections of the city known as Uptown received only wind damage, sparing the homes of blue bloods, academics, corporate leaders, trial lawyers, and real estate developers. Worried more about looting than flooding, the Uptowners decided to create their own community watch. Private police were shipped in to provide protection for the district. (They knew better than to count on the NOPD.) In the cases of most residents, the personal hurricane stories concerned dramatic escapes to Aspen and Sun Valley or how they flew in private helicopters, evacuating their poodles, de Kooning paintings, and old family portraits of soldiers in Lee's army. To many Uptown residents, dislocation meant staying at their second home in West Palm Beach or Cape Cod or Natchez longer than planned. The relatively unscathed Uptown and French Quarter sections became known as the "sliver by the river."[34]

IV

Just after 3 P.M. on Tuesday, Senator David Vitter participated in a nationally televised news conference from Baton Rouge, during which he made one of the more ill-advised statements heard in the aftermath of Hurricane Katrina. (Unfortunately, many remarks would vie for that inglorious designation.)

Vitter, a forty-four-year-old Republican serving his first year in the U.S. Senate, had been wandering around the Baton Rouge EOC since Monday, with essentially nothing to do. The other politicians at the press briefing—Governor Kathleen Blanco, Senator Mary Landrieu, Lieutenant Colonel Pete Schneider of the Coast Guard, and Bill Lokey, the coordinating officer for FEMA—took turns describing the worsening conditions in southeastern Louisiana. "I know there's been a lot of concern about the levee breaches," Schneider said. "We had a conference call just about an hour ago with Colonel Wagner with the Corps of Engineers. He has been up in a helicopter surveying the entire situation, and they're diligently working on a plan that is going to close these breaches. . . . They realize the gravity of the situation. They're not sparing resources on getting this fixed. And we're confident that the Corps will come up with a solution to this problem quickly."[35]

Senator Landrieu made the salient point that the flood reports pouring into the EOC were ninety minutes old. Everything was in flux, and an hour was an eternity. The water level, she said, had probably changed significantly since the last report. As she trailed off, Vitter clumsily leaped in, wanting to be part of the media action, brimming over with the last vestige of Louisiana's damning obliviousness to reality. "In the metropolitan area in general," Senator Vitter said, speaking of the floodwater, "it's not rising at all. It's the same or it may be lowering slightly. In some parts of New Orleans, because of the 17th Street breach, it may be rising and that seemed to be the case in parts of downtown. I don't want to alarm everybody that, you know, New Orleans is filling up like a bowl. That's just not happening."[36]

How wrong could a United States senator be? What planet was he on? New Orleans *was* filling up like a bowl. Lake Pontchartrain was gushing into the streets, hundreds of thousands of gallons of water every second. If you went to Lakeview, the water was twelve to fifteen feet high. Neither the levees nor the pumps were working. "A sunken city," John Zarrella of CNN had earlier described the situation. "Most of New Orleans is now under water. And it will likely be days, maybe weeks, maybe months before the Big Easy dries out."[37] The delicate bowl analogy had come to fruition. The water was omnipresent in many of Orleans Parish's wards. After offering his disclaimer, Vitter listened as Governor Blanco responded to the next question by saying that in the vicinity of the Superdome, the water was knee-deep and steadily rising. Vitter could only sheepishly stand by, smiling like Gomer Pyle—clueless. "Truth to tell," Terry Ebbert, New Orleans's director of Homeland Security, explained to *Times-Picayune* reporter Dan Shea about the same time Vitter was in denial, "we're not far from filling in the bowl."[38]

Giving Vitter the benefit of the doubt, CNN deemed his erroneous remarks "an oddly discordant statement."[39] Just an hour before Vitter made his gaffe that New Orleans wasn't "filling up like a bowl," Dr. Greg Henderson of Washington, D.C., stranded in the city while at a convention, e-mailed his family explaining the dire situation along Canal Street, which was "underwater." Although he himself was fine, he lamented that Charity and Tulane hospitals, both just blocks away, had flooded, and patients were unable to receive proper medical care. "The city now has no clean water, no sewage system, no electricity, and no real communications," Henderson wrote. "Bodies are still being recovered floating in the floods." In harrowing detail, he described how looters had made Styrofoam rafts to float down streets, how armed cliques of renegades were discharging weapons, and how a French Quarter bar had been made into a makeshift clinic. "Infection and perhaps even cholera are anticipated major problems," Henderson said. "Flood and water shortages are imminent."[40]

V

Not only was New Orleans flooding but in neighboring Chalmette, in St. Bernard Parish, a "Wall of Water" had completely overwhelmed the community the day before and little had improved since then. Chalmette was the predominantly white, blue-collar counterpart of the Lower Ninth Ward. It was bisected by the Mississippi River–Gulf Outlet (MRGO), the man-made waterway that had brought most of the flooding to St. Bernard Parish. The waterways that once were considered the conduits of Chalmette's lifeblood—MRGO and the Intracoastal Waterway—ended up being a twin curse, bringing a 30-foot storm surge that almost completely destroyed the town in the span of fifteen minutes.

One of the great misconceptions about Katrina was that the destruction was Mother Nature at her worst; that's only partially true. The horrific flooding that occurred in St. Bernard Parish, New Orleans East, and the Lower Ninth Ward was the direct result of the man-made "improvements" to the region's waterways. As the MRGO and Intracoastal Canal met and formed a V, they became what the *Times-Picayune* called "a funnel which causes storm surges to rise higher and move faster" as they took aim at St. Bernard Parish. When that same funnel of water continued west and crashed into the Industrial Canal, the jagged breach that resulted doomed much of Orleans Parish. New Orleans City Councilwoman Cynthia Willard-Lewis appropriately deemed MRGO the "highway for tidal surge," and in the months following Katrina, residents of both New Orleans East and St. Bernard Parish would gather at rallies to chant, "Go, Mr. Go!"[41]

Although Captain Chad Clark of the St. Bernard Sheriff's office had done everything he could to evacuate the town, many Chalmetteans simply refused to budge on the weekend before Katrina hit. Diligently, Captain Clark had stocked the fortresslike St. Bernard Parish courthouse with drinking water, set up roadblocks to keep unauthorized vehicles from coming into the parish, and dispersed his police officers so they wouldn't all be stuck in one place. He planned for about 130 St. Bernard first responders to hole up at the courthouse with him. Still, he felt there was a conspiracy of nature in the air; his gut told him that something was out of sync. "I chose the courthouse to wait out Katrina because it's way above sea level," Clark recalled. "It's high ground, up a flight of seaman's steps, eight or ten feet above sea level. The building's been there forever. It went through Betsy, the whole nine yards. Meanwhile, I put our one hundred and thirty vehicles, our whole fleet, in supposedly safe places."[42]

As Katrina hit, St. Bernard Parish was awash in rainwater. Captain Clark was all elbows on the windowsills, as debris hurtled past his face and a howling roar filled the courthouse. "Holy shit!" he kept saying. The profanities became more intense when he peered out and saw a railroad train moving down the tracks. "I don't understand, because what moron is going to move a train in these conditions?" he recalled. "Now we're gonna die. I was certain of it. The damn wind was moving the trains. Trees were snapping, you could see power lines falling over. Water was coming in the courthouse."[43]

The situation only got worse. Clark received a call from a police station nearby, where ten detectives were holding down the fort. "We think the levee has broken!" said the panicked officer at the other end. "We got water coming." The Industrial Canal had broken and water was literally rushing down the highway. Chalmette was badly hit with both the Industrial Canal break and the MRGO topping. Within a matter of fifteen minutes, the courthouse went from two to ten feet of water. A decision was made—much like how the Bay St. Louis police saved the Waveland force—to go get those ten drowning detectives. Unfortunately, Captain Clark had positioned his high-water vehicles next to Jackson Barracks, on the theory that if it was good enough for the National Guard, it was good enough for St. Bernard Parish. Katrina made a fool out of such logic. "They went underwater in nineteen seconds," he said. "But we had two boats placed by the courthouse and when the wind got down to about 60 mph, and with water just pouring into the courthouse, we hopped in the little boats. Nobody has ever seen water come up that fast."[44]

The two boats to which Captain Clark referred were, in actuality, Sea-Doos, sit-down watercraft, wave runners that were ideal for reconnaissance but not large enough for more than two people. Conscious of the danger involved in using these small craft, Clark and Officer Pete Tufaro volunteered to set out because neither of them had children. They would survey the damage along the St. Bernard Highway, check up on the officers stranded at the station near the Orleans Parish line, and report back to Sheriff Jack Stephens. "We started up the wave runners and took off," Captain Clark recalled a couple of months after Katrina. "The water was so high you could touch the red lights. Some of them you had to duck. It was unbelievable how high the water got so fast. As I got a little farther toward New Orleans, we were going over waves that were five or six feet."

A woman from the neighboring town of Arabi was standing on the balcony of a two-story building on Lebeau Street, screaming for help and holding a baby above the floodwaters. Captain Clark jetted his Sea-Doo over to her. "What are you, nuts?" he shouted. "Put the baby back inside." But she kept crying, "I need help! I need help!" She was having what Clark called a freak-out. She was convinced her child was going to drown in the deluge. "Put the baby back," he kept instructing her, demonstrating what to do with his hands. "Put the baby back." His words cut through her panic, and the mad gleam, the black fear, left her face. Her voice was high-pitched, shrill, crackling with anxiety. "Stay put," he told her. "we're gonna be back." She looked at him desperately. "If you don't come back," she said, "we're gonna die." Clark tried to make her understand that he was a man of honor, his word was golden. "I can still see her eyes," he recalled, months after Katrina. "I said, 'I promise you I'll be back. I'm going to check on these stranded cops. You're on the second story, so you're good for a while.'" She now believed him. "Please come back," she pleaded, as he increased the throttle.[45]

Clark and Tufaro made it to the police station, which was full of water. Clark reassured them that Sheriff Stephens would soon be rescuing them—since the train wasn't an obstacle—and that they should just chill out for two hours, three maximum. Time was of the essence. Clark and Tufaro made a U-turn and started heading back to the courthouse. A detour, however, was made to rescue the woman and her three-month-old boy. She was relieved to see them back so soon. "Now, how in the hell am I gonna put this baby on a wave runner and bring it to safety?" Clark groused at his buddy. "This is too dangerous!" He loosened his life jacket enough to place the baby firmly against his chest. The mother got on back and wrapped her arms around him, as if on a motorcycle. "Then we hauled ass to the courthouse," Clark said. "We just cut through the water at record speed."

Once her baby was safe and dry, the mother calmed down completely. She also seemed sedated, as if once her maternal worry was gone, she could become almost a child herself. "Why did you stay?" Clark asked. "Why were you so foolish?" She hesitated for a moment. "Well," she said, "because I'm a dumb ass." Without chuckling, Clark walked away. "Good answer," he said. "Very good answer."[46]

VI

Out of the devastation, the St. Bernard first responders—not just police, but judges, the district attorney, and others—fanned out from the courthouse and went on to evacuate more than 10,000 people between Tuesday and Friday.

Katrina flattened and dissolved mile after mile of St. Bernard Parish. The families huddled together in the attics and on the balconies of Chalmette, Meraux, and Arabi, watching the floodwater levels rise and creep up on them, were hoping for salvation. Certainly one could read passages of Nevil Shute's *On the Beach* or Kurt Vonnegut's *Slaughterhouse-Five* to find language that could be applied to St. Bernard Parish. But the only work of art that honestly mirrored the pulverized landscape was *The Ant Farm*, a couple miles outside of Amarillo, Texas, where ten Cadillacs were buried (front first) in the turf. Everywhere in the parish, cars were overturned in similar fashion, only they were upturned in six to eight feet of water. "We're talking about the complete destruction of St. Bernard Parish as we know it," Parish Council Chair Joey DiFatta said. "Every neighborhood, every street, every home, every building, has water—lots of it."[47]

There have been fine history books written on American disasters, like David McCullough's *The Johnstown Flood* and Simon Winchester's *A Crack in the Edge of the World*, but nothing compared to the oral history testimonials of Katrina survivors. Many of Louisiana's worst-hit towns, such as Grand Isle, Delacroix, and Port Sulphur, were virtually submerged under water. They didn't have as many eyewitness survivors to tell what happened, since Plaquemines Parish and the fishing camps along Barataria Bay in Jefferson Parish had been almost fully evacuated. But Chalmations by the thousands had stayed in St. Bernard Parish, socked with the double whammy of the Industrial Canal breach and MRGO overtopping. They hadn't expected a "Wall of Water" beating down their doors. Many of them stayed behind to take care of ill or elderly family members who simply could not evacuate. Others didn't have the financial resources to flee. All had been reassured that the Mississippi River levee would hold out—it did. But the MRGO topped the levee, hence the deluge.

One Chalmette survivor was fifty-two-year-old Michael Brown (no relation to the FEMA director). Conscientious and dutiful, Brown understood that Katrina was a potential Category 5 storm. As a precaution, he sent his wife and children to Cajun country in western Louisiana for the storm. But his in-laws, both well into their eighties, simply refused to evacuate. They were stubborn and won a battle of wills with Brown. He couldn't leave them alone in the face of a major hurricane. When the winds hit 150 mph, he made them wear their life jackets—just in case of flooding. "Then my worst fears materialized," he said. "I looked outside and I saw this black water starting to move across everything. I saw this wall of water hit the back of these homes and it just exploded. All of a sudden, water was everywhere."

Like all the Chalmette survivors, it was the roaring cascade that stunned Brown, as though it were something out of the movie *The Day After Tomorrow*. Quickly, he ushered his in-laws to the tiny second-floor attic; then, with a last burst of strength, he gathered whatever water bottles he could scrape up. (One of the great ironies of flooding was that you couldn't drink the water. You died of thirst in the general bedlam, surrounded by brackish water with massive seawater intrusion, and loaded with greasy toxins and chemicals.) As six feet of water filled his house, Brown looked out the window and saw his BMW float away. "The water was ferocious," Brown recalled. "You could see the natural gas bubbling up out of the water everywhere you looked. As far as the eye could see, there was nothing but black water."[48]

To put the Chalmette flooding into perspective, the Mississippi River levees in town were 16 feet high, while the "Wall of Water" was over 30 feet high. No "trickle-trickle," as WWL's Garland Robinette explained, it was "boom-boom." That the water was poisonous made it all the more incredible that Governor Blanco, pulling a Vitter, described it at the Tuesday press conference in Baton Rouge, "I wouldn't think it would be toxic soup right now. I think it's just water from the lake, water from the canals. It's, you know, water."[49]

Everywhere Brown looked after the storm, live animals were floating by, struggling to ground themselves on some rooftop or building, rarely to any avail. For a second, he felt lucky it wasn't him. Then another crisis hit the Brown clan. A heavy odor of mildew filled the house; he was not sure where it came from. The attic bricks started popping out of the house like baby teeth. Nervously, he moved his in-laws to the second floor. They would all be stuck there for days, as no Coast Guard helicopters or boats came to the rescue. A neighbor noticed that they were stuck in the house and stopped by in his little motorboat with bread, crackers, and water. On Tuesday evening, as the clear sky was full of glowing stars, yet another sinister terror struck the house. "I looked and saw this movement in the water," Brown recalled. "We had water moccasins on the ground floor. They were everywhere—on throw pillows, everywhere. They bite you and they're dangerous." A decision was made then and there. Come daylight, they were leaving the snake-infested house—somehow, some way.[50]

VII

Lieutenant Governor Mitch Landrieu made it to St. Bernard Parish on Tuesday, courtesy of a Wildlife and Fisheries flat-bottom boat. His goal was to track down Parish President Rodriguez and find out details about the "Wall of Water" and an oil spill in the Parish. "Nita Hutter had reached me by cell phone about what had happened to St. Bernard," Landrieu recalled. "She, Junior Rodriguez, Joey DiFatta, and a bunch of officials were stuck in the St. Bernard courthouse and I was determined to find exactly what they needed. Eventually, I found my way into a boat to Chalmette, where Junior and those guys were heroically saving lives."[51]

With flabby jowls that shook when he made public statements, which was often, Junior Rodriguez was a political legend in Louisiana. Somehow, he was born without a verbal filter and words just tumbled out of his mouth without restraint. Like Huey Long, Rodriguez was a populist, truly believing that every Louisianan deserved a whole chicken in their pot. Now, he wanted Landrieu to find him MREs, Gatorade, medical supplies—all the basics. St. Bernard needed them. "Junior was in his underwear and a T-shirt," Landrieu recalled. "I said, 'Junior, I came here. What do you need? Because we can't communicate with you, we can't talk to you. What the hell is going on? I want to get everything you need from Baton Rouge.'" Rodriguez handed the lieutenant governor his wish list and then jumped to the primary problem. "We have only one shelter," he said. "It's going to be the port. We have 2,000 people that need evacuating from Chalmette. Tomorrow, it will be 5,000 people."

Rodriguez wanted a ferry that could transport St. Bernard residents to dry land at Algiers, to be bused from there to appropriate shelters. Already, Rodriguez knew of sixty deaths in his parish. He didn't want any more on his watch. Unique among politicians in the first post-Katrina days, most of whom stood hat in hand, Rodriguez was a fireball of defiance. If Landrieu *really* wanted to help St. Bernard, Rodriguez intimated, he should have brought in explosives. If Rodriguez's constituents had to suffer much longer, he had a radical plan to dynamite a hole in the levee and drain the scum water out of *his* streets. "Junior kept saying, 'Should I blow the levee?'" Landrieu recalled. "'I just might have to do that.'" Such blasphemous, blunt talk startled Landrieu, largely because he knew Junior meant what he said. Rodriguez's threat had yielded a positive consequence: the National Guard got supplies and ferryboats to the Chalmette slip. "We never got explosives, though," Landrieu said. "But we got him just about everything else!"

Landrieu walked around the Chalmette slip, hugging refugees and talking to them. Scribbling down notes, Landrieu made sure a very sick, elderly woman was immediately medevaced out of St. Bernard to a hospital. (Her daughter later wrote Landrieu a long letter of gratitude.) For the first time they realized that they mattered, that perhaps the trauma would end soon. Collectively, they were saying prayers, like Charley Patton's song "High

Water Everywhere," about the 1927 Great Mississippi Flood, when folks chanted, "Lord, the whole round, man, is overflowed."[52]

With an understanding of the situation in Chalmette, Landrieu shifted his attention to New Orleans. He wanted to find out directly from Mayor Nagin why the Regional Transit Authority (RTA) buses and the yellow school buses were out of commission. Why wasn't Nagin getting people out of the Superdome? Not finding the mayor at City Hall, Landrieu waded, as he said, "in the water, into the Superdome a number of different times, walked over and tried to find the mayor." Eventually, he located Sally Forman, the mayor's communications director, at the Hyatt. She was glad to see the lieutenant governor.

"I'm looking for [Chief Administrative Officer of New Orleans] Brenda Hatfield," she said to Landrieu. "We must find her. She is in Baton Rouge, somewhere."

"If I see her," Landrieu said, "I'll tell her—"

Forman interrupted him. "Because we're looking for the RTA bus keys," she said. "Well, we don't know if she has them, but she'll know where they are."

As bad as things were in St. Bernard, Landrieu realized, it was worse in New Orleans. At least Junior Rodriguez was commandeering ferries and contemplating dynamite; it was, rightly or wrongly, a plan of action. Nagin was in hiding, looking for the RTA keys. This very fact meant that Mayor Nagin had been tricking the American public, saying he couldn't have used the buses to evacuate folks because he couldn't find drivers. In truth, he didn't even know where the keys were. He was willing to scapegoat hundreds of bus drivers to save, in his vernacular, *his* political ass. "So I went up to the top of the Hyatt and saw the mayor," Landrieu recalled. "He was sitting in a room, trying to pick up information from the TV and radio." It was a sad scene: twenty-seven stories below the mayor, flood-ravaged people were congregating at the Superdome, looking for food and water. Nagin's aloofness was chilling. "Mr. Mayor, is there anything you need?" Landrieu asked. "I just met with Junior Rodriguez. I'm about to go see Aaron Broussard. I'm trying to collect information." All Nagin would say, staring straight ahead, as if in a trance, was "We're looking for a command and control structure."

Mitch Landrieu looked twice at the mayor. At least in St. Bernard, there was a fighting spirit. For whatever reasons—exhaustion or fear?—Nagin was in low gear, unable to get jump-started, afraid of the poor people below. "From his hotel, I just walked back into the Superdome," Landrieu said. "They wanted to see you. They wanted to see somebody. They were always thankful that you were there. We were loading people into the back of trucks leaving the Superdome. The people that we took and put on the foot of the St. Claude Bridge, I actually ran into at that location, because there was a triage unit set up in back of the Superdome. An old man named Uncle Willy, whom I had taken off his Lower Ninth porch in a Wildlife and Fisheries boat, was in that triage unit." The sight of him alive and smiling brought tears to Landrieu's eyes.

"Hey, Uncle Willy," he said, "you're going to be just fine."

Landrieu helped load Uncle Willy onto an emergency truck headed to a proper hospital, somewhere out of the bowl. "I don't know where they took him," Landrieu said. "Or where he ended up. But thank God Almighty he was gone."[53]

VIII

The *Baton Rouge Advocate*, which considered itself the state's newspaper, felt an obligation to cover every stricken community in its pages. One especially harrowing story, written by Sandy Davis, told of the "Wall of Water" that came crashing into the Chalmette home of Dorothy Hingle and her son, Russell Embry. Terrified as Katrina blew into town, Hingle lit a candle, giving their modest brick home on Roselta Street a grotto feeling. She held her handicapped son in bed. Because Embry was a quadriplegic, the Department of Social Services was supposed to evacuate them. But no one came. They were stuck in Chalmette with prayer as their only safeguard. And while we'll never know Russell and Dorothy's last thoughts or actions, we do know, through the *Advocate*, that Katrina swallowed the mother and son without remorse. Five days after Katrina, Acadian Ambulance and Med Air still had not picked up their corpses. But the St. Bernard Parish Fire Department had come around, putting a huge orange X on the house adjacent to the numeral 2, to indicate two dead bodies.

In Chalmette, as well as the adjoining town of Meraux, regular citizens suffered something even worse than a hurricane and worse than a flood. The entire community was washed away, not only by up to twenty or thirty feet of water, but also by oil. Lots of oil.

Murphy Oil had evacuated its St. Bernard facility between 3 and 4 P.M. on Sunday, August 28. Unfortunately, the staff didn't follow procedure. When a hurricane threatened, they were required to fill any empty or half-empty crude-oil tanks with water. If a tank wasn't full, it could float away, causing spillage. That was exactly what happened to tank 250-2. The dispute was whether Murphy Oil did this intentionally to save money or whether it was an unintentional mistake due to human error. When the floodwaters came, the seven-foot-tall tank sloshed away and headed west into a residential area of Meraux. Big oil companies like Gulf, Shell, Exxon, and Texaco knew to avoid setting up shop in such a dangerously low-lying zone, but Murphy Oil, which had a crude-oil refinery in Meraux, probably couldn't resist the lower costs and they took the gamble. After Katrina destroyed the Chalmette area in a matter of thirty minutes, the firm's confused response exacerbated the suffering of thousands.

Shortly after Chalmette and Meraux were transformed into a watery tableau, a Coast Guard investigator reported that approximately 1,000 houses were poisoned by a crude-oil spill courtesy of Murphy Oil. More than 25,100 barrels of crude oil were dumped on these modest houses—based on a 2005 report from the Environmental Protection Agency. The Coast Guard eventually did a good job trying to clean up the Meraux streets—for example, sand was poured on roads to prevent oil slicks and lessen the possibility of auto accidents—but it was a lost cause. "It's all oil," Murphy Oil employee Shepard Brown, who lived only blocks from the refineries, told *USA Today*. "You're never going to get that clean."[54]

The health hazard created by Murphy Oil was immense. Not only does the crude oil burn skin but when inhaled in high concentrations, it can cause central nervous system damage, depression, convulsions, and a loss of consciousness. One thing was certain that Tuesday, as the residents of Meraux struggled with both oil and flooding: that Murphy Oil was going to have a class-action suit levied against it. (On January 30, 2006, a federal court did indeed consolidate twenty-seven lawsuits into one class-action suit, which was still pending as of this writing.)

"For a week after Katrina, all I could see was oil," Captain Clark recalled. "I'm a hunter, duck and deer. I owned a fishing camp in Delacroix. So it broke my heart to see animals trying to get out of the oil. Birds, nutria, small animals. We had one dog that walked up to us, looking for some kind of help. We should have put him out of his misery. Just covered in oil. It was pretty bad. But at that point it didn't even faze me because I had more things to worry about than this stinking oil spill."[55]

Within a week, a multiagency unit charged with responding to incidents of maritime oil and hazardous materials pollution in coastal Louisiana would arrive in St. Bernard Parish. It had the unwieldy name of the ESF-10 Maritime Pollution Response Forward Operating Base Baton Rouge. It would grapple with the Murphy Oil spill, staffers wearing anticontamination suits and special boots and gloves. But the citizens trapped without proper ventilation in the aftermath of the storm had no space suits upon which to rely, no oil-spill coordinator to turn to for health tips. All they had were fumes that wouldn't subside and a false belief that Murphy Oil had properly filled all of their container tanks with water.

A few days after pulling people out of the oil muck, Captain Clark noticed a rash on his stomach, something resembling a series of spider bites. It turned out that he and about ten other officers had contracted a severe staph infection from the filthy water. They were all treated in a MASH-like medical unit at Chalmette High School. "We knew that water wasn't healthy," Clark said. "We knew it probably would affect us healthwise."

Just talking about the Murphy Oil spill made Clark upset. Why didn't the company do right by the people? "I think we'll get sick," he said. "I remember going into that oil area one day during one of the missions. It would burn your eyes and burn your face. I said, 'There is no way this can be healthy.' It was benzene, they said. We never knew the difference at the time. I didn't think nothing of jumping off the boat into the water and going to rescue somebody. And I'd probably do it again tomorrow if I had to. People are walking around getting sick and dizzy and throwing up and diarrhea. That's just not normal. We all get sore throats every couple of days. I'm sure we're going to get sick."[56]

IX

On late Tuesday afternoon, the big worry of Governor Blanco, as well as of senators Vitter and Landrieu, was the lack of federal assistance. FEMA had promised 500 buses, but where were they? To the extent possible, without authorization from the Executive branch, the U.S. military started to get directly involved in Katrina relief. The USS *Bataan* was returning from a training mission in the Gulf of Mexico when Katrina swept through. The 844-foot Navy warship weathered the storm in fine shape, and then sent helicopters from its decks to assist in the rescue operations in New Orleans. Captain Nora Tyson elected to set sail for the Louisiana coast, where she could offer aid in the form of her ship's 600-bed hospital and its full store of food and water. On Tuesday, she sent a 135-foot landing craft on a 90-mile scouting trip up the Mississippi River. The sailors on the landing craft, which also carried some relief supplies, were among the first to see Plaquemines Parish in the aftermath of the hurricane. The sights were unbelievable, Navigator Rodney Blackshear recalled. "We saw a lot of dead animals, dead horses, floating cows, dead alligators. And a lot of dogs that had been pets. But no people." Dogs on the shore, driven mad by hunger and the traumatic experience of the hurricane, had turned vicious and kept the men from docking for a closer look. Near the town of Boothville, they saw a construction crane that had been left by the storm on top of a house.[57] The landing craft was within a few hours of bringing critically needed supplies to New Orleans and Chalmette when it was called back to the mother ship. Likewise, the *Bataan* itself was preparing to offer medical care to Katrina victims when it was suddenly, inexplicably, ordered to the Mississippi Coast—a region that had problems, to be sure, but which was easily accessible by truck. The *Bataan* was not a unique resource there, as it would have been in Louisiana.

One way or another, some people were being saved on Tuesday. Amantine Verdin, the Pointe-au-Chien woman whose granddaughter had checked in on her on Saturday, was flooded out of her home in St. Bernard Parish. She and her mentally retarded son, Xavier, were finally retrieved from their neighbor's barn on Tuesday, and taken to a shrimp boat that was sitting empty along the Mississippi. It was thought that in the boat they would be safe from the water. They were, but there was no food or water onboard. The elderly woman and her son had no choice but to accept what shelter the boat could provide. Lee Walker, the Poydras resident who couldn't afford to evacuate, was still in his attic with his three dogs on Tuesday, clinging to the only dry place that he could reach. With his bad back, he couldn't extricate himself or even signal for help. Like the Verdins, he was among the tens of thousands of people who were out of immediate danger on the day after the hurricane, but who were going hungry, in danger of dehydration and heat exhaustion.

North of Lake Pontchartrain, the flooding also continued on Tuesday, with water having filled streets from a few inches to twelve feet or more. Wherever the land poked through, residents were out in the heat to see what was left and, more important, who was left. Springing into action, deputies from Slidell and the rest of St. Tammany Parish rescued hundreds of people during those first days after Katrina hit. In Slidell, as in towns all over southeastern Louisiana, communication with the outside world had ceased. This was troublesome because all the fishing camps along the lake in Slidell—including Senator Mary Landrieu's family bungalow—were gone.[58] The only way to find lost relatives was to wander around, asking questions. "We're looking for my uncle," said a man sitting forlornly in the debris that had once been his uncle's house near the Oak Harbor Marina. "He told my mother the night before the storm that he wanted to stay home and watch over his house. Now we can't find him. We've picked through what used to be his house, but so far, we haven't found him."[59] After a short rest on a chair salvaged from the mess, he moved on. There was nowhere to turn for help. Gasoline was not to be found in the town, and even emergency vehicles were without fuel. "The hardest part is going to be going back later on and finding the casualties, although I hope there are none," Slidell Police Captain Rob Callahan said. "Until the water recedes we can't get to those bodies. . . . Imagine your worst nightmare and quadruple that 100 times."[60]

One Slidell resident who lost everything in Katrina was the legendary blues guitarist Clarence "Gatemouth" Brown. Ever since 1947, when Brown replaced an ill T-Bone Walker at a Houston club gig and thrilled the audience with "Gatemouth Boogie," he was one of the top performing artists in the Gulf South. He soon had hits with "Oakie Dokie Stomp" and "Ain't That Dandy." With his distinct blend of juke-joint blues, fast hillbilly, Texas swing, and Cajun boogie, Brown rapidly emerged as one of the three or four powerhouse guitar players America ever produced. He also played violin, harmonica, mandolin, viola, and drums. By the time Katrina arrived, the Stetson-wearing Brown was celebrated as an international music superstar, stubbornly loyal to his distinctive style. He had recorded with Frank Zappa, Bonnie Raitt, and Ry Cooder. In 2004, at the age of eighty, Brown, a chain smoker, had been diagnosed with lung cancer and was battling heart disease. For Katrina, his family wisely evacuated Gatemouth to East Texas. He watched on television as the community he loved—Slidell—was partially destroyed. The bayous he particularly loved had become one vast water hole, together with New Orleans. "I'm sure he was brokenhearted, both literally and figuratively," his booking agent, Rick Cady, said. "He evacuated successfully before the hurricane hit, but I'm sure it weighed heavily on his soul."[61] Just days later, he underwent angioplasty at a Port Arthur hospital. He was released to his grandniece's home in Orange, Texas, and died.

X

During the same Tuesday afternoon news conference described earlier, Governor Blanco addressed another plaguing issue: the myriad problems at the Superdome. One of the only state officials to have seen the facility in person, Blanco had to fight back tears as she said that Louisiana was looking for ways to move the terrified people out of the stadium.[62] "They're putting more and more survivors into the Superdome, and . . . the conditions there are very difficult," Blanco said. "But we're worrying first about the medically needy. So we have to set up shelters and make sure that their medical needs can be taken care of. Then in the next phase, we'll be looking for places to evacuate the rest of the folks at the Superdome. It's not a very comfortable situation right now. You can imagine: there's no power; it's hot; difficult to get food to them. . . . There's water lapping at the foot of the Superdome now. . . . I saw people walking in about knee-deep [water] as they were trying to get into the Superdome from the ground floor. And so we're, you know, that is definitely going to be phased in as we go through these next few hours and days."[63]

As Governor Blanco made clear, the Superdome was no longer a safe haven, the shelter of last resort that had been promised. It was a squalid holding tank. What Governor Blanco did not admit at the news conference, however, was that there were no concrete plans to evacuate the facility. Helicopters could transport a few of the most distressed people. The rest would have to wait—wait for the long-promised FEMA buses that had no way to navigate the waist-deep water blocking the exit route.

The absence of these buses was a continuing outrage. Day after day, as conditions changed radically in New Orleans, only two things remained constant: that people were in critical need of buses and that the buses were not running. Besides FEMA, the mismanagement of the bus transportation also fell in the lap of Mayor Ray Nagin. "We tried to start a few buses," Terry Ebbert recalled. "They just coughed, dead."[64] Nagin failed to follow his own Comprehensive Emergency Management Plan, which stated that city buses be operable and put on alert to evacuate citizens. As for the fleet of school buses owned by the city, they sat in a flooded lot a mere 1.2 miles from City Hall. A picture of the Metropolitan Street parking lot full of school buses, snapped from a helicopter by David J. Phillip of the Associated Press, stunned the nation. With no startable buses the mayor was in the throes of some kind of meltdown on Tuesday, unleashing profanities at anybody within earshot and constantly sobbing. Although he would put on a good public face, deep down he must have known just how delinquent he had been in preparing New Orleans for a major storm.

At least for a while, Mayor Nagin had one loyal ally: Councilwoman Jackie Clarkson. When she wasn't sleeping in the Hyatt ballroom, she spent much of her time in strategy sessions with Nagin. At the crack of dawn on Tuesday, Clarkson, with deputy Fitzgerald Hill as her guide, hopped into a police car and drove around Algiers. Refusing to stay holed up with Nagin at the Hyatt any longer, Clarkson then waded over to the real Command Center at City Hall and talked with Joseph Matthews, director of the city's Office of Emergency Preparedness. "Tell me what's needed to be done," Clarkson asked. She was referred to Chief Compass, who was back at the Hyatt. "Chief," Clarkson said, "put me in a boat. I'm a swimmer. I've been swimming all my life. Let me go out and save people." Chief Compass broke into a broad grin. "Mrs. Clarkson," he said, "I know you're a swimmer, but I'd feel like I'd have to use more police to protect you than to save lives." To which Clarkson said, "Fine, I won't bug you anymore."[65]

Clarkson asked herself two questions: Where can I be of the most help? What skill sets do I possess that are unique? Her answer was the telephone. Everybody said there was a "communications breakdown," but she was a master of getting a cell-phone connection when no one else could. After all, if she could reach her daughter at the Venice Film Festival, she could reach Washington, D.C., or Baton Rouge. "I found out Entergy had three satellite phones that were working [at the Hyatt]," Clarkson said. "Nothing else was consistently working. So I got on these phones and started calling places to find out flood information and who hadn't been evacuated. And people started calling my cell and leaving messages like 'So-and-so is still on the corner of Esplanade and Royal' or 'There's a nursing home on such-and-such street—we don't know if it's evacuated.' I was able to break through the communications breakdown. I found that not only could I get the message but that I could go to the satellite phone, call the Office of Emergency Preparedness and they'd send search and rescue people out to where I said. They got people out before we lost lives."

Although she stayed optimistic, Clarkson didn't like what she saw. Her city had "not done right" by the elderly. "They didn't realize they'd be trapped with no air, with no water, no food," Clarkson later said in tears. "I just wish I had known. Next time we'll know more. Next time I'm going to bring my entire database with me full of information about the elderly." Her biggest lament, however, was that Army troops weren't the first responders, in search and rescue as well as in maintaining security. "The Army would have done the job right," she said. "Give me the Army, Marines, and Navy anytime. If they had come in, everything would have been better."[66]

With the floodwater waist-deep by late Tuesday afternoon, Clarkson could no longer walk back and forth between City Hall and the Hyatt. She realized for the first time the magnitude of the disaster. She kept kicking herself that the elderly had not been sent out of New Orleans on boats, planes, trucks—anything. "Next time we've got to get all the sick and elderly out of here," she said. "A No One Left Behind Policy."

In some parts of New Orleans that afternoon, the hurricane was remarkably enough not regarded as a hardship. Looting just kept on increasing. It was a once-in-a-lifetime "shopping" opportunity. A couple of thousand people were wandering around, like sinister Santa Clauses, with bags of stolen merchandise over their shoulders. In areas without flooding, looters were having a rare field day, smashing windows on Canal Place, Riverwalk, and other shopping areas in order to carry away goods. The Wal-Mart on Tchoupitoulas Street opened its doors for local emergency personnel and police to collect materials needed in their work, but this generosity backfired when word raced through the city that the store was open—with everything for the taking. At first, people took only food and drinks and toiletries for themselves and their neighbors. Then other long-held desires took over and the electronics and jewelry departments were stormed. "It's a f——hurricane," said a fireman, watching the scene in disgust. "What are you going to do with a basketball goal?"[67] Nonetheless, some of those participating in the free-for-all were policemen. One was seen pushing a cart containing a Compaq computer and a 27-inch flat-screen television set. "The police got all the best stuff," said one looter. "They're crookeder than us."[68]

At the Walgreens store in the French Quarter on Canal Street, the thieves were on a rampage, grabbing whatever they could, when the police arrived to save what was left. A boy, stationed at the door as a lookout, cried out, "Eighty-six, eighty-six!" when he saw the cops ("eighty-six" was radio-speak for "police"). The store emptied instantly.[69] "It's like this everywhere in the city," said a longtime New Orleans policeman, as frustrated by the many absentees on the force as by the lawlessness in the midst of so much human misery. "This tiny number of cops can't do anything about this," he continued. "It's wide open."[70]

To their credit, some NOPD officers did help a group of doctors "commandeer" medical supplies from the Walgreens. Among the conventioneers who hadn't gotten out of the city in time were more than a hundred infectious-disease specialists. Rather than sit idly by inside the Ritz-Carlton, the doctors sprang into action, helping everybody they could. "Our biggest adventure today was raiding the Walgreens on Canal under police escort," Dr. Greg Henderson reported in an e-mail to his extended family. "The pharmacy was dark and full of water. We basically scooped the entire drug sets into garbage bags and removed them. All under police escort. The looters had to be held back at gunpoint."[71]

All sense of law was expunged from New Orleans within hours of the hurricane's passing. Some people sensed it, some seemed to have waited for it. With gangs and crime a way of life in the best of times in New Orleans, renegades continued coming out in large numbers. In fact, the proportion of criminals to the general population rose dramatically with so many of the more grounded residents having left. A great many of the looters were average citizens, overwhelmed by the temptation to grab what they wanted from neighborhood stores. The more serious criminals moved on the opportunities for major larceny. At Sewell Cadillac-Chevrolet, for instance, 250 cars were stolen from the lot. The biggest surprise was just who was in on the haul.

NOPD's Third District was probably the most heavily criticized police unit during Katrina. Because the Moss Street Station, located along Bayou St. John, flooded, a large contingent of the officers had to take refuge at the LSU Dental School. Everything went haywire for them. The rooftop generator was blown down, a sergeant tumbled down the stairs and fractured his ankle, and rescue boats passed by, refusing to evacuate them. When they eventually were rescued by Louisiana Wildlife and Fisheries boats, they were in need of vehicles. So they simply stole dozens of Cadillacs, some with "2004" stickers still on them, an estimated $3.7 million worth, from Sewell Cadillac-Chevrolet on Baronne Street in the Central Business District.[72] Warren Riley later insisted that most of the police stole the Cadillacs to help Katrina victims; perhaps most of them did. Of course, if all they wanted was transportation, they could have taken the Chevrolets. And there were NOPD officers like Willie Earl Bickham who was arrested in Houston driving a stolen Sewell vehicle. He had quit the NOPD a week after Katrina and drove his stolen car to Texas. "Bickham was booked in Houston with impersonating an officer and unauthorized use of a vehicle," the *Times-Picayune* reported in early 2006. "In the federal indictment . . . he is charged with interstate transport of a stolen vehicle, punishable by up to ten years in prison and a fine of up to $250,000."[73]

Perhaps more than any of the other indiscretions, the looting of the Cadillacs hurt the NOPD's image most. All Riley kept saying for months afterward was that it was "under review."

A major political split occurred between Louisiana Republicans and Democrats over looting. Governor Blanco, for example, feeling sorry for the poor who needed supplies, suggested that hotels and stores put an IOU pad at the checkout. Simply by signing their names and listing their goods, honest looters would be transformed into borrowers. It was a well-intentioned but impossible notion. When Senator Vitter heard this idea, he surmised that Blanco was out of touch with reality. "I knew we were in big trouble if the pad was our security plan," Vitter recalled. "At the same time, Governor Barbour of Mississippi made it clear that looters would be shot."[74]

There were those, however, who *did* leave IOU notes for using provisions during the deluge, like Blanco suggested. Sixty-two-year-old Skip O'Connor was part owner of six hotels in Greater New Orleans. As Katrina was approaching, O'Connor decided to leave his Marriott Courtyard on St. Charles Avenue open as a shelter. It was a civic gesture. A large church group noticed the door was unlocked and used the inn as their micro-Superdome. They slept in beds and consumed food from the kitchen. When evacuation buses finally arrived in New Orleans on Friday, Al Thomas, one of the Marriott Courtyard dwellers, left an IOU note which read, in part: "I'm writing to say thanks for the use of your hotel. Under these conditions, we had no other choice. We did not destroy anything, but we did find food to eat and water. . . . When leaving we took nothing but some food and some essentials because we did not come here to steal any that did not belong to us. . . . I will send something back finance-wise to compensate the use of your facilities. . . . There's nothing phony about what I'm saying because this is coming straight from my HEART. About sending money for the use of the building." In signing off, Thomas identified himself as "One of Katrina's Victims."[75]

The most dangerous criminals didn't bother with commercial looting for mere financial gain. The pathological criminal also sought domination and the chance for violence. To that end, a kind of "wilding" occurred in the wake of Katrina. Just waving a gun at helpless survivors somehow empowered miscreants. They barged into private residences, with the occupants home, weary and frightened from the storm. The crooks took what they wanted at gunpoint, and the residents were fortunate if they were still alive when the robbery was over. Nothing kept the crooks from murder, just for the heck of it. Patrick Wooten, a roofer, was still in his house in the Algiers section, trying to keep his wife and two teenage sons fed in the days that followed Katrina. The water had receded, but that didn't mean that there was anywhere to go, or any way to get there. One night, a group of thugs started trying to open the kitchen window. Wooten picked up a tire iron, the only weapon he had in the house, and shouted for his boys to take their mother into the bedroom and stay out of the way of gunfire. Alone against the intruders, Wooten hit the window frame with the iron. "Bam; get back!" he said, recalling what he did and said during his ordeal. " 'Bam; get back!' I said, 'Man, would you all please get back, man?' I said, 'We don't have nothing.' Then they hear me crying, you know. . . . [My sons] just stayed around their mom. They protected her. God must have been on my side, you know, to [have the robbers] not shoot in the house."[76]

Later in the week, Wooten's sons were outside when they spotted three corpses in the backyard of an adjoining house and heard a thug bragging that he had killed them in the course of robbing their house. Wooten had absolutely nowhere to turn. For days on end, his only choice was whether to be killed by the bears at the window or the lions in the street. The few police officers he and his family saw made it clear that they had orders to shoot to kill and that nothing would stop them from doing just that. Wooten stayed indoors night and day.

Many of the thugs were specifically looking for guns, hoping to trade whatever firearms they had for ever bigger ones. Gunfire was so rampant in the streets that the rumor quickly spread that rescue helicopters were under attack. That was not the case, but each time the story spread, rescue helicopters throughout the city were grounded for at least an hour.[77] "The rumors were way over the top," District Attorney Eddie Jordan recalled. "As a public official you couldn't believe what you heard. You had to be careful not to mislead people."[78]

Unfortunately, not everybody on the NOPD was as wise as the DA. By afternoon, the police had more serious problems than televisions and makeup kits disappearing from store shelves. "To my amazement and astonishment, armed crowds started shooting at rescue personnel," said Police Superintendent Eddie Compass. Instead of rescuing people, Compass's thin ranks had to try to protect rescuers.[79] Well . . . not really. It didn't happen. Caught up in the hysteria of the moment, Compass had exaggerated. He was in part responsible for floating rumors to the national reporters as hard news. Well-intentioned, Compass couldn't stop chatting up the press. He even held interviews with media from Brazil, Japan, and Italy, presenting himself as "Steady Eddie," the good cop on the beat. He seemed to bask in the limelight.

Warren Riley's girlfriend was in Houston when she heard the news that snipers were firing at rescue helicopters and that officers were being killed. "Warren, you've got to get out of the city," she cried. "Get out of there." Trying to comfort her, lambasting the grapevine, Riley explained that his duty was to stay in New Orleans. He couldn't leave. She was insistent, saying, "You're gonna get killed. They're killing everyone in New Orleans. They're shooting at helicopters, they're shooting at police, they're killing people. They killed Chief Compass." A rattled Riley knew this was bull, only the wildest of rumors being promulgated by irresponsible journalists. "What are you talking about?" he said. "They didn't kill Chief Compass. He is sitting right here. Do you want to talk to him?"[80]

Riley looked into all sorts of rumors. All the television stations were running stories about NOPD officers being killed by looters. He later explained, "All those stories about people shooting at helicopters weren't true. There was one guy [who] shot a helicopter in Algiers—a SWAT team went in and got him. [New Orleanians] were in their attics for days, seeing the [rescue] boats and they were shooting, though their shots were to say, 'We're here.' Our SWAT team would ax a hole in the roof and get them out. Imagine being in your attic for days with ten feet of water below you. The shooting was about people saying, 'Hey, somebody, help me.' [As for] people shooting at helicopters, that was bull."[81] Other prominent rumors were that DA Jordan had been murdered by prisoners and that Jefferson Parish Sheriff Harry Lee had died of a heart attack. All the rumors added to the turmoil—and frightened FEMA away from helping out the beleaguered.

With FEMA and the Red Cross staying out of New Orleans, and the Coast Guard and National Guard shorthanded for a disaster of such magnitude, journalists also found themselves in the role of emergency relief workers. Whether it was handing out pallets of water, rescuing people out of floodwaters, or finding rides out of New Orleans for the sick, reporters joined the effort to help. Not only did they bring international attention to the Great Deluge via newspaper dispatches and television reports, but they goaded city, state, and federal first responders to do more—much more.

At times, New Orleans police roughed up journalists in an attempt to prevent them from filming fellow officers who were mistreating citizens. Two *New York Times* photographers, Tyler Hicks and Marko Georgie, were taken from the front porch of the house at which they were staying, shoved to the ground, and handcuffed—all for getting fresh air during curfew. Jessica Willey of Houston's KTRK-TV was barred by the NOPD for filming looters (perhaps the police were worried that fellow officers would be captured in the act of committing a felony). Recognizing the police were using overt force, Michael Christie of Reuters compared certain New Orleans blocks to a war zone. "If I ever felt threatened, it was by the realization that there were an awful lot of back country sheriffs there itching to pull their triggers," Christie explained. "Having seen U.S. troops in action in Afghanistan and Haiti, where they went to shoot anything that moved, I felt rather nervous."[82]

XI

Along the Gulf Coast in Mississippi, the surging waters had been gone for more than a day as of Tuesday. However, many people were still in trouble. In addition, tons of debris remained, and the cleanup had not yet begun. Fleets of boats were found in scrub woods, and rooftops were separated from their houses. Holbrook Mohr of the Associated Press was in Gulfport on Tuesday. He heard of two people who had been carried out to the Gulf with the receding storm surge. They were rescued after clinging to floating wreckage for twelve hours. Mohr heard of other rescues, but was haunted by the endings that were neither happy nor tragic, only terribly confusing. "I met a man on the street," Mohr said. "He came up to me and asked me if I could help him. He pointed to a Shell gas station along the beach that was totally demolished and he just broke into tears asking, 'Where's Debbie? Where's Debbie? She's the love of my life.' People are pretty distraught here."[83]

Many historic Gulf Coast landmarks were simply gone when people came out on Tuesday to take a look. "The devastation down there," Governor Barbour told *Today* on Tuesday, "is just enormous."[84] Green Lawn Hall, which dated to the 1830s, was obliterated, only its foundation left. The ground floor of Beauvoir, Jefferson Davis's home, was gutted. Pelicans perched on the structural remains of the 800-foot Coliseum Pier in Biloxi. Pass Christian's harbor, full of shrimp boats, was deemed a total loss. That afternoon, police and other officials still couldn't even reach the once busy town of Moss Point, just north of Pascagoula—it was submerged under twenty feet of water.[85]

Governor Barbour had already reported that at least eighty citizens of Harrison County, which included Biloxi and Gulfport, had been confirmed dead.[86] Two hundred forty National Guardsmen rushed into the county Tuesday afternoon. Approximately the same number went to the other two coastal counties.

Although it was nothing like New Orleans, pillagers had come out in the broken communities of the Mississippi Gulf. Some were relatively innocent—for example, the castaways poaching the casino slot machines that were washed up all through the Biloxi waterfront, hoping to find some coins; or teenagers breaking into an amusement park that had been spared and riding the bumper cars for free. In other places, the criminal activity was serious—but not as serious as the opposition that arose to fend it off. Nanette Clark had invited her friend Jayne Davis to stay at her Pascagoula house during the hurricane—a fortunate idea, since Davis's home in Biloxi ended up being washed away. Clark's home, a dainty pink house with gingerbread trim, suffered only minor damage. On Tuesday, the two women sat on the balcony of that cheerful home with loaded guns, watching for looters or highwaymen. When they saw a suspicious-looking group of potential freebooters approach, they fired rounds into the air to scare them off.[87] The police could do little about looting, not only because they had rescue work to perform, but because there were no local jails left. All over the churned-up coast, homeowners took up arms and stood ready to defend their possessions with firearms. Some, like Clark and Davis, seemed almost eager to blast a pirate back to sea. It was a singularly American response—more specifically, a Southern one. Among the Confederate flags hung outside some homes and the American flags that were displayed everywhere, signs sprang up warning, almost excitedly, that trespassers would be shot. Fortunately, humor was also a reassuringly American response. Next to a pile of debris and wind-strewn lumber in Diamondhead, a high-end, golf-retirement community near Bay St. Louis, was a hand-lettered sign reading "House for Sale (Some Assembly Required)."[88]

The whole Gulf Coast region was, in the words of Mississippi Senator Trent Lott, "on its knees." Lott became a real critic of the slow relief and recovery response; ignoring the GOP talking points, political caution had been removed from his speech. On Tuesday, his home gone, Lott was just another Mississippian—an individual in trouble, in search of White House leadership, in hope of help. "The people of Mississippi are flat on their backs," Lott said in a public appeal to the President. There was something in his tone that was angry and smoldering. "They're going to need your help."[89] (A few days later, furious at FEMA's ineptitude, Lott charged that Director Michael Brown was "acting like a private instead of a general."[90])

XII

At the end of the day, Michael Chertoff returned from Atlanta to Washington, where he finally turned his attention to the relief effort along the Gulf Coast. In his capacity as Homeland Security Secretary, he issued a memo declaring the hurricane damage an "incident of national importance."[91] With that, Michael Brown, as the designated "principal federal official," could override local and state officials as he saw fit, and coordinate aid from nearly every federal department that could be of some use, including the military. Even so, Chertoff still failed to attach any real sense of urgency to the declaration, even delaying a public announcement about it until the following day. Judging by his actions, Chertoff seems to have been reluctant to rush the White House response. He consistently deferred to the President's initiative to form a special task force. At best, he was feeling his way through unfamiliar territory. At worst, his priorities were those of a political animal. The first instinct of a true emergency-response professional would have been to waste as little time as possible and use any and all powers available (according to the National Response Plan).

That Tuesday night, Ray Nagin was trying to exert the power just accorded to Michael Brown. In an evening news conference, he complained that his orders to use available National Guard helicopters to drop sandbags into the breach in the 17th Street Canal had been ignored, thus allowing the breach to widen. He complained that there were too many chiefs calling the shots—and he was probably right.[92] There were either too many chiefs or not enough effective ones. Nagin said that his priorities were to rescue people, fix the canals, and evacuate people from the Superdome. "We're not even dealing with dead bodies," he said. "They're just pushing them on the side."[93] Nagin used his mayoral power to declare martial law in New Orleans on Thursday. All residents who were able to were ordered to leave, and that included non-emergency police officers.

In the evening, having accomplished little and grounded in Baton Rouge, Michael Brown recognized that the magnitude of Katrina had outmatched FEMA. Chertoff was acting imperious and Brown's overtures to the White House were falling on deaf ears. The agency *did* manage to bring in eight disaster medical-assistance teams, with thirty-five members each. Beyond that, there was nothing much to boast about, just things like crews going out to inspect oil rigs for spills. Everything else was in the future. Medical personnel were being summoned through the Department of Health and Human Services, and food was being requisitioned through the Agriculture Department.[94] Someone at FEMA had been in touch with the U.S. Northern Command (NORTHCOM) of the Defense Department all day, trying to arrange for the use of two helicopters. That might have sounded good, except that the helicopters were needed, according to the request, for flyovers by FEMA officials.[95] With a little flexibility, they could have used the Coast Guard helicopters instead. Brown later testified against Chertoff before the Senate Homeland Security and Governmental Affairs Committee that his boss was "irate" because he tried to conduct field inspections.[96]

At the same time, NORTHCOM also began setting up a task force to be based at Camp Shelby, near Hattiesburg, Mississippi. Among other things, it would oversee deployment of regular army assets and personnel, if orders were received to that effect. The National Guard presence was at least showing some improvement. As of Tuesday, 7,500 Guardsmen were on the ground along the Gulf Coast.[97] That number was clearly inadequate, but NORTHCOM promised that 98,000 troops were available. "We have some unique capabilities such as airlift and amphibious vehicles that FEMA doesn't have," said a NORTHCOM spokesman, Michael Kucharek. "I think there's a realization that the devastation is so widespread that they are going to need more support than they can provide on their own."[98] He was certainly right about that.

By Tuesday night, Michael Brown was placing urgent, even panicked calls to the White House. "Guys," he recalled saying, "this is bigger than what we can handle. This is bigger than what FEMA can do. I am asking for help."

Brown reflected later, "Maybe I should have screamed twelve hours earlier. But that is hindsight. We were still trying to make things work."[99]

Chapter Ten

THE SMELL OF DEATH

> The moment an undertaking begins to shape up, it becomes ipso facto the butt of a thousand hostile, treacherous, subtle, and untiring intrigues. . . . Nobody can say different. . . . A tragic fatality penetrates its very fibers . . . slowly lacerates its warp, so profoundly that, when you come right down to it, the shrewdest captains, the snootiest conquerors can only hope to escape disaster, to keep from cracking up, by some cock-eyed miracle. . . . Such is the nature and the burden, the true upshot of the most admirable ventures. . . . It's in the cards. . . . Human genius is out of luck.
>
> —Louis-Ferdinand Céline, *Death on the Installment Plan*

I

SARA ROBERTS WAS FURIOUS. A partner at Dunn, Roberts, & Company in Lake Charles, Louisiana, and a practicing certified public accountant since 1987, she just couldn't stand watching New Orleans turn into a Pompeii or an Atlantis as she passively stared at her television set. Her husband, André Buisson, was the ninth of ten children from a New Orleans family. They had, as she put it, "tons" of friends in the city. Word eventually reached her in Lake Charles—the seat of Calcasieu Parish and the urban hub of southwest Louisiana—that Katrina had flattened the homes of two of her brothers-in-law, in Shell Beach, near Chalmette. To add insult to injury, Roberts was a member of the Superdome Commission and couldn't believe reports that over 20,000 people had congregated there. "The Dome is not built to be a shelter," she later said. "It's just not; it's too open. You have to be able to separate certain groups, for example, the elderly, children, and families. The Dome is not really made for evacuation. We had many, many meetings regarding that. And the one thing we were all, as commissioners, very interested in, was making sure that we made it as safe as we possibly could. That we had appropriate provisions for a couple of days."[1]

On Tuesday, conditions at the Dome weren't improving much; a human rights crisis still was on hand. Out of Roberts's concern for New Orleans was born the Cajun Navy. The armada originated with a telephone call. On Tuesday, Sam Jones, the governor's liaison to parochial and municipal governments, had telephoned her, saying FEMA and federal troops were slow in coming. "Let's call out the volunteers," Jones said. "We've got to get those people out of New Orleans."[2]

Jones wanted Roberts to procure twenty boats in Cajun country, hitch them to pickup trucks, and caravan them eastward to New Orleans to join the Louisiana Department of Wildlife and Fisheries (LDWF) boatmen. Without hesitation, Roberts said, "I think I can do it." Her first act was to telephone a client named Ronny Lovett, the owner of R & R Construction, a lumbering, bespectacled entrepreneur with thinning hair. A country boy from Leesburg, Georgia, he had over six hundred employees in his contracting firm, mainly no-nonsense types, who helped build chemical plants. After Lovett agreed, which didn't take long, he started hunting down employees with boats. "My guys don't back off of anything," Lovett said, "and we all take care of each other. So, when Sara called, we all acted."[3] The Cajun Navy started with the donation of his own Bay boat.

Roberts had iron-straight blond hair and a lighter complexion than most of her fellow Cajun colleagues. In addition to her full-time career, she was busy raising two kids, one in the first grade and one in the third. But she was "stoked" to suddenly be corralling boats for such a worthy cause. Born into Scotch-Irish and Cajun lineage, Roberts's mother had hailed from Rayne, Louisiana, the second-largest city in Acadia Parish, also known as the "Frog Capital of the World." Roberts quickly rounded up boats from friends. Meanwhile, the Lovetts started telephoning their employees in tiny Lake Arthur, Louisiana, population 3,007, located in Jefferson Davis Parish, for rescue boats. Ruth continued to score. Virtually every employee she asked said, "Count me in."

The Lovetts offered any employee who joined the citizens' navy triple wages for the time spent in New Orleans (they wound up spending about $200,000 of their own money to fund the Cajun Navy). Roberts also corralled boats from friends. Call it a humanitarian monetary incentive. And with fellow Louisianans in need, and FEMA a missing-in-action joke, this citizens' navy was rapidly materializing. The three magnets that pulled all the filings together were Lovett, Roberts, and Buisson. It was like an American Civil Defense effort, something akin to the Battle of New Orleans of 1815. "Once we got thirty-five people and eighteen boats on Tuesday morning, we were ready to go," Roberts recalled. "My husband and I were going to lead the caravan to New Orleans."[4]

Because Lovett had helped the victims of Hurricane Andrew in 1992, he knew what provisions were needed. Each boat had to have spotlights, gas-powered chain saws, picks, axes, life jackets, oars—every tool, in short, that might make a difference in saving a life. "Ronny also made sure that we had a fuel truck that went with us and a tool truck," Roberts recalled. "We had mechanics and electricians. Most of our crew members were skilled laborers, and we purposely assigned tasks according to skills."[5] Sara, using her own talents as an accountant, tallied and organized the supplies as they were packed onto the vessels. The shallow-water boats on her ledger sheet came in all shapes and sizes, including tunnel hull boats, aluminum-welded boats, a Sea Fox, and homemade johnboats. "I even got us a gooseneck thirty-foot trailer," Lovett said. "A one-ton truck was brought in full of water. Everybody we encountered was going to get a drink."[6]

Somehow, as Cajun folk watched New Orleans flood on TV—while FEMA abandoned the stranded and then treated them as if they were human driftwood—the old ghosts of the past stirred in the hearts of southwestern Louisianans. Many of these Cajuns worked in the oil and gas exploration business but their hearts had never left their pirogues cutting through the bayous at dusk. Many had never been to the city of New Orleans. As the Cajun Navy rendezvoused at the Wal-Mart parking lot in Jennings, these Ragin' Cajuns were ready to earn their overtime. "The gathering looked like the beginning of a weekend duck hunt or fishing trip among a band of brothers," Andy Buisson recalled. "On this trip, though, the beer stayed at home as did the guns . . . well, most of them. Thank goodness not all of them."[7]

He and his wife, Sara Roberts, a Blanco supporter, dressed in blue jeans, T-shirt, and baseball hat, made sure all the pickup trucks pulling boats were fully fueled and ready to roll. Together they made quite a sight. Andy was six feet three inches tall with plenty of bulk. By contrast, Sara was five feet one inch in her stocking feet, barely registering in at 115 pounds. He drove the lead vehicle, a white Chevrolet pickup, while she handled communications on her cell phone. "We were in a huge caravan," Roberts recalled. "And people on I-10 would just pull over and let us pass. They'd see our boats and give us a thumbs-up. They knew where we were going. They cheered what we were doing."[8]

For Buisson, a smudge of oil on his face, this ruddy afternoon was bittersweet. A Lake Charles lawyer, he was the scion of Confederate royalty, a proud descendant of General P. G. T. Beauregard's sister.[9] However, it didn't escape Buisson, a student of history, that Beauregard, the "hero of Fort Sumter," might have deemed the idea of a citizen flotilla to New Orleans to rescue African Americans from floodwaters ludicrous. But Dixie had changed. A litigation lawyer who mainly represented small businesses and local government, Buisson championed equal rights. To think in racial terms at such a dire moment as the Great Deluge was not in his makeup. Reflective, detail-oriented, and observant, Buisson ended up keeping a diary of their frenetic experiences that week. Sam Jones, from the governor's office, had instructed them to proceed to the LDWF staging area at Causeway Boulevard and I-10 in Metairie.

"When we all arrived at Causeway, there were literally hundreds of boats in a line," Roberts recalled. "There was quite a bit of confusion. Wildlife and Fisheries was trying to get some order to this because all of these people just showed up in boats."[10] The caravan was then directed to the Sam's Club on Airline Drive, where Roberts was handed two cans of Day-Glo paint and instructed to mark each house searched with an alphanumeric code. However, because there was an overabundance of rescue boats at that site, LDWF told the Cajun Navy to line up on Canal Street, alongside Harrah's Casino, and wait there for an assignment. The Cajun Navy's original plan was to deploy boats along the Industrial Canal for rescues in the Lower Ninth. But a group of NOPD officers, concerned that New Orleans East was being neglected, persuaded this citizen flotilla to head out to Chef Menteur Highway in New Orleans East the next morning.

At dawn on Wednesday with a police escort, the Cajun Navy tried to head east on I-10. However, they were stymied by the flooding and the crowds. They were going to deploy in a residential neighborhood, where most of the NOE houses were originally built after World War II to absorb G.I. Bill families. The convoy trip, particularly the scene on I-10, overwhelmed Roberts. "I've never seen so many people," she said. "They were all on the elevated portion of the interstate and many were sitting in the sun in wheelchairs waiting for buses. They were all obviously sunburned, obviously dehydrated, very listless. At that point I text-messaged Sam Jones, Leonard Kleinpeter, and Ty Bromell, all three of them, to get these people off the interstate. It was horrific. I've never seen anything like it. You talk about the Dome being a disaster? The real disaster, in my opinion, were the people left on that interstate. It was horrible. People literally crying for water. Begging us for water."

The Cajun Navy eventually made its way, via some slick maneuvering, to Chef Menteur Highway. They created a staging area at Crystal Palace, a large reception hall located only about a half-mile from Lafon Nursing Home and the Sugar Bowl Lanes. Nearby was a largely flooded-out branch of the New Orleans Public Library. A man suddenly appeared at the lead pickup truck, screaming for help. His pregnant girlfriend was going into labor. "Our baby is breached. Can you please help us get to the hospital?" he begged. Roberts and Buisson had been instructed by NOPD officers on hand not to leave the caravan, their armada's mission being to get people out of the water. Roberts tried to make the man understand that they weren't an ambulance service. "I was devastated," she recalled. "It was very hard for me to tell him that I would get the [NOPD] captain and try to get some help. But we couldn't take them out of the area. We couldn't leave the caravan.

Meanwhile, a new problem presented itself. As Buisson explained, the Cajun Navy discovered there was a martial atmosphere in the area. "It became obvious that our lightly armed band of hunters and fishermen might have come under-equipped," he wrote. "Wearing Kevlar vests, out-of-parish deputies brandished short-barrel shotguns, automatic rifles, and [a] weapon my unsophisticated eye could not identify, and which I would surely not know how to fire. Unlike our good-natured volunteers, with the occasional sidearm on their hips, concealed by their t-shirts, these others exhibited their weapons for all to see. Pumped up, they were ready to go hunting. I suddenly felt naked, wishing I had pulled my one shotgun out of the closet, or wherever I had left it after I used it last."[11]

An NOPD officer gave the Cajun Navy the rules of engagement. "If you encounter a dead body, don't touch it," the officer instructed. "Leave it alone. That will be handled later. We are here to help people. There have been some reports of rescuers being shot and boats being taken. If it gets too rough in there, we're getting out. Bring those who will leave, but don't force them. No pets. Do not travel alone. Go out in pairs. As far as weapons, carry what you need for protection."

By 7 A.M. that Wednesday, five or six boats were deployed down each of four highly populated roads off Chef Menteur Highway: Read Boulevard, Ridgefield Drive, Crowder Boulevard, and Bullard Avenue. "My husband and I worked the same area," Roberts said. "When we first put the boats in, the first thing I noticed were the police officers, because there was no one evacuating right there where we had launched the boats. It didn't take me long to figure out that these officers were so incredibly damaged. They were looking for their own relatives. They worked with us, side by side, pulling people out of the water, but they were most interested in trying to find out where their own family members were. Another one of our boats took a police department member with us. He wanted us to find his father. They ended up going to his house and they found his father dead."

No matter where Sara Roberts looked that day, there was somebody who needed attention; the number of sick and elderly just floored her.

The first person Roberts rescued was a middle-aged woman with her husband and mother. She never forgot how the bedraggled woman kept waving a picture in front of her, moaning, "I've lost my best friend, see? My sister." While being ferried back to the Crystal Palace on Chef Menteur, the evacuee clutched the color photograph. "I think the floodwaters got her," the woman groaned. "I've lost my best friend." She was hysterical. When Roberts dropped her off, she tried to convey to the woman that she needed to take care of herself. Just because she was on dry land didn't mean her evacuation journey was over. There would be long, hot hours waiting for buses. Then it would be weeks or months at a shelter. But Roberts knew she didn't need to hear about all that for now. What she needed now was comfort and reassurance. "You're safe," Roberts said firmly. "We're going to get you some attention. You're going to be okay."

Later that afternoon, as Roberts was helping another group out of a rescue boat she did a double take. She was grabbing the arm of the "sister"—the woman in the photograph! "You're the sister!" she exclaimed. "Guess what? I know where your sister is. I'll bring you together." It was the high point of a grueling day. "It was the sweetest get-together imaginable," Roberts said. "They each thought they had lost their sister."

The crying, screaming, praying, dragging, and alternating between hope and despair continued. On another rescue, as Roberts helped a woman out of her flat-bottom boat, she accidentally got some foul-tasting water in her mouth. She thought of the bacteria now in her body. Meanwhile, the woman being rescued pleaded, "Please, don't let them leave my boy!" A concerned Roberts asked, "Ma'am, where *is* your son?" The response was chilling. "He died in the attic," the woman sobbed. "He had cancer and he died while we were in the attic. It was so, so hot. He couldn't breathe. I just can't leave his body. He's the only thing I have left."

Each Cajun Navy boat was packed to the gills with refugees, and Roberts couldn't accommodate a corpse. "You're safe," Roberts said, stroking the grieving mother's head. "We're going to try and get you into a safe place. We can't take your son, but you're going to be okay." As the rescue boat pulled away, the mother, reaching out for the receding silhouette of her attic, knew she would never see her child again.

On one run down Read Boulevard, Roberts saw two African Americans coming out of a Rite Aid with big bags full of merchandise. She had seen looters on her TV back in Lake Charles and now they were in front of her very eyes. Her blood boiled. Creeps! Swine! Degenerates! People all around were dying and they wanted *things*. It made her stomach turn. She later confessed to being extremely "judgmental," convinced that their bags were full of radios, CDs, and cameras—unessential things not needed to survive. "Later that day I saw those same black guys in a fifteen-story high-rise," she recalled. "They had taken those supplies and were passing them out among these elderly people that had been left behind, people that were desperate for Gatorade, energy bars, and medical supplies and things of that nature. It made me very ashamed. At that moment I realized they were taking care of the people they loved. I put myself in their position: if I had an elderly mother or partner or friend or child that was in trouble, I would stop at nothing—nothing—to help them. How wrong was I to judge all of these people as looters. I never again thought anything negative about anyone just trying to survive Katrina and the whole bloody aftermath."

The high-rise Roberts was talking about was Forest Towers East Apartments at 10101 Lake Forest Boulevard located just off Read Boulevard in a cul-de-sac. It was a generic-looking senior-citizen residence, not a nursing home. It had no legal obligation to evacuate their residents. One man waved his empty oxygen bottle out of a window like a railroad lantern, hoping to catch the Cajun Navy's attention. It worked. Roberts's boat, followed by two others, pulled up to the front door of Forest Towers. The debris-filled entrance looked like a hand grenade had just gone off. A soaked picture of the Sacred Heart of Jesus, walkers, artificial plants, plastic furniture, dead animals were just some of the items in the floating morass. "We drove the boats into the dining room. We carried those in wheelchairs downstairs to get them in our boats," Roberts recalled. "We had to get them to a stairwell where we had to get them through the dark water. In the stairway were mounds of dead rats, frogs, mice—it was disgusting. Then we set up tables so we could lift residents onto the tables and get them into the boats—that's how deep the water was."[12]

The residents told Roberts and Buisson that on Saturday they had been told by the management of Forest Towers that they would have to fend for themselves. "We're evacuating," the infirm were told. "You must call your families. You're on your own." According to Roberts's estimate, about 70 percent of the residents left behind were in wheelchairs, many needing insulin or dialysis treatment. They were helpless. Some of the residents did have family who came and rode out Katrina with them. Their loyalty to kinfolk saved lives. The able-bodied people, ignoring mandatory evacuation orders, tried to be caretakers for all the seniors.

What truly amazed Roberts, however, was the number of people in the high-rise with pets. Many of the elderly pleaded with her not to leave a beloved pet behind. The police had instructed the Cajun Navy not to rescue animals—in their view, things were difficult enough without snarling dogs and pissing cats on overheated evacuation buses. But by this time, Roberts was fed up with NOPD rules. "I simply refused to tell these people who had been through so much that they couldn't take the only loved one they had," she said. "I lifted many a fat cat and put it into the boat."

Besides Forest Towers, the Cajun Navy came to the rescue of the Metropolitan Hospice on Read Boulevard. As in so many Louisiana facilities, the weakest and most frail beings had been left to stew among bits of decaying food and putrid water. However, a chaplain, Bible in hand, had refused to abandon them.

Months after Katrina, Ronny Lovett admitted that he had forgotten the African-American chaplain's name, but he didn't forget his deeds. After entering the facility, the Cajun Navy quickly loaded up three boats, their last run of the day. Lovett offered the chaplain a ride, but he refused. There were seven women in their eighties and nineties, simply too ill to be evacuated by boat and left on a street corner to wait for a bus. "So he stayed with those seven white women," Lovett recalled. "He just wasn't going to leave them. I think about that chaplain a lot. I just wonder what happened to him and the others at that hospice. He was a man worthy of respect."[13]

As twilight turned into nightfall at the Crystal Palace staging area, Roberts stood in line to take the antibiotics that were being handed out by the NOPD. Surveying the hundreds of refugees, she noticed many had open sores, a whitish-yellow pus oozing out of their skin. "I don't know if, during the storm, when they were trying to escape the floodwaters, they were gashed," Roberts recalled. "Also, many of the people were diabetics, desperate for insulin." Struggling with hyperglycemia, which could cause dehydration and ketoacidosis, many were at risk for gangrene, which could lead to the amputation of toes, feet, or even legs. There were also people with heart and kidney ailments, intestinal ulcers, glaucoma, and other serious medical conditions. "These people needed medical aid fast," Roberts recalled. "There just wasn't much I could do to help."

At the end of the day, the Cajun Navy loaded its boats onto the trailers, re-formed the caravan, and headed back to dry land downtown. Seemingly docile evacuees, tired of waiting for phantom buses, jumped on the boats, pounded on vehicle windows, and encircled the caravan. Warped by fear, they wanted a lift. A riot nearly broke out. "We somehow made it through the human barricade," Roberts said. "We got back on I-10 and we exited at the Convention Center. That was a madhouse. They were angry, they wouldn't move off the freeway exit. They were blocking the exit. They were shouting and the street was very tense. They were hungry, thirsty, mosquito-bitten, and sick. They were exposed to the elements. They were hot. The police managed to get us past that. We lined up on Canal Street."[14]

The Cajun Navy spent the night in their pickups, just three blocks from the tortured Convention Center. They were, as Andy Buisson put it, "sitting ducks, barely armed with nowhere to go."[15] A SWAT team of forty or fifty armed men patrolled the area. Their primary objective was to make sure those congregated around the Convention Center didn't suddenly turn into a riot army. Meanwhile, Ronny Lovett worried about his Cajun Navy troops. Should they leave or stay? Things were getting hairy. Sara Roberts had fallen ill, with dry heaves and an excruciating headache. She blamed her condition on the antibiotics.

Suddenly, an NOPD officer appeared alongside Buisson's truck, asking for help. Apparently, six SWAT officers hadn't returned from a rescue mission around Esplanade Avenue, between Broad Street and City Park. The police wanted to take the Cajun Navy boats to go look for them. "Ronny was brought into the conversation," Buisson wrote. "After ten minutes, he noted in frustration that the officer had told him three different stories while trying to engage our help. Tensions were high. Unknown dangers, whether real or imagined, appeared to be all around. Now some of the group was being asked to carry three boatloads of tactical specialists into the dark bowels of 'New Venice' to rescue six armed officers, whose fate was unknown. The choice was soon removed when the desperate officer declared Martial Law, commandeering the boats."

The NOPD took three Cajun Navy boats and three R & R Construction workers with them, and disappeared for nearly five hours. When members of the Narcotics Division of the NOPD found out about the boat heist, they were livid, furious that fellow officers had infiltrated the Cajun Navy and "stolen" people. Eventually, at 2 A.M., the boats returned. With reports of violence breaking out at the Convention Center, Lovett made a decision to get out of New Orleans. "By 6 A.M. Thursday morning, we arrived home," Buisson recorded. "We bathed and crawled into bed thinking we would sleep away the day. I awoke three hours later, rested but restless. I began to think about New Orleans again."[16]

Although the Cajun Navy took Thursday off, they were back in New Orleans East on Saturday, this time with thirty-five boats. They worked all weekend long. It's estimated that this patchwork outfit rescued close to 4,000 people in the flooded neighborhood.

II

On Wednesday morning Reverend Willie Walker tried to make it to Central City from his wind-damaged home in Kenner. His big worry was that Noah's Ark Missionary Baptist Church was underwater—which it was. He knew from radio reports that South Saratoga Street was swamped, but that didn't deter him. From his minister's perspective, grief and concern easily trumped pragmatism in the wake of Katrina. He hatched a plan. He would sneak past police barricades in his Mustang in order to get into New Orleans to rescue those left behind, particularly parishioners from Noah's Ark. Somehow, by the grace of God, he would find a boat. Off he went. Driving through largely white Metairie, it saddened him that de facto segregation still reigned in Louisiana. He couldn't help but think about race as he drove past the old office of ex-Klansman David Duke, wind damaged by Katrina. At least the Lord did something right, he chuckled to himself.

Realizing he would need provisions for the stranded, Walker made a U-turn and headed north to Dorseyville, a hamlet in the petrochemical area known as Cancer Alley. You could almost smell the carcinogen called vinyl chloride. Liver cancer was widespread in the area, particularly in the African-American and Hispanic communities. He stopped at his cousin Deedee's trailer home on the outskirts of town, which still had power. On the television, he caught his first look at the sad, abysmal situation at the Superdome. It spurred him to accelerate his rescue mission. Rushing to a Wal-Mart in nearby Donaldsville, he began to purchase as much food and water as possible. The air-conditioning felt good, but he was worried that he was moving too slowly. After buying a shopping cart full of basic provisions, he sheepishly asked the store manager to donate the goods, showing his church business card. "She gave me thirty more gallons of water for free, saying they considered it a worthwhile church donation," Walker recalled. "They also gave me turkey, some cookie packages, anything that would give some energy, like Cheese Nips and Red Bull."[17]

As Walker took the old River Road route past Ochsner Hospital, he stumbled upon fifteen or twenty NOPD officers operating a checkpoint with a few out-of-state policemen. He wondered why there were so many police in such an out-of-the-way locale. He quickly answered his own question: the perimeter was just the place to "work" if you didn't really want to, a refuge from the chaotic demands of the rest of the city. This group of NOPD simply shied away from duty, transforming the city's limits into one big Dunkin' Donuts coffee break.

Reverend Walker wanted to offer up his services to the Jefferson Parish Response and Rescue Team, which was in charge of the rescue boats based at Causeway and Airline. "Who's in charge?" Walker asked a checkpoint guard. A few men pointed to a sergeant. Each officer had on a bulletproof vest and was carrying three or four guns. "I thought to myself, this was search and rescue?" Walker recalled. "It was like a bear hunt." The sergeant rushed at him and sized him up, making sure he wasn't drunk or an Al Qaeda operative or a member of a Memphis gang looking to seize turf. Under such scrutiny, Walker felt like a peyote smuggler from

Juárez—frustrated cops could make even a righteous man quiver. Nonetheless, Walker passed the profiling drill. It paid to look clean-cut and offer up an Uncle Tom smile. But in spite of that, the sergeant flat out rejected the reverend's offer to help; the fact that Walker's church was named Noah's Ark didn't impress.

Refusing to leave the launch site, Walker approached a group of officers from Monroe, Louisiana, who agreed to take him along. He brought all the Wal-Mart provisions down to the dock and filled a flat-bottom boat with them. "Grim, grim, grim," Walker said about his first boat rescue. "You'd see bodies hanging on fences, on railroad tracks. The smell of death was everywhere. Everywhere. The wind blew and if there was a body two miles away, you could still smell it."

Reverend Walker bonded with the Monroe Police Department. There were six of them in a boat and together they saved dozens of people. But the scene back at the staging area, at the highway overpass, was dolorous. The old-fashioned priorities of rescue—women, children, handicapped people, and the elderly—was in play. Helicopters were landing on the overpass, whisking away the critically ill and infants at regular intervals. After one drop-off, Walker walked to the top of the overpass, hoping to get a view of the city. "I could see to my right Xavier University submerged," he said. "Tulane Avenue was submerged in water."[18]

All told, Walker would go on more than one hundred boat rescues that week, helping folks down rickety staircases and collapsed porches and fire escapes. Some houses were gutted, nothing left but joists and studs. Sometimes out of courtesy Walker would go up to a partially flooded house and knock. He knew nobody would answer but it was still the civilized thing to do. And it wasn't just the levees that had breached: the water had lifted asphalt parking lots and dumped piles of gravel into the middle of streets, making some roads impassable for flat-bottom boats. For the next ten days Walker, usually wearing a flotation vest, never complained about the privations of his mission. He figured, What better way to show your love of God than to help the destitute? Just being in a rescue boat, ready to pull in the forgotten and forsaken, was ennobling.

At one point on Thursday a Coast Guard helicopter hovered over the boat he was in, its rotor blades almost causing the vessel to tip over, the stagnant water blowing into the faces of those on board. It was blinding, but the pilot had a good reason for hovering so low. He was pointing to an apartment complex behind the roughed-up Carrollton Shopping Center. Walker gave the helicopter pilot an okay sign, and the boat headed for the apartments. It looked like a cheerless harmonica, pocked by gaping holes—the doors kicked out. "We knew something was terribly wrong as we approached the apartments," he said. "Katrina was bad, but it didn't blast open every second-story door of an apartment complex."

The traumatized people Walker and his company encountered there were looking around for fragments to save from rubble. Many were in shock, silent, refusing to move, still feeling the brunt of the storm in their heads. The evening before, a coterie of thugs had raided the apartment complex, stealing from the storm-weary everything they had: TVs, jewelry, kitchen utensils, and CDs. They took things for a simple reason: they *could.* Oddly, the building did have phone service and desperate residents were able to send an SOS to which the Coast Guard responded after a long wait. "The sun was out," twenty-eight-year-old Nicole Kelly recalled from the Cajundome shelter in Lafayette after escaping the ravaged apartments. "But the water started coming right up the street and it kept rising. Then we heard Governor Blanco saying on the radio that they couldn't stop the water and everybody should just get out. So we started calling 911, but we couldn't get any help."[19]

By Thursday night, Walker was weary. He felt good about saving dozens of people from New Orleans floodwaters over the previous two days. But he still hadn't made it to Noah's Ark. He just couldn't walk away with people suffering from dementia or psychotic illness. He took special care of the meek, making sure they found rides to Ochsner Hospital or Algiers Naval Base. He tried to be stoical. But he was furious. Where were FEMA and the Red Cross? Why wasn't the U.S. Navy on the scene with airboats? Why had the levees broken? All he saw were NOLA Homeboys with recreational boats and waders, pulling people out of the bilge while officials loitered around the staging area. Incredibly, two policemen in the area had told Walker not to worry about saving the rather pathetic people left in New Orleans; in their opinion the real crisis in the city was automobile theft in high-rise parking garages. Throughout his rescue effort, authorities kept asking to see Walker's ID, implying that he had no right playing "Bigshot Cavalry." The pecking order of the law, originally designed to serve those in need, was topsy-turvy. Ex-cons were now lifeguards while NOPD were naysaying birds on a wire. Sometimes Walker heard the police come right out and say that those left behind were nothing but "fire ants" or "dumb niggers." This blatantly racist attitude was even held by some black cops who looked on the stranded black folks with self-loathing. In 1981, Jean-Michel Basquiat painted *Irony of a Negro Policeman,* depicting an African American torn between his heritage and his job. This, in many ways, was the dilemma some NOPD officers faced. They were torn as to whether to run away or help out or ignore rescues altogether and smoke cigarettes at checkpoints out of harm's way.

Not all the flooded areas reeked with dangerously polluted water; on many streets, it wasn't toxic at all, just Lake Pontchartrain overflowing. Nevertheless, many police officers were unwilling to risk contracting hepatitis, salmonella, or other diseases, just because the city had not bothered to equip the NOPD with Hazmat protection. For whatever reason—old tensions or storm-induced confusion—the police were often part of the problem after Katrina, not the solution. Many stranded African Americans found that every time they met a police officer in the days following the hurricane, a gun was either drawn or touched or insinuated. It was part of the white-knuckled reality of a city in which laws had washed away—for citizens and police officers alike. To get anything done, Reverend Walker had to be smart, even using wardrobe trickery to his advantage. "My jeans and sport shirt weren't working well," he recalled. "I had no special permit to be in the flood zone. One night I met a buddy of mine, Mike Powell, who slipped me a Red Cross T-shirt. So the next morning, on my way to the [South] Carrollton [Avenue] boat launch, I slipped it on. If the Red Cross couldn't make it to New Orleans, I'd take it upon myself to use their good name to keep pulling people out. And guess what? It worked. But having the shirt didn't mean much when you were wading in debris."[20]

Before Katrina, the word "debris" had a friendly, culinary ring to it in New Orleans. At Mother's Restaurant on Poydras Street, the house specialty was the Debris Po'-boy, sliced roast beef with plenty of fat, drenched in a rich brown gravy, placed on French bread and loaded down with mayonnaise, pickles, lettuce, tomatoes, and, for the real connoisseur, plenty of Crystal Hot Sauce. Such delights were part of the old New Orleans. While Mother's survived Katrina, and continued to serve their marvelous po'-boys, every time the word "debris" was uttered, people thought of smashed glass or blasted furnaces, sopping wet clothes or splintered church collection boxes, fallen Our Lady of Lourdes statues or deflated soccer balls. When you saw somebody picking through debris, looking for a forlorn memento to snatch out of the wreckage, an awkwardness came over you. It was too humbling to watch; you had to turn away.

III

At 1:45 A.M. on Wednesday, August 31, FEMA officials formalized a request known as a tasking assignment to bring vehicles into New Orleans—specifically, 455 buses and 300 ambulances. The action was days late, but at least the bureaucracy had finally budged. The request was duly submitted to the Department of Transportation (DOT). Another full day went by before anyone at FEMA found out that, as one of the officials phrased it, "the DOT doesn't do ambulances."[21] That was only the beginning of a day that might be considered the eye of the Katrina aftermath. By Thursday FEMA buses started arriving in small numbers at the Superdome.

The misery of Katrina didn't wane on Wednesday and Thursday with the nightfall. Darkness made life in the bowl doubly worse. In the bleak, subtropical hours of the early morning, people in New Orleans and the surrounding parishes felt supra-stranded, still frightened, and still waiting. "People had to stay in groups," said Lieutenant Lawrence McLeary of the Louisiana State Police, "huddle in groups and operate and function as groups, because they didn't have communication with anyone else. And I don't know that people can even understand how dark the city got. I mean, no lights at all. It was just pitch-black. There was nothing, no sound at all."[22]

In the aftermath of Katrina, 1.1 million buildings in Louisiana were without power. In some Gulf Coast areas, the blackout was only temporary, but in others, the situation was chronic—for the simple reason that it couldn't get better. "It's catastrophic," said David Botkins, an official with Dominion Virginia Power, a company that sent 200 workers to Louisiana and Mississippi on the day after the storm. "The entire grid system in these areas is completely ruined."[23] Through the grid system, utility companies transfer power between power plants and ancillary stations; they also transfer it for sale to other power companies and deliver it to their customers. On each level, electricity demands balance between the source and the receiver. In making repairs to the mighty grid after Katrina, the local utility, Entergy, had one terrific advantage: it had prepared for just such a hurricane.

Devastating as Katrina was, it didn't send the power company into a frenzy, as it had FEMA. The flooding did, however, slow the process. Half of the forty-two substations in the city and its environs were flooded,[24] damage that would involve major repairs. In the midst of the flooding, linemen did their best, even climbing from boats onto towers to make repairs and reattach fallen wires.[25] But such electrical work was dangerous and time-consuming. The looting also hampered the restoration of power. Crews could not work with thieves circling them and their equipment, often with guns cocked and aimed. Within a week, the company would have power restored to more than 647,000 customers in Louisiana and parts of Mississippi; it would even manage to turn the lights back on in most of the French Quarter.[26] The rest of New Orleans, waterlogged and decaying, would wait months for power. Entergy, one of the nation's largest nuclear generators, sustained $1 billion in storm damage. It could survive financial loss, but not the lack of customers in Greater New Orleans. With so many places, including New Orleans, having lost the majority of its population, the local subsidiary, Entergy New Orleans, Inc., filed for bankruptcy in the third week of September.[27]

The efforts of Entergy were called "nothing short of amazing" by *Transmission and Distribution World,* an electric industry trade magazine.[28] All of Entergy's nuclear facilities in the Gulf Coast states—Waterford 3 in Taft, Louisiana; River Bend Station in St. Francisville, Louisiana; and Grand Gulf Nuclear Station in Port Gibson, Mississippi—suffered only minor, insignificant damage.

The Gulf South region was crumbling, but not merely in the physical sense: the mental strain of seeing one's hometown wiped away could wear down even the strongest personality. The whole thing might have seemed like an adventure on Katrina's first day (Monday), at least for those who survived without injury. The second day (Tuesday) was perplexing; a time to complain, loudly, to anyone who would listen. But by Wednesday, it was no longer possible to say what was real in the disorder left by storm and disappointment.

At 4:30 A.M., Eddie Compass, the police superintendent of New Orleans, was running up and down the halls on the fourth floor of the Hyatt Hotel, where some NOPD officers were headquartered. "Where are all the men?" he shouted. "We need all the men downstairs, now!"[29] He rushed from room to room, grabbing anyone he could find, from policemen to frightened hotel guests, insisting that they rush to the first floor. The reason, he said, was that looters were attacking the hotel. There was going to be a battle and he needed troops. When the makeshift army arrived in the lobby, though, they didn't find any looters at the doors. That Eddie Compass had overreacted was obvious. He desperately needed sleep. At that hour, the hotel was in the process of receiving a load of supplies, including food and gasoline. Compass was so concerned at the thought of losing the commodities—which were practically priceless under the dire circumstances—that he seemed on the verge of cracking.

Bogus rumors aside, Chief Compass *never* abandoned his post during the Katrina crisis. He worked harder than anyone to get New Orleans functioning in a coherent fashion. He did not flee to Austin or Galveston to be with his grown-up children. He *was* asked by Nagin's communications director, Sally Forman, to fly to Newark somewhat later, to flip the coin at the Saints versus Giants NFL game, along with former presidents George H. W. Bush and Bill Clinton. "Forman asked me to go on behalf of the mayor," Compass later complained. "I didn't want to go. Not at all. Later, [Nagin] used it as a reason for me to be dismissed."[30] Besides that official trip to Giants Stadium at Forman's request, Compass claimed he left New Orleans only once, to attend a meeting in Baton Rouge, sleeping for one night in Denham Springs. Another reason it seemed that Chief Compass had skipped out on New Orleans was that he offered the NOPD free five-day vacations to Las Vegas on September 4. "When you go through something this devastating and traumatic," Compass said, "you've got to do something dramatic to jump-start the healing process."[31]

The idea that slot machines and roulette wheels were considered the "healing tonic" only further embarrassed the New Orleans Police Department in the eyes of the American public.

IV

Back at Drew Elementary School in the Ninth Ward, Charmaine Neville was trying to hold herself together after being raped. Her nerves shattered like a broken glass jar, she had to put her own pain away and refuse to let the rape mess with her self-esteem. Helicopters flew overhead, their pilots aware that there were hundreds of evacuees to get out of Drew. But they were painfully slow coming back, usually just basket-lifting one or two people at a time. "So that Wednesday," Neville recalled, "we all decided to get out of Drew, forget about a mass rescue and we walked to Canal Street to get out of the floodwaters." While they were all standing around, wondering what to do, Neville commandeered a bus that was parked near the French Quarter. Nothing was more valuable in New Orleans on Wednesday than a bus, not even a diamond as big as a bus. If it could transport people, it was priceless. At that exact moment, no one was tending the bus, it was just parked. In an instant, Neville decided to take it. As if singing the Curtis Mayfield song "People Get Ready," she blurted out, "People, get on board," standing at the front door and waving them toward it.

Neville knew that the West Bank had only suffered wind damage—no flooding. Getting across the Crescent City Connection bridge was the ticket to freedom. She plopped into the driver's seat, flipped on the engine and put her right foot on the gas. The only thing on her mind was that bridge. If she made it over, she could work her way toward Baton Rouge, where there would be plenty of provisions and medical attention. She didn't know how to drive a bus, but she did the best she could. "I was doing well, but all of a sudden, a damn police car pulled in front, making a kind of roadblock," she recalled. "And all of the poor people around the Convention Center were looking to get out. I just couldn't take them. My bus was full. I still have nightmares about leaving them behind." Around New Orleans in the aftermath of the storm, police frequently set up roadblocks, in order to control just who went into and out of the city. With a little moxie, a driver could sail on through most of them. Neville had a lot of moxie. "Instinctively," she said, "I didn't slow down, I just put on the gas. . . . I just kept going. Everybody was screaming, hiding. It was just wild and scary and crazy. I was tryin' to turn on the windshield wipers and couldn't. But no way was I going to stop for those clowns." Seeing it was a city bus, and assuming that it was on government duty, the police didn't give chase.

It was a scene straight out of *Smokey and the Bandit.* Anybody watching would have naturally cheered Neville onward. And drive she did, getting her people out of the New Orleans bowl. By the time she got to Donaldsonville, however, her adrenaline was depleted. At a diesel filling station, a truck driver, seeing her with pump in hand, asked if he could help. He bought her fuel and directed the bus to the Shekinah Glory Full Gospel Baptist Church. Although the church had only been open eight months, it was ready to take care of evacuees by the hundreds. Pastor Ronald Harbin greeted Neville's exhausted busload at the door.[32] For each person, stepping into the church was like falling into heaven. Clean clothes, hot meals, and beverages were provided. Those in need of medical attention quickly received it. Neville herself was able to call her best friend, Wendy Haydel, in Hawaii. The word then hit the extended Neville tribe—scattered all over America—that Charmaine was okay.

Donaldsonville provided Neville with a chance to catch her breath. In counseling with Pastor Harbin, she described the horrors of the rape for the first time. The national media was clamoring to talk with her, to hear first-hand about her daring bus escape. A devout Catholic, she decided to tell her story on television, with Archbishop Alfred Hughes of New Orleans. As she entered the studio in Baton Rouge, a TV executive came up and asked, "What are you, another refugee?" Offended by the remark, Neville shot back, "Honey, I'm a survivor." Her interview with the round-faced archbishop was extremely controversial, for not only did Charmaine recount her rape, she criticized the Bush administration and told the far-fetched story about seeing alligators preying on people in the Ninth Ward. "I felt blessed to be talking to the archbishop. I wanted people to know that we weren't crooks," Neville later explained. "We weren't looting. We weren't on crack. But yes, I took a bus and if that makes me a looter, then guess what? I am a looter. And other women were attacked. That's why I went public."[33]

V

Early in the day on Wednesday, Governor Kathleen Blanco was keeping her calm, despite her frustration. First thing in the morning, she tried to put through a telephone call to the President, in order to tell him that "expected and promised federal resources still have not arrived."[34] Having spent Tuesday night in Crawford, Bush was preparing to leave Texas for Washington at the time. But at the Crawford ranch—as in any presidential limousine or on Air Force One—the facilities were such that he could speak on the telephone as easily as he could in the Oval Office. Blanco had every reason to insist that she needed to speak with him at once. As former New Orleans Mayor Sidney Barthelemy explained on CNN that morning, "The President of the United States is the only person who has the resources to coordinate, to bring in the troops, to make this city a safe place and solve the problems, particularly the breach in the levee. [People in New Orleans] are losing hope."[35]

In response to Blanco's call, the White House did not make the President available. Nor could she speak with Chief of Staff Andrew Card. After this futile attempt to reach the nation's leaders, Blanco joined Archbishop Hughes in ceremonies designating the rest of the day as an official day of prayer in Louisiana. In her remarks, the governor admitted humbly that with monumental work before the state, Louisianans needed the help of a "higher power." She then returned to her office and tried to reach the President again. After a short delay, her call was transferred to a low-level bureau, the White House Office of Intergovernmental Affairs. That was, of course, a dead end; a silent graveyard of inaction. The governor of the state most affected by the worst national disaster since the San Francisco earthquake and fire of 1906 was being handed around like a small-time lobbyist in search of a favor. During the course of the morning, Blanco received calls from presidential surrogates, including one from Andrew Card, who was then on vacation in Maine, but who listened to her insistent request for buses—as many as 5,000. According to Blanco's recollection, Card didn't exactly promise to help. "He affirms that he believes he can help with this," she noted.[36]

To keep up the morale of her constituents, Governor Blanco did not admit publicly that the White House was stonewalling her. Instead, at a little after 10 A.M., she announced that she was in touch with the White House and that buses were headed to the Superdome to begin the evacuation. Blanco's primary concern was reducing the roiling population at the Superdome. One FEMA medical team had arrived on Tuesday to augment the medical staff that had been on duty since Sunday morning. FEMA also sent a single "Emergency Response Team Advance," which joined Marty Bahamonde in attempting to satisfy the basic needs of those on hand. "Each day," Bahamonde said, "it was a battle to find enough food and water and get it to the Superdome."[37] Somehow, the skeleton crew succeeded in providing the crowd with at least two meals per day.

The hundreds of National Guardsmen on duty at the Dome kept order. They were, however, outmatched by the sheer numbers. Rumors that murders were committed at the Superdome were untrue, but rapes had been perpetrated. Some of the people in the Superdome, according to Terry Ebbert, were addicts. The exact number will never be known, but hundreds there were thrown into withdrawal by the lack of alcohol, crack cocaine, heroin, or some other substance. "Did you have people at the Superdome in withdrawal?" Ebbert later asked rhetorically. "Yes, that was a real problem. The drug addicts were in physical distress. The tremendous heat. All that went into creating a very, very bad situation."[38] And yet those who were managing the disaster operations were continuing to direct rescuers to drop off hurricane victims at the Superdome. Hotels, like the Ritz and W, that were closing suggested that guests find shelter at the stadium. The population there had actually increased by 5,000 since Tuesday evening, when the governor had called for the mandatory evacuation of the Dome.

President Bush knew what was going on, or at least he should have. He started the day with a teleconference from Crawford on the subject of hurricane relief, speaking with officials spread by vacations all over the country. The participants were Michael Chertoff, Deputy Secretary of Homeland Security Michael Jackson, White House Homeland Security Advisor Frances Townsend, Michael Brown, Dick Cheney, Karl Rove, Deputy National Security Advisor J. D. Crouch, Andrew Card, and White House Counselor Dan Bartlett. White House Press Secretary Scott McClellan later summarized the discussion:

> The meeting began with an operational update from Mike Brown. Mike—well, they discussed the options for an evacuation of the Superdome in New Orleans, and the people that have been—that were moved there originally. They also discussed the issues relating to the flooding going on in New Orleans, and Mississippi, as well. And Mike talked about the work going on to fix the breaches in the levees. And so there's a good bit of discussion about what they were looking at doing for the levees, to fix the levees, and they're working with the Corps of Engineers in that regard.
>
> They also talked about the security situation. As you're all aware, marshal [*sic*] law has been declared in Mississippi and Louisiana. And they talked about the National Guard response to that, as well. And then they talked about the coordination of the response efforts within the federal government. And the President wanted to make sure that Mike was getting all the cooperation he needed from all the different agencies within the federal government on the ground. And Mike expressed that he was getting good cooperation within the federal government.[39]

President Bush raised two concerns: the short-term need for food, water, and shelter in the affected areas and the long-term need to rebuild communities. While that much may have been obvious, the devil was in the details that would answer those universally held concerns. Significantly, the President was flanked at the teleconference by Karl Rove. To such a master politico, Louisiana was known as the only Democratic bastion in the South, the region where he had devoted much attention during the previous decade to building a solid Republican coalition. "Wednesday,

Karl Rove woke up and said, 'Hey,'" Blanco recalled, "'why are the national talking heads criticizing my president? Why not the governor?' They have a huge network and we weren't trying to manage anything like that. We were trying to save lives."[40]

By 2005, if Louisiana was a holdout, it was a wobbly one. In the blue (Democratic) versus red (Republican) paradigm, Louisiana was purple, turning redder every month. The partisan politicking of Katrina kicked in. Alabama and Mississippi were red, and, if played right—if Blanco got scapegoated—Louisiana could soon be as bright as a fire engine. While David Vitter had been elected two years before as the state's first Republican senator in over a century, Democrat Kathleen Blanco had managed to win the governor's office. The TV political pundits weighed in with an unprovable theory: For any political animal on Bush's side of the fence, the best-case scenario in the aftermath of Katrina would be the use of federal resources to help the people of Louisiana, without lending one iota of assistance to the state's beleaguered Democratic governor. Even more tempting, the theory went, was the notion that Katrina could be used as a tragic but convenient excuse to reshape the state politically: gerrymandering by bulldozer. All such plots and subplots undoubtedly occurred to those at both ends of the phone lines connecting Baton Rouge to Washington. Every time Governor Blanco telephoned that week, she was fully aware that she was a Democrat calling an unfriendly Republican White House. And so the struggle for survival that was facing hundreds of thousands of people in southeastern Louisiana on Wednesday was played out in its own way at the highest levels of government. "They were trying to blame Louisiana for all the problems," Blanco said. "And I just wouldn't let that happen. The White House just flat out made [up] stuff."[41]

VI

At 11:20 A.M., not long after Michael Brown had finished telling the President and others involved in the teleconference that "he was getting good cooperation within the federal government," the FEMA director received an e-mail from Marty Bahamonde. "Sir," it began, "I know you know the situation is past critical. Here some [*sic*] things you might not know. Hotels are kicking people out, thousands gathering in the streets with no food or water. Hundreds still being rescued from homes. The dying patients . . . are being medivac [*sic*]. Estimates are many will die within hours. Evacuation in process. Plans developing for dome evacuation but hotel situation adding to problem. . . ."

Bahamonde heard back from the director. "Thanks for the update," Brown e-mailed back. "Anything specific we need to do or tweak?" Three hours later, Brown's press secretary issued an e-mail to several FEMA officials, noting that "it is very important that time is allowed for Mr. Brown to eat dinner. Given that Baton Rouge is back to normal [*sic*], restaurants are getting busy. He needs much more that [*sic*] 20 or 30 minutes. We now have traffic to encounter to get to and from a location of his choice, followed by wait service from the restaurant staff, eating, etc."

A colleague forwarded the message to Marty Bahamonde, who composed a response that showed just how hard it could be to remain committed to any effort in which one's leaders seem indifferent. "Just tell her," he fired back, "that I just ate an MRE and crapped in the hallway of the Superdome along with 30,000 [*sic*] other close friends so I understand her concern about busy restaurants. Maybe tonight I will have time to move the pebbles on the parking garage floor so they don't stab me in the back while I try to sleep, but instead I will hope her wait at Ruth's [steakhouse] is short. But I know she is stressed so I won't make a big deal about it and you shouldn't either."[42]

In a February 2006 interview, Brown explained that the debacle at Morial Convention Center—which on Wednesday started filling up as an alternate shelter to the Superdome even though there were no provisions or triage—wasn't FEMA's fault. He pointed his finger at City Hall. "The Convention Center, from my perspective, came as a surprise," Brown said. "The Superdome was supposed to be the primary evacuation center of last resort. I began to receive some e-mails that the hotels were evacuating and telling people to go to the Convention Center. And those poor people, the tourists and others from the hotels, began to show up at the Convention Center and somehow the doors become open [*sic*]. . . . So I'm in Baton Rouge trying to figure out if this is true or not because we're getting conflicting information.

Yeah, a shot was fired somewhere. But does a shot fired at someone evolve into widespread civil disturbances? And we had come to the conclusion sometime on Tuesday that it really wasn't that bad in terms of civil disobedience. But Marty called me at some point and said, 'The medical teams and all of us feel unsafe and we're going to evacuate.' And I said, 'Well, Marty, you're on the ground. You know better than I do. You need to make your own decision about what to do and if that's the case then go ahead and get out of there.' Then we started hearing stories about other people going into the Convention Center and it's starting to get worse. So sometime Wednesday we began to realize there really were people showing up at the Convention Center. Who are they? Is this an organized evacuation or is this just an ad hoc thing that's occurring? Then I found out that there was absolutely no planning, no indication from the mayor, nothing that this was in place. . . . We weren't prewarned in Baton Rouge that we were going to have to deal with people in the Convention Center. It was a total surprise to us."[43]

No one, however, had a more frustrating day Wednesday than a man named Peter Pantuso, the head of the American Bus Association, a trade group that represented over 950 intercity bus companies, including the giant Greyhound. America's bus companies were ready to help; Pantuso knew who they were and how to organize them. He wanted to help, too. Instead, he spent the whole day trying to get through to the people at FEMA who were coordinating the search for buses. "We never talked to FEMA or got a call back from them," said Pantuso.[44] Only later did he learn that the agency had contracted all transportation needs in case of disaster to Landstar, a trucking firm that happened to be headed by Jeffrey Crowe, former chairman of the U.S. Chamber of Commerce, which funneled funds to the Republican party through its political action committee.[45] (The contract was originally worth $100 million, though a post-Katrina modification raised it to $400 million.) As a trucking company, Landstar didn't have much familiarity with bus logistics. After making a few general inquiries on the weekend before the hurricane struck, it dallied, making no immediate preparations. In fact, Landstar didn't fully address the problem until Tuesday morning and then only by assigning a subcontractor, Carey Limousines, to find buses. "They found us on the Web," said a vice president at Carey.

After that, two days went by. It need hardly be said that the urgency for quick transportation out of New Orleans during those two days was a matter of life and death to some of the stranded Katrina victims. Late on Wednesday, Pantuso learned from friends on Capitol Hill that Carey Limousines was in charge of finding buses. He immediately called the company, which was headquartered near his own office in Washington, and spoke to the vice president working the problem. She said that she had meant to call Pantuso but didn't have his telephone number. (It was listed in the local telephone book.) Finally apprised of the needs, Pantuso could start contacting his member companies with specifics. However, the full complement of buses wouldn't arrive until the weekend—five days late—due to mismanagement on the part of the federal government and its contractors.

VII

On Wednesday morning Tony Zumbado and Josh Holm woke up early in their Hallmark motor coach, which was parked on Canal Street. The night before Zumbado, a good cook, helped prepare sandwiches for about fifty crew members. Now at the crack of dawn, they cleaned up the mess. Quickly they ate a bowl of Total and put out boxes of breakfast bars for their just-rising colleagues. They plotted their day. They were joined by Denny Miller, an NBC engineer. All around them outside were people wandering about, dazed and exhausted, sloshing through the gurgle of water coming up out of the gutters. A curfew had been issued for the hours between 6 P.M. and 6 A.M., but it wasn't really enforced. "We were just starting to walk to our van when out of nowhere this African-American guy named Dwayne Jones just popped in front of us," Zumbado recalled. "He was dripping in sweat and his eyes were wide. He looked like he was strung out, if you can imagine that, and I say this with all due respect. I'm not stereotyping, but that's the image that I saw in him."[46]

Startled, Zumbado worried briefly that Jones might be aiming a weapon. "I need to talk to somebody," Jones said. "You need to talk to me. I need to speak to somebody. There's people dying in the Convention

Center. No media's been over there. They told us to go over to Canal and find somebody." As Jones rambled on in this fashion, many people might have written Jones off as a bum detached from reality. Miller, in fact, got between Zumbado and Jones, fearful that some act of violence was about to occur.

Zumbado, however, could take care of himself and he waved Miller away. "Look," Zumbado said to Jones, "do you want some coffee? Do you want something to eat?"

"No, no, man," Jones said. "I don't want nothin' to eat. I don't want anything. I want you to come with me."

A streetwise Zumbado believed that Jones was telling the truth. "Let's go," Zumbado said. "Show me."[47]

The Morial Convention Center was only four blocks away, but Zumbado drove. They had a lot of camera gear, too much weight to trudge around like pack mules. He told Miller to let Heather Allan know he went to investigate something and would be right back. Part of Zumbado's rationale for driving Jones was in fact to get him away from the NBC trailers. It was a safety measure. Stories of floating corpses, gang fights, and hijacked supply trucks were all over the wires. "It was chaotic and it was unsafe and everybody was starting to want our provisions. The human tide would, sooner or later, turn against us, so when Dwayne and I left, I told those watching to please go back to your motor houses and eat your breakfast," Zumbado said. "The previous night people had come up to our camp asking for a cup of coffee, bottled water, a Band-Aid, you name it. And we were like, 'Oh my God, this is no good. We're catering to the cops, to everybody.' But we just couldn't say no to anyone. It was just horrible."

So Zumbado headed down Convention Boulevard with Holm and Jones, parking a block from the center. A group of people was behind the Convention Center, standing on the Mississippi River levee, hoping to hijack a blunt-nosed barge or tourist side-wheeler. Others were fanning themselves in the early-morning heat. "What a scene," Zumbado recalled. "People sleeping in the street. People lying in the gutter. Dead people. People chanting, 'Help. Help. Help.' Kids crying. Dogs running around. Trash all over the place. It smelled atrocious. There were maybe thirty seniors, special-needs individuals, on the side of the sidewalk in wheelchairs. They had empty canisters of air, plastic tubes still hooked up to their noses. They hadn't gotten out of their wheelchairs for days, sitting in their shit, basically." Zumbado raised his trusty PD-150 Sony handheld camera. "We're goin' to get the footage," he told Holm. "Stay behind me."[48]

The stench inside the Convention Center was so awful that Zumbado threw up his breakfast. Pale, sweating, nauseated by the horrific scene, Zumbado made his way through the throng of people, many of whom insistently pulled on his shoulders, asking, "Why won't you help us?" or "Where's the government?" or "Look at my mother, can't you see she needs insulin?" or "Why are kids dying and people getting raped?"

How did the situation deteriorate this badly? Zumbado wondered. "Everybody was yelling and telling me stuff," he recalled. "And Dwayne was taking me to where the dead bodies were on the side of the building. Eventually they were moved into the deep freezer, where dead chickens and cattle rumps were kept frozen before cooking. And he showed me where a person was dying on the sidewalk. Where a lady had died in a wheelchair, where another man was dead in a lawn chair over there and was just left in the sun. All this horrific stuff. I'm videotaping this stuff and I'm totally, totally overwhelmed."

Not only had Zumbado lost his stomach, he was losing his head. His body took on a funny odor. He just pushed forward and filmed the bad dream. The bathrooms were overflowing with feces. Bloodstained stairs and muddy wet, clothes tossed everywhere. Flies were buzzing around babies. Dogs ate dirty diapers. Dead cats were kicked into piles of debris. It was barbaric, medieval. Staggering about were a few drug fiends, looking for heroin to snort or jab. "I mean everything vile and deranged imaginable was going on," Zumbado said. "It was like a bad Woodstock."

With Jones continuing to point out the vilest attractions, and Holm following to capture the shrieks, they recorded the hellish scene. "See, I told you I was going to bring somebody," Jones kept telling people. "See." That line only infuriated some of the people. "Well, he ain't got no goods and he ain't got anything," one woman charged, with a group nodding in agreement. Jones defended Zumbado. "Well, yeah," he said, "but he's the media and he's going to bring us help." With an angry wave, the woman dismissed the notion. "He ain't goin' to do shit for us," she said. "He's like all the rest." This time Zumbado spoke directly at her, forgoing interpreter Jones. "Ma'am," he said. "I promise you, if you let me videotape this and you let me get out of here with this, I'll bring you help."

Worried his presence might spark a riot, Zumbado did a last round of shooting and then drove back to the NBC compound to find the dysentery-consumed Brian Williams and overtired Heather Allan. "Look," he told them, "you gotta look at this tape." Both were flabbergasted, cringing at what they saw and heard, and disgusted that City Hall had let the situation get this wildly out of control. "Williams wanted to run it all, but knew we couldn't," Zumbado said. "He knew the guys in New York running a family network wouldn't let him do that." They all quickly agreed to show some of Zumbado's footage and play some of Holm's sound bites. But Williams and Allan wanted Zumbado to go immediately on MSNBC and, as an eyewitness, report the degradation to the world. "So I went on and talked about it," he recalled. "Williams and Allan were able to get some of the worst scenes on the air without making families watching vomit like I did."[49] Wearing a blue NBC ball cap and green work shirt, Zumbado told about the Convention Center with articulate passion. In cable TV terms he was "on fire." Reporting that there were no C-rations or water bottles, Zumbado insisted that the victims "needed help yesterday."[50]

NBC News, who had already broken the story of the Superdome roof holes and the first wave of looting, had now had another exclusive. Hanging around the network trailers that morning was New Orleans native Harry Connick Jr., the singer. No sooner had Zumbado finished talking than Connick, whose father had been district attorney for twenty-nine years, pleaded to be taken to the Convention Center. "Connick had been sleeping in his car," Holm recalled. "When he caught the live segment he said something like 'This is my town and I have to go see my people.'"[51]

Allan okayed the idea of a filmed Connick tour of the Convention Center. Weary and depressed, Connick was miked and, with Zumbado and Jones as guides, headed back there. "We basically did the whole thing again," Zumbado said. "Connick just wanted to witness that it was that bad. They thought he was a FEMA official when they saw me walking with him. They're saying, 'Oh, look, he did bring somebody.' And he's like,

'No, no, no,' and then somebody yells, 'Hell, no, that's Harry Connick. He ain't no official.'"

Trying to be the voice of reason, Connick stood up on a chair and launched into a flat speech. The air was heavy with the smell of vomit and cigarette smoke and urine. "I'm here to witness what you're going through," Connick began. "I'm going to get help. I'm going to talk to somebody. I will help you. You're my people. I come from here. I just could not pass the moment to see how bad this was."

The speech did not go over well. "Man, you can see," somebody shouted. "Get the hell out of here: go get us help." According to Zumbado, Connick, overcome by a sudden dizziness, "just couldn't take it" inside the Convention Center. "I've had enough," he told Zumbado, glancing for the exit. "This isn't good." On the way out the door, a man died at Connick's feet, the victim of a massive stroke. "Look!" Zumbado shouted, videotaping the man. "Look!" A panic-stricken Connick couldn't look at the fate of "his people" when they were dying like dogs. "Let's get out of here!" Connick said. "Now!"

Zumbado and Holm kept returning to the Convention Center that day. People were still congregating, without permission—or any protection from one another. Weapons abounded and crime was unchecked. Almost as large a number huddled outside, afraid to go in the building, a formless, irresolute mob. As Zumbado and Holm soon found out, stories of the so-called Death Freezer had made the rounds. They themselves filmed around five corpses. Muffled screams could be heard from one end of the center to the other. Ninety-three-year-old Allie Harris was one of the anguished. She had been evacuated from the New Orleans East home she had shared with her husband, Booker. During the evacuation ride in a hot panel truck, Booker died. In the New Orleans created by Katrina, the driver simply unloaded the old man's body, propped the corpse in a folding chair, and left. Someone covered the remains with a yellow blanket. On Wednesday Allie sat in front of the Convention Center, once a focal point for millions of visitors, guarding her husband's body. She sat next to the corpse and slowly ate crackers, "seemingly unaware of all the tragedy unfolding around her."[52] Losing her husband was hard enough, but she wasn't going to allow him to be dragged away and tossed into some meat locker. She wasn't going to let her beloved husband's body become debris—as the saying goes, "over her dead body."

VIII

Ivory Clark got very little sleep at the New Orleans Grand Palace Hotel early Wednesday morning. His seventy-two-year-old aunt, Yvonne Green, coughed continually; her asthma was definitely getting worse. He had tried to take her to Charity Hospital but was turned away at gunpoint. He had to keep hitting her on the back, which helped open her bronchial passages. If her nebulizer—a machine that turns medicine into a breathable mist—didn't work soon, she would die. She was in desperate need of electricity or a generator. At the crack of dawn, Clark headed out to Canal Street, hoping to flag down a helicopter or boat. For two or three hours, he waved at every vehicle that moved, to no avail. Then, just as he was about to give up, a man in an inflatable raft paddled toward him. "Please, mister, help me," Clark begged, "I've got a really sick woman and a ninety-year-old woman who can't walk. They're not going to make it." Without hesitation, the stranger said, "Bring 'em on."[53]

Clark ran to his hotel room, telling his family, "Let's go! Pick up your stuff—we're leaving." Unfortunately, his mother-in-law, Sedonna Green, could barely move, so it took nearly a half hour to get her downstairs to the back parking lot. Eventually, the Clark family—all six of them—made it outside, and to Ivory's great relief, the boat owner hadn't paddled away. The raft was soon so weighed down with Clarks and Greens that it almost sunk. But with Clark pushing from behind, and the stranger pulling in front, they waded through the floodwaters, arriving at a ramp that led to I-10. This was the waterline. From this point Clark would have to walk his family to the Superdome, where the medics would be able to tend to Grandma and Auntie. "I told that guy thanks," Clark recalled, "I was halfway to the Superdome and, most important, out of all of the flooding."

However, the Superdome was still a quarter mile away, and neither Grandma nor Auntie was in walking condition. Out of the corner of his eye, he saw a woman with two children standing on the sun-scorched interstate with a stolen stock cart from a Winn-Dixie. It was a large cart, with two tiers. He wanted it. With all the charm and desperation he could muster, he approached the sheepish woman, telling her his plight. Two elders might die if he didn't get them to the Superdome as soon as possible. The woman looked over at the Clark family, where Auntie, as if on cue, was having a coughing attack. Her first answer was no. But Clark insisted, invoking God and Jesus and Stevie Wonder and pointing out that her family could still walk. In Christian terms, his family situation was "meeker" than hers. At most, this woman might lose possessions, not human lives. Her defiance melted. "She started mumbling and unloading her cart and gave it to me," Clark recalled. "She wished us well . . . sort of. . . . We put Grandma and Auntie in the cart, one on each tier, and started 'walkin' to New Orleans,' as Fats Domino used to sing."

On any other day, Ivory Clark pushing a cart with two old people lying on it down the middle of I-10 would have attracted no little attention. But not on the Wednesday after Katrina. The entire highway was now a staging area for evacuees, with people screaming at every helicopter, babies crying, and people passing out from heat exhaustion. Water was a rare commodity and people took deep swigs and then passed along the plastic bottle. Every twenty yards, Clark would stop the cart and whack Auntie's back. She was having wheezing fits and Clark kept glancing at his wife, Donna, afraid that at any minute Auntie might be a goner. You didn't have to be a doctor to realize that the elderly were "passing" all over the Gulf South. Eventually, the family made it to the Superdome. Clark ran around outside looking for a Red Cross volunteer or a nurse, anyone who could help. He had imagined there would be FEMA booths—like at church fairs—where teams of disaster experts were dispensing medical help, washing out cuts with disinfectant, gurneying the truly needy onto ambulances. How wrong could he have been? At the Dome he might as well have taken a number like at a bakery—one that read "27,363." It seemed everyone in sight had medical needs, and virtually none of them were receiving any help. The officials whom Clark approached only shouted at him, "The Superdome is closed. Go to the Convention Center."

Clark told his family the bad news. But, as always, he put on a sunny mask. "Kiddos," he said, "we're going to the Convention Center." Nobody heard the cursin' under his breath.

Down I-10, the Clarks saw mayhem swirling all around them. Eventually, they got to the Tchoupitoulas exit. It was a downward ramp, so Clark stood in front of the cart, his feet acting like brakes, making sure it didn't become a runaway train. It was impossible to measure how much Clark sweated that muggy afternoon. His T-shirt was drenched, as if he had gone swimming for a day in Lake Pontchartrain. In a sense, he had. His eyes were stinging from the sweat, and every time he spoke, beads would roll into his mouth. It was like drinking salt water. His lips were cracked. He pulled off a piece of dead skin. He thought of the Ray Charles/Gladys Knight song, "Heaven Help Us All," a cut on a duet CD he had recently purchased at Starbucks. He wanted to fall down on his knees and say a prayer. Instead, he kept pushing onward like a plow mule, saying, "Lord have mercy" and "Keep the faith," even though, for the most part, he wasn't a religious man. Jesus had a way of coming to the forefront of his mind only when he was in a jam. "This constituted a jam," Clark said. "Yes, indeedy."

When the Clarks finally arrived at the Morial Convention Center, they were shocked. Everywhere they looked, there were piles of rubbish. Most people were afraid to go inside the building. Clark decided that in order to find a generator for Auntie's respirator, he "had to assert myself," he said. "You had to look for an opening."[54]

IX

With no effective barrier, Lake Pontchartrain had been draining into the city for more than two days. The good news on Wednesday afternoon was that the floodwaters stopped rising in New Orleans. The water stabilized and the lake no longer had anywhere to go. They were just two adjoining bodies of water—Pontchartrain being a lake and New Orleans being a huge pond. The bad news was that with the parity between the two, the floodwater could not recede from the pond. The only hope was to re-erect the barriers. To that end, the Army Corps of Engineers and the National Guard started using helicopters to drop fodder into the breaches, starting with the gaping levee of the 17th Street Canal. To block the 200-foot-long, 25-foot-high hole, the military dropped sandbags and concrete highway-lane barriers. These sandbags were not the 50-pounders used to fill sandboxes for kids. They weighed 3,000 pounds each, but even so, it looked as though the helicopters were dropping peas into the gaping breach. Cynics predicted that it would be three years before the holes were all fixed and the water could be pumped out. Engineers said that it would be three weeks, but in New Orleans on that Wednesday, years and weeks, hours and days all seemed to be the same.

At 1:45 P.M., Air Force One was flying over Louisiana, taking the President from Texas to Washington, D.C. George Bush was onboard, along with Karl Rove, Deputy National Security Advisor J. D. Crouch, other aides, and a gaggle of reporters and photographers. As usual, President Bush remained in the forward compartment of the 747, away from the members of the press and most of the staff. Unlike noisy airplanes, Air Force One was a relatively silent, meditative chamber, in which the President could collect his varied thoughts. The President's bedroom, bath, and office were in the nose of the plane. A little farther back was a small lounge and then a galley, skirted by a passageway and a row of chairs and sofas along the window. That seating area was usually used by the Secret Service, who killed time playing games on their BlackBerrys or flipping through magazines. Just to the back of the galley area was a conference room that doubled as the dining area. Farther toward the rear of the plane was a work area for presidential aides. Finally, there was a seating area, looking much like the first-class cabin on a commercial jet, for members of the press. President Bush, a private man, remained in his section of the 747, generally avoiding the press, in their section.

Somewhere over Louisiana, the presidential staff ordered the chief pilot to make a detour over the Gulf Coast, so that Bush could survey the hurricane damage, two full days after Katrina struck. Air Force One slowed as it approached New Orleans from the northwest, descending gently from its cruising altitude of about 37,000 feet to just 2,500 feet. The pilot remained in contact with ground control in the area, to make sure that the jet didn't intrude on the work of rescue helicopters. President Bush sat in a seat across from the galley, the area normally used by his security detail, and peered out the window. His hands were curled into fists, an unusually grim expression seemed frozen on his face.[55] Rove, Crouch, and Press Secretary Scott McClellan hovered nearby, pointing out landmarks. One of the Air Force pilots onboard stopped by to indicate towns, submerged or barely protruding from the floodwater. Bush himself spotted the Superdome first, its roof tattered by the wind. As the plane continued east and reached the Ninth Ward, the President said, "It's devastating. It's got to be doubly devastating on the ground."[56]

As the presidential plane flew over New Orleans, it dipped lower, to 1,700 feet, and people who had been stranded on rooftops for forty-eight hours easily heard the roar of the engines. The city was so eerily silent, except for the whir of helicopters, that the engines of the President's 747 filled every corner of the air for miles around. The blue-and-white jet swooped past the Superdome; the President could see the displaced people clustered around the arena. Likewise, those onboard Air Force One could all but see the desperate people who were still waiting on rooftops, trapped on highways, or wading through floodwater in search of safety. Those in the luxury jet, surrounded by immaculate furnishings, were assured by an aura of order and security. They were 2,500 feet above and a world apart from the victims below.

The destruction in the wake of the hurricane was still fresh. In two days, thousands of people had been rescued, but none of the damage to buildings had been repaired. And the water just sat: stubborn, ugly, and dangerous. For those who experienced Katrina's flooding, the foul smell would linger, and probably stick with them to the grave. The media continually referred to the waters as a "toxic gumbo." But if you stood on a porch or balcony on Wednesday and tried to inhale and exhale normally, your entire nasal cavity rebelled. The floodwater wasn't for the fainthearted—or even the brave, for that matter. It was a rotten stench, as bad as that of decomposing flesh.

Many first responders could take one whiff in a house and tell whether an animal or human had died inside. Some knew the difference between a dog or cat. The poisoned New Orleans floodwater, by contrast, was the death smell, plus dozens of other odors commingled with the overflow from the sewage system. It was as if the saucer that was New Orleans had been turned into a giant petri dish, where some mad chemist had tossed every wretched waste product imaginable into a stirring, stagnant pot. With temperatures in the nineties, this hell broth cooked. One whiff and your eyes burned, your face cringed, and your lungs pulled up their defensive drawbridges, hoping to survive the onslaught by partially shutting down. From Claiborne Avenue came smells—seawater and carbolic acid, rubber tires, dead plants, linen and common house paint, gasoline and dead rats. All you could do was cover your nose and search for a surgical mask. And then, just as your senses were poised to adjust, a slight breeze brought with it the smell of mildewed refrigerators, sour cheeses, and curdled milk. In post-Katrina New Orleans, the evil aroma attacked your sinuses, causing them to swell in minutes.

Soon, you developed "Katrina cough," a bronchial hack that left rescue workers feeling dizzy and heavy-headed. Anybody in or near the water naturally worried about the long-term health risks. That's what it was like in the bowl for everyone: the victims and the first responders; the powerful and the needy; Democrats and Republicans; NOLA Homeboys and the Cajun Navy; the Uptown rich and the Ninth Ward poor. Anyone who was in the region knew what that smell was like, but wondered if anyone else did. Could President Bush really know the smell from a limousine in the sky?

As Air Force One continued on its thirty-five-minute tour of the hurricane-ravaged coast, the President himself spotted the shocking state of Pass Christian, Mississippi. "It's totally wiped out," he said. As he correctly observed, the bridge across St. Louis Bay was gone. All of Harrison County, in fact, was in pieces. Every juke joint and eating house along the coast was gone. "There wasn't a whole lot of conversation going on," Press Secretary McClellan said about the mood on Air Force One. "I think it's very sobering to see from the air. And I think at some points you're just kind of shaking your head in disbelief to see the destruction that has been done by the hurricane."[57]

President Bush opted not to put the flashlight in his face, saying, "This is your president," as LBJ had done during Hurricane Betsy. He chose a more fastidious approach. While it was true that President Johnson's flashlight trip was grandstanding, he had made Betsy survivors feel significant. The

President of the United States *cared.* President Bush, while emotionally distraught, chose the remote approach instead of an on-the-ground inspection. It was a blunder. "The President pointed out this one church that was still standing," reported McClellan, "but all the homes around it there were completely wiped out. There's a causeway we saw that was in pieces that the President pointed out."[58]

Around this time Karl Rove made his way to the back of Air Force One and invited photographers to the forward compartment to take shots of Bush surveying the damage. Some reporters tagged along as well. The invitation surprised the members of the media, who were rarely allowed access to the President onboard Air Force One.[59] It was obviously intended as a photo op, a depiction of the President as a compassionate man, connecting to the tragic region. It backfired. No one expected the President to go to New Orleans, pick up a bucket, and start bailing water, but detouring over it in a jet was a meaningless gesture at the other end of the spectrum. Howard Fineman of *Newsweek* described the resulting pictures in terms of "president as tourist, seemingly powerless as he peered down at the chaos."[60] Bush, already two days late in seriously addressing the Katrina situation, was finally entering the recovery foray from an unfortunate springboard of high-altitude disconnect. He should have smelled the death. He should have touched the floodwaters. He should have showed he cared a bit more. White House political advisors admitted later, "It looked like he didn't know what was going on."[61]

Chapter Eleven

BLINDNESS

> The Government is fully aware of its responsibilities and hopes that those to whom this message is directed will, as the upright citizens they doubtless are, also assume their responsibilities, bearing in mind that the isolation in which they now find themselves will represent, above any personal considerations, an act of solidarity with the rest of the nation's community.
>
> – José Saramago, *Blindness*

I

ONLY A SURREAL ALLEGORIST, like Portuguese Nobel laureate José Saramago, could describe Katrinaworld, a denuded black hole of double-talking gibberish where the bureaucrats hid behind the white marble walls of statistical procedure and partisan politics. Starting on Wednesday, August 31, Americans were awash in Katrina statistics: Category 3 or 4; 20,000 or 25,000 at the Superdome; 40,000 troops needed; 120,000 people without their own transportation; 38 percent of New Orleans living in poverty; a chilling 2,430 children separated from their families;[1] 1,700 people still trapped in New Orleans hospitals; 40,000 National Guard needed in the disaster zone; 1,150 Red Cross shelters opened in twenty-seven states; more than 100 tons of sandbags already dropped by the Coast Guard to repair the levees; 90,000 square miles of the Gulf Coast declared official disaster areas. Hope evaporates when a person feels like a number.[2]

What Saramago wrote in his novel *Blindness*—"blind people need no names"—was the attitude that the authorities took on Wednesday. New Orleans, in particular, was in body-count mode. Perhaps 1,000 or 3,000 or 10,000 or 100,000 dead. The politicians spoke of the dead with greater certainty than they spoke of the living. But they didn't really know anything about either. You called 911 and heard an operatic beep. Baton Rouge was a civilization away. Washington, D.C.—to use cell-phone lingo—was out of range. Not even a sputtering little Toyota was on Esplanade Avenue. You want soldiers? The soldiers were holed up in the Superdome. You want light? Strike a match. The thirty-five St. Charles line streetcars were silent—no more clackety-clack. The Napoleon House bar was boarded up. Decent coffee or brandy wasn't being served. Water was scarce and the music was stilled. No braying of trumpets around Jackson Square.

On Wednesday, Homeland Security spokesman Russ Knocke announced that his boss, Secretary Michael Chertoff, had declared Katrina an "incident of national significance." That designation, an understatement, granted the department the right to provide "operational and/or resource coordination for federal support to on-scene incident command structures."[3] In other words, Homeland Security was supposed to expedite relief by serving as a central command post. According to the National Response Plan, however, the designation stopped short of allowing the federal government to assume overall management of the affected areas. That was left to those "on-scene incident command structures," or in plain English, local governments. The Bush White House was determined to attain full federal control, though, of troops. While speaking to reporters Wednesday, Knocke pointedly observed that while "state and local officials have not formally declared that they can no longer manage the disaster on their own, that is the case."[4] Therein lay a point of contention that would dog and delay the effort to deliver help to Louisiana. "Dying," Saramago wrote, "has always been a matter of time."[5]

Every time the Bush administration and the State of Louisiana hesitated, lawyered-up, and read the fine print on Homeland Security procedure, an

American died prematurely. One of the biggest lessons of Katrina was that in times of disaster, bad bureaucracy plus presidential hesitation equals corpses. Meanwhile, Governor Kathleen Blanco believed that she *could* manage the rescue and relief effort and needed only resources from the federal government. From the point of view of the governor and most local officials, the federal response had already been monumentally mismanaged, with the military getting a late start and FEMA in a daze. As former New Orleans mayor Marc Morial said, FEMA needed to be "completely rebuilt" to keep with "the needs of twenty-first-century disaster response."[6] From the other side, though, to send the cavalry instantaneously to Louisiana under the aegis of "local officials" was to risk making heroes out of political enemies. Politics aside, the White House and Homeland Security were reluctant to hand over extraordinary assets to Louisiana officials whom they regarded as ill prepared and inexperienced. "Blanco reminded me of an aunt I have whom I love to pieces," FEMA Director Michael Brown said. "But I would never trust this aunt to run a state or be a mayor. She was just a wonderful human being. I just see Blanco as this really nice woman who is just way beyond her level of ability."[7]

If Brown saw Blanco as doddering, by Wednesday he was beginning to see other local leaders in an even darker light. According to the testimony Brown gave before the Senate Homeland Security and Governmental Affairs Committee on February 10, 2006, he continued informing his boss, Chertoff, of the breached levees in New Orleans, starting with their conversation on Monday evening. In addition to the telephone call to Chertoff, Brown delivered his message in video-teleconferences (VTCs) with Chertoff aides on at least two occasions over the course of that time span. In his testimony, he called the Department of Homeland Security's later claims that it was not informed of the levee breaches, "bologna." E-mail messages supported his version of events.[8]

When Chertoff seemed unresponsive, Brown also contacted presidential aides to clarify the extent of the flooding and the dire state of those left in greater New Orleans. For twenty-four hours, he saw little or no concern in either camp regarding the message that he and by that time others were attempting to deliver in either camp. According to

Brown in his congressional testimony, the problem was that Hurricane Katrina was a natural disaster, rather than a terrorist attack. As he explained it:

> It's my belief that had there been a report come out from Marty Bahamonde that said, yes, we've confirmed that a terrorist has blown up the 17th Street Canal Levee, then everybody would have jumped all over that and been trying to do everything they could; but because this was a natural disaster, that has become the stepchild within the Department of Homeland Security, and so you now have these two systems operating—one which cares about terrorism, and FEMA and our state and local partners, who are trying to approach everything from all hazards. And so there's this disconnect that exists within the system that we've created because of DHS. All they had to do was to listen to those VTCs and pay attention to these VTCs, and they would have known what was going on. And, in fact, I e-mailed a White House official that evening about how bad it was, identifying that we were going to have environmental problems and housing problems and all those kinds of problems. So it doesn't surprise me that DHS would say, well, we weren't aware. You know, they're off doing things, it's a natural disaster, so we're just going to allow FEMA to do all that. That had become the mentality within the department.

That was, at least, one explanation. The Government Accountability Office (GAO), the nonpartisan investigative arm of Congress, concurred that the delay in Chertoff's response was critical, and it even went further. In a preliminary report submitted by Comptroller General David M. Walker issued in February 2006, the GAO noted that "the DHS secretary designated Hurricane Katrina as an incident of national significance on August 30—the day after final landfall. However, he did not designate the storm as a catastrophic event, which would have triggered additional provisions of the National Response Plan (NRP) calling for a more proactive response." The report concluded, "In the absence of a timely and decisive action and clear leadership responsibility and accountability, there were multiple chains of command, a myriad of approaches and processes for requesting and providing assistance, and confusion."[9] When the White House finally stepped in to take an active role in disaster response, it only added to that confusion.

Just after noon on Wednesday, Louisiana Senator David Vitter acted as go-between for fellow Republican Karl Rove, bringing a message to Governor Blanco. According to notes taken by Terry Ryder, the governor's executive counsel, Rove took it as a given that Blanco wanted President Bush to assume control of the evacuation. In the meantime, Rove directed the governor to put New Orleans under martial law, "or as close as we can get."[10] The response to the White House overture was icy on Blanco's part. "We had our two senators, who decided to camp out in the EOC, which was probably a big mistake," Blanco said. "They were both emotionally involved because their homes and their home territories were directly impacted. [Vitter and Landrieu] were *trying* to be helpful. I would definitely say that. Vitter would try and pass a message to me from the White House. They came on their own just because they were upset. After I told the President that I felt like I needed 40,000 troops—now, I didn't mean all federal troops, but a combination of National Guard and federal troops in any kind of assortment. When I looked at the magnitude and saw how many parishes were involved and saw how much work there was to do, I knew I needed bodies. I needed manpower to help us stay alive, in many, many kinds of ways."[11]

At 2:20 P.M., Blanco telephoned the White House again. This time, President Bush took the call. Blanco described the desperate state of affairs in the New Orleans region—which he had seen for himself a few hours before—and she requested the 40,000 troops. Blanco later admitted she "just pulled a number out of the sky," though it ultimately proved an accurate assessment of the troop strength needed. In any case, if federal troops were to be in her state, then Blanco expected that they would officially be under her control. She was not ceding the sovereignty of Louisiana or the foothold she had as its governor. Significantly, she told the President, as she later testified to Congress, that she wanted "to continue to be his partner in a unity of effort as is called for under the National Response Plan."[12] Whether the White House was looking for the same type of partnership with her remained to be seen. "Karl Rove asked

David Vitter, what did I actually mean by asking for federal troops," Blanco recalled. "Did I mean to federalize the situation? And when David Vitter brought that message to me, I picked up the phone and I called the President and I clearly stated [my position] on Wednesday. 'Mr. President, I definitely need more troops, but I do not want to federalize the troops. What I want is some soldiers with equipment that can help us. We have a lot of work in the aftermath of our lifesaving operations.' " She also asked for medical assistance. "The hospital ship would have been nice to have early on. I don't know if it ever came. At least the *Iwo Jima* came."[13]

Blanco's communications director, Bob Mann, was one of those giving the governor media advice. A decision was made to leave the media to fend for itself. No embedded reporters. No preening for the media; this was a tragedy. There was a thin line between grandstanding and mourning, and the governor wasn't going to cross it by staging too many press conferences or photo ops. By Wednesday, however, Mann realized that the White House had launched a public relations onslaught against his boss, blaming Blanco for *everything.* "It was pretty obvious on Wednesday what was happening," Mann recalled. "We were getting telephone calls from reporters who were citing the White House as their source. It was just coming at us in waves. Every reporter that Wednesday said, 'The White House is saying . . . White House officials say . . .' They were sticking it to us. But then at the exact same time they were publicly saying, 'This is not the time for a blame game.' "[14]

While the White House was blaming Governor Blanco, the media lit into President Bush. Some journalists recalled a prescient point Bush the candidate had made at one of the 2000 presidential debates. When the moderator, Jim Lehrer of PBS, observed that perhaps 90 percent of a president's job was dealing with unexpected emergencies, an eager Bush jumped in: "That's the time when you're tested not only—it's a time to test your mettle," he said. "It's the time to test your heart, when you see people whose lives have been turned upside down. It broke my heart to go to the flood scene in Del Rio [Texas] where a fellow and his family just got completely uprooted. The only thing I knew was to get aid as quickly as possible, which we did with state and federal help, and to put my arms around the man and his family and cry with them. But that's what governors do. Governors are often on the front line of catastrophic situations."[15] So are presidents, of course.

II

With the declaration that the hurricane had been an "incident of national significance," the Department of Defense organized Joint Task Force Katrina, which could draw on all of the branches in response to the hurricane. According to the National Response Plan, requests for military assistance were to come directly from FEMA. The first requests on Wednesday were for about fifty helicopters. The Navy sent the hospital ship *Comfort,* with its one thousand beds and extensive medical facilities.[16] It was due to leave its berth in Baltimore in two days and arrive in New Orleans in twelve. This was so late as to be useless.

That Wednesday, California Governor Arnold Schwarzenegger stepped into the fray. He mobilized eight Swift Water Rescue teams—the Sacramento City Fire Department, Oakland Fire Department, Menlo Park Fire Department, Los Angeles City Fire Department, Los Angeles County Fire Department, Orange County Fire Authority, Riverside City Fire Department, and the San Diego City Fire Department. Task Force 5, for example, from Orange County, was deployed to New Orleans. This seventy-six-member team of firefighters was among America's best units for water and boat rescue operations. Usually, this crew team worked regular hours at firehouses in Anaheim, Santa Ana, and Orange City. "But during a crisis, they get the signal to mobilize as one specialized subgroup of the seventy-member task force," reporter Lynn Armitage explained in *OC Metro: Business Lifestyle Magazine.* "They were one of the eight swift water rescue teams in California deployed to New Orleans—the first responders."[17]

Task Force 5 flew out of March Air Force Base in Riverside, California, on a National Guard military transport plane that Wednesday. It would turn out to be an eighteen-day deployment. The task force brought along utility trucks, a tractor-trailer loaded with rescue boats, an SUV, and plenty of food, radios, water, batteries, helmets, and other rescue equipment. From their rubber rafts, which held up to ten people each, these firemen saved over two hundred New Orleanians, but it wasn't easy. Two of the firemen, Dave Baker and Dave Thomas, were knocked out when a Chinook helicopter accidentally blew a wood crate in their direction. Baker received thirty-seven stitches from the mishap. "Without the helmet," Baker said, "I'd be dead now."[18] But injuries aside, pulling seniors out of hospitals, saving a couple out of a church loft, and extracting children from a collapsed shotgun all made their efforts meaningful. "If I hadn't been able to go," Chris Boyd said, "it would have killed me."[19]

Governor Schwarzenegger also sent more than 500 members of the California National Guard and hundreds of medical workers. Speaking in front of a banner reading "Donate Money, Time, Blood 1-800-HELP-NOW," Schwarzenegger gave the kind of emotional appeal that President Bush didn't. "The devastation caused by Hurricane Katrina has challenged the limits of despair," Schwarzenegger said. "Entire communities destroyed, people driven out of their homes, maybe thousands of people dead. I know that the people of California have shown tremendous compassion and generosity because of this disaster . . . please continue to be generous. . . . We know that the recovery is going to take months or maybe even years. But this disaster is no match for the resolve for us to stand side-by-side with our fellow Americans that are struggling right now, that are hurting, that need our help."[20]

National Guard troops, called up from the affected states, numbered about 6,000 when the storm hit the coast on Monday morning. With up to 40 percent of their respective forces already on duty in war zones overseas, however, those states had limited troops left to call as the disaster expanded in scope. On Tuesday, the number had increased to only 8,000.[21] Before more could come from other states, authorization had to come from the National Guard bureau at the Pentagon. That formality had become another roadblock. And then there were the federal troops, the active-duty soldiers. Before they could be deployed, federal approval had to be granted. Initial approval came with the designation of the hurricane as an "incident of national significance," but ultimately the President would have to give them their final orders. The plan put in place on

Wednesday called for 20,000 troops, both National Guard and active-duty, to be on the ground by the next morning in the damaged areas. Then they had to actually be deployed. Oddly, Task Force Katrina bypassed the 8,000 regular soldiers available at Fort Polk, which was located in Leesville, only 270 miles from New Orleans. Instead, it called on personnel from Texas and California, in addition to the 82nd Airborne Division from North Carolina, legendary for its ability to reach any destination in the world within eighteen hours.[22] That was fine—but the U.S. soldiers from Fort Polk could have been in New Orleans in half that time. Anyway, until the President signed off on it, none of them were going anywhere.

Lieutenant Colonel Bernard McLaughlin, who was in New Orleans on military duty for the first twelve days of the Great Deluge, kept a detailed diary of his shifts at both the Superdome and the Convention Center during the critical periods. A Louisiana State University alum, McLaughlin was a professional litigation mediator from Lake Charles. He took his obligation in the National Guard just as seriously as Jimmy Duckworth did the Coast Guard Reserve. Stocky, with a Marine-like crew cut, he was a real fitness fanatic. Although mediating legal disputes before they went to court was his forte, he had also graduated from the U.S. Army Judge Advocate General School in Charlottesville, Virginia, as well as the U.S. Army Combined Arms and Services School and the U.S. Army Command and General Staff College, both at Fort Leavenworth, Kansas. "Whatever I could do to better myself," McLaughlin said. "My interest was in being of service to my country."[23]

McLaughlin had been assigned to Task Force 134/Multinational Force Iraq, their operations headquartered in the Republican Palace on the banks of the Tigris River in southwest Baghdad in the Green Zone. Their court building was located in the Red Zone of Baghdad in a landmark building known as the Clark Tower, formerly Saddam Hussein's personal museum in which the dictator displayed lavish gifts and antiquities that he received from other countries. "We converted it into a court building," McLaughlin recalled, "complete with chambers for the Iraqi judges, offices for prosecutors' courtrooms and holding cells. We tried several hundred insurgents and terrorists for various crimes including murder, terrorism, illegal weapons possession, etc."[24]

When McLaughlin went to Iraq, he carried the "royal purple" and "old gold" LSU flag with him, along with a copy of *The Collected Poems of Rudyard Kipling*. Down the streets of Fallujah and the alleyways of Balad and the roundabouts of Baghdad, there was McLaughlin waving his collegiate banner. He would fly it down to Checkpoint Two, the most dangerous road in the world. "Everywhere I went," he recalled, "the flag went with me." It was a humorous prank, aimed at goading Longhorns and Aggies and Razorbacks and Crimson Tide. Nobody knew what the Iraqis thought of the strange flag. Did it belong to some inconsequential Coalition of the Willing ally like Eritrea or Albania or Uzbekistan? Was it the pennant of some secret CIA covert division?

McLaughlin was awarded the Bronze Star in addition to the Iraq Campaign Medal for his sterling service. "When I returned from Iraq, I never dreamed that I would serve on military duty in New Orleans six months," McLaughlin recalled. "As it happened, I was an invited speaker on mediation of employment law cases at a Louisiana State Bar Association conference seminar at the Loew's on Friday, August 26, and stayed over in NOLA, as I had a mediation in a personal injury case scheduled in the city for the next morning. I did the mediation and drove home to Lake Charles, which took me about nine hours." The next morning he was called to active duty and told to report to Jackson Barracks in the Lower Ninth Ward.[25]

When Katrina hit, McLaughlin was stuck at the barracks as it flooded. To watch the flood grow from one inch of rainwater to a five-foot wall of water in about twenty minutes, was startling; it was particularly wrenching for him to watch his car succumb to the storm. It wasn't until Tuesday morning that the Jackson-based Guard contingent realized how badly flooded the rest of New Orleans was. The stranded troops moved by boat from their headquarters to the Mississippi River levee a couple long blocks away. They were soon airlifted by Blackhawk helicopters to the Superdome for the start of what McLaughlin called "our real mission."[26] They joined more than 300 Guardsmen already on duty.

The Jackson Barracks Guardsmen, there with both Old Glory and the Louisiana State "Pelican" flags sewn on their shoulders, began their duty by securing the various food and water stations. Discarded piles of clothes, MRE wrappers, and water bottles were scattered all about. Each MRE contained a 1,200- to 1,400-calorie meal usually with the following items:

1. Soda crackers and a spread, like cheese or peanut butter.
2. A main course, like beefsteak, chicken and noodles, turkey breast.
3. A side dish, like fruit, rice, soup, or potatoes.
4. Some type of dessert, snack, or candy.
5. An accessory packet: matches, spoon, wet wipe, salt, pepper, coffee, and sweetener (some may include an alternative beverage powdered drink too).
6. Military versions also include a mini roll of toilet tissue.[27]

The Guard wanted everything to be "orderly"—that was the main word with them. With just enough MREs and water left, as well as concession food left over from the recent Saints game, they fed the crowd as best they could. Provisions were rationed instead of being handed out at will. At one juncture a few disgruntled Domers got in McLaughlin's face. "We want a hot meal," they carped. "We don't want MREs." McLaughlin's mind was racing. "You ungrateful SOBs," he thought. But he kept his composure. "Well, I'd like a hot meal too. But this here MRE is good enough for me and the troops around me—it ought to be good enough for you."

The Guard had armed patrols inside and outside the Superdome. A large medical/triage station was set up with plans to transform the Arena, next door to the Superdome, into a permanent hospital. "Most folks were appreciative of our efforts," McLaughlin recalled, "especially the older folks. Many of the younger set, however, wearing sagging pants, tattooed, street gang types, were surly and rude. I felt sorry for lots of the kids, as they had no control over what family they happen to be born into."

On Wednesday, McLaughlin's mandate was to keep law and order in the Superdome. He also had the difficult job of letting the stranded know that it might be another day before buses arrived. His fellow Guardsmen ranging in age from eighteen to twenty-eight started working a series of twenty- to twenty-two-hour days, taking only the occasional catnap. McLaughlin's diary entry shows how hard the Guard's work was:

> The Dome situation is deteriorating rapidly—the crowd is in a bad mood, many want to leave the Dome, we tell them they can go if they want but they will probably miss the buses which we are staging to take them to Houston. Quite a few are thankful for our presence. . . . My mediation skills are handy for letting people vent and then redirecting their focus to solutions rather than conflict. Young thugs are harassing and threatening the weak and sick, our food distribution area is mobbed and almost overrun, as people push, shove, curse and fight. I disarm a number of punks armed with sticks and metal rods, including taking a hammer from one and a socket wrench from another, which are obviously being carried as weapons. It is a hot, brutal day—the Dome is reeking of sweat, feces, urine, discarded diapers, soiled clothing, discarded food, and the garbage strewn about is almost frightening in its sheer volume. People are openly cursing and fighting, many are openly angry with Mayor Nagin. Our focus is maintaining order and keeping things in control until the buses arrive.
>
> In late afternoon, I accompany MP Teams doing a sweep of the box suites, as looters have infiltrated that area, we arrive and find every single box suite has been looted with the doors or ventilation grills kicked in and stripped of all liquor, in total, we take down about twenty looters, confiscate liquor (which some of [them] tell us they brought with them to the Dome, we are talking $50 and $60 bottles of scotch, bourbon, etc); the looters are flex-cuffed and turned over to the NOPD; they have not only ransacked the suites, they have crapped on the furniture and rugs. . . . [It's a] senseless, wanton destruction of property. As our [rules of engagement] limit us to self-defense, not much we can do to deter looters, unless we are in the immediate area. The public bathrooms on Levels One and Two are trashed—urine and crap on the walls, stalls, counters, a fetid, reeking stench permeates the air.
>
> As evening descends, we have various groups chanting very negative things, primarily along the lines of destroying the Dome and the

> city. The night is a descent into hell. . . . One of our MPs is shot, several fires are started, we continue to confiscate and destroy liquor, which is fueling much of the violence. Yes, there are those who respond to our acts of kindness as we try to make sure the kids and the elderly are taken care of. I place 6 British university students, pretty young ladies, in a safe area, along with 5 Polish students who had worked in Six Flags. The big problem is sorting out the troublemakers from those who are behaving well. I stay up all night, crisscrossing the Dome in hundreds of mini-missions, each a human vignette of its own.[28]

The Louisiana National Guard did a pretty good job of maintaining order at the Dome on Wednesday. But they weren't alone in New Orleans. Other states were nimble in offering assistance to the ravaged states, none more so than New Mexico. Its governor, Bill Richardson, telephoned Governor Blanco on Sunday, concerned that the hurricane was indeed going to be the killer storm so long dreaded along the coast. "We had offered our assistance," recalled Richardson's spokesperson, Paul Shipley, "and told Governor Blanco and her people that we'd be ready to help, and we had already put our National Guard on standby."[29] The next day, when Richardson realized how damaging the storm had been, he ached. "The New Mexico National Guard had extensive experience in disaster relief," said Governor Richardson on Monday, "and after my discussion with Governor Blanco, I have ordered two hundred members of the New Mexico National Guard to leave immediately."[30] Unfortunately, they would not leave immediately.

The Pentagon, which was supposed to track, but not oversee, such National Guard deployments for up to four days, claimed that certain forms had to clear before permission could be granted for the transportation of troops across state lines.[31] If the Pentagon had been purposely keeping the troops from Louisiana, it could not have done a better job of causing delays. Just why the Pentagon wanted to delay the arrival of help was a mystery. But New Mexico's troops were ready when they were finally allowed into the state on Thursday, September 1. "We were the first responding—outside of Louisiana—National Guardsmen on the ground there, of all the states," said Sergeant First Class Jim Lee of the New Mexico National Guard. "We beat the active-duty Army—we beat the 82nd Airborne."

The amount of concern the New Mexico National Guard exhibited was inspiring. "Sergeant Lee and his units were sent to Plaquemines," he said, "over 50,000 left, not homeless as in 'they're homeless and they're going to go back,' but homeless forever. Their homes are destroyed. Or they are not on the same property they started, and only frames are there, a rooftop here or there. People who lived in homes that were built by their great-grandfather were going back to nothing. Nothing." Sergeant Lee shared his memories of the operation with a New Mexico journal, the *Desert Exposure*. A straightforward storyteller, Lee recalled that the New Mexico troops went to Louisiana with a very specific directive:

> Our leadership told us: "We're not running the boat. We are there to help the people of Louisiana." So those of us in a leadership position, when we went to our counterparts in the civilian community, it was almost scripted to the extent that I walked up to my counterpart and said, "Hi, I'm Sergeant First Class Jim Lee from New Mexico, and we're here to help you. What can I do?"
>
> I went to my counterpart, Gene Fox, who was in charge of all the freshwater and sewage facilities in southern Louisiana. He breathed a big sigh of relief, and said, "Man there's a million ways you can help. I'm so glad to see you." Later on, he said—I'm using his words—that "we were sitting there looking at each other. We were devastated. Then in the door walks Jim Lee with the New Mexico National Guard and says, 'What can I do to help?' And you just tell me, 'We're here, we're gonna do it!' That spirit that came through my door changed our whole attitude."[32]

III

At the Lafon Nursing Home of the Holy Family on Chef Menteur Highway in New Orleans, the supervisor, Sister Augustine McDaniel, made the decision before the storm that moving her 103 residents would be traumatic for many of them. The home was a sturdy, two-story brick building; as Sister Augustine lay in a supply of water, food, and medicine, she assumed that if there was any flooding at all, the residents could be moved to the second floor. She also made sure that there would be plenty of staff members on hand during the hurricane to assist the residents.

When the storm hit, and water rose on the first floor, she and the others moved residents to the second floor, even though hauling those in wheelchairs up the stairs or carrying people, some of whom were hefty, wasn't easy. McDaniel was the commander, cool under pressure, but seeing the Herculean race against the water to save the residents, recalled one of the employees, "she kept saying, 'I'm so sorry, I'm so sorry.'"[33] Nonetheless, she and the staff saved every one of them. When the winds died down, the residents were fed and cleaned. The air on the second floor was fetid and the temperature shockingly high, but staff members fanned the residents and tried to keep them cool. On Monday night, the water receded and by Tuesday morning, a trickle of traffic appeared on Chef Menteur Highway, the staging area of the Cajun Navy rescues.

Lafon employees went out to the street to flag down someone who could help evacuate the residents. A Department of Wildlife and Fisheries truck pulled over just long enough to say that they were only rescuing people from flooded areas—there was no one to help the largely dry residents of Lafon. All day Tuesday, the resourceful employees tried everything they knew to draw attention to the plight of the ailing residents. Nothing worked—including the telephones. On Wednesday, the story was the same.

The residents at Lafon were among the many people to suffer from the fact that the rescuers on the ground in New Orleans were undermanned and overwhelmed. Officials assigning priorities could not see the 103 residents at Lafon, lined up in hallways, gasping for breath, sniffing the mildewed air. No one, it seems, could see them. That Wednesday the Cajun Navy rescued the destitute at Forest Towers East and Metropolitan Hospice, but they missed Lafon. Lafon would have been better off if it *had* flooded. In Washington, which still could not locate a fleet of buses in the whole United States of America, they knew nothing of Rosalie Daste, a one-hundred-year-old woman, about whom Anne Hall and Doug Struck wrote in a rich *Washington Post* article on Lafon. Officials in Washington who thought there was time for more meetings didn't know "that she had never missed a Southern University football game, that she was famous for her shrimp and okra, and that she made her grandchildren pick up pecans in the hot Louisiana sun because she wanted them to know what life would be like without a college education."[34] She was one of the old and weak, yet irreplaceable people, who had no voice in the cacophony that followed the hurricane.

By Wednesday, some of the residents at Lafon had died; their bodies had been taken to the chapel downstairs and laid out respectfully. One employee's cell phone gained a signal just long enough for her to call her brother, Irvin Boudreaux, who lived near Atlanta. He had no luck, either, in eliciting help from any official source, and so he heroically made his way to New Orleans to assess the situation. A day and a half after his sister's call, he finally managed to hire a bus from a company outside of New Orleans, at a cost of $1,000 of his own money, and collect forty of the residents for transportation to another nursing home in Houma. When he tried to return for a second group, though, the bus driver heard an explosion in the city and flatly refused to go. Irvin Boudreaux, the only hope for Lafon, had reached a dead end. It was Friday before two FEMA employees happened to pass the facility and see someone who told them of the crisis inside. By then—five days after the storm and four after Sister Augustine had started trying to call for help—twelve residents were dead, from the heat or the lack of medicine. Within an hour, a convoy of helicopters had evacuated the residents. A grateful Sister Augustine, soaked with sweat from her tribulation, was standing in the parking lot in a T-shirt.

On Wednesday afternoon, people outside of the Gulf South coast were shocked when they realized that absolutely nothing was getting better in New Orleans and almost everything was getting worse. Americans were not used to seeing their country in ruins, their people in want. They began to feel ashamed that their country was either unable or unmotivated to provide the basic necessities of food, water, and shelter within its own borders. As Terry Ebbert, head of local homeland security, later told a Tulane University audience, "for the first time in my life I was ashamed to be an American."[35] Senator Mary Landrieu wasn't ashamed, but she was frustrated. She was standing in front of the Superdome on Wednesday trying to find buses. She borrowed Marty Bahamonde's official telephone in order to call her office in Washington, D.C. "It didn't work," she said. "I thought to myself, This isn't going to be pretty."[36]

All through the day on Wednesday, as television film crews focused on the dramatic rescues, still photographers fanned out across New Orleans. They may have been trained in the tradition of Walker Evans, the renowned photorealist who spent a few years during the Depression documenting the city in a time when "the weather was balmy and the pace of life unhurried."[37] However, the most incredible images of Katrina seemed to have been influenced more by Man Ray. As the *Washington Post* headlined a story, "Houses Walk and the Dead Rise Up."[38] The second most commonly bandied-about word in response to the post-storm scene—after "debris"—was "surreal." And photographs are the permanent testimonial of why.

The work of Kathy Anderson, staff photographer for the *Times-Picayune,* was a case in point. As she wandered around Gentilly, the Lower Ninth Ward, and the Garden District, snapping color photographs, she developed a truly surreal portfolio, which was collected together as a feature entitled "The Ways of Water." The captions gave an indication of the subject matter: "Mattress on Telephone Pole," "Refrigerator on Roof," "Door on Wire," and the favorite of every posthurricane photographer with a lens, the proverbial "House on Car." Her two classics, photographs that would make a flaming Dadaist proud, were "Barge on Bus" and "Rocker in Tree." When asked to explain her art, Anderson had a minimalist answer: "They captured the storm in all of its terrible and absurd power."[39]

In the immediate response to the attacks of September 11, 2001, those affected were at no loss for courage and dignity, but Katrina was a different story. As the events in southeastern Louisiana were shown all day on TV and described in detail in newspaper reports, Americans came face-to-face with a weak, pathetic, and perhaps even shoddy side to their national disposition. Across the country, whether sitting by a lake in upstate New York, at a theater in Chicago, or in a pristine park in downtown Portland, Oregon, Americans were aware of an ill-fitting guilt: their fellow citizens were at that very moment stuck in grinding misery, and yet it was a benign, carefree summer day elsewhere. Viewers used to seeing natural disasters on television are aware that they unravel with a certain rhythm and the first day or so is bound to be chaotic. By the second day some sense of order should emerge. After that improvement runs on a course and generates an aura of hope. No matter how distressing the original crisis, the expectation in the human heart is that it will be followed by a sense of renewal. But when the government fails, each American fails, too, and feels it inside.

At that juncture, the government agency most people were livid about was Homeland Security, particularly its FEMA division. What people didn't yet fully comprehend was that the overall disaster, the sinking of New Orleans, was a man-made debacle, resulting from poorly designed levees and floodwalls. Although Congress had authorized Louisiana hurricane protection in its 1965 Flood Control Act, the Army Corps of Engineers had failed to live up to its provisions. "Perhaps not just human error was involved [in floodwall failures]," Raymond Seed, a civil engineering professor at the University of California–Berkeley, said ominously of the floodwall failures to a Senate committee. "There may have been some malfeasance."[40]

IV

Novelist Richard Ford, sitting at his home in East Boothbay, Maine, was watching New Orleans flood on television. He temporarily put aside work on his sprawling sequel to *Independence Day.* East Boothbay was a long way from New Orleans, 1,687 miles to be exact, but Ford had lived in the Big Easy for years and it had stayed with him, especially the eccentric people. There was his friend Curtis Wilke, a former *Boston Globe* reporter with a gray beard to make Whitman proud, and a whiskey drawl to put Jim Beam to shame. Sometimes Ford and Wilke would dine on seafood at Galatoire's, talking about racehorses, boxing, and dog breeding. But Wilke had moved to Oxford, Mississippi, and Galatoire's—long unique to Bourbon Street—used the hurricane as an excuse to open up a new branch restaurant in Baton Rouge. Nothing was sacred, even culinary conceits, as the people of New Orleans were finding out.

Then there was Rick Barton, the potbellied provost of the University of New Orleans and movie critic for *Gambit Weekly*. "Rick the Barbarian" was the kind of guy who won buffalo-wings-eating contests. Ford wondered if Barton had gotten out of the bowl. There was Ken Holditch, who knew more about Tennessee Williams than the playwright had ever known about himself. There were also the folks at the Garden District Bookshop—Britton Trice, Ted O'Brien, Deb McDonald, and Amy Loewy—who used to blow up Ford's dust jackets to poster size and have him autograph them with a felt pen. Ford, who had bought a Bourbon Street home in 1990, and then one in the Garden District a decade later, was a yellow-dog Democrat and a friend of trial lawyers—their class-action suits often cheered him up.

One of Ford's political friends was Marc Morial, the former mayor. Back before 9/11 they would sometimes talk about street crime, cultural trends, or down-home restaurants. Morial—who moved to New York with his wife, Michelle, to become head of the National Urban League in 2003—was a voracious reader; Ford was a literary hero to him. When Katrina hit, Morial was heartbroken. He started working the phones and e-mailing friends. No one in the media was able to express his frustration until he read Richard Ford's op-ed piece in the *New York Times* on Sunday, September 4. The prose of the stunning, beautiful "A City Beyond the Reach of Empathy" spoke directly to Morial: "For those away from New Orleans—most all of us—in this week of tears and wrenching, words fail. Somehow our hearts' reach comes short and we've been left with an aching, pointless inwardness."[41]

Watching TV that week with Richard Ford was his wife, Kristina. Having worked closely with Mayors Sidney Barthelemy and Marc Morial while she was executive director of New Orleans' City Planning Commission, Kristina Ford was irate by Wednesday. She couldn't believe Mayor Nagin was refusing to walk to the Superdome, grab a bullhorn, and reassure the frightened crowd. "My own opinion is that he has no empathy for them and the poor blacks of that town," she said. "And the black people knew from the get-go that he didn't care about them. He just made it really plain on this issue. Sidney was a social worker and Marc cared deeply about the poor. You couldn't have kept Sidney or Marc from going inside. They would have been down there with a megaphone to say, 'Help is coming. We're doing everything we can.'"[42]

Ben Jaffe, the son of the founder of Preservation Hall and a fantastic stand-up bass player, was also suddenly a member of the expatriate music community. He had driven out of New Orleans for Lafayette on Tuesday. Then he headed to New York City. He and his wife, Sarah, dug into their own pockets to help relocate great local musicians. Other New Orleans music legends who lived in the Lower Ninth—like Oliver "Who Shot the La-La" Morgan and Al "Carnival Time" Johnson—lost everything they owned. Ben was worried about the Lower Ninth Ward, and not just the stranded people—as the brothel song goes, "There is a house in New Orleans. . . ." The status of a particular Lower Ninth Ward building haunted him. On Sunday afternoons as a kid, he used to go over to Sister Gertrude Morgan's house on North Dorgenois Street, right near where the Industrial Canal breached. The combination house and church was called the Everlasting Gospel Mission. "She used to preach to my parents while I picked four-leaf clovers in her front yard," Jaffe recalled. "We were mesmerized by them. My brother and I would pull them up by the root and transplant them to our yard. They never grew." Sister Gertrude had died in 1980 and was buried near the airport; her home was sold. Nonetheless, he prayed the flood hadn't destroyed her home. "She profoundly affected me when I was growing up," Jaffe said. "She was always exuding love, care, courage, warmth, and strength. I was a Jewish kid going to synagogue, but she didn't care. She was the most liberal person. Her Jesus was about love and I never forgot it."[43]

Sister Gertrude, a self-taught painter, was a French Quarter soap-box evangelist who interpreted the book of Revelations as if it were emblazoned in her soul by a bolt of Mississippi lightning. An Alabaman by birth, she had arrived in New Orleans in 1939, christening her Lower Ninth "back-a-town" house the Everlasting Gospel Mission, and told people that she was a bride of Christ. She wrote poems about the apocalypse, strummed folk songs like Odetta, and illustrated passages from the New Testament on toilet paper rolls, lamp shades, and detergent boxes. She was, in *artiste* parlance, a practitioner of "outsider art." There was no holy water jive in Sister Gertrude's repertoire, no snake oil pretense of having the Jordan River bottled up in a vial. She delivered her sermons with a cardboard megaphone, her satisfied mind spreading the gospel with a wildcat rebuking Hallelujah! The Lower Ninth was her New Jerusalem and she was its wild-eyed African-American spiritualist, bouncing a tambourine with one hand while painting a self-portrait titled *Jesus Is My Airplane He Hold the World in His Hand He Guide Me* in the other. She was one of the Elect—"Oh, Sister Gertrude"—an apocalyptic seer whose Throne of God art would compensate for all her youthful indiscretions. As she inscribed one of her paintings: "G is for a girl a female child O don't Be so frisky and wild."[44] She constantly warned people to beware of "conjurers," aka Lucifer or the Devil. Her heart was wide open, however, to "angel children" in trouble; her home-church-studio always had room for an orphan or seeker or wanderer. She was safeguarding them from New Orleans which, in her mind, was "the headquarters for sin."[45]

Jaffe's parents early on realized the untutored Sister Gertrude was a primitive art genius—one who eventually had a major retrospective titled "Tools of Her Ministry" at the American Folk Art Museum in New York. Although her paintings went for $100 during her lifetime, at the time of Katrina, Sotheby's was selling them for upwards of $5,000 apiece. But Ben didn't know about the rising value of her art; he just loved her radiant colors and folk wisdom. He remembered her Everlasting Gospel Mission as an exotic theme park dedicated to Jesus; the walls were illustrated with brightly colored red camels, blue angels, and green churches. Art filled up every nook and cranny of her ranch-style house.

What Jaffe recalled most, however, about Sister Gertrude were the four-leaf clovers that blanketed her lawn. Right there in the middle of the Lower Ninth Ward, a short walk from the Industrial Canal, were fields of these shamrocks which surrounded Sister Gertrude's home studio. While Ben's parents were listening to her sing "Precious Memories" or watching her paint a self-portrait on her guitar case, Ben would pick four-leaf clovers, treating them gingerly so as not to have bad luck.

Driving to New York, leaving the bowl behind, Jaffe kept thinking about Sister Gertrude's lawn. He imagined it underwater. Even though somebody else now owned the church, he still considered North Dorgenois Street sacred ground. And then there was the keepsake, one of her self-portraits that Jaffe inherited when she died, of Sister Gertrude in an all-white nurse's uniform, carrying a bag, fleeing the flooded Lower Ninth. It was titled *Lord, I Don't Want to Be Buried in the Storm.* Jaffe made sure her post-Betsy artistic vision of the Big One stayed high and dry throughout Katrina, but he was haunted by whether her house survived the deluge. "I knew that when I got back to New Orleans," Jaffe said, "I'd go check up on Sister Gertrude's place, to see how it fared."[46]

When the Jaffes returned to New Orleans weeks later, Ben solemnly drove through the Lower Ninth. He saw the giant rusted *ING 4727* barge, owned by the Ingram Barge Company of Nashville, Tennessee, which had gotten loose from its anchor in the MRGO during Katrina and had burst through the Industrial Canal levee, perhaps causing the breach. With a length of over 200 feet, a beam of 35 feet, and a height of 12 feet, *ING 4727* could carry 1,877 tons of cargo. He grew numb. For the first time ever, Jaffe felt God's presence lurking in his car. He turned off St. Claude Avenue, pulled in front of the Everlasting Gospel Mission, turned off the ignition, and froze. The flood had lifted Sister Gertrude's house off its foundations and moved it ten or fifteen yards away. But it was standing—the Lower Ninth Ward was gone, but the house was there. For decades Sister Gertrude had talked about Judgment Day—now it had arrived. "I never have really felt that alone in New Orleans, sitting on her stoop for hours on that Sunday afternoon," Jaffe recalled. "Nobody was around. It used to be a busy block—everybody out and about in church clothes. Everything now was still. It was the first time that I cried. . . . It was cathartic, though."[47]

Why would God do such damage? Jaffe wondered. All around him were desolate, uprooted, trashed, cracked, and broken homes. What a messed-up, vengeful world, he thought. And then he looked down. All around him were four-leaf clovers. They had survived the deluge. Everything else in the Lower Ninth was mud-gray, and lifeless, smashed-concrete desolation. But like sunflowers blooming in junk heaps, the field of clovers had survived. He bent down on one knee and passed his hand lightly over the clovers. Carefully, he plucked one, stood up, and walked to his car. He had never bought into Sister Gertrude Morgan's notion of New Orleans as "the headquarters of sin"—if it was, then he was a terminal case. The four-leaf clovers, however, were a sign that Sister Gertrude was all right in heaven. He made a pact with himself to try and save her house. Or was it just a pipe dream? The City of New Orleans had allowed the federal government to destroy Storyville to build I-10; back in 1964 they even razed the birthplace of Louis Armstrong. The wrecking ball, however, wasn't going to claim Sister Gertrude's house-church. Not if Jaffe could help it. Preservation Hall was going to get into the preservation business.

V

Jazz trumpeter Wynton Marsalis was another New Orleans expatriate glued to his television that week. "It's as if your parents have died," he told David Fricke of *Rolling Stone*. Marsalis was safe in his New York home, but he wasn't comfortable. "All your memories, your ancestors, are underwater. My great-uncle's house is probably gone. I go back to that house in my mind all the time. It was a shotgun house. I think of the fan they had in the front room, the shed with all the newspapers he put up for wallpaper—old yellow paper from the 1920s; the little plot of land he used to cut with a hand mower. His front porch—we put that down ourselves. He was a stonecutter; I was five or six."[48]

Marsalis had been born on Louisiana Avenue in Flint-Goodridge Hospital, right across from the Magnolia projects and just down the road from Reverend Willie Walker's Noah's Ark Missionary Baptist Church. He loved everything about the birthplace of jazz. He wasn't alone in having New Orleans memories etched into his imagination. Ex–New Orleans residents scattered all over the world were in the same sentimental situation, watching the city's demise on television. No matter whether they lived in New York City, Denver, Milwaukee, or Seattle, they possessed a New Orleans state of mind. All the Big Easy expatriates would surely agree with Marsalis when he wrote, post-Katrina, "Our city is still alive. It's generations of us who are still here, and will get our city back together. There are things that are tragic losses that will never be recovered, but I feel like the most valuable thing is the people, the spirit of the people, the will of the people, the mind, and the hearts . . . that's not lost. That's not even close to lost."[49]

Mark Broyard of Los Angeles counted himself among their ranks. He had been born in New Orleans—one of the first "colored" children ever born at Ochsner Clinic—and lived in town until he was five. His father and grandfather had their own bricklaying company, working mainly in Harahan and Metairie. With white flight to the suburbs in the 1960s, the firm thrived, but the racism became too much to bear. "My brother was scheduled to start first grade," Broyard recalled. "They didn't want him to be the first colored kid back there and have things thrown at him and have him abused at all. He was already kind of a sickly kid, asthmatic and all that. So with my dad's two sisters taking the lead, we all moved to Los Angeles. But they never really left New Orleans, it was always their home. I think now, even forty years later, every time, I'm with my mother, at some point, she'll say, 'We should have never left New Orleans.' "[50]

Every summer during Broyard's childhood, the family would hop in the car, get on I-10 in Los Angeles, head east for the Franklin Avenue exit in New Orleans, near Bayou Sauvage. The London Avenue Canal helped the various residential neighborhoods—Gentilly Ridge, Filmore, Gentilly Terrace, Gentilly Woods, among others—drain water into Lake Pontchartrain. "There were oyster and clam-shells in the street," Broyard recalled, "so when you turned onto Franklin, there'd be a dip and the wheels of the car would start grinding in the dirt in the oyster shells. That sound was like music to me. I knew that I was home, just hearing the sound of the car on those white streets full of broken-up oyster shells." The Broyards spent the summer months in Gentilly, not far from Lake Pontchartrain, and it was a magical place. Using his grandparents' house as a base, Mark befriended all his neighbors. He remembered eating piles of crawfish and corn on the cob off old *Times-Picayune*s with his Uncle William. Both of them were sprayed down with Off!, to keep the bugs away. Not too far from St. Gabriel the Archangel Church, where his family often attended mass, was Grandma's bungalow, where once a week they had a trout or buffalo fish supper. He could still smell the butter, corn dumplings, turnip greens, and cayenne pepper decades later.

Now, as Broyard learned that Gentilly had been destroyed by Katrina, and all of his relatives' houses filled with floodwater from the London Avenue Canal breach and MRGO topping, the precious memories came pouring back like a brass trumpet titillating with pain. There was Teddy's Grill, where the dressed roast beef po'-boy sandwiches were overstuffed. They were his absolute favorites. He ordered them with delicious little lemon pies for dessert, washed down with a Barq's root beer.

Broyard was carefully monitoring the Seventh Ward from his Los Angeles home. The little boy who used to run the streets of Gentilly was forty-eight, director of the choir at St. Jerome's Church and cantor for Catholic churches all over Los Angeles. He was part of an expatriate New Orleans community there, always singing at weddings what poet Langston Hughes called "lucky hymns," for other former New Orleans residents. What most concerned him on Wednesday, however, was the fate of his godmother, who was in Lafon Nursing Home in New Orleans East. He had called Sister Augustine McDaniel as Katrina approached and he didn't like what he heard. "What evacuation plan is in place?" Broyard asked. "Have you evacuated everyone?" The response was, "No, no, everyone is asleep." Broyard said, "You are aware that a Category 5 storm is on the way?" Again, the response was inadequate. From California, all Broyard could do was shake his head. "It just blew my mind," he recalled, "but sometimes, that's just the New Orleans mentality about things. 'Oh, there's a Category 5 on the way, fix me a sausage sandwich and let's go pick up a beer. . . . Everything's goin' to be all right.' Katrina, particularly the breached levees, just caught everybody by surprise."[51]

Twenty-two residents of Lafon Nursing Home died, but Broyard's godmother wasn't among them. She had weathered the storm on the second floor and on Thursday was among the forty residents boarding Irvin Boudreaux's bus. "My godmother thinks she did something wrong and they put her in jail," Broyard said. "Isn't that terrible? I talked to that manager [Sister Augustine], and just chewed her out. I just couldn't believe that they'd be so inept and so incompetent to leave those little old people stranded like that and that some of them would die and that my godmother, at ninety-six years old, thinks at this stage of her life that she did something wrong, so that she's now in jail. That's just horrible." Sister

Augustine didn't need anyone to tell her that it was horrible. She knew firsthand.

Two days after helicopters ferried the rest of the residents away from Lafon, Rosalie Daste died from the strain. Broyard and his friend Roger Guenveur Smith, an actor who'd appeared in Spike Lee's *Get on the Bus,* had already created and performed *Inside the Creole Mafia,* a play to honor New Orleans culture. They now reconfigured the dark-humored show to reflect post-Katrina angst, to remember the Rosalie Dastes of the Big Easy.[52]

When Broyard was asked how he felt about Gentilly now that the Tambourine Choir was shut down, the delicate jazz silenced, and his dreamy-eyed summer playground transformed overnight into a dead neighborhood, he paused and fought back tears. He gave the matter serious consideration. He put his finger on his chin to meditate. He was fishing for a verdict worthy of Matthew or John. "Well," he said forlornly, "Teddy's is gone. So that means no more roast beef po'-boys for me."[53]

VI

Expatriates were one thing, but nearly one million displaced residents of the ravaged areas were also watching the coverage of the storm, or reading about it. "I just heard that in the area where I live—people are on top of their roofs," said Aaron Parker, who was at a Red Cross shelter in Beaumont, Texas. Parker managed a Walgreens in New Orleans.[54] He had no way of knowing what had become of his house or his store. Nearby at the shelter, a woman clutched a cell phone, hour after hour, hoping it would ring with a call from her son, who had remained behind in New Orleans. She and Parker were among the many who had left their homes believing that they were escaping the winds but expecting to be home in a day or so. With little in the way of clothing or other belongings, they were stuck in a kind of limbo: safe in body, but literally clueless in regard to their own lives, past and future. They were the Children of the Exodus.

Towns across Louisiana were burgeoning with evacuees. The city of Lafayette, for instance, population 110,000, took in 40,000 of them. Many, if not most, stayed with friends or relatives. It was not unusual for a family of four to take in eleven people. Had the hurricane been only a very bad storm, that might have been akin to a holiday gathering for the hosts and guests alike. But Katrina was a vicious and unending disintegration of society's fabric. According to academics such as Carl A. Brasseaux, a professor of history at the University of Louisiana at Lafayette, the family networks that had remained strong in Louisiana were more important than any government aid in helping to ease the impact of the disaster. "If family structures in Louisiana," he said, "had eroded to the point they have in many parts of the country, many refugees here would face a very long and bleak road ahead."[55] Aside from the fact that every foldout bed across inland Louisiana and Mississippi was in use, the storm ensured that hotels and motels in the same regions were filled to overflowing. You couldn't even rent a room at the Fountain Motel, the Shades Motel, Riverbend Motel, and scores of roadside accommodations, whose hallmarks were vacancies. Rest stops along highways were also filled, with people unable to find or afford a hotel and living out of their Oldsmobiles and Grand Cherokees. All fees were waived at state parks by Lieutenant Governor Mitch Landrieu, so evacuees could camp at Sam Houston Jones Park north of Lake Charles, Chicot State Park in Ville Platte, and elsewhere.

One Louisiana facility running on overdrive was the Pete Maravich Assembly Center on the LSU campus in Baton Rouge. If the Superdome had become "hell's island," and the Convention Center "the pit," then the Maravich Center was "purgatory central." The 14,164-seat multipurpose arena was the holding tank for those bused out of New Orleans. It also became a makeshift hospital, a triage hub for medics, volunteers, and a constant stream of patients. The center did have air-conditioning, electricity, and drinkable tap water. Although at times it became too crowded, and there was a definite disorder in the air, the medical teams, including those from the Red Cross and FEMA, did a fine job. Ambulances and helicopters kept dashing in and out, dropping off Katrina victims needing insulin injections, stitches, or IV bags. "It was miserable, miserable, so miserable out there," Pearl Johnson recalled when she at long last got dropped off in purgatory. Her husband, just grateful to be at the Maravich Center, added, "But we're here! At least we made it here!"[56]

Fats Domino was among the hundreds of Lower Ninth Ward residents who eventually wound up at the Maravich Center. Domino and his wife had been rescued from their home on Monday and taken to the Superdome. Sweating profusely, his blood pressure soaring, and confused, the seventy-seven-year-old Domino thought he was there to sing. Sherry Watters, a Department of Social Services caregiver at the Dome, worried that Fats might die. She tried to get him to eat a vegetarian MRE and drink water, but he was uncooperative. He preferred the greasier fare served at the Popeye's Fried Chicken near his home—in fact, the company, in an act of generosity, had given the music legend an all-you-can-eat card, good at any of its franchises. "He walked around the Superdome, shaking hands," Watters recalled. "He was really nice, but he just refused to eat because he thought he was about to perform."[57]

Called "Saint Sherry" for her nonstop efforts to take care of diabetics and high-blood-pressure patients at the shelter of last resort, Watters made sure Domino got over to the Arena, the basketball facility next door to the Superdome that had been pressed into service as a triage area. After that Fats Domino, like everyone else at the Dome, waited to escape the bowl. On Wednesday, Domino and his wife, Rosemary, along with their large extended family, boarded a bus outside the Superdome bound for Baton Rouge. Once they arrived at the Maravich Center they waited patiently to register just like any other evacuee. Refusing to trade on his celebrity, Fats checked in as Antoine Domino. He was then taken to triage where medics could calm his heart palpitations and monitor his blood pressure. A volunteer named JaMarcus Russell, a student at Louisiana State University (and the quarterback of the LSU Tigers), recognized the rock 'n' roll legend. His girlfriend was related to him. Russell was thrilled to approach one of his idols, introduced himself, and shook his hand. Seeing that Domino was sick and exhausted, and wanting to help, Russell offered to let the Dominos sleep at his apartment. Eventually over twenty Katrina refugees ended up "crashing" at Russell's place. Fats, like everybody else, was glued to CNN and the other news channels, trying to monitor events. A bed was found for him, though at one point he chose the floor.

Because there had been reports that Domino had died, the news that he was alive and well in Baton Rouge was one of the cheerful moments of a gloomy week. "The important thing is that Fats and his family are all right," Russell told the press. "I'm not sure where they are headed, but I feel better knowing that they're okay." Although Fats refused to do interviews, an LSU spokesperson conveyed a heartfelt message: "Tell the people of New Orleans that I'm safe," Domino said. "I wish I was able to still be there with them, but I hope to see them soon. I want to express my sincere gratitude and appreciation to JaMarcus for opening up and sharing his home with us."[58]

On Friday, September 3, Fats Domino, Rosemary, and some of his kids piled into cars and headed toward Dallas. Once they crossed the border, Fats became part of another official statistic: one of 230,000 Katrina evacuees to arrive in Texas.

Fats Domino eventually made it to Arlington, where he was housed in an air-conditioned high-rise. Get-well greetings soon came pouring in from all over the world. His blood pressure stabilized and his bouts of sweating ended. All things considered, he was doing well. But he was homesick.

In the middle of October, after living in Bedford, Texas, Fats returned to the Lower Ninth Ward. Both his Marais Street home and the adjacent office on Caffin Avenue looked like they had been dumped into a washing machine for days. A big satellite dish sat in the front lawn, but the pink and yellow trim on the house hadn't lost its pastel appeal. And the flowered tiles on his front porch were still in place, muddy but not cracked. As he wandered around the premises looking for salvageable items, a wave of despair threatened to consume him. He shook his head in disbelief at the eight-foot watermark. But he put on a valiant front for the TV crew on hand to document his return. A favorite shirt was found and a handsome bust unearthed. Suddenly Fats cheered up more—a lot more. His world-famous smile beamed on his face. Three of his twenty-one gold records hadn't washed away or been looted: "Rosemary," "I'm Walkin'," and "Blue Monday." For the first time since Katrina, the Fat Man was a Happy Man.[59]

Scattered about Fats Domino's compound debris was a white Steinway piano missing its legs. As artifacts go, this one was filled with symbolic meaning. It was unclear whether Fats ever wrote any songs on the grand piano. It probably didn't have any import at all. Also found in the rubble were an electric Wurlitzer piano and another Steinway. But staring at the thick gray dust on those desecrated piano keys, laying decommissioned on the old wooden floor, it was hard not to believe something profound had perished. The Lower Ninth Ward would never be the same. Something new would take its place. The jamboree, however, was over. Salvaging scraps from the past was all that was left to do. In early 2006, the Louisiana State Museum salvaged the legless piano, and someday it would be under glass on display. Schoolchildren would study the artifact as a Katrina learning tool. But it was perfectly reasonable to believe those months after the hurricane that Fats's piano symbolized just one thing: the day the music in New Orleans died.

VII

Many Louisianans had long regarded Texas as a tall, dark shadow over their state, dominating business, reaping the benefit of Louisiana's cornucopia of natural resources, and even pulling many of its brightest minds away. Texas, it has been said, looked at Louisiana as a Third World territory, wide open for exploitation. In the bleak humor of the lower Mississippi, the capital of Louisiana was not Baton Rouge; it was Houston. However, it was Texas that opened its doors and surely its hearts the widest to people made homeless by Katrina. Throughout the day on Wednesday, Governor Blanco had extensive telephone conversations with Governor Rick Perry of Texas, who wanted to join in the official effort to give relief to the evacuees. Blanco's immediate problem was finding a place to send the 30,000 people in the Superdome and at least twice that many who were stranded in other venues in and around New Orleans.

Perry was a fifth-generation Texan, a rancher from Abilene who was elected as a Republican state legislator in 1985. In 1999, he was George Bush's running mate as lieutenant governor, and when Bush left Texas to move into the White House in January 2001, Perry took over the governorship. A tough, fiscal conservative, Governor Perry made his mark in tort reform, legislative vetoes, and an abiding belief in making health care more readily available to children. At the time of Katrina he was a popular governor, albeit one facing a reelection challenge in 2006.

In responding to the hurricane disaster, Governor Perry reached out to officials in Texas's biggest cities, finding some support in all of them, but a true ally in Houston's mayor, Bill White, a Harvard-educated attorney who had spent most of his career as a businessman, running energy and construction companies. White, a Democrat, had been mayor of Houston since January of 2004. He had been raised in San Antonio. His father was a leader in registering Mexican Americans to vote. Fighting to eliminate poll taxes and a staunch advocate of "one man, one vote," he was elected to the State Senate when Bill was thirteen years old. "We did voter registration drives in the west side of San Antonio," White said. "That got me started. By 1975, I was a legislative assistant to a U.S. congressman."[60]

White was obsessed with studying energy consumption issues. A true "wonder boy," when the OPEC embargo of 1973 hit, White, not yet twenty years old, headed to Washington, D.C., to write energy legislation pertaining to fuel efficiency standards and strategic petroleum reserves. Upon graduating from the University of Texas School of Law in 1979, White remained in Texas and opened a public interest law firm in Houston. His reputation soared, and he was admired by Democrats and Republicans alike. Along the way, he became heavily involved in the oil business. Short, bookish, and soft-spoken, with big ears and a Howdy Doody grin, White was the antithesis of a Texas cowboy; everything about him spoke of Georgetown salons and Cambridge think tanks. Married with three children White taught Sunday school at St. Luke's United Methodist Church in Houston and attended, seemingly, every social event in town.

What differentiated White from most other politicians of his generation was his nonpartisan nature. Like Perry, White responded to Katrina with pragmatic generosity. Both intended to accommodate the refugees of the storm, but to do so in an organized way that would be accountable financially as well as socially. In opening Reliant Park—a convention, sports, and entertainment complex consisting of the Stadium, Center, Astrodome, and Arena—for the evacuees, Perry and White were undaunted by the fact that the Superdome had deteriorated under a similar weight. A catering company was engaged to serve three hot buffet meals per day. Medical personnel began to set up stations to screen for those who needed hospital treatment. The Red Cross arranged for cots and extra toilet facilities. Metropolitan buses were put on a schedule to ferry people to shopping malls, parks, and other sites around the city. According to this plan, Reliant Park was projected to be a temporary stop, until more permanent housing became available; if necessary, though, it could handle evacuees for up to ninety days.[61] At the same time, Perry made sure that Texas schools prepared to receive children displaced by the storm. "No community is equipped for such an enterprise," White later recalled. "It's just a reality we had Americans in need and we were the closest urban area with the capacity."[62]

VIII

Texas stepped up, but the federal response continued to be muddled on Wednesday afternoon. At about 1 P.M., Michael Chertoff hosted a press conference that included five other cabinet members. "We are extremely pleased," Chertoff said, "with the response that every element of the federal government, all of our federal partners, has made to this terrible tragedy." He did not explain why there was so little to be seen for it all. He also did not mention the name of FEMA Director Michael Brown once, although he made reference to many other colleagues and seemed to be in a generous frame of mind, when it came to extending praise for whatever it was that he thought had been done by his department.[63]

Americans, however, did not share Chertoff's good humor, and they were decidedly not "extremely pleased." Secretary of State Condoleezza Rice was on vacation, shopping at the Salvatore Ferragamo store on Fifth Avenue in New York, when a fellow shopper spotted her and shouted, "How dare you shop for shoes while thousands are dying and homeless?" That evening, Rice attended the Broadway musical *Spamalot.* Word went through the audience that she was present, and when the lights came up after the show, she was roundly booed.[64] The next day, she ended her vacation early and returned to Washington, D.C. At least one administration official had finally gotten the message: Katrina was a crisis that could not be ignored.

On Wednesday afternoon, Governor Blanco finally issued the executive order that allowed the state to commandeer every school bus in Louisiana. In fact, many people from throughout the country were hankering to help, but most found that it wasn't that easy. The road of good intentions led only to frustration and madness on the way to New Orleans. In an effort to secure the area, police or Guardsmen were stationed at most of the roads leading into southeastern Louisiana. They had orders to keep potential looters out. They also had to keep home owners from returning and becoming a danger to themselves in the midst of the crisis. That was also understandable. But when rescue crews arrived from municipalities all over the nation, fitted out in uniforms and carrying specialized equipment, it should have been clear that they belonged in a different category. One reason given for turning away rescue crews was that they would need to be fed and housed—and the city had no way to help anyone, not even those coming to help.

While rescuers were being turned away under the absolute lockdown, reporters and cameramen seemed to have no trouble moving in and out pretty much as they pleased. Even a fake press pass was an invitation into the bowl. People watching the coverage were perplexed to hear from a veritable battalion of reporters freshly arrived in New Orleans that the federal government couldn't find any way to bring help into the flooded city.

Assuming that the trouble in absorbing volunteer rescuers was just part of the fog of war—the war on the hurricane damage—it was still hard to explain why FEMA was so irresponsible with the expert help that was proffered. In the most egregious case, a group of more than 100 first responders who came in from all over the country were diverted to Atlanta on Wednesday for training on rules against sexual harassment. With the equivalent of a sinking ship going down in New Orleans, no one would think of sensitivity training as the absolute top priority—no one, apparently, except FEMA and the federal government. In any case, these rescuers, including firemen, experts in search and rescue, Hazmat, and paramedics, were told that their only job would be handing out flyers containing the FEMA toll-free number. They were not to interfere further; no matter what they were asked, they could only respond by repeating the toll-free number. "There are all these guys with all this training and we're sending them out to hand out a phone number," said a firefighter from Oregon, who was thinking of the storm's victims. "They are screaming for help and this day was a waste."[65]

Some people wondered why local firefighters had been on duty nonstop, with time off only for catnaps. Under the leadership of Superintendent Charles Parent, the NOFD demonstrated extraordinary courage putting out huge infernos. "Down there in the Ninth Ward we thought there was a toxic chemical fire," Terry Ebbert recalled. "The fire department, perhaps risking their own lives, put it out. And then there were all the gas fires flaring up because of the gas lines that were severed, and we were working with folks to get those shut down."[66] Mayor Nagin used one of his television appearances during the day to solicit more help from the nation's fire departments. The situation was that dire, which was why the firemen sitting through the FEMA seminar in Atlanta were perplexed and angry. The Oregon firefighter was right: people *were* screaming for help.

One New Orleanian still needing help on Wednesday morning was chef extraordinaire Austin Leslie, who had been trapped in his attic for two days. Rescuers would finally arrive that afternoon. At seventy-one, Leslie was the city's reigning expert on cooking fried chicken. With his muttonchops and trademark white yachting cap, he was well-known in the restaurant world, both in the back, at the fryer, and out front, hobnobbing with guests. "There's nothing like cooking Creole food in New Orleans," he said in an interview in 2004. "That's your toughest audience, your best one."[67] The two days in the attic heat had placed an unbearable strain on Leslie, however. After persevering through an arduous trip to the Morial Convention Center, and eventually to Atlanta, he was unable to shake a fever and died on September 29.[68]

IX

On Wednesday, officials in Texas rushed to prepare the Reliant Park complex and other venues to receive evacuees. At the same time in New Orleans, Colonel Jacques Thibodeaux of the Louisiana National Guard took up his new assignment: organizing the people inside the Superdome for evacuations onto buses. He and his top aides, including McLaughlin, spent the afternoon on the scene, devising a practical plan. They settled on a short walk to Loyola Boulevard, which had remained dry and where the buses could load passengers and pull away for a relatively easy trip out of town. The only problem was, as Colonel Thibodeaux said, "You wanted to have 1,000 buses to put your hands on."[69] And there weren't any. Blanco had put out the call for Louisiana school buses, and her trio of Ty Bromell, Leonard Kleinpeter, and Angele Davis was corralling them at mile marker 209, but the governor and every person stranded in New Orleans still waited for the Federal Emergency Management Agency to produce hundreds. FEMA, however, barely even knew who was looking for the buses on its behalf. "But eventually FEMA delivered some," Kleinpeter recalled. "They did start helping to evacuate the Dome."[70]

Meanwhile, at 4 P.M., President Bush convened the hurricane task force at the White House. An hour later, from the Rose Garden, he addressed the nation on live television. An editorial the next day in the *New York Times* worried, "Nothing about the president's demeanor yesterday—which seemed casual to the point of carelessness—suggested that he understood the depth of the current crisis." The President did not seem to be engaged in the subject, glancing at his notes and delivering a speech that smacked of the press release issued by the Department of Homeland Security that same afternoon. The speech ran:

> I've just received an update from Secretary Chertoff and other Cabinet Secretaries involved on the latest developments in Louisiana, Mississippi, and Alabama. As we flew here today, I also asked the pilot to fly over the Gulf Coast region so I could see firsthand the scope and magnitude of the devastation. The vast majority of New Orleans, Louisiana, is under water. Tens of thousands of homes and businesses are beyond repair.
>
> A lot of the Mississippi Gulf Coast has been completely destroyed. Mobile is flooded. We are dealing with one of the worst natural disasters in our nation's history. And that's why I've called the Cabinet together. The people in the affected regions expect the federal government to work with the state government and local government with an effective response. I have directed Secretary of Homeland Security Mike Chertoff to chair a Cabinet-level task force to coordinate all our assistance from Washington. FEMA Director

Mike Brown is in charge of all federal response and recovery efforts in the field. I've instructed them to work closely with state and local officials, as well as with the private sector, to ensure that we're helping—not hindering—recovery efforts.

This recovery will take a long time. This recovery will take years. Our efforts are now focused on three priorities. Our first priority is to save lives. We're assisting local officials in New Orleans in evacuating any remaining citizens from the affected area.

I want to thank the state of Texas, and particularly Harris County and the city of Houston and officials with the Houston Astrodome, for providing shelter to those citizens who found refuge in the Superdome in Louisiana. Buses are on the way to take those people from New Orleans to Houston. FEMA's deployed more than 50 disaster medical assistance teams from all across the country to help the affected—to help those in the affected areas. FEMA's deployed more than 25 urban search and rescue teams with more than 1,000 personnel to help save as many lives as possible.

The United States Coast Guard is conducting search and rescue missions. They're working alongside local officials, local assets. The Coast Guard has rescued nearly 2,000 people to date. The Department of Defense is deploying major assets to the region. These include the USS *Bataan* to conduct search and rescue missions, eight Swift water rescue teams, the *Iwo Jima* Amphibious Readiness Group to help with disaster response equipment, and Hospital Ship USNS *Comfort* to help provide medical care.

The National Guard has nearly 11,000 Guardsmen on state active duty to assist governors and local officials with security and disaster response efforts. FEMA and the Army Corps of Engineers are working around the clock with Louisiana officials to repair the breaches in the levees so we can stop the flooding in New Orleans.

Our second priority is to sustain lives by ensuring adequate food, water, shelter, and medical supplies for survivors and dedicated citizens—or dislocated citizens. FEMA's moving supplies and equipment into the hardest hit areas. The Department of Transportation has provided more than 400 trucks to move 1,000 truckloads containing 5.4 million meals ready to eat, or MREs; 13.4 million liters of water; 10,400 tarps; 3.4 million pounds of ice; 144 generators; 20 containers of pre-positioned disaster supplies; 135,000 blankets and 11,000 cots. And we're just starting. There are more than 78,000 people now in shelters.

HHS and CDC are working with local officials to identify operating hospital facilities so we can help them, help the nurses and doctors provide necessary medical care. They're distributing medical supplies, and they're executing a public health plan to control disease and other health-related issues that might arise.

Our third priority is executing a comprehensive recovery effort. We are focusing on restoring power and lines of communication that have been knocked out during the storm. We'll be repairing major roads and bridges and other essential means of transportation as quickly as possible. There's a lot of work we're going to have to do. In my flyover, I saw a lot of destruction on major infrastructure. Repairing the infrastructure, of course, is going to be a key priority.

Department of Energy is approving loans from the Strategic Petroleum Reserve to limit disruptions in crude supplies for refineries. A lot of crude production has been shut down because of the storm. I instructed Secretary Bodman to work with refiners, people who need crude oil, to alleviate any shortage through loans. The Environmental Protection Agency has granted a nationwide waiver for fuel blends to make more gasoline and diesel fuel available throughout the country. This will help take some pressure off of gas prices, but our citizens must understand this storm has disrupted the capacity to make gasoline and distribute gasoline. We're also developing a comprehensive plan to immediately help displaced citizens. This will include housing and education and health care and other essential needs.

I've directed the folks in my Cabinet to work with local folks, local officials, to develop a comprehensive strategy to rebuild the communities affected. And there's going to be a lot of rebuilding done. I can't tell you how devastating the sites were. I want to thank the communities and surrounding states that have welcomed their neighbors during an hour of need. A lot of folks left the affected areas and found refuge with a relative or a friend. And I appreciate you doing that. I also want to thank the American Red Cross and the Salvation Army and the Catholic Charities and all other members of the armies of compassion. I think the folks in the affected areas are going to be overwhelmed when they realize how many Americans want to help them.

At this stage in the recovery efforts, it's important for those who want to contribute to contribute cash; to contribute cash to a charity of your choice but make sure you designate that gift for hurricane relief. You can call 1-800-HELP-NOW or you can get on the Red Cross Web page, redcross.org. The Red Cross needs our help, and I urge our fellow citizens to contribute. The folks on the Gulf Coast are going to need the help of this country for a long time. This is going to be a difficult road. The challenges that we face on the ground are unprecedented. But there's no doubt in my mind we're going to succeed.

Right now, the days seem awfully dark for those affected. I understand that. But I'm confident that, with time, you'll get your life back in order. New communities will flourish. The great city of New Orleans will be back on its feet. And America will be a stronger place for it.

The country stands with you. We'll do all in our power to help you. May God bless you. Thank you.[71]

Where was the passion in President Bush's speech? Could anybody find a single memorable line? His delivery had all the drama of a reading from the phone book or an IRS form. "George W. Bush gave one of the worst speeches of his life yesterday," commented the *New York Times,* "especially given the level of national distress and the need for words of consolation and wisdom. In what seems to be a ritual in this administration, the president appeared a day later than he was needed. He then read an address of a quality more appropriate for an Arbor Day celebration: a long laundry list of pounds of ice, generators and blankets delivered to the stricken Gulf Coast. He advised the public that anybody who wanted to help should send cash, grinned, and promised that everything would work out in the end."[72]

As the *Times* noted, nearly all of the help that the President described was still in the future, many weeks away, a fact that he neglected to mention. Meanwhile the situation was only getting worse with every hour. Because soldiers were not yet on the ground in large numbers, for instance, security was eroding.

On Wednesday evening, Mayor Nagin ordered the majority of New Orleans police officers to stop rescuing people. Looters were considered a higher priority. "They are starting to get closer to the heavily populated areas—hotels, hospitals," Nagin said, "and we're going to stop it right now."[73] In the beginning, anyone might have been a looter. Those who needed food or water during the clutches of the disaster had some intrinsic right to take it from a store or office. The spree escalated from there, however, to the point that two very different scenarios were playing out at once in the city. People were seeking relief, still standing on rooftops or struggling with the problems at the shelters. Simultaneously, bands of lawbreakers were running rampant. They took the contents of stores and even automobile dealerships, preyed on hurricane victims, and used guns with abandon, shooting at the sky or threatening other people. The rescue effort might have continued alongside the looting, but when it developed into wanton violence, Nagin ordered the police to turn their entire attention to regaining control. It was a steep challenge. There was nowhere to put those arrested except the temporary jail at the Greyhound bus station. More commonly, the police could do little but put ruffians in line for evacuation.

The descent into *Lord of the Flies* anarchy was the third tier of the Katrina disaster. The first tier had been the storm. The second had been the flooding, due to the levee failure. The third tier was the violence that swept away laws, due largely to the absence of federal troops. There was a fourth tier, too, awaiting in the city and it scared everyone, good or bad. Floodwaters stagnated in the steamy hot region, creating a breeding ground for infectious disease that would soon take whatever lives the wind and the flood and the violence hadn't. Or so it was feared. The dragonflies helped, but the fear of malaria or tetanus or worse persisted.

As the Cajun Navy and the Louisiana National Guard at the Superdome had found out, at night New Orleans was ruled by "anything goes." The U.S. government had sent armored vehicles to haul out the money at the

Federal Reserve Bank. Likewise, other banks had private security teams safeguard money in the cover of darkness. Slowly, but surely, buses were starting to transport some Superdomers to Baton Rouge, Lafayette, and Houma. The mass exodus to Houston was yet to begin, however.

That Wednesday at 6:00 P.M., most of the national media were being kicked out of downtown hotels. "Vans were now pulling up to the W Hotel; the entire hotel was under a forced evacuation," Jeff Goldblatt of Fox recalled. "My bosses told me it was time to leave. All I had was my rental-car, no trailer to sleep in."[74]

Just as he was doing his final report from the W, however, Goldblatt saw a vagabond submerging a plastic jug into an ornamental fountain out front. Ernest Hemingway had written just such a scene on the first page of *To Have and to Have Not,* only it was a "bum" in Havana with a cup. "I pivot around and I notice this bedraggled, filthy man with mangy grayish black hair in his fifties going on about eighty, dip his jug into the slimy water and then drink," Goldblatt recalled. "As he put the jug to his lips I turned around and said something like 'Sir, what are you doing? You can't drink that!' His gesture went right to my spine; it wasn't just a tingle, my whole back went rigid and I said, 'Stop, we'll get you some water.' "

Broadcasting live, Goldblatt's audio technician walked over and handed the needy man some water. It was a coalescing moment for the reporter who hadn't slept in days. He had been hoping to find the libertine side of Katrina, i.e., Johnny White's bar serving beer or Bourbon Street revelers waving Mardi Gras beads. But suddenly Goldblatt turned deeply pessimistic about New Orleans's immediate future. When the homeless were drinking contaminated water out of scummy fountains and hotel owners were booting out media, New Orleans had reached the forbidden hour, the moment when you felt like a lemming about to walk without protest over a cliff. Things were going to get worse, not better. The trickle effect had arrived. All that was left was panicked improvisation. "Getting out of town was insane," Goldblatt said. "But you've got to know when it's time to go."

Right then and there, Goldblatt decided he was going to improvise his way out of town, back to Chicago, where his wife was waiting and Morton's steakhouse was open. "I hope people didn't think that the fountain drinker was staged, fabricated to get ratings," Goldblatt later said. "This guy was the real thing. Although I had seen many distressing signs of depression that Wednesday, that was the icing on the cake. That's when I knew the entire urban grid was gone. I mean, I heard about dead bodies rotting in the street. I heard that a couple of blocks away at the Convention Center there were bodies. I knew that if I walked over there I would have seen them. But to me the drinking fountain was the moment when the bottom had fallen out."[75]

Chapter Twelve

THE INTENSE IRRATIONALITY OF A THURSDAY

To see the sufferation sicken me
Them suit no fit me
To win election they trick we
And they don't do nuttin' at all.
—Damian Marley, "Welcome to Jam-Rock"

I

ON THURSDAY MORNING, SEPTEMBER 1, President George W. Bush consented to an interview on *Good Morning America* to discuss the catastrophic sequence of events in the Gulf South. In his talk with Diane Sawyer, which took place in the Roosevelt Room of the White House, the President was low-key. "I fully understand people wanting things to have happened yesterday," he said. "I understand the anxiety of people on the ground. . . . So there is frustration. But I want people to know there's a lot of help coming."[1] At times, he was vague, as when Sawyer asked him to compare the devastation he saw at the sight of the World Trade Center after the terrorist attacks in 2001, and the Gulf Coast after Katrina. He said that New Orleans suffered greater physical damage. It was an unimaginative answer that suggested he hadn't contemplated or processed the full implication of Katrina. "Nine-eleven was a man-made attack," he offered, "this was a natural disaster." The only surprising comment from the President was "I don't think anybody anticipated the breach of the levees. They did anticipate a serious storm." At first, it seemed consistent with his generally tepid remarks that morning. And yet protests cropped up that FEMA had conducted the Hurricane Pam exercise in 2004 and it had indeed predicted a failure in the levee system. The Army Corps of Engineers, worried about potential hurricanes, had also warned of the potential for breaches, making it known that it wanted to enlarge the levees and improve the drainage. These were not small points.

By "anybody," the President was presumably referring to people around him in the government; that is, he was saying that no one in the federal government anticipated the breach in the levees. If that was the case, and the extent of the flooding came as a surprise, then it might explain the total confusion that reigned for the first four days. After all, even a superpower can be caught by surprise. President Franklin Roosevelt had been "caught by surprise" when the Japanese bombed Pearl Harbor. Harry Truman had been half-asleep at his home in Independence, Missouri, when his Secretary of State Dean Acheson telephoned with the startling news that the North Koreans had crossed the 38th Parallel on June 25, 1950. And presidents being "surprised" had antecedents during past natural disasters; nobody blamed William McKinley for being surprised when a hurricane struck Galveston in 1900 or Theodore Roosevelt when an earthquake destroyed San Francisco. For the time being, the American people, thanks to the Sawyer interview, had some sort of answer to their question: the United States couldn't respond any more quickly, because the federal government had been caught by surprise.

However, Terry Ebbert of New Orleans Homeland Security immediately contradicted Bush. "This is a national disgrace. FEMA has been here three days, yet there is no command and control. We can send massive amounts of aid to tsunami victims but we can't bail out the City of New Orleans?"[2]

As time passed, even more pointed evidence emerged to refute Bush's claim of innocent surprise. In January 2006, the *Washington Post* obtained two documents that proved that the White House had been forewarned. On the Saturday before the hurricane struck, FEMA concluded that Katrina would likely cause as much damage as was described in the Hurricane Pam exercise. On Sunday, at the teleconference, NHC head Max Mayfield warned the President that the levees might be topped. More urgently, the White House Situation Room received a forty-one-page analysis early on Monday morning (at 1:47 A.M., to be exact) from the National Infrastructure Simulation and Analysis Center, a component of the Department of Homeland Security, and it specifically warned of breached levees, in addition to other manifestations of damage from the monster storm.[3] With the storm only hours from landfall, the report offered a very small window of opportunity for preparation—but the White House Situation Room was designed to operate under just such tight parameters. Clearly, there was no truth to the President's assertion that no one knew that the levees might breach. The warnings were part of the record, and part of the process of responsible prediction that keeps the White House informed of probable crises.

On Thursday, as Bush talked with Sawyer, the nation was confronting a disaster potentially even bigger than that brought on by Hurricane Katrina on Monday. With little progress for the hurricane victims in sight, Americans throughout the fifty states entered into a crisis of faith, uncertain about anything except that the nation they thought they knew was nowhere in sight. Callous, ignorant, inefficient—the words that described the relationship of the United States to its maimed Gulf Coast were usually used to describe Third World countries in the midst of a disaster. From experience, Americans knew their country to be generous, organized, and absolutely unstoppable in the face of disaster. Even the most cynical knew that America could do anything and do it faster and better than any other country—if it wanted. Watching the scenes of New Orleans still enduring want and danger four days after the hurricane was tearing the whole nation apart. There were only two possible ways to understand what was going on besides the "surprise" alibi: (1) the U.S. government couldn't respond, being woefully weak and confused; (2) it didn't want to respond, being disinterested in the people and their plight.

That Katrina unveiled a very different bolt in the American fabric was apparent to anyone who tuned in to Katrina coverage on television. The same comments were heard over and over in conversations across the country. "It doesn't look like America," people said. "It looks like a Third World country," they said. A common question echoed Ebbert's comment: "How come we could send aid for the tsunami victims on the other side of the Pacific in no time and we can't get anything to the Gulf Coast?" By Thursday, in fact, Katrina had become one of the half-dozen moments in American television that not only revealed events but actually defined the community of millions responding to it on television. It joined the Army-McCarthy hearings (1954), the Kennedy assassination (1963), the Apollo 11 moon landing (1969), the Watergate hearings (1973–74), and the attacks on September 11, 2001.

In each case, Americans had been drawn together by the shared response of living together through a searing event, moment by moment, in the intimacy of their living rooms. We were Americans because we lived together through the sense of emptiness after the Kennedy assassination or the betrayal during the Watergate hearings. In the case of Katrina, the shared response was *shame*. The word was blasted across the headlines of magazines and newspapers. It was at once a confession of culpability and a rejection of the seeming indifference and incompetence that was bringing so much misery to so many Gulf Coast residents. On the day that Americans pondered why the government couldn't or wouldn't do more, representatives of the Bush administration gave no comfort. The President came across on *Good Morning America* as remote, unable to articulate the comforting words to rally the nation in its hour of peril. Former New York Mayor Ed Koch summed up Bush's dilemma: "I learned that people want you out there," he said. "They want you to suffer a little with them, they want you to convince them that you will protect them as part of your family, they want you to be an extension of them."[4]

II

For three days, six-year-old De'Mont-e Love was stuck inside an apartment at 3223 Third Street in New Orleans with his family and some neighbors. From the roof of the redbrick, three-story building, you could see flooded Memorial Medical Center and the torn side of the Superdome roof. All the neighborhood haunts, like the Piggly Wiggly, Chicken-in-a-Box, and the New Mount Era Baptist Church, were inundated with water: so was the shotgun called Smothers's Grocery, where De'Mont-e bought his Big Shot soda. "We stayed on the second floor and the water was almost to the second floor," said De'Mont-e's mother, twenty-six-year-old Catrina Williams. "So we didn't have to move to the third floor. We thought we'd be all right. But we occasionally checked out the roof."[5]

Up until the hurricane, Williams, a Jehovah's Witness, had led a fairly routine life in New Orleans. For six years she had worked the night shift at the Quick & Easy Food Store in Kenner. Her husband, a carpenter, traveled a lot. Her rent was $513 a month. During the days leading up to the storm, everybody lampooned her first name—Catrina. "Yes, they all made a joke about it," she said. "I don't find it too funny. But you know, I would laugh it off because, after all, my name is with a C, not a K. Don't get that wrong, you understand?"[6]

When Katrina hit, Williams's apartment building was packed with people, including her niece Zoria and nephew Tyreek. With no electricity, Catrina started up a portable grill and cooked ground meat, smoked sausage, and chicken, so that, as she put it, "it wouldn't go bad." Cousins, half brothers, aunts, and uncles were all crowded into her two-bedroom apartment, which had become the hub of the entire building. "We started panicking," she recalled, "trying to keep everyone together. At first, I was like, 'We're not going to make it through this here.' And then I put all of that aside . . . you know, stopped thinking the worst and tried to think the better."

Having survived the storm, Catrina turned to the big question on Tuesday, which was how to evacuate. A middle-aged couple from the apartment complex became sentinels, sitting on the roof and waving sheets to try to grab the attention of Coast Guard or National Guard helicopters. There were five mothers and seven children in the building. Before long, everybody started congregating on the fiberglass roof. Catrina brought an umbrella with her to shade De'Mont-e and his younger brother, Da'Roneal, from the sizzling heat. "It was so hot on the second floor," Williams recalled. "At least you could breathe a little bit at night on the roof."

Catrina, as it turned out, was a good helicopter spotter. All day Wednesday, if a chopper was even vaguely heard, she would jump up and down and scream, "Us . . . us . . . us!" She wanted her baby boys out of New Orleans, away from the smelly floodwaters and the violence. "On Thursday, I actually flagged a helicopter down," she recalled. "It landed on the roof. I was like, 'Okay, God answered our prayers.' So we all waited in line, and the Guardsman said he's only going to take the children. When he said that, I freaked out. They only took one mother because she had a baby that was actually disabled. He was one of the seven and he required extra attention, so she felt like she was going to go with her baby. We could not go, they was like if we don't let our kids go, there's going to be a lot of trouble."[7]

Catrina put De'Mont-e onboard and handed him his five-month-old brother. For a mother to give up her two sons to an unknown man in a helicopter was clearly a distressing situation. But the rescuers insisted that they would be back in twenty-five minutes. Williams carefully told her son what to expect and how to act; not to be afraid. He was young but bright as a button, and he listened to everything Mom said. "Don't let him out of your sight," she instructed De'Mont-e, referring to the baby, "clutch him always." Two cousins and three other children were also on the first helicopter flight, making seven kids in all. "It was really loud," De'Mont-e recalled of his chopper experience, "and when I looked down, I saw all the houses under water. The little kids were crying a lot. But I didn't cry."[8]

The National Guard pilot dropped De'Mont-e, Da'Roneal, and the other five children off at the Causeway Boulevard/I-10 junction, a prospect that would have been confusing even for an adult. People were aimlessly wandering around and no one was in charge. De'Mont-e shepherded his flock to a safe spot. Mom had given him the order to be a "little man," so he had to step up and become the protector. Holding tight to Da'Roneal, as Mom had instructed, he told the other children to hold hands. "We formed a circle chain," De'Mont-e later explained. "Nobody was gettin' away from anybody."[9] As the hours passed, his mother didn't arrive; the Guard helicopter had never returned to the roof for the other adults. "We waited for five hours and no helicopter," Catrina Williams recalled. "We thought something terrible happened to the kids. You'll never know the pain we had, having our children stripped from us like that."[10]

Later in the day, a National Guard helicopter picked the I-10 children up from the overpass and took them to Louis Armstrong International Airport, where Robert Clement, a medic with Acadian Ambulance Service, took charge. De'Mont-e carefully gave his name and spelled it out, before reciting his address and telephone number, just as his mother had taught him on the roof. Remembering her instructions, he explained that his father had put them on the apartment roof to avoid floodwaters and looters. Clement took one of the other children, a two-year-old named Gabby, into his arms. He was afraid that their parents were dead. Eventually, a Social Services caseworker from Baton Rouge arrived to take the group of seven children away, loading them into an ambulance. Clement felt ill as he handed over tiny Gabby, fearing for her future. "When I gave her away on the ambulance," Clement said, "I cried all the way home."[11] They were placed under the supervision of National Guard Major Steven Trisler.

When the National Guard helicopter finally came back to 3223 Third Street to pick up a second group, it only took the women; Catrina had to leave her husband behind. "They took us to somewhere on dry land," she recalled. "Some Navy base in the West Bank. We spent the night there, outside on the water, along the Mississippi River. Then they took us to Kenner airport. We was trying to get ourselves situated. We weren't going anywhere until we found where our kids was at." Fortune shone on Catrina at that airport. Across the terminal, she saw both her husband and her father, who had just been evacuated from Third Street. "When we had got to where we was at in Kenner, we just prayed on it and everything," Williams recalled. "Just hoped and prayed we'd somehow find De'Mont-e and Da'Roneal. That's all we could do."[12]

Taken by bus to San Antonio, Catrina Williams and her family were assigned to the Kelly U.S.A. Shelter. She was grateful to be fed, even though the "beans looked like soup" and the lettuce was brown. Williams spent every waking hour trying to locate her boys through the Red Cross, FEMA, the National Guard, and anyone else who would listen. She was frantic. Over a week after Katrina hit, she got a telephone call from the National Center of Missing and Exploited Children. With De'Mont-e's help, social workers there had traced Williams to San Antonio on Sunday, September 4. "They had found my boys," she explained. "They were in Baton Rouge. We hopped in a car and started driving. Our reunion was at Baton Rouge airport. That was the best moment right there. That was all the headache gone, being able to try and pick up your pieces once you're reunited with your kids, all the hugs and kisses. . . . That night there was just unexplainable, the relief right there just to be with my children. I thanked God because I got them back."[13]

De'Mont-e, who had kept the other children safe, entered first grade at Harmony Hills Elementary School in San Antonio. He later received a national award for his courage in protecting the others. Former state senator Joe Neal of Nevada, a political commentator, concluded that the heroism of De'Mont-e Love proved "that a six-year-old demonstrated more leadership than the President of the United States."[14]

With children like De'Mont-e Love feared lost, FEMA in disarray, and the politicians trying to fling blame off their shoulders, one Louisiana businessman, Richard Zuschlag, stepped into the leadership void and saved lives, seven thousand of them. Zuschlag, a Pennsylvania native, living in Lafayette, Louisiana, was cofounder of Acadian Ambulance Service. At the time Katrina hit, Acadian was the largest privately held ambulance service in the nation, with two hundred ambulances, two thousand employees, and seven air ambulance helicopters. "I founded Acadian Ambulance because I thought it was a shame that cement trucks and beer trucks had better two-way radios than what the ambulances had with the hospitals," Zuschlag recalled.[15]

In fact, communications was Zuschlag's obsession, which made him a bit of an oddity in the world of private-sector emergency rescue work. Not only did his company have its own large communications tower in Gray, Louisiana, he had satellite telephones by the thousands and, as *Inc. Magazine* explained, "his back-ups had back-ups."[16] Bold, unflinching, and on the job while FEMA and Homeland Security had a command breakdown, Zuschlag evacuated chronically ill people and newborns from six hospitals and opened a triage center at the Causeway Boulevard/I-10 junction, using his fleet of two hundred ambulances. His helicopters worked as frenetically as the Coast Guard did. "I tried the best I could to keep a medic with a satellite phone on the roof of every hospital," Zuschlag said. "I had a difficult time, because the numbers kept perplexing me. It was one thing for a hospital like Tulane to have 300 patients, but I had no idea that they had 500 employees and another 1,500 family members."[17]

Employees who witnessed Zuschlag in action compared him to General Douglas MacArthur retaking the Philippines. Nobody in the Gulf South had reliable communications but Acadian Ambulance; all those years of preparation had paid off. Joking, in a letter, Zuschlag claimed that the Acadian Ambulance was the "Cajun Air Force," and they were. He was, as *Inc. Magazine* rightfully noted, "the anti-FEMA."[18] Every second wasted on poor logistics infuriated Zuschlag, a Republican who believed Governor Blanco did a first-rate job of keeping the rescue efforts alive. "She was impossible to reach, you only got through one out of eight hundred times, but she encouraged our efforts," Zuschlag said. "One of the reasons that [Acadian Ambulance] may have been successful was because the New Orleans 504 area code got knocked out. Over here in the 337 area code, we kept on going. Nobody could get through to anybody else, so they'd call me."[19]

III

By Thursday, all sorts of "Out-of-Towners"—like James Cardiff of McKinney, Texas, a sixty-one-year-old corporate tax management executive—were arriving in New Orleans to help rescue Katrina survivors. A great admirer of President George W. Bush and of all Republicans for that matter, he was a man of substantive wealth. Reverend Billy Graham was his hero. A deeply committed Christian, Cardiff, a member of the nondenominational Stone Wire Community Church, believed he had a moral obligation to help the storm victims. "I don't like being called a fundamentalist," he said. "I'm not a zealot. I just try to live by the Bible, to not be a hypocrite."[20]

Cardiff was watching TV on Wednesday and couldn't believe that the federal government wasn't sending in the Army and the Marines. He was worried about the stranded. People on life-support machines had only a day to live without electricity. Dialysis patients had only 72 to 106 hours. FEMA had supposedly already sent 25,000 body bags to the region. "So I said to my wife that I can't just do nothing, what kind of Christian would I be?" Cardiff said. "Immediately, I packed a bag, flew to Houston, and rented a huge white van there. I packed it with gallons of water. I kept thinking about what Jesus told his disciples: 'What you do to the least, you do unto me.'" His smartest move was also getting 150 gallons of diesel fuel, the rarest of all commodities in New Orleans. And he had printed an official-looking sign that read RESCUE WORKER. "God has always given me great intuitiveness," he said. "My first idea was to hook up with Harry Heinz in Baton Rouge. He was working around the clock at the LSU shelter, and he showed me a map of where people needed saving in New Orleans, the really bad places below sea level where people were just left. That man worked so hard saving people that his heart went out and he died a few weeks after the storm, once all the forgotten and forsaken were all right."[21]

Upon arriving in New Orleans on Thursday, he stayed at the home of Chaplain Henry "Hy" McEnery III on Laurel Street in Uptown. Besides being a chaplain, McEnery was a sergeant in the Mississippi National Guard 20th Special Forces Group 2nd Battalion. A warm, engaging man, large in stature, and a good talker, he was bitter that decades ago the Louisiana governor had abolished the Louisiana Special Forces. So, in protest, he simply joined Mississippi Special Forces. What he had in common with Cardiff was a penchant for humanitarian relief, and the Maker. They were on a mission to save lives. "I love him, but he talks too much," Cardiff said of McEnery. "But we made a good team."[22]

Their first baptism by fire occurred when Cardiff heard that a seventy-nine-year-old man named Alcede Jackson, in their neighborhood had died in his easy chair on August 31. They found the nearby house where his corpse was, offering their services to perform a funeral. The widow, a distraught, extremely ill Violet Jackson, received them, in desperate need of dialysis. The problem was she wasn't going to abandon her husband's body, which was decomposing in a chair like a ghost from William Faulkner's short story "A Rose for Emily." You could smell the rot. A mentally challenged son was also in the house. "How can I help you to leave?" Cardiff asked Violet in the most reassuring voice imaginable. "How long were you and Alcede married?" She broke into a soft, helpless smile, as if she were an orphaned child. "Forever," she said, "forever."[23]

A negotiation began. Cardiff promised to give Alcede a proper funeral, to wrap him up in blankets, along with some plastic garbage bags to stop the insects from congregating on the corpse. They would hold a proper Christian memorial service on the porch. She, in turn, after the service, would allow McEnery to rush her to Ochsner Hospital. After some easy arm-twisting, she agreed to these terms.

Lifting the corpse was tough duty, gas being dispensed in a long smelly run. The odor was atrocious. "By coincidence, when I was taking provisions for a stay earlier, I had grabbed all the roses that the looters had left behind," McEnery said. "Now I had them for the service. And I made a homemade headstone with a quote from John. So we had a front-porch service." Refusing to drop anybody off at the Superdome, the Convention Center, or I-10, because they were "abominable" according to Cardiff, they loaded Violet into McEnery's car, along with a few other sick people and whisked them off to Ochsner. While McEnery was going to get Violet dialysis, Cardiff was going to look for food and a rescue route to Memorial Medical Center, where he heard that doctors, nurses, and patients were stranded. Unfortunately, Ochsner couldn't take Violet or the others; the hospital had a huge overflow of patients. What to do? McEnery decided the only other option was Louis Armstrong International Airport. "The airport was crammed with people, medical teams working in rapid-fire fashion to help the truly sick," McEnery said. "I rushed up to a doctor and told him that this woman, Violet, was going to die if she didn't get dialysis now."[24]

Without even a second of hesitation, the doctor dropped everything he was doing, leaving dozens waiting in the triage line. "She's bad," he said. "But I'll take care of her. She'll be okay. You can go." McEnery had grown attached to Violet. He hated to just leave her at the airport, but he had a Christian rescue calling, and there were more New Orleanians to save. "I wondered about what had happened to her," he recalled. "And then weeks later, out of the clear blue sky, a special nursing center gave me a courtesy call from Alexandria, Louisiana. Violet had just come through, the nurse told me. She was all right." As McEnery said, "What a blessing."[25]

Meanwhile, Cardiff went to Interstate 12 and somehow found a food warehouse. He bought bologna and peanut butter and passed it out among displaced persons along the roadside. Then it was on to the corner of Napoleon Avenue and St. Charles, the driest place to park if you wanted to get to flooded Memorial Medical Center. Money once again spoke. He essentially rented a johnboat to take him to the hospital. He was relieved to see a group of Acadian helicopters evacuating doctors and nurses from the rooftop. But all around the hospital, in the streets of Central City, people were begging to be evacuated. For the next week Cardiff would save hundreds of New Orleanians by boat, living the teachings of Jesus. "There was a hateful NOPD officer who tried to stop my efforts," he recalled. "But I wasn't going to let the Devil stop me from doing what was right."

Asked why he'd driven *into* New Orleans from Texas when martial law dictated that he stay out, Cardiff had a pat retort. "How could you leave all those poor people behind?" he asked. "How could you have resources and not help?"[26]

IV

For those who were rescued from their garrets in New Orleans, plucked out of the mire, or picked up off trash-strewn streets, the sense of relief was overwhelming—and short-lived. First responders had neither the time nor the means to offer any further comfort to Katrina victims. Though it sounds callous, survivors were dropped off as quickly as possible on the nearest dry land and left to fend for themselves. Only those in *very* grave medical condition were given expedited transportation out of the dragonfly zone to Ochsner or Armstrong International Airport. For thousands of others, the struggle to survive continued and sometimes failed. "I stayed in the control center for eight straight days, working sixteen hours a day in Lafayette," Zuschlag recalled. "To my knowledge, we got everybody out who was sick. We didn't turn down nursing homes. We did turn down some individual requests from people who needed out. But we just couldn't do it; we didn't have enough manpower."[27]

That Thursday, in the rest of the country, while New Orleans suffered,

Americans went about their business. The Boston Red Sox began the day two and a half games ahead of the New York Yankees in the American League East Pennant race. The debate over Supreme Court nominee John Roberts heated up, with the NAACP, the Mexican American Legal Defense and Educational Fund, and the National Women's Law Center coming out against him. His Senate Judiciary Committee hearings were slated to begin the following Tuesday. The previous day 950 people had been killed in a stampede at a Shiite pilgrimage in northern Baghdad, caused by the hysteria of people claiming a suicide bomber was in the crowd. The number one box-office movie was the raunchy Steve Carell comedy, *The Forty-Year-Old Virgin.* But in New Orleans and its vicinity, the days of the week ceased to have meaning. Monday was the day of the hurricane; that was the day no one would ever forget. But Tuesday, Wednesday, and Thursday were eerily all the same to the people trapped in the devastation left by Katrina. Louisianans lost all sense of time, as dozens of rescues turned into hundreds and then thousands. Governor Blanco continued finding evacuation buses from parishes like Sabine, Bossier, and Natchitoches. Familiar Louisiana sights like Six Flags (New Orleans East), Ruth's Chris Original Steak House (Mid-City), Dooky Chase's Restaurant (Tremé), and the New Orleans Fairgrounds (where the New Orleans Jazz and Heritage Festival took place every spring) were half-submerged in water. Logic was suspended and no explanation justified the devastation that people saw in New Orleans.

The most stunning examples of surreal New Orleans were the sights that greeted survivors looking for respite from the storm and its aftermath: the hospitals, the bridges out of New Orleans, the Superdome, and the Convention Center. In normal times, each was nothing if not reliable and strong. Much of Interstate 10, which ran east-west across the heart of the city, was an elevated expressway. In Center City, the highway was flooded in places, but for people on the northern and western sides of town, the roadway was dry land. Once they climbed up an access ramp, they joined the throng walking along the empty lanes. The highway led west through Jefferson Parish and eventually to Baton Rouge. The march started Tuesday and was still strong on Thursday, yet few buses arrived to help; few trucks brought water or food. On the other side of New Orleans, to the east, I-10 was now a dead end: the road had crumbled away at a bridge crossing the Industrial Canal.

Before the storm, New Orleanians who chose to remain in hurricane-proof, high-rise buildings, like the Hyatt, Sheraton, or Hilton, called their strategy "vertical evacuation." The same might be said of those newly homeless who found themselves on the elevated roadway that was all that was left of I-10 in central New Orleans. They climbed up because they would be dry there and because they would no longer be suffering alone; they would be part of a community. In the absence of any sense of protection from the government, there was some small security offered by association. Rickey Brock managed to get his family out of their flooded home on Dorgenois Street in the eastern part of the city, with the use of a rowboat. They headed for the Superdome but found it closed to new evacuees. For two nights, the Brocks slept nearby, under the interstate. Driven away by the corpses lying all around them, they climbed a ramp to a section of I-10. The bodies, Rickey Brock said, "smelled so horrible we came up here."[28]

Ten-year-old Ernest Smith tried to comfort his elderly grandparents as they sat in the trash and dust of the shoulder of the road. Eventually, the three of them hoped to reach Atlanta, where Ernest's mother was living. The boy was mature for his age, and reflective about the circumstances, but he doubted that they would ever see his mother again. "I would tell her," he told a *Times-Picayune* reporter, looking around the road that had been his home for days, "I love you. Please come get me, and I don't want to be out here no more."[29]

Kache Grinds, an eleven-year-old, had been on the interstate for two days with her grandmother, who was nearly blind. They had been evacuated from a housing project and deposited there. On Thursday, she was standing at the end of the water, where the highway descended into the flood, trying to gain the attention of the emergency rescue boats that occasionally passed by. She was going to offer them her life savings of six dollars to take her grandmother to safety, but none of the boats paid her any attention.[30] Babies were getting sick on the interstate; elderly people were dying of heart attacks or strokes, but very little help arrived.

The heat of Thursday was so brutal that Ceci Connolly of the *Washing-*

ton Post reported seeing "one group of more than a dozen stretched out in single file across I-10 to squeeze under the narrow band of shade from highway signs."[31] The people huddled together, making a home of the barren roadway, holding their belongings in a pillowcase or on a cart, looked like refugees from some war-torn area overseas. The term "refugees" offended some Americans, though. Because the crowds awaiting rescue were overwhelmingly black, the term seemed to imply that they were strangers in their own land. Indeed, they were refugees, in that they sorely needed refuge of some sort, but in the American perception, shaped by centuries of good fortune, refugees were necessarily foreigners. Americans *gave* refuge; they didn't have to seek it. Katrina changed that. A whole population fit the description, if a refugee was someone who had nowhere to go—but went anyway—and no one to care about them. And so, although Reverend Jesse Jackson said, "It is racist to call American citizens refugees," it was only the truth.[32] Oprah Winfrey led the anti-"refugee" crusade. "I think we all in this country owe these people an apology," Winfrey said. "We still don't know how many of our fellow Americans lost their lives in the Katrina catastrophe. . . . They are not refugees. . . . They are survivors and we, the people, will not let them stand alone."[33]

Truth be told, they were evacuees and refugees and survivors. All three. And racism was in play, to some degree. If thousands of storm-ravaged citizens were stranded in Boston's Back Bay, caught on some portion of the Massachusetts Turnpike during a flood and if they were white, you can be sure it wouldn't have taken days for them to be evacuated. Whatever the conditions, it wouldn't have taken officials four days to rescue them. With that assumption—that there were certain populations in America that would not have been left in primitive conditions for so many days—the question is whether the I-10 gathering was ignored because it was predominantly African-American or largely poor. A third possibility was that the treatment stemmed from the fact that the hurricane occurred in the state of Louisiana: rescue was complicated enough in isolation, but apparently impossible when it was at the mercy of an unresponsive White House.

One can conjecture whether a poor white crowd would have fared better in New Orleans or whether the black population of another state would have been treated so shabbily. For those on the I-10, there was no question about why they were being left to fend for themselves. Watching one another weaken in the heat, many of them came to believe that, when push came to shove, they were regarded as second-class because they were black. The truth was not in the comforting supportive words of the government but in its deeds. Thousands of African Americans were cast back to primitive conditions, and in the scheme of the response to Katrina, the attitude seemed to be that it was all right that way. Four days later, the President's parents, George H. W. and Barbara Bush, visited Reliant Park in Houston. They found conditions that were clean and orderly, but hardly homelike. It was a goodwill tour that was much appreciated until Barbara Bush expressed her opinion that "so many of the people in the arena here, you know, were underprivileged anyway, so this, this is working very well for them."[34] No one would blame the President's mother if she misspoke in the fray of an interview, but many people leapt to her defense. And that indicated hers was a broadly held opinion and therefore worth pondering, in view of the racial questions raised by the slow response to Katrina.

"Anyone who has visited the most deprived parts of America's cities," wrote Gerard Baker in *The Times* of London, "rather than merely empathized with them from afar, would have no difficulty whatsoever with the proposition that the inhabitants would prefer an air-conditioned sports stadium with all the food they can eat, the country's best medical attention and the benign security of National Guard protection to the hunger, sickness and lawlessness in which many of them live."[35]

No one would deny that in the poorest neighborhoods of New Orleans—and other American cities—conditions were poor. Nonetheless, the underlying implication went too far, holding that since Katrina's displaced people were "underprivileged anyway," they looked upon life in the shelter as a step up—that they were grateful for the orderliness of their new life. That line of thinking turned them into something far worse than displaced. They were perceived as beggars. Jack Shafer, a first-rate columnist at Slate.com, came to Mrs. Bush's defense, saying, "She probably got it right. The destruction wrought by Katrina may turn out to be 'creative destruction,' to crib [economist] Joseph Schumpeter, for many of New Orleans' displaced and dispossessed."[36] Gaining something more than they "deserved," they were supposed to be elevated to a higher way of life, and to be grateful for it. That attitude stripped the refugees of their dignity, first by presuming that all of them were the same, and then by implying that being "under-privileged," they lived in squalor and were used to it. Standing around on an interstate for four days was just more of the same. When they got more, it would be in the form of largesse on the part of the government, providing *clean* beds, *hot* meals, and *safe* conditions.

As he had the last few days, Brian G. Lukas, the chief cameraman of WWL-TV, set out early Thursday to film the grim conditions on I-10. Everything now was making him angry, disgusted, and almost physically ill. In his diary he described the displaced persons crisis he encountered that week.

> Causeway and I-10 has become a massive staging area for the evacuees for New Orleans. There are thousands of people here. . . . In this heat, they are just trying to survive. Many are looking for displaced family members . . . all are suffering terribly. They seek the camera out crying for help. Many are old . . . many of the evacuees are sick. This is a makeshift refugee camp . . . the sanitary conditions are deplorable. I cannot believe that I am in the United States of America . . . the richest and most powerful nation in the world. Government—local, state, and federal—have failed these people. Helicopters are landing and scatter the debris. . . . The helicopter wash [also] provides a momentary relief from the heat. A lady is crying to me . . . telling me that they are not being treated as human beings and asked why doesn't anybody help us. . . . This was supposed to be an area of rescue . . . they were going to be bused out of the area, but there are too few buses. Instead of the hundreds of buses promised, they came in groups of five or six. There are just too many people so they have to wait in the hot sun. . . . Helicopters shuttle some of the sick out . . . many of the elderly are rolled from their wheelchairs into the transport door of troop-carrying helicopters. However, other helicopters are [bringing] additional evacuees into the area. This is . . . tough to witness.[37]

There was no doubt that the urban poor and the African-American population of New Orleans had social problems. But what outsiders forgot all too easily was that those forms on the overpass weren't abstractions, they were families. Jerome Wise, standing in despair on the I-10, had saved only one possession from his house near Hayne Boulevard, a few blocks from the lake. He didn't take food or money, but a photograph of his family, who had evacuated in advance. Jerome stayed behind to be with his son, who had to work a shift at McDonald's over the weekend. After four days, Wise and his son were still separated from their family and without them, he felt he had nothing at all. In that, he was no different from the vast majority of other refugees—and other Americans of all races and incomes. "I can't go home no more," Wise said. "Nobody wants us. Nobody wants to help New Orleans."[38]

Oprah Winfrey wanted to help these people. More than any other television personality, her appeals were full of deep emotion; she communicated an almost tangible pain to tens of millions of viewers. "Just as the storm moved off the coastline, pretty soon these stories will no longer be a part of the headlines of our lives," Winfrey said. "I feel deeply that we owe it to every single family who has suffered to not forget and to not let them stand alone."[39]

V

The other principal highway west out of New Orleans was U.S. Route 90, which sprang out of I-10 just north of the Superdome and then wended its way south out of the city. Crossing the Mississippi River at the 3,000-foot twin bridges known as the Crescent City Connection, it led to the West Bank city of Gretna. First settled in 1836 by German immigrants, it was originally named "Mechnikhan." Gretna's historic district included numerous nineteenth-century traditional Louisiana state structures, like the Kettie Strehle House and the German-American Cultural Center on Huey P. Long Avenue. On Thursday, a group of two hundred dehydrated refugees walked along Route 90 to the bridge, on the way to Gretna, where hurricane damage was minimal. Electricity was out and water was unreliable at best in the suburb, but the streets were dry and the buildings largely intact. For people from New Orleans, it may as well have been the promised land. If they made it to Gretna, food and water would be plenty. According to Gretna Mayor Ronnie Harris, more than six thousand evacuees tried to flood across the Crescent City Connection begging for help.

The group included many residents, as well as displaced persons from New Orleans hotels, most of which closed abruptly on Wednesday. One could hardly blame the managers: without sanitation and plumbing, a hotel can maintain living conditions for only a day, at best. A San Francisco couple named Larry Bradshaw and Lorrie Beth Slonsky were in town for a paramedics conference when the storm hit. On Wednesday, they were told to leave their hotel. They were a resourceful group. With the other guests, they collected $10,000 for transportation and called an out-of-town bus company, which sent ten buses. On the way in, however, the fleet was commandeered by the National Guard. The guests were disappointed, of course, but accepted the fact that severely ill people at the Superdome needed the transportation even more than they did. "The thing that gets me," Bradshaw said, "is that if we could get on the phone and get ten buses, why couldn't FEMA make that call?"[40]

On Thursday afternoon, the group of hotel guests had been turned away from both the Superdome and the Convention Center. A few of the hotel guests took the advice of an official who suggested they walk across the Crescent City Connection to Gretna. Bradshaw and Slonsky were in the group, which was described, according to an estimate by one member, as being 95 percent black. As they walked onto the bridge, two police officers from Gretna blocked their path, brandishing guns. As a few members of the throng approached to inquire about the problem, the officers discharged the weapons, firing over the heads of the people before them. Bradshaw put his hands up, holding his paramedic badge where the police officers could see it. By way of explanation, he was told only that "there would be no Superdomes in their city."[41] That was shorthand for the fact that there would be *no disorder* in Gretna. But the people on the bridge were quite obviously harmless: frightened tourists with their luggage and garment bags, parents with babes in arms, and elderly people, some in wheelchairs. No one was allowed to come into Gretna—not from New Orleans. That was the direct order from Mayor Harris. Physically exhausted, and with nowhere to go, the group of refugees set up an encampment on the edge of the bridge, surrounded by liver-brown sludge. They thought they would spend the night there and try to cross again the next day. The Gretna police came back at the group in force. "Get the fuck off the bridge," the police shouted. When the people didn't respond immediately, the officers pointed their weapons at individuals.[42] Ultimately the Gretna police called in a helicopter, which hovered low over the encampment, its rotor blades blowing the makeshift shelters away and pelting the people with dust and debris kicked up by the downdraft. The group retreated to New Orleans.

Sheperd Smith, anchor of *The Fox Report,* was ideally situated to witness the Gretna Bridge Incident. He had arrived in New Orleans on Sunday, and he and Jeff Goldblatt had broadcast live from the French Quarter once the Katrina winds died down. Deemed by a *TV Guide* poll as America's "second most trusted news anchor," Smith had logged a lot of miles as a reporter, covering Operation Iraqi Freedom, 9/11, the Columbine school shooting, and President Clinton's impeachment trial. Nothing about him was squeamish; in fact, he served as a witness to the execution of Timothy McVeigh in June 2001. But nothing he ever witnessed affected him like the Gretna Bridge Incident. As a Southerner, Smith knew instinctively that the remnants of Jim Crow still flourished in the Gulf South. While other TV commentators were describing the Superdome in racist terminology, Smith balked; it was more about urban poverty and bad planning than anything else. "I was based in the Royal Sonesta Hotel," Smith recalled. "When the 17th Street levee breached we were doing our reports from the balcony. I couldn't get through to New York except by text message."[43]

When the hotels were evacuated, Fox News decided to move their satellite trucks to the Crescent City Connection. Not only could they see the Superdome and the New Orleans skyline from there, but the vantage point afforded them a bird's-eye view of the Morial Convention Center, which had suddenly, due to NBC's heart-wrenching footage, become another focal point of the disaster. The Gretna Bridge Incident happened on Fox's doorstep. "Suddenly these cops came driving over on the bridge," Smith recalled. "We saw them blocking the road. Rumors were flying that there were armed gunmen roaming about, and we did hear gunfire. So I assumed these police officers were tracking down some criminals. There were helicopters flying overhead."

Smith, in his good-natured way, walked up to the Gretna police. "What are you guys doing?" he asked. To which the Gretna officer said, pointing to a group of largely African-American refugees trying to cross the bridge, "We just want to keep them fucks out of there." The Gretna cops were furious that much of the Oakwood Center Mall, located on the Gretna-Terrytown border, approximately a mile from the bridge, had burned Thursday afternoon. An unidentified group of thieves had scaled onto the roof using a ladder. Like cat burglars, they entered through a blown-out vent. They piled up store merchandise and lit a bonfire. Using the blazing fire as their light, they ransacked Lady Foot Locker, Underground Station, and Adler's among other stores. The fire soon spread out of control. It raged for hours. Jefferson Parish firefighters, with inadequate water pressure, futilely fought the inferno. Crowds gathered to watch the mall smolder, among them a group whom Parish Councilman Chris Roberts called a "massive congregation" of refugees from the Superdome and Convention Center. They had walked across the Crescent City Connection earlier in the day.[44] The Gretna cops may not have *known* who had broken into the mall and started the fires, but these New Orleans outsiders—nearly all African-American—had raised the cops' suspicions. It was under these tense circumstances that a perplexed Smith asked, "Are you telling me that these Convention Center people can't go over the bridge? That you're forcing people to not evacuate?" He got an affirmative nod. "That's absolutely crazy," Smith said.[45]

Smith was truly "dumbfounded." A short while later Fox's Geraldo Rivera, reporting from the Convention Center, confirmed what Smith had reported: stranded New Orleanians were "not allowed out" of the bowl. That day Smith became one of the leading American voices of indignation. "A continuing cycle of gloom was occurring," he recalled. "The Gretna Bridge thing really ticked me off. . . . The next day the Gretna police offered to show us the mall where the police substation had been burned. The last thing we had time to do was go to the mall."[46]

The next day, on Friday, an unrelated collection of people had the same idea: to use the Crescent City Connection as a route out of New Orleans. Jill Johnston, a Canadian health care executive on vacation in the Big Easy when the hurricane struck, was among those who made the long trek to the bridge. "We all walked the five or so miles up onto the bridge," she said, "and the police start shooting at us."[47]

Mayor Ronnie Harris and Police Chief Arthur Lawson Jr. of Gretna did not try to deny that they had told the police to use necessary force to keep people from New Orleans from using the bridge. They claimed that their own city was in a lockdown, as part of a mandatory evacuation order, and that they had no provision for handling refugees. In addition, they seemed to resent the fact that no one at City Hall in New Orleans had even tried to coordinate movement of people with them. "All of this was crashing down on all of us who were in charge," Harris said to CBS News. "[We] had to make decisions in a crisis mode." In any case, they deeply resented being labeled as racists for not allowing the (mostly black) refugees into their city, which has a racial mix of about 60 percent whites and 40 percent blacks. "It was the right thing to do at the time," Harris told MSNBC's Brock Meeks. "The charges of racism in my community are absolutely off the mark, absolutely rubbish, and that's a nice way for me to say it, okay?"[48]

Perhaps Gretna's officials were not at all racist. One might as well take them at their word on that. And perhaps they did not have much in the way of water, food, or personnel. That was possibly true as well. Yet that was not the issue. The point was that they had far more provisions than those who came to them on the bridge. If they thought about it, then of course, they came up with somewhat logical reasons to block the way. The impulse to help would have come from the heart, long before the thinking began. That Gretna's officials, with the support of the residents, condemned their fellow humans, of whatever color, back into the unlivable conditions of New Orleans gave yet another biblical cast to Katrina—the unwanted New Orleanians walked back off that bridge, wandering like refugees in the book of Exodus.

The Gretna Bridge Incident was a clear example of callousness. Sadly, it wasn't the only one. Governor Blanco inexcusably ordered National Guardsmen to other bridges and points of exit around New Orleans in order to keep people from walking out. The refugees inside the city could see lights and dry land across the river, but the guards were keeping them from attempting to leave. As the days passed and the week wore on, many believed that they were being held prisoner and that the government was trying to kill them. Neither conclusion was entirely unreasonable under the circumstances.

VI

Many of the people who tried escaping to Gretna had already trod the route of the hopeless in New Orleans: dropped off at the Superdome or told to go there, they found that it was closed to new entrants. The next stop was typically the Morial Convention Center. Only after seeing what it was like, inside and out, did some determined people keep plodding on, in search of safety in Gretna.

The Convention Center offered shelter, if endless rooms running with open sewage can be termed "shelter." Crime was a fact of life—and death—there. Dereece Bailey, a medical technician, was at the Convention Center on Thursday. "In one of the bathrooms," she recalled, "you had a little girl. She could have been maybe thirteen. Her neck was slit. Her clothes were like ripped, so she could have been—I can say could have been raped. But she was lifeless and a lot of people were saying that she had been in the bathroom for maybe a day or two."[49] While stories of violence and crime at the Superdome were either exaggerated or fabricated altogether, the horrors at the Convention Center were definitely not exaggerated. Corpses were left in plain sight both inside the building and outside, as Tony Zumbado of NBC had documented.

Many survivors, like Natalie Rand who lived on Melpomene Street in the Tenth Ward, have vivid memories of what transpired. The National Guard arrived at her door on Thursday and told her, guns pulled, "Get out now!" They gave her family a lift to the Convention Center—just dumped them off. No water or MREs, just the hot pavement. She had her three teenaged cousins with her. Early on Saturday morning, "around 2:00 A.M., for the first time in my life, I saw a man get shot," Rand said. "A NOPD car pulled up to the Convention Center, shining a light on us. They had broadcast not to move. A scared man simply moved and they shot him dead and drove off. They left him on the ground; it was a drive-by shooting. Somebody put a blanket over his body."[50]

The dead man Rand was referring to was perhaps forty-five-year-old Danny Brumfield, lean and taut like a young Chuck Berry. Since Tuesday he had camped outside the Convention Center protecting his five grandchildren from lurking drug addicts. Rumors of rape were widespread and Brumfield knew that if one of the kids wandered off, he might never see him or her again. He didn't need to learn from the media that more than 2,000 registered sex offenders in the Gulf South had disappeared, their whereabouts unknown; he could sense it. A nondrinking, decent family man, Brumfield was a success story who had risen out of the Desire projects, married his high school sweetheart, and worked in construction. "The lady next to us, she had like ten to fifteen children with her and nobody to help her," Brumfield's daughter Shantan recalled about their nights stuck outside the Convention Center. "He [Danny] helped her feed the children."[51]

Brumfield had built his house with his own hands, so he didn't want to abandon it for Katrina. When the floodwaters rose inside, however, he was forced to the second floor with his wife, Deborah, and his diabetic mother-in-law, Ruby Augustine. Like so many others, this was followed by an escape through a hole in the roof. The Coast Guard came and did their job just right. They plucked Deborah and Ruby from the roof and eventually transported them to Reliant Park in Houston. But Danny refused to leave New Orleans without his daughter, Shantan. That was a hard fact. The family grapevine said she was at the Convention Center, so that's where he went to look. He found her in front with her kids huddled around her in the stifling humidity. Next to them was a man with infected ulcers on his legs, the pus so bad all you could do was turn away in horror. Shantan later talked about the "screaming" that was occurring outside the Convention Center starting at around 1 A.M. the evening her father was shot. Others use nouns like "wails," "groans," and "shrieks." Clearly the Convention Center evacuees, tired and thirsty, were cracking under the strain.

At that moment, on cue, like in some X-rated version of *Cops,* NOPD officer Ronald Mitchell and his partner arrived on the scene in a squad car. Because Danny Brumfield had become a Convention Center leader, a reassuring magnet of strength for the weak, he started walking toward the police car. His primary motivation was to get a handful of people medevaced out of the area before they died. To gain the attention of the police, he started walking right in their headlights. Brumfield wasn't going to let the police car leave the Convention Center area without seeking emergency help from the officers. Somehow he had to convey to the NOPD that *people were dying* while they hovered around Harrah's Casino like disorganized pigeons. According to family members, the patrol car slammed on its brakes, actually hitting Brumfield, who quickly put his hands on the hood in a quasi-spreadeagle to demonstrate that he was not armed. The squad car backed up and then hit Brumfield a second time. Again he fell on the hood. Responding reflexively, he pulled a pair of garden shears his niece Africa had given him earlier for self-protection. This gesture triggered a lethal response. The two officers, apparently angry and scared, felt threatened. Officer Mitchell shot Danny Brumfield dead. One minute, an evacuee leader, the next, another corpse in front of the Convention Center oozing blood. The NOPD dismissed family complaints as being emotional. Brumfield had lunged at them with garden shears; they had a right to protect themselves. Point in fact, Brumfield had obstructed official police business. That was the public police line. What kind of society would it be if pedestrians would just stop police willy-nilly with a "brick-wall" hand. As of April 2006, an investigation of the circumstances of Brumfield's death was ongoing.

A wave of despair swept over the crowd following the shocking death of Danny Brumfield. Nobody knew what to do. Africa stepped up, checking her uncle's pulse and then writing his name and birthday on gray duct tape. She wrapped the identifying tape around his right arm. One family member, in the midst of the shrieks, had enough composure to write out biographical details about Danny, about how he was a great father and a hard worker. The note was then sealed in a Ziploc bag and taped to his body, which was now lying in a pool of blood. A stranger found a blanket and placed it over him. Too much blood for young girls to see. Prayers were said. The grandchildren were shooed away. As the night wore on and Danny Brumfield's body started decomposing, maggots arrived. In the morning the humidity increased, and as the sun rose toward its zenith, gnats and flies landed on the exposed surfaces of his brown skin. At 10 A.M. Saturday, buses finally showed up at the Convention Center, three blocks from the corpse. The Brumfield clan had to evacuate; they had to leave Danny Brumfield behind. "My mother walked up to him," Africa Brumfield later told the Associated Press. "She said her good-byes there. She said as much as we didn't want to leave his body there, we really didn't have any other choice. Then we ran because our lives depended on it."[52]

On another occasion a few NOPD officers were standing around the Convention Center, guns drawn, scanning the crowd. Desperate to get a bus out of New Orleans, Conia Sherman approached them for transportation information. She had her young son at her side. Surely they wouldn't shoot a woman with a child? "Excuse me, Officer," she said, "do you know when the buses are—" Before she could finish, he cut her off. "Fuck you, bitch," he shouted, pointing his gun at her head. "They told you hos to get out and you didn't listen." Sherman meekly offered, "Officer, I'm just worried about my boy." The officer shot back: "You better keep your nigger boy with you, because if I had my way, I'd shoot you."[53]

Inside the sprawling Convention Center complex, gangs marked out territory and defended it with guns purloined from stores. All the while, they were drinking liquor (stolen as well). And yet, while some of the gangs were the source of violence, something unexpected happened to others. They realized that there was no authority—or that they were, by virtue of their very organization, the only authority in the whole building. In the human drama that was unfolding at the Convention Center, they stepped up to protect the weak. Denise Moore was an eyewitness: "They were securing the area," she said. "Criminals. These guys were criminals. They were. But somehow these guys got together, figured out who had guns and decided they were going to make sure that no women were getting raped, because we did hear about women getting raped in the Superdome, and that nobody was hurting babies. And nobody was hurting these old people. They were the ones getting juice for the babies. . . . They were the ones getting clothes for people who had walked through that water. They were the ones fanning the old people, because that's what moved the guys, the gangster guys, the most, the plight of the old people."[54]

It rankled Moore to hear the gang members at the Convention Center called "animals" in later press reports. She didn't see anyone from the government—city, state, or federal—trying to help anyone there. The help of the "good" gangsters didn't go very far, though. Most of the young men who were armed were threatening one another or anyone who got into the way. People were knifed, robbed, and sexually molested.

The Louisiana National Guard had arrived at the Convention Center on Tuesday, not long after those people turned away from the Superdome had broken into the riverfront facility. The 247-troop detail were, in effect, squatters, just the same as the other intruders. Representing the 769th and 527th Engineering Battalions, they were specialists in debris removal. They did not even try to restore order; they ducked into a hall and locked the door, appearing only rarely after that.[55] Perhaps they couldn't have maintained the law in the entire facility, but they could have laid out some section as a safe zone, without weapons or violence. Instead, they did nothing. On Thursday, they realized that the place was out of control, and they left for good. The estimated 15,000 people inside—and the 5,000 in the area surrounding it—were left to their own devices for food and water. Late that morning, help seemed to arrive in the form of a SWAT team that entered the building in formation, weapons drawn. Those who thought that they represented the forward detail of a peacekeeping force were disappointed, however. They were only looking for one person in particular, the wife of a Jefferson Parish government official. The woman had sent out word that she had ended up at the Convention Center with a relative. With the SWAT team as an escort, the pair were taken from the squalor and any hope that order would be restored exited too. It did not help that the two women so highly valued by the SWAT team were both white.[56]

People who were afraid to go inside the Convention Center, or disgusted by the fetid atmosphere in there, hovered around outside, surveying the desperate masses along the street. Even there, the surroundings had become so filthy, without sanitation, that when night came, it was impossible to find a clean place to lie down. Every so often, an official with the National Guard would yell for the people to line up for the buses. All the while, they were told that if there was any pushing whatsoever, the bus would be waved on. That was a good rule, for the sake of order. Except for one thing: there weren't any buses. During the first four days, none came. And yet the people, exhausted and ill, had to keep getting into long lines. Some people realized why: it was a cheap, backward way to maintain dominance. If officials let on that there were no buses coming, then the crowd might become truly uncontrollable. The drill of lining up and then sitting them down was a means of holding power over the helpless.

According to Convention Center general manager Warren Reuther, there were anywhere from 18,000 to 25,000 evacuees who had congregated at the center. Nobody was providing food or water. And although Michael Brown of FEMA claimed these storm victims were receiving medical attention, he was misinformed.[57] "It was as if all of us were already pronounced dead," said Tony Cash after three nights there. "As if somebody already had the body bags. Wasn't nobody coming for us."[58]

Eight of those gathered at the Convention Center were Ivory Clark and his family. All Wednesday night, Clark guarded his clan, who were forced to sleep on the curb in front of the Convention Center. For good luck he kept rubbing a crucifix he had purchased at Mr. Goldman Jewelry. After seeing a couple of corpses wrapped in blankets, perhaps one of them that of Danny Brumfield, Ivory decided he must move his family to a safer location. On Thursday he found a quasi-safe haven at the Harrah's Casino parking lot. Grabbing a few cardboard boxes, Clark ripped them in quarters, using them for his kids to sleep on. He had found Grandma a wheelchair, but Auntie was still in desperate need of an electrical outlet for the nebulizer that kept her asthma under control. "Things were real bad," Clark maintained. "We hadn't eaten in a couple of days. So, as the man, I had to go find food."[59]

Clark, with his son, Gerald, clinging to his side, decided to walk down the trolley tracks a short distance. Clark worked as a cook for a family at One River Place, a high-rise condominium complex perched on the Mississippi River. An NOPD beat cop tried to stop him from entering. "Look, I live here," Clark said, even though he didn't. "I'm just trying to get back to my condominium." The cop eyed him suspiciously, pointing his rifle at Clark and Gerald. Just then two of the One River Place guards recognized Clark from his days as a private chef there. "Hey, Ivory," Robert Celestine shouted down. "Come on up." The other guard, Chris Turnball, waved the guns away and invited Ivory and his son to trudge up the hill to the lobby of the high-rise.

Both Celestine and Turnball put their shotguns down and hugged Clark. During the pre-Katrina years, they used to stand around the lobby and gossip about the Saints. Now they were engaged in a combatlike situation, swapping very different kinds of stories.

Before long the two security guards waved Clark and his son into the lobby, where they took the elevator to the fifteenth floor. A generator was supplying basic power to the building. Clark had a key to the suite. Five nights a week he had cooked everything from lamb chops to crawfish etouffée for an elderly Jewish couple. They had gone to Europe for a long cruise, leaving Ivory behind as caretaker. For a ravenous man, the sight of two huge freezers of food was like finding a diamond mine. Pure ecstasy—and habit—took over. He immediately began cooking up the perishables. He and Gerald tore the cellophane off the ground pork, knockwurst, bacon, and hot dogs. Using the gas stoves, Clark started cooking like a fiend. Bottles of Perrier, Coke, Tab, and orange juice were also rounded up. "After cookin' it all up with Gerald, we headed back to Harrah's," Clark recalled. "I was real proud."[60]

"Wow!," "Holy cow!," and "Look at that!" were the kinds of greetings Clark received when he started passing out home-cooked food first to his family, then to the others camping out in Harrah's parking lot. "We ate well," Clark said. "Friday morning, I went back up to One River Place, climbed the stairs, and made a huge batch of grits. We had it goin' on."

At the Convention Center the authorities were obsessed with the chairs. They viewed the bringing of them outside as looting. "I'd have Grandma in a wheelchair and Auntie lying down on the cart, and we'd stand in them buses' lines for hours," Clark recalled. "Friday night we were told 100 percent, absolutely they were coming. They never did. We ended up sleeping at Harrah's garage again. But come Saturday morning the National Guard arrived by the hundreds, full of evacuation orders. They acted like we were the Vietcong or something, maybe Indians in a damn concentration camp.

They kept shouting for everybody to give up their weapons. Problem was, nobody really had weapons, although a few guys coughed up penknives."[61] That was the Convention Center: something bordering on a prison camp.

VII

New Orleans had nine major hospitals. Six of them—Lindy Boggs Medical Center, the Ochsner Foundation Hospital, Memorial Medical Center, Tulane University Hospital, Methodist Hospital, and Children's Hospital—were privately owned. The federal government operated the Veterans Administration hospital in Bywater. Louisiana State University Hospital was also a government operation. And so was the largest of them all—Charity Hospital, a nonprofit institution created to provide care for poor people.

Children's Hospital and the Ochsner Clinic, which were on the relatively high ground of Uptown, near the Mississippi, remained safe from flooding. The other hospitals were much more vulnerable. Memorial Medical Center was located toward central New Orleans. Lindy Boggs was in Mid-City. Methodist Hospital was on Read Boulevard in Lakeland, only a couple of blocks from Lake Pontchartrain. The other four—the VA, Charity Hospital, University Hospital, and Tulane Hospital—were all located in the medical district, about six blocks north of the Superdome, and under the shadow of Interstate 10.

As Katrina approached, a hospital seemed to be a smart place to spend a hurricane. The buildings were securely constructed, with auxiliary generators and plenty of supplies. Nonetheless, Charity Hospital evacuated its stronger patients during the weekend before the storm.[62] That left only the critical cases, those least able to sustain any stress. During Katrina, the wind damage at the hospitals was by and large light. Even after the power failed in New Orleans on Monday, the hospitals were still viable, relying on generators to power the lights and lifesaving or life-sustaining equipment, such as kidney dialysis machines, respirators, ventilators, and defibrillators. With the flooding on Monday and Tuesday, however, the hospitals lost most of their power: generators had been placed in the basements, which filled with water early on. Portable generators could operate machines for the critically ill, but not indefinitely. By Tuesday night, helicopters were evacuating the most serious cases from Tulane Hospital, using a parking garage as a landing area. Such evacuations were not in and of themselves chaotic; hospitals practice for emergencies regularly, honing the art of triage and planning for the type of equipment that would have to be transported with each particular type of patient.[63]

No hospital, though, was prepared for the conditions that met New Orleans as Tuesday turned to Wednesday. The staff at Memorial Medical Center, which was surrounded by six feet of water, couldn't properly care for its 260 patients. When people swam or paddled to the hospital seeking help, they were turned away. Memorial was in fact trying to evacuate its own patients, even though, as CEO Rene Goux would later write, "there was no sign of any organized rescue effort, just these people who came from out of nowhere."[64] Remarkably, no government agency made any effort to canvass the hospitals or coordinate an evacuation effort. Memorial staff members were reduced to hot-wiring a boat parked in a nearby garage, so that they could evacuate a critically ill 400-pound woman. About one hundred patients were sent to safety by Wednesday night, largely through the efforts of Acadian Ambulance Services. The Memorial staff tried to care for those who were left, but the level of care was primitive. As the death rate rose, bodies were placed in the chapel. With temperatures high, the stench from the decay was overwhelming. Angela McManus, who had weathered the storm in a bed next to her mother, Wilda Faye Sims-McManus, a cancer patient on the seventh floor, said, "The sewer lines had all backed up, and we were down there in all that stifling heat and this odor was horrendous. People were trying to get into the hospital just to get to higher ground, and they weren't allowing that . . . so they boarded the doors up, and we were just in there smothering all night long."[65]

Most of the New Orleans hospitals ultimately bypassed the government and its inadequacy and hired their own medevac helicopters. The going was slow, but at least something was being accomplished. At Charity Hospital and University Hospital, though, there was no money for private evacuation services. They would have to await government help, but there was no sign of it as Wednesday came to a close. At Charity, the staff operated ventilators by hand, pumping steadily to keep patients alive, for two days.

On Thursday, the situation was desperate: no one was coming to evacuate Charity and many patients who ought to have lived were about to die for lack of proper care. The situation at Tulane Hospital was slightly better. CEO Jim Montgomery and his colleagues at the Hospital Corporation of America (which owned the hospital) had arranged for small helicopters to remove the last patients. Once that was accomplished, Montgomery lined up his staff and their families—and their pets. "Then a situation developed," he wrote. "A frantic Medical Director of Critical Care showed up by boat from Charity Hospital. Major problem. Charity was in a meltdown. He had 21 critical care patients, many being hand ventilated for two days and he couldn't get any help from the state. Can you help me, he asked? This was a tough question, but it had only one answer."[66] Montgomery assigned the small helicopters to help Charity. His staff, exhausted as each member was, would wait for another helicopter.

One doctor who wouldn't leave New Orleans was Dr. Christopher Wormuth of Ochsner Clinic. Flood victims were arriving at Ochsner's emergency doors in a steady stream. Wormuth lived on South Carrollton Avenue with his wife, Laura, and their two kids. Early on they made a "family decision" to stay. They were going to protect their property with a shotgun. At dawn on Wednesday they drove to Lafayette, purchased six generators and seven gallons of gasoline, and felt fortified as they returned home. But Thursday the looting intensified and gunshots echoed throughout the night. "Then on Friday a policeman came up to me and said, 'What are you doing in town?'" Wormuth recalled. "I explained my job at Ochsner and how I had a gun and was protecting my family. He said, 'You think you're protecting your family? Really? I could have just shot you dead, you didn't know I was a cop. If you care about your family, get them the hell out of here.'" Wormuth was rattled enough by this NOPD officer that he loaded his SUV and had his wife drive the kids out of town. "I stayed at Ochsner," Wormuth said, "but I gave up on that whole 'community watch, protect my property' thing. New Orleans had descended past that."[67]

For a day or two, difficult to accept as it was, the general feeling was that the hurricane had landed a powerful punch and they would all have to help overcome it. By the fourth day, Thursday, people in the hospitals couldn't understand why the government had forsaken them. At Memorial, there was a palpable feeling of hopelessness when word ran through the ranks that the institution was low on a list of state priorities for evacuation. Hospitals weren't used to being dead last on anybody's priority list, nor should they be. Awful things happened to patients as a result. "Yesterday was the first day I think that I only cried once," Dr. Randy Roig, director of a rehabilitation hospital on the north shore of Lake Pontchartrain, said ten days after the hurricane struck. "You don't really have much choice, and every now and then, I just get into a corner and get to myself and let my emotions run rampant. But then, I have to get right back to work."[68]

Oscar Zavala was a patient in a New Orleans hospital on Thursday, suffering from a bacterial infection. "What hurts in my soul," he said later, "is that we waited for the government response and it never came. My government is the first one there whenever something happens in a foreign country . . . and here they failed from the very beginning."[69]

AT CHARITY Hospital—or Big Charity, as it was popularly known—food had run out that Thursday and no assistance was forthcoming. The busiest trauma center in New Orleans was itself in critical condition. Founded in 1736, Big Charity, according to *New York Times* reporter Adam Nossiter, was "a symbol of a social commitment to the poor." When Charity celebrated its 250th anniversary in 1986, Allen Toussaint wrote about the legendary hospital in a song titled "Charity's Always There." But Katrina turned Toussaint's lyrics upside down. The hospital was in lockdown, forced to turn people away.[70] Dr. Norman McSwain, the chief of trauma surgery, waded from Charity to his office at Tulane Medical School, where he was a professor. He called the Associated Press. "We have been trying to call the mayor's office," he explained, "we have been trying to call the governor's office . . . and have tried to use any inside pressure we can. We are turning to you. Please help us."[71] His hope was that the media would become the cavalry. Every time they reported a heart-wrenching post-Katrina story, some private sector mogul stepped up to help out.

AIDS specialist Ruth Berggren was another doctor still at Big Charity on Thursday. Born in Boston, she was raised in Haiti, where her parents were public health physicians. In rural Haiti, Berggren witnessed how disease could consume entire villages, leaving rummaging shadows where people had once trod. After specializing in infectious diseases during her medical education, Berggren eventually wound up working with AIDS patients at Charity. "We had 7,000 HIV-infected individuals in New Orleans," Berggren recalled, speaking of conditions on the eve of Katrina. "We ranked very highly in terms of the incidence of the new cases nationwide, which was a phenomenon that had been noticed in the southeastern United States. Both New Orleans and Baton Rouge had an alarming number of new cases, and it particularly appeared to be affecting adolescents and young black females."[72]

When Katrina was headed toward the Gulf Coast, Berggren's conscience dictated that she ride out the storm with her eighteen AIDS patients. Sneaking in a quick five-mile run along the Mississippi River levee before Katrina arrived, Berggren had been the assigned teaching physician on call at Nine West for the entire month of August. "I knew that I was going to stay for the duration," she said. "My husband and my son were going to stay at Tulane University Hospital, just across the street from Charity. My husband was the chief of oncology there. He was not on call, but he made the decision to stay because I was staying and he wanted to take care of patients, and to make sure that his research cell-lines wouldn't be damaged in a power outage."[73]

Berggren's patients ranged from twenty to sixty-five years of age. She was extremely worried about two patients on dialysis: one had advanced AIDS and the other was a young woman who "was responding very beautifully" to HIV therapy. "I was really desperate to get them out," she recalled. "I was especially focused on the young girl because I felt like her life wasn't in imminent danger and because she had a lot of life ahead of her."

When Katrina hit Berggren was a steel magnolia, all steady nerves and tranquil compassion. She ate a little overripe fruit and reassured all of her patients that hurricanes came and went and life went on. She caught a little sleep Monday night on an air mattress in the conference room. But when she awoke on Tuesday and looked out the window, she saw slow-moving floodwaters lapping at the front steps of Charity. "That was the moment that I knew that this wasn't what we thought," Berggren said, "that things were going to get very different around here. I assumed, however, that we would get evacuated within the next twenty-four hours. I never imagined we would be there as long as we were."[74]

With the flood putting the hospital in grave danger, a quick organizational meeting of doctors and nurses was called. There were over 250 patients in the building, scattered over all thirteen floors. The staff started categorizing patients according to the severity of their illness and their priority for evacuation, whether they were ambulatory or wheelchair bound. Although Berggren went along with the color code that resulted, she wasn't very keen on it. "People got confused as to what the colors meant in terms of whether a patient could walk or not," she recalled. "I had a guy in there who had a spinal fracture, who really wasn't very sick; he could have been the last person out, from a medical standpoint, but he needed full support to be carried out. On the other hand, the patient I was most worried about, the girl on dialysis, could actually walk out, but her life was in the most imminent danger, because she needed dialysis. So there was a lot of confusion. I think this was a problem with the evacuations of all the hospitals and nursing homes. National Guard rescuers or police, for example, when they showed up would say that they 'were only taking patients who can't walk.' They disregarded that there was another person who, if they weren't airlifted out in twenty-four hours, would die. They just weren't receptive to what doctors and nurses had to say. They had a fixed idea."[75]

Imbued with a deep commitment to Nine West, and an abiding passion for her eighteen patients and the loyal nursing staff, Berggren decided that they were going to stick together as a group. Quiet, respectful, and full of fear, they were now dependent on Dr. Berggren. They had all become friends because of Katrina, and they pledged to leave nobody behind. Katrina, or the blunders of Bush, Brown, Chertoff, Blanco, and Nagin, weren't going to shatter the esprit of Nine West. Pulling for Dr. Berggren to evacuate them from the bowl became their monomaniacal cause. Somehow the chemistry gelled and Nine West adopted an us-against-them attitude.

Meanwhile, mayhem had broken out in the hospital. Armed security guards were turning away storm victims at gunpoint. Locals, believing the hospital was a safe haven, had started arriving through the floodwaters en masse. While the occasional exception was made—for instance, an NOPD officer who was coughing up blood—the guards weren't letting anyone in. To cope with the growing security menace, Charity doctors deputized three of their residents to protect the roughly 800 people in the hospital. An iron-clad rule was set: Let no new patients inside. "We couldn't be a refuge to the public," Dr. Peter DeBlieux, director of the emergency medicine residents, recalled. "We couldn't even get our own patients out. We were turning people away from medical attention. You had people floating in little wading pools, you know, Grandpa's had a stroke, being floated by his family in a wading pool to Charity Hospital, and we're saying, 'We're not taking anybody. You can't come to the hospital.' "[76]

Dr. Berggren didn't like guns. She was a Brady Bill advocate, a staunch antigun mom. Over the years, going in and out of Charity's trauma center, she had seen too many gunshot wounds that had crumpled young men into permanent unconsciousness. So, as the flooding of Charity continued, she was less interested in National Guardsmen with long-nosed rifles pointed at looters than in rafts, flatboats, or helicopters for her marooned patients. Rumors of urban sniping, however, had turned the Guard on high alert, slowing down rescue efforts, causing choppers to abandon basket-dropping. On Wednesday there was National Guard support; on Thursday they left Charity due to snipers. Here was the rub; the Guard was not, by mandate, supposed to shoot back at fellow Americans. Posse comitatus was the law of the land. "The Guard said they couldn't shoot," she recalled. "That was fine by me. But we were left alone and that was particularly devastating."

The Posse Comitatus Act of 1878 was passed to remove the U.S. Army from engaging in law enforcement during Reconstruction. The term literally meant "the power of the county." It was devised as a way to reassure former Confederate states that the federal government wouldn't abuse them anymore. Basically, the act severely hampered the ability of the Army—and later the Navy, Air Force, and Marines—to become law enforcement officers against American citizens. But in truth, posse comitatus had evolved into a procedural issue. National Guard units—under the authority of the governor—could bypass posse comitatus in an emergency. During the Los Angeles riots of 1992, for example, U.S. troops were used to end the cycle of street violence. In addition, the Coast Guard was exempt from the act. So doctors like Peter DeBlieux and Ruth Berggren were right to be angry that the National Guard and Coast Guard refrained from evacuating Charity Hospital. Certainly the National Guard and Coast Guard had a legal right not to engage. But didn't they have a moral obligation to save the sick? Did one or two guys with a Saturday night special really stop an entire military operation?

When the National Guard pulled out, abandoning doctors and patients alike, on late Wednesday and early Thursday morning, Berggren had an ace up her sleeve. For some strange reason, even though the entire hospital was without phone service, one line near the Nine West ward was working. Call it blind luck or divine design, for some unknown reason, Berggren had a phone with juice. "I expressed a lot of anger about the fact that our evacuation had been stopped and that the National Guard had left," Berggren recalled. "At one point I was berated by this guy from the media, who said, 'What's your problem? Can't you see there's 15,000 National Guard in New Orleans right now? Can't you see them?'" Berggren couldn't see anyone outside her window except some guys in padded fatigues. She got off the telephone, found the chief of security, and had him escort her to every door and exit in the hospital. She was right. The only protection around was a few NOPD officers or security guards. "There was no National Guard presence during the hours when the sniper shut us down and we resumed."

Making matters even more insufferable at Big Charity was the flooded morgue. Over a dozen corpses floated around, with five of them piled up like logs in a beaver's dam, blocking a stairwell. "Our morgue at Big Charity," said Don Smithburg, CEO of the LSU hospital system, "is full and it is underwater."[77] Everything was a race against time. The last unevacuated remaining patients were struggling from depravations. The insulin supply was running out. Intravenous fluids were nearly nonexistent.

Sickly newborns were in incubators which had become airless suffocation tanks. The National Guard and Acadian Ambulance Service *had* to rescue these folks *fast*. There was no time anymore for the luxury of worrying about snipers. But the Coast Guard remained hesitant. "We'll be doing rooftop extractions as soon as possible," Coast Guard Lieutenant Commander Jeff Cater told *USA Today*. "There are some areas where shots were fired. We are avoiding those until we get some assurance our crews can fly in there. As much as we want to help those people, the safety of our people has to be a concern."[78]

Pushing the issue, Dr. Berggren accused the National Guard and Coast Guard of fussy cowardice in the face of an evil piffle. They had an army. Why were they terrified of one lone sniper? To her mind, it was wrong for Guardsmen to halt their mission over a single renegade and floodwaters. Realizing that the situation was going from very bad to worse, Tyler Curiel, her husband, arrived at the ER loading dock in a canoe he borrowed from a surgeon at Tulane. "I'm here to help you," he told his wife. He also reported that their twelve-year-old son had been choppered out of New Orleans in the care of family friends.[79]

For Berggren, it was a moment of great relief. Her boy, Alex, was all right and her husband was at her side. Together they would be able to evacuate the AIDS patients from Nine West. Things were looking up. Beyond that, Curiel had come to Charity with the intelligence that helicopters *were* evacuating Tulane University Hospital patients from the adjacent Saratoga parking garage roof. "I needed a boat to get to Charity, to check up on my wife," Curiel recalled. "There was a canoe padlocked to some pipes in the parking deck. An orthopedic surgeon had decided to bring a canoe to Tulane University from home. He also kept life jackets and oars in his office. So I grabbed those. I now had access to a canoe to see my wife. Unfortunately, paddling my way to Charity people in the water shouted obscenities at me, and they splashed in the water after me trying to steal my canoe. It felt horrible not picking them up, but I was on a mission to Charity and I wasn't going to be hijacked."

Not long after Dr. Curiel arrived, two National Guard Humvees arrived at Charity's ER dock. They were ready to take patients to the Tulane University helipad. The first drop-off went without a hitch. Like a dutiful scoutmaster, Curiel left the parking lot with the guard trucks to evacuate another five or six Charity patients via canoe and bring them to the helipad. All of a sudden, as they arrived back at the Charity ER dock, gunfire rang out. Curiel felt his back tense. National Guardsmen emptied out of the trucks and bolted inside the hospital. Berggren said, "I'd been assigned a sheaf of papers and I'm tallying which patients we've gotten out from which ward. I'm standing right there and people are rushing by with gurneys and there's a feeling of frenetic activity, but it's also hopeful—we're finally getting out of here. It's Thursday morning, we've been here since Sunday. All of a sudden, people start screaming, 'Sniper! Sniper!' They're charging in from the dock. I grab my papers and flatten myself up against the wall because I thought I was going to get trampled by the panicked crowd stampeding. They were like buffalo. I felt that the danger to those of us in the hospital was not the flood or even the temporary lack of food and water; it was actually crowd behavior and people with guns."[80]

Berggren rushed back upstairs to the working telephone near Nine West with her husband. They reported the sniper attack to CNN. Their profound SOS was heard around the world. But in the coming days, when it became clear that reports of gunshots and violence in New Orleans were overdrawn, some press questioned whether the sniper story was apocryphal, a creation of histrionic doctors, blurry-eyed nurses, and jittery Guardsmen. "My response to that is anger," Berggren later stated. "There's no question we were being shot at, and you have to tell somebody you're being shot at or you won't get any help. For reporters to come back to us at Charity and say, 'Maybe you were messing it up or maybe you were so anxious and overwrought, you just *thought* someone was shooting at you, and gee, you shouldn't have done that because it shut down the rescue operation,' is outrageous."

Dr. Peter DeBlieux confirmed that Charity doctors and other emergency personnel were sitting ducks for a rooftop sniper on Wednesday and Thursday. "What looked like a steady stream of patients getting evacuated got shut down," the doctor explained. "Snipers were firing at people rescuing folks from the Tulane Hospital. We got to witness people throwing things off the interstate at people in boats . . . so there were some pretty awful things going on."[81]

By Thursday, the U.S. government started to downplay reports of violence out of New Orleans. The motivation for a response from Homeland Security was obvious: the government was being embarrassed internationally in New Orleans. If the United States couldn't control New Orleans, could they handle the streets of Baghdad or Kabul? Therefore, there were two ways to get control of the anarchy: send in federal troops *and* start claiming the sniper reports were largely false. Both responses were put into effect. The record shows that many reported sniper stories *were* false. The U.S. government was right to quash some of the exaggerations. But there was also no reason to dismiss the claims of Charity and Tulane doctors that somebody took potshots at them while they were on the ER loading dock. They were credible eyewitnesses.

Late Thursday afternoon, Berggren received an uplifting phone call. Libby Goff, a friend in Texas, managed to dial into Charity on a landline and speak to Berggren directly. Goff sometimes spoke in a hippie slang, using terms like "groovy-doovy," but she was loyal beyond words. She was overjoyed to hear Berggren's voice. "What do you need to do, Ruth, to get out of there?" Goff asked. Not wasting time with idle chitchat, Berggren told Goff how worried she was about her patients with tuberculosis; the air from their quarantined rooms was no longer able to circulate through a special decontamination air filter. Staff had already put masks on the faces of those with TB, but tuberculosis *could* be spreading throughout the hospital. Goff immediately launched into action from her University Park home in Dallas. In quick succession she called her congressman, Pete Sessions, Jack Vaughn, vice president of Vaughn Petroleum, and entrepreneur Ross Perot Jr. From the combined efforts of those three wealthy Texans, Goff was able to secure a helicopter from Tenet Healthcare to pick up Drs. Berggren and Curiel from Charity. The question remaining was how to get choppers to Charity in New Orleans, where the airspace was temporarily shut down because of the sniper reports. A private jet was also procured.

Eventually, Goff called Berggren back with good news: a jet from Sterling Aviation was going to first send a helicopter to the parking garage roof on Friday afternoon and then have a private plane waiting at Louis Armstrong International Airport that evening. "Thanks for the jet," Dr. Berggren said, "but you can't just take me and Tyler. I've got eighteen patients and I'm not leaving them!" Goff understood and promised she'd have multiple helicopters and a big private jet, one that could hold up to thirty or forty people. Cheers rippled through Nine West where they heard that transportation and food were on the way. At long last they were getting out of Charity. All they had to do was hold tight one more day. "We had found a plane in Chicago that would get them," Goff recalled. "Everything was coming together serendipitously. No matter what, I was getting Ruth out of there."[82]

When patients and medical personnel were evacuated from New Orleans hospitals, they were usually taken to Louis Armstrong International Airport, located about fifteen miles from downtown. People with health problems from other points in the city were also dropped at the airport. On Thursday, very few people made it there, and as a field hospital, the place was practically empty. But as evacuations finally accelerated on Friday, Saturday, and Sunday, the airport was full of people, each of whom had a different medical need. Patients were laid out on the floor, and even on the baggage carrels. The overcrowding was shocking. "Remember the scene in Atlanta in *Gone With the Wind,*" said a paramedic named David Spence from Texas City, Texas, "That's what it was like."[83] Doctors and paramedics from all over the country scurried around in a rush, trying to tend to the ill amid the arduous conditions. Working a twenty-hour shift was the norm. The operation wasn't as efficient as it might have been, and supplies were limited, until the U.S. Air Force took over the job of deliveries from FEMA. The sanitation was poor, with the signature odor of Katrina—overflowing toilets—in the air.

A seventy-eight-year-old man named Roy Britz was also taken to Armstrong airport. On Thursday morning, a bus came to his nursing home, which was without electricity or plumbing, but safe enough. On the way to the airport, the driver had to take evasive action when someone fired a gun at the bus. Sitting in a chair in the middle of a terminal, Britz told anyone who would listen that all he wanted was a cold drink. He could hardly swallow the lukewarm water he'd been handed. It would certainly help combat the dehydration that was a problem throughout the "wards" at the airport. But no one gave Britz a cold drink. He had been through a lot that day and seen a lot of the response to Katrina. "It's been nothing but stupidity," he said, "and hell."[84]

VIII

While President Bush was talking on *Good Morning America,* fifty-five-year-old Larry Dixon was still trapped in his Upper Ninth Ward two-story house. He had planned to evacuate for Katrina. He had even dutifully sent his wife to Houston in advance on the Sunday before the storm, but he had stayed to help his Mazant Street neighbors board up and tie down. Exhausted, he had dozed off in a La-Z-Boy chair during those key hours late Sunday night, when escaping the bowl was still a possibility. He woke up to find the Upper Ninth flooding, debris flying everywhere. Like tens of thousands of others, he used the freight-train analogy to describe the loud noise causing his parlor to rattle nonstop. The water rose to four feet in his living room in half an hour. "I was supposed to leave," he said. "Now I was thinking, I've got to survive."[85]

Calmly, without even the slightest tremor, Dixon grabbed bottled water, fruit, and his portable TV and climbed into his attic. A little water wasn't going to unnerve an ex-Army man like him. Now he was going to become an attic survivor, with plenty of stories to gas off about at the nearby We-Got-It Café over breakfast. Without question, Dixon was an Upper Ninth man, whose countenance and body language reminded one of a middle-aged James Baldwin. He spoke with a steady, unflappable conviction that his life mattered, that he was important. As a boy, his hero was Sidney Poitier, whose character in *Guess Who's Coming to Dinner?* he tried to emulate in both wardrobe and demeanor. Hats, for example, were important to Dixon; his favorite was a light brown fedora he'd purchased at Meyer the Hatter shop on St. Charles Avenue: it was a Dobbs with a black band and a short red feather sticking up. "My mother wears hats," he said, "and so do I. And so did Malcolm X and Martin Luther King and Harry Belafonte and Sidney Poitier."[86]

Dixon wasn't a rich man, but he made a good living. His family owned

Cars carrying residents left downtown New Orleans ahead of Hurricane Katrina on Sunday, August 28, 2005. Hundreds of thousands of residents were ordered to flee on Sunday as the hurricane strengthened into a rare, potential Category 5 storm and barreled toward the vulnerable below-sea-level city. MICHAEL AINSWORTH/DALLAS MORNING NEWS/CORBIS

Satellite imagery showed Hurricane Katrina over the Gulf Coast making landfall as a Category 3 storm at 1:15 P.M. (EST) on Monday, August 29. NOAA VIA GETTY IMAGES

A mother and son rested amid the chaos of the Superdome, where thousands of New Orleans citizens sought refuge from Hurricane Katrina's wrath. Mario Tama/Getty Images

Survivors took food and supplies from a drugstore in New Orleans on Tuesday, August 30, as the city was cut off from the outside world, submerged by rising floodwaters, and troubled by anarchy in the streets. James Nielsen/AFP/Getty Images

The vast flooding of New Orleans was shown in this aerial view taken by *New York Times* photographer Vincent Laforet on Tuesday, August 30, a day after Katrina passed through the region. Vincent Laforet/AP

The entire fleet of much-needed New Orleans school buses was flooded and rendered useless due to breached levees in the aftermath of Hurricane Katrina. Phil Coale/AP

A man held himself on his porch in the flooded Lower Ninth Ward of New Orleans on Monday, August 29, after Hurricane Katrina slammed Louisiana as a Category 3 storm. MARKO GEORGIEV/ GETTY IMAGES

A family waited on their porch to be rescued while holding on to supplies for their survival in flooded New Orleans on Tuesday, August 30. RICK WILKING/ REUTERS/CORBIS

People on Canal Street made their way to higher ground as water filled the streets on Tuesday, August 30, a day after Hurricane Katrina hit the city. MARK WILSON/GETTY IMAGES

A corpse was tied to a tree to prevent it from floating away in the great deluge that engulfed New Orleans. Paul Sancya/AP

A U.S. Coast Guard boat patrolled a neighborhood in New Orleans. The Coast Guard did a flawless job as first responders to the deluge. Christopher Morris/AP/VII

Evelyn Turner cried alongside the body of her common-law husband, Xavier Bowie, after he died in New Orleans, on Tuesday, August 30. They decided to ride out the storm because they had no means to evacuate the city. Bowie, who had lung cancer, died when his oxygen supply ran out. Eric Gay/AP

President Bush surveyed the damage from the window of Air Force One while flying back to the White House on Wednesday, August 31. Christopher Morris/AP/VII

Rhonda Braden walked through the destruction in her childhood neighborhood in Long Beach, Mississippi, on Wednesday, August 31. Rob Carr/AP

Survivors walked along a beach-front road in Bay St. Louis, Mississippi, which was reduced to rubble by Hurricane Katrina. David Rae Morris

Heavy flooding in New Orleans forced families into their attics, and ultimately through their rooftops as they awaited rescue on Tuesday, August 30. Vincent Laforet/Reuters/Corbis

An airboat pulled up to Memorial Medical Center while Acadian Ambulance Service medevaced doctors, nurses, and patients from the hospital. Many of the elderly and infirm were left behind to die. Bill Haber/AP

Survivors of Hurricane Katrina were stranded on the elevated Interstate 10 in downtown New Orleans on Wednesday, August 31, as they sought refuge from the deluge. They waited for days to be evacuated to Baton Rouge, Lafayette, or Houston. One group sought shelter in nearby Gretna, Louisiana, but were turned away at gunpoint. This became known as the Gretna Bridge Incident. Rick Wilking/Reuters/Corbis

Residents in New Orleans, forced to their rooftops to escape the rising floodwaters, were rescued by helicopter on Thursday, September 1. David J. Phillip/AP

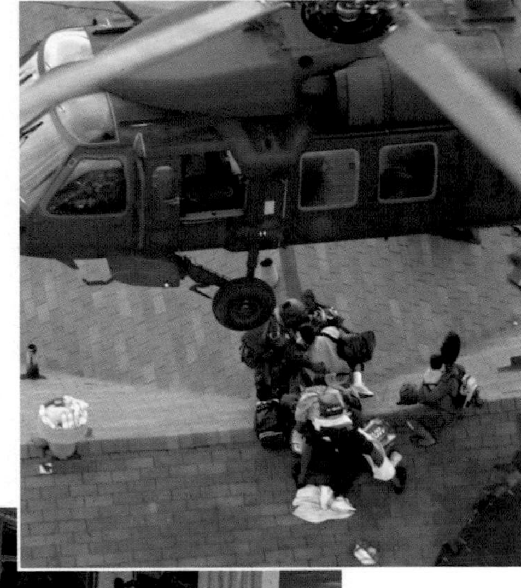

Louisiana Lt. Governor Mitch Landrieu aided in the rescue effort in the Lower Ninth Ward of New Orleans following Hurricane Katrina. Working side by side with Louisiana Wildlife and Fisheries, they saved hundreds of people. Arthur Lauck/The Advocate

Stranded victims of the hurricane waved umbrellas with distress messages written on them from a second-story New Orleans balcony behind a flooded cemetery. Mario Tama/Getty Images

In front of the Convention Center, a bystander views the body of Danny Brumfield, allegedly killed by the New Orleans Police. Ron Haviv/AP/VII

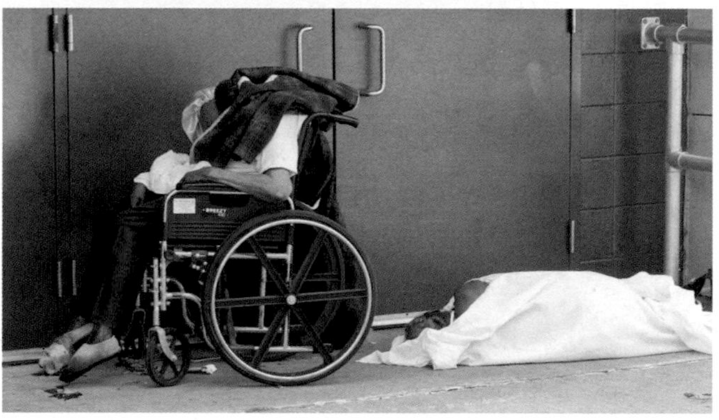

The body of Ethel Freeman, ninety-one, and that of an unidentified resident of New Orleans remained at the door of the Convention Center on Friday, September 2. The previous Sunday Mrs. Freeman's son was forced to board a bus and leave his mother's body behind. He didn't see her again until he saw her photograph in newspapers and on television as an unidentified casualty of the storm. Robert Sullivan/AFP/Getty Images

Bottled water was tossed to dehydrated survivors of Hurricane Katrina in New Orleans as they waited in long lines to be evacuated from the city. Stephen Azzato

Thousands of flood victims walked from the Superdome to a staging area outside the Hyatt Regency, where many city officials remained throughout much of the crisis. The angry crowd complained that Mayor Nagin would not come out and address their concerns. BILL HABER/AP

Twenty-year-old Jabar Gibson commandeered a bus to carry hurricane evacuees to the Houston Astrodome. It was the first to arrive there at around 10 P.M. on Wednesday, August 31. He was greeted by Harris County Judge Robert Eckels whose "can do" attitude made Houston proud. © HOUSTON CHRONICLE

Residents of New Orleans were airlifted by a U.S. Coast Guard helicopter from the misery of the Convention Center to an evacuation staging area at the Louis Armstrong International Airport on Saturday, September 3. ROB CARR/AP

Sara Roberts of Lake Charles, Louisiana, helped organize the Cajun Navy, a loose consortium of private citizens with boats from southwestern Louisiana that rescued residents of New Orleans East. Along with her husband, André Buisson, Roberts carried flood victims to the safety of Chef Menteur Highway. Eric Gay/AP

U.S. federal troops began to arrive on Thursday, September 1. Their primary goal was to establish law and order in the streets of New Orleans. Lindsay Brice

FEMA marked searched homes with DayGlo paint, but the recovery of bodies was slow. This body of an unidentified man remained exposed below Interstate 10 at Earhart Expressway more than a week after Hurricane Katrina devastated New Orleans. Lindsay Brice

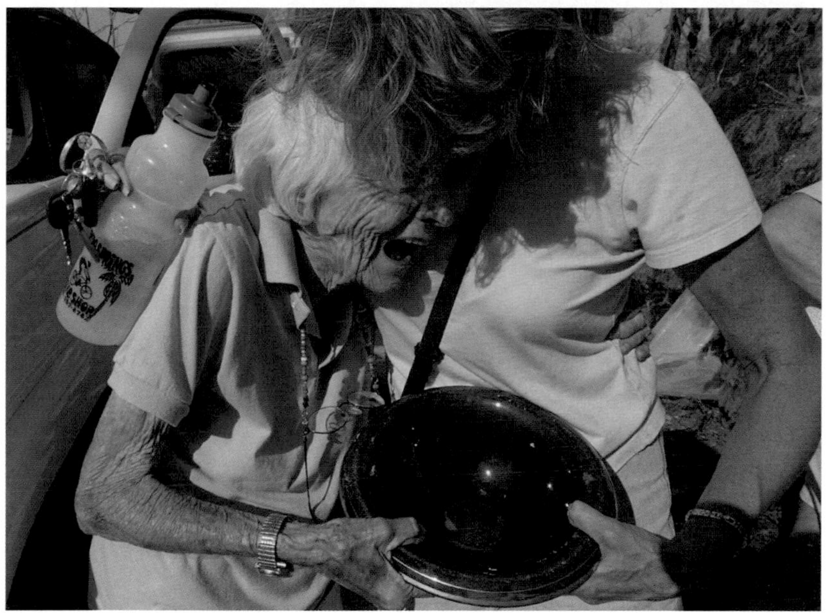

Lucretia Fly, ninety, of Bay St. Louis, Mississippi, broke down crying on the shoulder of her daughter, Nyshie Seymour, when she found a family heirloom, a glass bowl over one hundred years old, in the sand behind where her home used to be. Chuck Liddy/The News and Observer/AP

Glenda Thomas sat in despair on the steps of what was the front of her home in Gulfport, Mississippi. Many residents along the Gulf of Mexico lost all of their worldly possessions. Ross Taylor/Getty Images

Gulf South residents expressed their frustrations by "street blogging" on boarded-up buildings in the days and weeks following Hurricane Katrina. Anja Niedringhaus/AP

A Red Cross volunteer comforted a hurricane victim in Houston's Reliant Center Astrodome on Friday, September 2. More than 65,000 evacuees were processed through Reliant Center Park. The Astrodome itself housed 17,500 persons, while the Reliant Center and Arena housed over 9,000. Andrea Booher/AP

Lt. General Russel Honore *(left)* talked on a cell phone while waiting for a helicopter as an aide took notes. Honore provided leadership in the Gulf South in the days following Hurricane Katrina. Rob Carr/AP

The largely evacuated Uptown neighborhood of New Orleans was left littered with muck and debris in the ghost-town aftermath of Hurricane Katrina. Lindsay Brice

Shotgun row houses on Elysian Fields in New Orleans remained flooded one week after Hurricane Katrina. The U.S. Army Corps of Engineers dropped sandbags around the clock to bolster the breached levees. Lindsay Brice

Members of the D-Mort team carried the body of Leola Lyons, seventy-two, from her New Orleans home weeks after the hurricane. A month after the storm hit Louisiana, more than 972 victims in the state lay nameless in a morgue on the outskirts of Baton Rouge. Ann Heisenfelt/AP

A wheelchair covered in mud and debris remained in a room at St. Rita's Nursing Home in St. Bernard Parish, where thirty-five people died. The owners apparently neglected to evacuate the residents, resulting in their drowning in the Wall of Water. Justin Sullivan/Getty Images

NBC News videographer Tony Zumbado discovered bodies lined up before the altar in the second-floor chapel of Memorial Medical Center in New Orleans. Photographer Tony Carnes of *Christianity Today* later took this photograph. Tony Carnes

Tony Zumbado fearlessly followed leads and captured images that shocked the world. He was the first videographer to film the Convention Center mayhem. Much of his footage was deemed too graphic for broadcast. DWAINE SCOTT

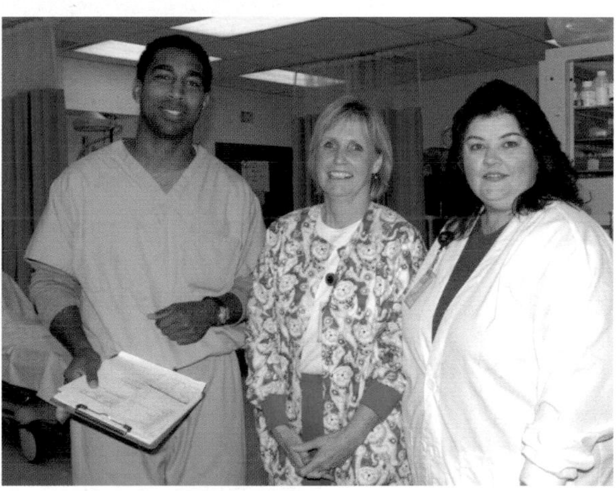

The Hancock Medical Center in Bay St. Louis, Mississippi, did not turn away a single patient despite the rising waters and limited resources within the hospital. Pictured here are Dr. Fredro Knight and nurses Sydney Saucier and Angie Gambino, who worked nonstop for days treating nearly one thousand hurricane victims. LINDSAY BRICE

Ruth Berggren landed in Fort Worth, Texas, and was embraced by longtime friend Libby Goff after having made sure all of the patients and staff from the Nine West Ward at Charity Hospital were rescued together. RICHARD GIBBE

Rug merchant Bob Rue raised "street blogging" to an art form at his St. Charles Avenue emporium, where he threatened looters and invited tourists to return to New Orleans for Mardi Gras. CNN's Anderson Cooper frequently used this storefront as a backdrop for his reports from New Orleans. LINDSAY BRICE

Michael Knight *(left)* and Jimmy Deleray, two of the heroes of what became known as the NOLA Homeboys movement, met for a beer at Vaughn's in the Bywater. Bob Marley aficionado Knight saved hundreds of people in Tremé in his personal boat. After warding off looters in the Riverbend section of New Orleans, Deleray became a leading rescue boatman in the flooded neighborhoods of the city. LINDSAY BRICE

Larry Banks, U.S. Army Corps of Engineers chief of Watershed Division for the Mississippi Valley, examined the breached section of the 17th Street Canal levee in the Lakeview neighborhood of New Orleans. CHARLIE RIEDEL/AP

Dixon's Central, a snack food company headquartered in Carthage, Mississippi. They manufactured bags of Pork Skins and B-B-Q Flavored Pork Skins, delicious with ketchup or hot sauce and an alternative to potato chips. Dixon made sure every mom-and-pop store in the Upper Ninth stocked their cured chips, void of carbohydrates. And, due to his salesmanship, every Winn Dixie in the Greater New Orleans area was carrying his family's two-ounce bags right next to big-shot snacks like Fritos and Lay's. "I had boxes and boxes of my pork skins on the first floor when the water came in," Dixon recalled. "Unfortunately, they all washed away."

Every morning Dixon started his day with a bag of his pork skins. They were better than doughnuts, he'd say. At night Dixon, his Dobbs hat hiding his bald spot, would walk up to Sweetwater's Bar, play Sam Cooke's "You Send Me" and Al Green's "Love and Happiness" on the jukebox, and have four or five bags of pork skins, washed down with Crown Royal and water with a twist of lemon. Just seeing that pale, copper whiskey bottle made him grin from ear to ear. For well over a decade, Dixon had owned three retail shops, specializing in 45s, called Gemini Records. What a grand time he had, a good run while it lasted. But Gemini Records was now defunct and the Upper Ninth was underwater. His two thousand warehoused boxes of pork skins in New Orleans East were also now mush. "To kill time in the attic I thought about all the people I loved, how I'd be a better man once the Katrina mess was over," Dixon said. "No more scolding my wife."

Like most of the attic people, Dixon had to knock out his vent to allow air to draft in. He spent most of Monday, Tuesday, and Wednesday on his roof checking up on neighbors. According to Dixon, the attic dwellers of the Upper Ninth had created their own communications system "like in Africa." Standing on the eaves, they made slight hand gestures, as if in a prison yard; for example, holding up three fingers, followed by two, meant that three houses down, there were two people stuck inside. National Guard and Coast Guard helicopters flew over the Upper Ninth, but never dropped a basket. Dixon surmised that their infrared, high-tech maps probably showed that the Lower Ninth Ward was in much worse shape. It was. He didn't blame them for not stopping. "They were just busy tryin' to get as many people out as possible," he said. "Remember I served at Fort Polk. I knew what they were trying to do and they were doing it very well."

By day Dixon busied himself with business books. He had a dream that someday his pork skins would be stocked in every A & P and Kroger from San Diego to New York. Maybe even Wal-Mart would carry them. Every person in the Delta—his home turf—could testify that they were good, they were damn good, ten times better than all these New Age snacks like Cool Ranch Doritos or pizza-flavored Pringles. His snacks had that real pig-grease flavor, straight from the slaughterhouse, not saturated with salty chemicals concocted in a laboratory. What the hell did those fools know about food? "I kept thinking about new marketing plans up in the attic," Dixon said. "Ways to quadruple my distributors. I had a few business books with me and they were helping out."[87]

At night, Dixon would sit on the roof and study the galaxy. He was looking north for Polaris, which during the Dark Ages of slavery had been a guiding star which had directed 70,000 to 90,000 of his brothers and sisters to the Promised Land. How little things had changed, he chuckled. Here he was sitting on a roof, in an all-black neighborhood, trying to follow the drinking gourd in the dark sky, while he saw flashlights talking in code up and down Mazant Street. There was no Peg Leg Jones, whispering directions, but the Underground Railroad was alive and well; blacks were still trying to escape Alabama, Mississippi, and New Orleans. Nobody could argue that point. Just another black migration moment was the way Dixon wrote off all the Katrina hullabaloo. He simply didn't believe the U.S. government was going to lift him out of the debris. African Americans had to take care of each other, without the "Caucasian race." To think that President Bush was going to lift the black man out of the muck was absurd. Self-reliance was the ticket for survival. So by Thursday morning, Dixon had stopped waving his fedora at the choppers like a beggar. The best way to deal with whites was to walk past them, not bringing any undue attention to yourself. They'd pick you up in a chopper just to arrest you for unpaid parking tickets.

Given this pessimistic attitude, there was no jubilation when he popped out of his roof vent and saw his forty-seven-year-old cousin Arthur "June" Thomas rowing down the street, engaging in low-grade rescues, a person here, another one there. June wasn't trophy-hunting, like some of the NOLA Homeboys, claiming two or three hundred saves. The only reason he was rowing a sixteen-foot aluminum boat was that his house was a junk heap. He had stayed for the storm, hanging out in his attic, when his house had moved off its foundation, causing the gas pipes to break. His entire downstairs lit up like a blue fireball, Thomas recalled, like an enormous baked Alaska being served up to the Jolly Green Giant.[88] All over the Lower Ninth, "people were hollerin'," and he didn't try to save them. He swam for his own life. Eventually he was picked up by a Wildlife and Fisheries boat and dropped off on the St. Claude Bridge with everyone else. June, however, wasn't the kind of man to sit around. He went and found himself this aluminum boat and now was pulling his Uncle Larry off his roof. "He was in no real hurry," Thomas said of his uncle. "He just tried to stay all dignified."[89]

Without question, Dixon was glad to be saved. But it was a long way from Mazant Street to Houston. All he had were his walking shoes and credit cards. He asked June to dump him off at the St. Claude Bridge and he'd walk the 4.3 miles to the Morial Convention Center. Little did he realize that the massive center, the very place he had sold pork skins a few years ago at a snack convention, had become uninhabitable. Upon arriving, he surveyed the situation, wiped his brow, and went looking for water. That was no easy chore. His lips were as swollen as a brass band player's. But he recognized an NOPD officer smoking near Harrah's and went up to him. They had known each other for twenty years. "You can have a bottle," the officer told him. "But don't tell anybody where you got it from."

Quietly, as if invisible, the still well-groomed Dixon, now hydrated, entered the Convention Hall and shook his head in disbelief. Clearly this was the wrong place to be. So he moseyed back outside, walked across the Convention Center Boulevard to the foot of Canal Street, and deliberately sat down under an awning; shade was the key for surviving. Shade was as important as a canteen. He sat there all Thursday night. He didn't even get up for a nature call. He just sat and stared as if in a trance. He missed his wife, Joanna, terribly. "'Gotta get to her in Houston,' I kept thinking," he recalled. "We were going to reunite. No doubt about it."[90]

At 5 A.M. on Friday, Dixon stood up, dusted himself off, and made sure his clothes weren't rumpled, that he looked respectable, snappy enough to have lunch with Ralph Bunche or Eartha Kitt at Commander's Palace with nobody thinking he was a second fiddle. He had on his favorite shoes and he started walking to the Crescent City Connection, the bridge that the Gretna police had blockaded. And he just kept walking, one step at a time, head held high, a man not in a rush. As he approached the police barricade, he just saluted and walked right past them. They must have believed that he was a city councilman or private eye. At least, he hoped so. He had purposely exuded the aura of a man who owned a pocket watch.

Although he had plenty of opportunities to hitchhike in his youth, hopping on the back of cotton trucks headed to the levee market in Memphis, he had always declined. He wasn't a tenant farmer or a day wager. He was a Mississippi businessman, one of the Crackling Kings of Leake County. But Katrina had messed him up. Out his thumb went, hobo style. He was headed west. Another displaced man, about his age, walked up to him and talked. He seemed like a decent fellow, with a strain of Jamaican or Trinidadian blood in him. They decided to hitchhike together. Dixon never really got his name. For about an hour, no vehicles would stop. Then again, there were virtually no cars on the road. As luck would have it, however, they stumbled upon a pickup truck wedged in a drainage ditch. The driver's misfortune turned out to be their ticket out of misery, or so they thought. "We helped pull him out of the hole," Dixon recalled. "And he, in return, gave us a ride."

Out of all the Katrina-related experiences Dixon encountered, this ride with a drugged-up, manic gangbanger was the most harrowing. Even though there was debris littering the road, Mr. Pickup Truck insisted on driving 75 or 80 miles per hour. Dixon was in the passenger seat and kept glancing back at his hobo partner scared silly in the truck bed. "You've got to slow down," Dixon pleaded with the driver. "He's goin' fall out."

The drug addict dropped them off. At least he had gotten them past Gretna. They were now hoofing it down Highway 90 to the hamlet of Boutte in St. Charles Parish, sixty-seven miles from Baton Rouge. Out the thumbs went again. Before too long, a pickup pulled right beside them, the driver sizing them up with his eyes. They passed muster. A deal was struck. He would give the hitchhikers a lift to the Baton Rouge airport if

Dixon filled up his gas tank and the other man gave him twenty dollars. "Sounded fair to us," Dixon recalled. "And you know what? He took me right to where I wanted to be."[91]

Once at the airport, Dixon caught a break. For some strange reason, there was a burgundy Malibu still available on the rental lot. He tossed down his credit card, showed them his Louisiana driver's license, signed a form, and took off headed west over the Mississippi River. Houston had by now taken on a mythical dimension in his head. Dignity was still his lodestar. The Astros looked play-off bound and he imagined himself in the Homeplate Bar and Grill, watching the baseball games in air-conditioned glory just a few weeks ahead. He imagined his wife's warm embrace followed by a carefree shave, baby powder—the whole nine yards.

And the big reunion came. He parked the car on Pebbledown Street, walked up to the appropriate door, and knocked. Joanna answered. There was no joy in her face. She just stared at him and him at her. A few awkward moments passed, and she gave him a perplexed look. "Damn," she said, with arms folded, full of disappointment. "You still livin'."[92]

Chapter Thirteen

"IT'S OUR TIME NOW"

We acknowledge some forms
of considerations.
We open for those
who adhere to our one rule
endure.

—Louise Erdrich, "Asiniig," *Original Fire*

I

SOME PEOPLE CALLED IT the "Battle of Oak Street." It lasted from Monday to Thursday and forty-eight-year-old Bruce Foret—a white, middle-class business owner—was in the thick of it. Foret lived in a 3,500-square-foot house on Finch Street in Lakeview, a Lake Pontchartrain neighborhood tucked between the 17th Street Canal and the London Avenue Canal. His family business was an Ace Hardware Store, located along a somewhat dilapidated shopping strip just off Carrollton Avenue, only blocks from the Mississippi. The store, at 8338 Oak Street, opened in 1929 and Foret's father had acquired it in 1964. In 1970, they purchased the "Ace Hardware" franchising rights. "At one time, before the big-box stores, we were a good-size hardware store that specialized in service," Foret recalled. "We knew fifty percent of our customers by name. It was a family business with guys that worked for us for anywhere from twenty to fifty years. New Orleans always had an underlying racial tension. I'm no expert. Probably sixty percent of my pre-Katrina customers were black. New Orleans was probably the most integrated city in America. We never had many problems, but it took something like Katrina to uncover the problems."[1]

Like most New Orleanians, Foret spent the weekend before Katrina boarding up his Lakeview house. He had evacuated his wife and their three children to Dallas, Texas, while his brother evacuated their mother to Jackson, Mississippi. That Saturday, as Katrina approached, Foret kept Ace Hardware open, selling nails, hammers, and plywood. "Business was brisk," he said. "But basically not a single African-American family came in. They weren't boarding up." They were, of course, but perhaps not in his area. Realizing that Lakeview would probably flood, Foret decided to spend the storm in Harahan, which lay above sea level, at his mother's suburban home. It proved to be a wise move. His Finch Street house took two feet of water and then burned to the ground when a gas line ruptured. He lost everything in the fire. "I woke up at around 6:30 A.M. Tuesday and something told me I better get down to the hardware store," Foret said. "I threw some clothes on, got in my truck, and started heading Uptown. It was pretty difficult. There were a lot of power lines down, a lot of trees. So I just meandered around, up and through old Jefferson Parish, up on the Mississippi levee where it was dry, all over. I finally got to my store and that's when I found out it had been looted."[2]

The word "looting," derived from the Hindi *lut* and the Sanskrit *lunt,* was defined by the *American Heritage Dictionary* as "seizing goods wrongfully or by force, especially in times of riot or war." Basically, looting was considered an opportunistic act, an open season to get "free" stuff during a crisis. Although the term had ancient antecedents, it had been used by the American media to describe a whole series of recent historical events. When Islamic fundamentalists bombed the World Trade Center on September 11, 2001, ATMs all over Manhattan were "looted." Then there were the unforgettable scenes from Baghdad of the Iraqi National Museum being looted in April 2003. In times of unrest, the fear of looting was a constant in every storekeeper's life, particularly when massive power outages occurred. At Ace Hardware, the possibility of collective theft had caused Foret to install a sophisticated alarm system. But it only worked with electricity. In a powerless New Orleans, his store would be wide open, like a free buffet, for looters to carry off his merchandise in droves. As a precaution, back in the late 1990s, he had taken out full-coverage liability insurance against theft. "Good thing I was covered," he said. "Man, I ended up needing the insurance."[3]

No sooner had the Katrina winds dissipated than one of Foret's employees, thirty-seven-year-old George Morales, part of New Orleans's Guatemalan community, ventured from his Dauphine Street house to Ace Hardware to check the store's pulse. The structure was in pretty good shape: a little wind damage to the electric Ace Hardware sign, which was still hanging, but nothing major. He took a quick drive to Kenner to inspect some rental property he owned and then headed back to Oak Street for a second glance. Things had changed. What he encountered at Ace Hardware made his blood boil: between twenty and twenty-five African Americans, many of them women and children, were looting the store. It was a bad dream. There was an old woman carrying out a power saw and a middle-aged man carting out a $600 power washer. Before being hired by Foret, Morales had been a bouncer at two French Quarter bars: Razoo's and Cat's Meow. Confrontation was part of his nature, flexing his muscles a component of his work regimen. "I parked, slammed my door, walked to the storefront, and told a looter to put the merchandise back," Morales recalled. "He didn't like that. He jumped all over me with 'Fuck you' and 'Who do you think you are?' I told him he was robbing my store. Then he got in my face. He broke my two-foot line, getting within eighteen inches of my face." Incited, Morales snapped and punched the looter with a quick right, knocking him backward three or four feet and cutting his lip. "He then ran off," Morales said, "and I went to stop the others from stealing. My mistake was I didn't have a gun."[4]

Shouting loudly at the looters had a positive effect. Many of them dropped the merchandise and quickly fled the scene. For a moment, Morales thought he had control of the bedlam. But then the looter he had punched returned, running at him, head high, waving a crowbar frantically in the air like a tomahawk. Morales ducked, the crowbar coming within inches of his skull. Using an old wrestling move, he laid his body into the looter's torso, struggling all the while for control of the weapon. Morales absorbed one blow to his arm, which became bruised and swollen, but was able to get the attacker in a headlock. He started squeezing the life out of the looter, choking him breathless. Before Morales broke his neck, however, he let go. He just shoved the thief aside. Some little voice told him that manslaughter equaled Angola State Prison. "I let him go, and he ran," Morales said. "Other merchants had now congregated on the street and all the looters left." Round one of the Battle of Oak Street was won by Morales. "That was just the beginning of a week of trouble," he said. "I owed my boss to defend his business. He'd been good to me. And any way you sliced it, looting expensive power tools was wrong."[5]

When Foret arrived at Ace Hardware on Tuesday morning and saw his front window bashed in, he realized that a turf battle was on for control of the Riverbend area. It was white versus black, merchants versus neighborhood residents. Foret encountered Morales pacing about, holding his hurt arm. A few other merchants, including Hank Staples from the Maple Leaf Bar, Ralph Driscoll of Driscoll Antiques Restoration, and John Burwick, who worked at Jacques-Imo's Cafe, were pacing up and down Oak Street. With the NOPD in disarray, they would have to fend for themselves. The white business owners of Oak Street spontaneously formed a so-called community watch. Morales had already grabbed a loaded pistol and an unloaded shotgun from Ace Hardware; he knew where his boss kept them hidden. An enraged Morales explained to Foret how, in addition to the crowbar bandit, about twenty-five African Americans had come marching up Baronne Street looking to steal merchandise. Many were waving guns. The word was out: Ace Hardware was an open store. According to Morales, he had aimed his shotgun at these potential looters and they had quickly turned around, full of curse words. Slapping him on the back, Foret told his employee he had done the right thing. The other white merchants nodded their heads in agreement. "I used to be kind of a gun dealer, so I had four or five pistols," Foret recalled. "I gave pistols out, so everybody was pretty much armed. Hank Staples had a .38 pistol that he keeps, his personal gun. The antiques dealer on the street, he had a 9mm which he kept hidden under his clothes. He had it on him that whole week."[6]

Both of Ace's front doors had been blasted out by the looters, and there was glass everywhere. Foret was determined to repair the front of his building before nightfall. He paid John Burwick fifty dollars to stand over him with a shotgun, to chase potential looters away, while he diligently and secured his storefront with wire mesh and plywood. There was also the back door to barricade. "It was so hot and miserable, and I only had cordless tools, which didn't have any charges, because they were new tools from my store," Foret recalled. "Luckily, one of my Oak Street neighbors had a generator, so he would charge them. I worked from eight in the morning to four in the afternoon."[7]

During the day, while Foret was boarding up his storefront, two pickup trucks full of young men kept cruising down Oak Street, taunting the merchants. "I thought we were going to get sprayed," Foret said. "I felt outnumbered and outarmed." Some of those on the street had been part of the group looting the store when Morales had arrived that morning. "They were trying to intimidate us," Foret recalled. "I never did see any guns with those young black guys, but they were just looking to see what they could get come nightfall. George kept saying, 'That's them sons of a bitch right there. They stole stuff.' Their brazenness was self-evident. They were mocking us, a bunch of old white guys."[8]

Quite predictably, given the psychological black-versus-white dynamic that had developed, racial gamesmanship engulfed the entire area. The looters again broke into Ace Hardware in the middle of the night, around three o'clock on Wednesday morning. Besides ransacking the hardware store, they stole a forklift from Maximilian's Rugs on Oak Street, which they ran through the front entry of the Rite Aid. Within an hour, over one hundred African Americans poured into the drugstore, stealing or commandeering everything in sight. When Foret returned at eight Wednesday morning, he was flabbergasted by the shattered storefronts. Oak Street had been vandalized beyond recognition. About five blocks of it looked like a chicken bone picked clean. Another "community watch" meeting was held. This was urban war. It had become a matter of principle.

Some of the city's richest residents took security into their own hands. In New Orleans's upscale Uptown neighborhood, well-heeled and well-armed property owners, sometimes with security guards to assist them, kept possible looters at bay, carrying firearms openly in their neighborhoods and looking after neighbors' homes and valuables—keeping a close watch on friends' irreplaceable art collections. Calvin Fayard—one of the region's major political fund-raisers for the Democratic Party, and the owner of the so-called Wedding Cake House, one of the city's grand mansions—would remain at home and on guard with a coterie of like-minded friends. Some would use their powerboats to rescue those who'd lost their homes. Their neighbors would dine on gourmet food from nearby specialty stores. Some would bathe in their stagnant swimming pools. One or two would take the opportunity to fly by helicopter to the office to shred potentially sensitive business documents—just in case the papers were to fall into the wrong hands, should law and order break down altogether. "We were, thank God, on high ground," Fayard recalled. "Most of Uptown just didn't flood."[9]

Unlike Uptown, all the residents of Hollygrove used to complain that their neighborhood flooded terribly even during a summer thunderstorm. The residents demanded drainage projects. Hollygrove recording artists like R&B great Johnny Adams and rapper Lil' Wayne sang about how nobody ever listened to their flooding woes. In 1996, however, a new pumping station came to Hollygrove. It was supposed to keep the streets dry. "Damn thing never worked right," Ivory Clark said. "Hollygrove still flooded from a few raindrops."[10]

When Katrina hit, the Hollygrove neighborhood took in up to eight feet of water. The residents lost everything they had. Anger welled up at white New Orleans, particularly the merchants on Oak Street. How come every African-American family in Hollygrove had gotten flooded while the white businesses were dry? Some in the neighborhood felt victimized, as though all their worst dreams were suddenly coming true. Stranded on rooftops or crammed into attics, the Hollygrovers felt justified in looting or commandeering. It was their way of lashing out against the backed-up sewers and failed pump house. "They just marched down our streets and stole whatever they pleased," Hank Staples recalled. "They had just lost all good sense. The flood triggered a deep resentment. Hell, we didn't cause their homes to flood. I run a bar and employed black musicians. Why make me suffer?"[11]

The most important thing, Foret decided, was that the forklift had to be decommissioned at any cost. He grabbed electrical clippers, about five or six merchants loaded their guns, and they marched together to the front of the Rite Aid, where the forklift had been abandoned in the doorway. A mob of African Americans was actively stealing merchandise, so Jimmy Deleray, standing next to Foret, fired a warning shot in the air from a .45. "This caused the looters to scatter," Foret recalled. "A little black girl thought Jimmy was shooting at her and she started screaming." Boldly, with anger in her eyes, the girl, undeterred by the guns, threatened Foret and Deleray, saying that she was going to get her brother and he was going to shoot the Oak Street merchants dead. "It was unbelievable," Foret recalled. "Then one of my black customers, a guy who was watching the showdown, said to me, 'She ain't kidding; her brother's a murderer. You better get out of there.' So I'm like, 'Oh my God.' We all got really concerned at that point. This was on Wednesday."[12]

Foret spent Wednesday evening at his mom's house in Harahan. The battle for Oak Street and, to his mind, for the future of the city, was brewing along the levee. There were no NOPD officers or National Guard to save the day. It was the OK Corral, and he was willing to risk his life for Ace Hardware. The dynamic that had developed was a slightly tilted version of Spike Lee's *Do the Right Thing.* Suddenly he was pitted against his former black customers. He didn't harbor them any ill will. He wasn't a racist. But he felt his honor was at stake. If he let hooligans own his store, then they owned him. "I got up early Thursday with my pistol," Foret recalled. "I had gotten used to having my pistol with me wherever I went. As I drove up to the hardware store, I caught a looter coming out the front with a handful of supplies." Spontaneously Foret swerved his Dodge pickup right over the curb, just ten feet from the looter. "Is that shit worth dying for?" Foret shouted at the looter. "Is it?" Terrified to be caught red-handed, the looter, an African-American man in his thirties, dumped the goods and took off running. Another African American, an unexpected ally, shouted out from across the street, "Man, there's plenty more inside!" This was the moment Foret dreaded. A panic swept over him. Cautiously, he crept forward. He pointed his pistol at the smashed storefront doors. Then, in a few steady, long paces, he walked into Ace Hardware. "Whoever is in here," he screamed, "better get the hell out!" He then fired his .38 automatic into the ceiling. "I figured that would scare them," he said, "shake 'em up."[13]

Everything was dark inside the store. Nothing moved. He kept aiming the pistol down the aisles just in case an armed intruder was hiding. Eventually he realized the coast was clear. The store was empty. There had only been one looter. Clamps, drill bits, chisels, handsaws, fuses, and screwdrivers were scattered all over the floor. The looters hadn't been content with just stealing; they had vandalized the store for sport. Foret's cash registers were flipped over, his computers smashed, and his office files in utter disarray. "Just malicious stuff," Foret recalled. "They had gone on a frenzy." He surveyed the interior. He walked down the aisles. Like Job, Foret started again: cleaning up the store and resecuring the doors. "They weren't going to get the best of me," he said. "They weren't."

The Battle of Oak Street even carried over into Houston, to which Jeff Amann, owner of a popular outdoor landscaping company headquartered at 8616 Oak Street, had evacuated. His business was responsible for the care of many exquisite yards in Uptown and the Garden District. Tourists would photograph his mazelike hedges and exotic swimming-pool areas. He owned five specialized service vehicles and dump trucks, painted with his logo. Amann was in Houston with his family when he received a telephone call from his friend Patrick Berrigan, a contractor based on the north shore of Lake Pontchartrain. Berrigan, who had also evacuated, was driving down a Houston street when he saw an Amann Landscaping truck full of African Americans. The logo had been spray-painted over but was still visible. Berrigan hopped out of his car, pulled his gun, and forced the truck to the side of the road at gunpoint. There were eight men in the box and three in the truck cab. According to Amann, all but three of them fled the vehicle. "Berrigan made a citizen's arrest," Amann recalled. "At first the Texas State Troopers were about to arrest him for holding a gun on I-10, but when they found out what had happened, how he saved my stolen truck, they congratulated him on a job well done."[14] Amann had kept all five of his trucks behind a barbed-wire fence on Oak Street. The looters had stolen bolt cutters from Ace Hardware and used them to break into Amann's property. "They turned over all my files and took shits on them," Amann recalled. "They stole everything from my office, including cell phones, checks, tools, office supplies, everything."[15]

Word on Oak Street was that three of Amann's other landscaping trucks were constantly pulling up to a boarded-up house on Willow Street, and merchandise was being unloaded. Eventually, a combined contingent of NOPD and National Guard conducted a raid on the house. Stolen merchandise was piled high in every direction, in sitting rooms, the kitchen, and the bedrooms. The place looked like a warehouse. "Those guys were using my trucks to cart off every kind of electrical equipment you could imagine," Amann said. "These weren't poor Katrina victims. They were hardened professional felons."[16]

Tensions in the Riverbend neighborhood were rising. Deleray, for example, ran into a clique of looters robbing thirty boats from the Aqua Marine Inc. store on Oak Street. The looters were carrying the boats out in rickshaw fashion, trailers and all. When Deleray decided to confront them, to challenge them eyeball to eyeball, he had two AK-47s pulled on him. "He backed off," Foret recalled. "I told him saving boats wasn't worth getting shot over."[17] But Deleray, besides patrolling the neighborhood with his gun, also turned first responder. Starting on Thursday, and for the next two weeks, he went out in a boat rescuing hundreds of New Orleanians, most of them African Americans. He was also hired by CNN to take reporters like Anderson Cooper and Karl Penhaul to newsworthy disaster sites.

That Thursday Foret had an encounter with another looter. It was an African-American gentleman who called him Mr. Ace. Although Foret didn't know him by name, he liked the cut of his jib. He was surprised such a seemingly decent man would participate in massive, wholesale thievery. "Why you doing this?" Foret asked. "Why? Because," the looter told him, "it's our city. It's our time now." Foret stared back at him. "Everything but Oak Street," he said. "Oak Street is still ours."[18]

II

Jefferson Parish Sheriff Harry Lee was determined to stop looters under his jurisdiction. His police officers, for the most part, hadn't left, and the flooding in the parish was minimal. Sheriff Lee ordered his officers to arrest or shoot anybody breaking into businesses or homes. The criminals needed to know that the law still ruled on the streets of Jefferson Parish.

He also ended up sending ammunition to Orleans Parish, trying to help the NOPD to reclaim the upper hand. "Jefferson Parish did a great job," NOPD police chief Warren Riley recalled. "Harry did a great job. We needed assistance from him and he came through. He got a helicopter for us, flew it in from Tennessee."[19]

At a 7-Eleven store on the West Bank, Sheriff Lee himself broke up a looting ring. Instead of being angry, he was more inquisitive. "Why are you stealing from this store?" Lee asked the thieves. "Because we missed out on the Wal-Mart," a looter responded. "We got there too late. Everything was gone. It made us mad."[20]

It was their time now.

Everyone was mad at someone in Louisiana. Governor Kathleen Blanco, for her part, was incensed at the seemingly gratuitous breakdown of order in New Orleans. In any crisis, it was taken for granted that criminals would be on the loose, but in the sections of New Orleans that were above sea level, like the Carrollton-Riverbend area around Ace Hardware, a significant proportion of those who remained turned delinquent. It was distressing that so many people were taking cruel advantage of the disaster circumstances. The robberies and marauding were paralyzing the progress, such as it was, of recovery. Gun violence—usually random firings into the sky—interfered with the rescue of tens of thousands of stranded people. Buses that straggled into the city drew anarchist gunfire. In response, bus drivers took evasive action: they just returned home. As any National Guard or Coast Guard first responder could attest, it was the sniping—or rumors of it—that slowed down rescues more than anything else. Because of snipers, Governor Blanco truly believed federal forces were needed in New Orleans to quell the unrest. As of Thursday morning they had yet to arrive, though. "They have M-16s, and they're locked and loaded," Blanco said of the fresh National Guard troops entering New Orleans at a Friday-morning press conference. "I have one message for these hoodlums: these troops know how to shoot and kill . . . and I expect they will."[21]

Embracing General Bennett C. Landreneau of the Louisiana National Guard as her top advisor, Governor Blanco could barely contain her anger at the lawbreakers, warning them that Arkansas National Guardsmen had just arrived on the scene "fresh from Iraq," and they were "more than willing" to kill anyone breaking the law in the city. Unfortunately, it was only three hundred Guardsmen who were arriving that day, hardly enough to take control of the tinderbox that was New Orleans. Governor Blanco vented her fury at the disintegration of society in New Orleans and then stalked out of the news conference.[22] No sooner had she left than state police superintendent Henry Whitehorn stepped to the podium. Among his announcements was the fact that NOPD officers were quitting the force and surrendering their badges and the city. It was old news with a new twist. "They lost everything," Whitehorn said, "and they didn't feel it was worth taking fire from looters and losing their lives."[23]

Using the Wal-Mart on Tchoupitoulas Avenue as the NOPD supply depot, Warren Riley blamed FEMA for the debacle at hand. At a time when the police department desperately needed support he tried to talk with FEMA officials but met only a bureaucratic blow off, a shrugging don't-bother-me attitude. "The first three or four days [after Katrina] FEMA couldn't answer any questions," Riley said. "They would listen but never commit to anything and then when they did, it's telling you what they can't do. That's certainly not the time to hear what *can't* be done. I said, 'Listen, just be quiet. Tell me just one thing you can do for this city. One thing!' And they just got quiet. They're on television giving us no guidance, no direction, they're not telling us anything, they're not supplying us with anything. Why are they here? And that was the most frustrating part of the entire storm for me. I guess it goes back to when I was a kid. I liked Westerns. I liked military movies. I liked police stories. So you believe in the government. I truly believed that the military helicopters would be here, would be flying in. I believed that we would see trucks coming in. It just wasn't happening."

Although Riley hated to admit it, he believed racism was playing a part in the federal government's slow response. He knew President Bush cared about black people, he just didn't think he cared enough. "I can assure you that if some storm hit Kennebunkport, Maine," Riley said, "it would have been a different story." That suspicion or ones like it occurred to many millions of Americans. In the meantime, Riley and the few hundred officers on whom he could rely had to keep working against all odds in New Orleans.

Some 890 officers lost their homes and 400 had themselves been trapped by the flooding. Riley could tick off the harrowing flood story of virtually every officer. Troilin Laos was stranded for days on her roof in Hollygrove with no food or water. Sergeant Michael Levassier had stayed in New Orleans East to take care of his wife, who was recovering from breast cancer surgery; a seven-year-old daughter; and his mother with Alzheimer's disease. "The roof just came off my house," he told Riley. "I saw a body just float by my house." Nevertheless, once the Coast Guard rescued the family, Levassier voluntarily reported for duty to save others stuck in the deluge.[24]

Also marooned in New Orleans East was Officer Kathy Carter. "The water's coming up, she lives in a town house, she's on the second floor, and she's saying 'Oh my God, snakes are coming up! Water moccasins! The snakes!'" Riley recalled. "We said, 'Cathy, do you have your gun? And she said, 'Downstairs.' Now I don't think the snakes were trying to get her, I think they were trying to get away like everybody else. But still, we had police officers that were on roofs for three or four days and the worst feeling in the world is for us not to be able to help, not even our own, because of the flood location they were in."[25]

For others on the force, the storm, the sense of loss, and the sheer shock of the sights, sounds, and smells was truly unbearable. Officer Paul Accardo killed himself on Saturday, September 3. Late that night, Accardo, whose St. Bernard Parish home was wiped out in the storm, drove aimlessly to Luling, a nearby town, and parked his patrol car in the deserted parking lot of a boarded-up restaurant. After writing a note with the police department contact information, he shot himself.[26] Lawrence Celestine also used his police gun, shooting himself in front of a fellow police officer, a relative of his, on Friday.[27] "Their deaths just broke my heart," Eddie Compass recalled. "A lot of my bad press moments, when I cried, or seemed broken was because I just couldn't handle their deaths."[28] Riley was also extremely close to both men. "They were like family, brothers of the shield. I got a phone call at about 6:30 that morning that Paul had committed suicide," Riley recalled. "I talked to Paul Accardo at about 1:30 that morning because we'd had a report of gunfire near the Convention Center. I was at City Hall, at the Emergency Center. I went into a room where we had some cots set up, where Paul was sleeping, and I went in there to get my radio. Paul Accardo was sitting on the side of the cot. He had just awakened and I said, 'Hey, man, I'm going out. There's been some shooting. It never stops.'" A downcast Accardo, engulfed in sadness, said, "Yeah, Chief, it never stops." That was the last time Riley talked with Accardo.[29]

Smart, kind, and hardworking, Accardo was more of a desk cop than a brawler. Celestine, on the other hand, was one of the NOPD's toughest street cops, a fearless go-getter, physically fit, arms bulging, never willing to back down. "After thirty-one years on the job, I never thought I'd have to go and put a police officer in a body bag," said Captain Bob Bardy, Celestine's supervisor. Bardy said he didn't want his officers—"the kids" as he calls them—to see the much-admired Celestine brought so low.[30] "He could run people down in chases and things like that," Riley recalled. "I was sitting at Harrah's and got a phone call that he committed suicide and that was in about the same twenty-four hours as Paul. Celestine was first, then Accardo. It was a shock. We didn't know why. There were all sorts of rumors as to why he did it. Nobody knows why [Lawrence] pulled the trigger."[31]

Given the NOPD's disintegration you would think Riley would have welcomed the additional help of the National Guard. Not so. Louisiana and Oklahoma Guardsmen wanted nothing to do with the NOPD. They viewed them as part of the problem, not the solution. Riley recalled trying to evacuate about six hundred stranded people from I-10 on Friday at around one in the morning. About fifty military trucks had arrived at the Convention Center. Riley, wanting a piece of the action, pulled up to a Guardsman. "We're NOPD," Riley said. "We're glad you're here. We've got six to seven hundred people on the interstate; they've had no water all day. We need ten to twelve of your trucks." The Guardsman told Riley, "I need to clear that through my general." About twenty minutes later the Guardsman reappears. "The general won't talk to me and the guy comes back with a little smirk on his face and says, 'You know we just drove in a long way. We can't help you.'" A flabbergasted Riley said, "What do you mean? We've got seven hundred people up on the interstate. It's one-thirty in the morning. They're soaking wet from being in the water all day." The Guardsman held his ground. "I'm sorry, sir," he said, "I can't help you." Riley exploded in disgust. "I don't believe this," he said. "I need to talk to the general." The Guardsman fired back with some attitude of his own. "The general," he said, "has already spoken."[32]

What Riley didn't understand was that to be seen standing next to an NOPD officer had become anathema. Like Brown in Baton Rouge, there was an informal three-foot rule for the National Guard. At all costs don't get into a loser picture with an NOPD officer. Reports of desertion, suicides, and stolen Cadillacs were overwhelming the newswires. To his credit, however, Riley, one of the heroes of Katrina, decided he'd find another way to bring the I-10 folks to the Convention Center area. He flagged down a guy with a scraggly beard who had a big truck. "Sir," Riley said, "we're going to have to commandeer your truck." The guy asked, "For what?" Riley and his fellow officers explained that seven hundred people were on I-10. "You don't have to commandeer my truck," he said. "I'll help." They shook hands. "This guy followed us there, loaded up his truck, probably made seven or eight runs until six in the morning," Riley recalled. "And not only that. The guy came back every day and helped. Never asked for a penny. Came, stayed with us. That's a hero to me, not the military. And that's when I say: Is America what it says it is if the military refuses to help its people?"[33]

Because the NOPD was in such continued disarray, the security onus of New Orleans continued to fall on the Louisiana National Guard. They were, for the most part, clustered around the Superdome. On Thursday, Lieutenant Colonel Bernard McLaughlin was still at the Superdome. On Friday, he was ordered to secure the Convention Center in a military action. His diary offered a record of the tumultuous situation in the bowl of New Orleans on the third and fourth days after Katrina:

> We continue the bus evacuation, we get reports of the events in the city (looting and bad guys shooting at us and the police and rescue workers), the Dome is still surrounded by water, and there are roving groups of troublemakers on the highway overpasses. Bodies are being found in the water. We are getting pretty worn out but are heartened [when] troops arrive from Arkansas, Nevada, and California. More reports of looting in the city, the Wal-Mart is looted of all firearms and ammo, and teams are looking for those looters. Lots of private help is on-scene now, doctors, paramedics, out of town police and sheriff

units. . . . Lots of private helos and lots of private boats as massive search and rescue ops are underway.

I find a bucket, quickly take a washcloth bath, wash out my uniform, and put it on wet (it dries in about forty minutes in the heat).[34]

III

The most devastating rumor of all in New Orleans, the one that caused the most distress among the displaced people at the Superdome, was that the buses were coming. Just around the bend. They were always just an hour away, day after day. The lies were widespread: FEMA had sent a fleet down from Vicksburg; the U.S. Army had them rolling down I-20 from Fort Knox; they were already at Jackson Barracks and would arrive momentarily. There were rumors that children would be taken from their mothers and put on separate evacuation vehicles. In truth, the Blanco buses had been slowly moving people out in a haphazard way. FEMA-contracted buses were also starting to arrive. Throngs of displaced people were jockeying for position in the disorganized bus lines outside the Superdome. Tempers flared. Shoving and yelling were the norm. "It wouldn't be until Thursday that we started to get everybody out in a logical way," Terry Ebbert recalled. "By then, Houston had opened its shelters."[35]

The city of Houston, under the leadership of Mayor Bill White, had stepped up in a matter of days with the perfect place to receive the newly homeless of New Orleans. Reliant Park had 45,000 cots, hot food, medical units, and plumbing that worked. Those factors met the immediate requirements of the displaced storm victims. Houston was also prepared with trauma counselors ready to meet long-term needs. Lists of available housing were prepared, schools made room for an influx of students, and employment bureaus prepared to place willing job seekers. "I had read both John McPhee's *The Control of Nature* and John Barry's *Rising Tide*," White recalled. "I was familiar with the issues surrounding the levees. As soon as I heard they had broken, then I knew that our lives would change for many, many months. No community is equipped to absorb 200,000, maybe 300,000 people. But it was a reality that Houston had Americans in need and so we made it happen. It wasn't just our shelters that performed, our churches, businesses, and hotels all worked together to help as many people as possible."[36]

More than any other city in America, Houston exuded the can-do, let's-figure-out-how-to-make-it-happen attitude. Although the Astrodome had been built in 1962, and the presence of NASA had made Houston seem like a space-age city, it was really the economic woes in the industrial Midwest (or Rust Belt) during the 1970s that allowed Houston to prosper. With people migrating southward from Chicago, Cleveland, Detroit, and other Midwestern cities, Harris County's population doubled to two million people. The challenge facing Houston was how to build an infrastructure that could grow into a larger metropolis. A wildly pro-business posture was taken by both Democrats and Republicans. Houston soon became the energy/petrochemical capital of the world. The port of Houston supplanted the port of New Orleans as the commerce vortex of the Gulf South. Coinciding with the growth of both the port and the oil industry was the Texas medical industry. A golden age of aggressive research and modern medical care was launched in Houston, at facilities including Texas Children's Hospital, Memorial Hermann Hospital, and Methodist Hospital. "We developed a big civic spirit," White said. "We didn't hesitate or wrap ourselves in red tape. We learned from the mistakes made up north."[37]

By Wednesday, Houston was all set and waiting for its chance to help. But there were more glitches. Finding buses was hard enough in the midst of the Gulf South disaster; engaging qualified drivers was just as tough. One brave wildcat bus driver, however, became an unexpected hero. On Wednesday morning, Jabar Gibson, a twenty-year-old African American from the West Bank, heard that a bus barn in Algiers Point contained Orleans Parish school buses. Algiers Point never flooded, so the buses were dry—although the barn was a pen for vehicles awaiting service. Gibson didn't know that; he just found the keys to bus 0235 and drove it to the Fischer Housing Project in Algiers, where he invited panicked residents to clamber onboard. Although Gibson didn't really know how to drive a school bus, he managed to control the commandeered vehicle on a twelve-hour trip to Houston, shifting gears only when absolutely necessary. Texas police had been warned that a "renegade bus" was on the way, and they were on the lookout, but Gibson pulled up safely at the Astrodome just before midnight on Wednesday; his was the first bus to arrive from New Orleans. The fact that a high-school dropout could find a workable bus when none of the city's officials could became the stuff of legend in the unreal world of Katrina. "Back in New Orleans," *Times-Picayune* reporter Josh Peter later wrote, "the first FEMA bus would not pick up evacuees at the Superdome until later [Wednesday]—those at the Convention Center would wait two more days—but Gibson, even with Texas law enforcement out to stop him, had safely evacuated a busload of New Orleanians to Houston."[38]

It was Harris County Judge Robert Eckels, the county's chief executive, who was on hand at the Astrodome greeting Gibson when he arrived. At the request of Jack Colley, the Texas governor's Emergency Management director, Judge Eckels had gotten Reliant Park ready for the Louisiana evacuees. "We don't need 2,000 beds like we thought," Colley had telephoned him. "We have 23,750 and we're going to evacuate the Superdome and transfer them to the Astrodome. Can you do it?" The idea of saying no never crossed Judge Eckels's mind. A Texas Republican, the judge, whose father had been a Harris County commissioner, was best known for promoting a high-speed light-rail system between Houston, Dallas, Austin, and San Antonio. Although Harris County had a reputation for cowboys, as everything branded Texas did, in truth, Harris County was only 50 percent Anglo. "We have a huge Hispanic population, a large African-American population, and the largest Asian population in this part of the country," Judge Eckels said. "We have more consulates than Los Angeles. The port anchors a lot of that. What we have focused on, more than diversity, is the commonality, the things that bring us together. We understand that there's more that unites us than divides us as a community."[39]

When Gibson pulled up to the Astrodome, Judge Eckels welcomed him to Harris County. It was an odd moment, the first of many buses the judge would greet.

"Gibson just jumped in the bus and drove," Eckels said. "He just grabbed a bus and drove here, picked up people on the highway and drove them to Houston. That was the first bus. We had other people showing up. He got off the bus and people saw him and asked, 'Where's the driver? We need to park this.' And he was like, 'What do you mean? I drove it.' "

Judge Eckels claimed that in the end none of the Superdome evacuees were problematic. In fact, he thought the reports out of New Orleans were exaggerated. The African-American poor he encountered at Reliant Park were bedraggled and scared. There wasn't even a scent of violence about them. "All day we were getting ugly reports from the Superdome," Judge Eckels said. "Governor Blanco called me and said, 'There are criminal-type people; there have been a lot of problems in the Superdome. You need to be ready for it.' What we found was just the opposite. The people that got here were tired, they were in poor physical condition. They had been without food or water for several days, picked off the roofs of their houses or from a bridge. They had a long bus ride, some of them twelve hours or longer coming from New Orleans. But they were nice, courteous, no problem at all really. We made sure 2,700 doctors came to see them from medical centers all over the country."[40]

IV

Jabar Gibson drove only the first of many Louisiana buses to come. By noon on Thursday, two hundred buses had left the Superdome for Houston. Many of these late evacuees had nothing but the clothes on their backs, but they were the lucky ones, to be leaving in the first wave. Regrettably the police and military continued to enforce their no-pets policy. Mary Foster of the Associated Press wrote about one little boy at the Superdome who was about to board a bus with his fluffy white dog, Snowball. The National Guard stripped him of the dog at gunpoint. "Pets were not allowed on the bus, and when a police officer confiscated a little boy's dog, the child cried till he vomited," Foster wrote. " 'Snowball, Snowball,' he cried. The policeman told a reporter he didn't know what would happen to the dog."[41] The little boy was forced to board the bus and Snowball was left behind at the Superdome. "When we heard about it, a few of our volunteers went looking for Snowball," recalled Laura Maloney of the Louisiana SPCA. "We even heard he was in the Superdome, but it was the wrong dog. We never found Snowball."[42]

An even more egregious case of forced pet abandonment occurred when Denise Okojo, blind and ill with cancer, was evacuated from her New Orleans East apartment. Although Okojo was in desperate need of chemotherapy and radiation treatment, she didn't want to leave her Seeing Eye dog, Molly, a Labrador retriever. For six years, Molly had been her companion, making life bearable. "The Coast Guard airlifted me from my roof to Lake Charles Memorial Hospital's oncology unit," Okojo recalled. "They said no animals and pulled Molly away from me. I screamed and yelled, but they lifted me into a basket and off I went."[43]

Maloney had set up LSPCA headquarters in the Lamar-Dixon Horseshow facility in Gonzales, a couple of miles from where Tony Zumbado had stashed his Hallmark trailer before Katrina. Her task was daunting. Over seven thousand people had contacted the national SPCA asking for help finding their stranded pets. Molly was just one of the animals she helped find. Within two weeks, the LSPCA saved over 8,500 pets. Even more impressively, Maloney, following Duke Ellington's maxim of "no boxes," allowed other animal rights activists to join in the rescue effort. They helped save an additional 7,000 pets. "We had a lot of inexperienced people helping us," Maloney recalled. "So we had a lot of bites."

In Lake Charles, Okojo told a nurse about having had to abandon Molly. Word was sent to Maloney. Determined to reunite Okojo with her companion dog, Maloney and Caroline Page, also of the LSPCA, joined forces with two volunteers from the ASPCA in a daring boat journey. It took the four women six hours in a flatboat to make their way to New Orleans East. All the houses were underwater and addresses were impossible to discern, but they eventually found Okojo's apartment building. Page took an oar and smashed a first-floor window. She swam in the murky water, calling, "Molly, Molly." Hearing a throaty bark on the second floor, she made her way upstairs and found the faithful dog alive, but, as Maloney put it, "afraid, hungry and cold."[44] They rushed her to an animal clinic for care. Then they took Molly to Okojo, who was still at the hospital. "We all cried," Maloney said. "It was so emotional for all of us."[45]

In too many cases, people stayed behind and faced death rather than leave their pets. The many tragedies involving people and their animal companions during Katrina forced officials to recognize that rescues *have* to take pets into account. In the immediate aftermath of the storm, United States Senators Rick Santorum (R-PA) and John Ensign (R-NV) asked President Bush to appoint an "animal czar" for the Katrina response. Although the suggestion was declined, the nation was beginning to understand that animals are part of U.S. society and can't be separated from their human companions in good times gone bad. "You have no idea how many people wouldn't get on boats or helicopters because their pets couldn't come," Maloney recalled. "When you're old, and alone, your dog or cat *is* your family. To ask them to leave them behind is cruel and dumb. We need to educate our government accordingly."

What frustrated Maloney the most was that when the National Guard (and others) finally came to evacuate humans off of I-10 and the Causeway, they forced people, often at gunpoint, to leave their pets on the side of the road. The LSPCA had found buses, taken them to Jefferson Parish, and pleaded with the authorities to allow the organization to shadow the human rescue buses with ones for animals. The LSPCA would do the corralling of the pets and be responsible for the paperwork. They were told to get lost. "After the storm, Homeland Security came to us and told us they were wrong," she recalled. "To their credit, they wanted to work with us in the future to create a database of humans and pets so they wouldn't get separated permanently due to natural disasters."[46]

V

Ron Forman, CEO of the Audubon Institute, was responsible for the hurricane preparation plan for the New Orleans Zoo, Aquarium, and Species Survival Center on the West Bank. He had a permanent staff in place to tend to the animals both during and after Katrina. "The staff that stayed lived in the grounds in the reptile house, of all places!" Forman recalled. "You wake up and you're sleeping diagonally and there's the king cobra looking at you and you're wondering, 'Is this the best place to be sleeping after a hurricane?' "[47]

Luckily, Forman had Dan Maloney (husband of Laura Maloney at the LSPCA) spearheading his zoo efforts. While their number one concern was to care for the animals themselves, they also wanted to protect the community. "Trees were falling all over the place," Forman recalled. "If a tree hits the tiger enclosure and a tiger's running loose on St. Charles Avenue, we've got a safety issue. Some of the more dangerous animals truly are of concern. You couldn't have a community recovering from a hurricane with a lot of animals running around."[48]

Because Forman's wife, Sally, was Mayor Nagin's communications director, he stayed at the Hyatt, where the Office of Emergency Preparedness was set up. Two of the Formans' three kids endured the storm with them at the Hyatt.

Along with a couple of Audubon employees with tranquilizer guns, animal nets, and other high-powered weapons, just in case an animal got loose, Forman arrived at the zoo expecting the worst. Debris was everywhere; enormous branches and even uprooted oaks littered the once lovely grounds. As he approached the elephant exhibition, he saw his huge, gray mammal friends, Panya and Jean. They trumpeted him, as if to say, "We've been waiting for you guys. Get over here." At that point, Forman knew the zoo had survived. Out of 1,400 animals, he lost only 2 otters and 2 birds.

The Aquarium of the Americas was an entirely different story. After the 17th Street Canal breached, Forman told his aquarium employees to leave New Orleans ASAP. He wasn't going to be responsible for the loss of human life. On Tuesday evening, he found himself standing in the aquarium with water starting to enter the building and only about three hours of fuel left in the generators. Once the generators stopped, the fish and marine animals would die. "I started walking through the water down Canal Street to the Marriott," Forman recalled. "At the Pelham Hotel, I knew the general manager, so I asked him, 'Where are you getting diesel fuel for your generators? Because to keep the animals alive, I need diesel fuel.' "[49]

The answer to the question was Patrick Quinn Jr., part owner of over twenty hotels in the city. Quinn took it upon himself to hunt down diesel fuel to keep hotels and hospital generators running. On Tuesday, he had Steve Carville, brother of Democratic strategist James Carville, send a flatbed truck loaded with ten-gallon drums of diesel. That was a start. By Wednesday, Quinn's son, Frank, had one-upped Carville, procuring a huge diesel dump truck with high-pressure hoses. It was filling up the generators at Bell South and Tulane University hospitals. It then came to the aquarium. "This fuel truck shows up with weapons on it," Ron Forman recalled. "It was like the Marines were coming to save the fish. Eventually we found out where the four thousand gallons of diesel fuel went. We were in business."

With that goal accomplished Forman went looking for NOPD officers who could protect the aquarium from looting. Forman told the highest-ranking officer he could find, "I have a building with a generator that's worth a hundred million dollars. I've got food. I need police." That gave the officer an idea. "Well," he said, "we lost our communication center in the French Quarter. Can we move the communication center in?" A relieved Forman was happy to oblige.[50]

Within an hour, truckloads of NOPD police equipment started rolling into the aquarium. Most of the officers hadn't taken a bath or shower in days. Forman broke into his souvenir store and started handing out T-shirts and tank tops. "They cleaned up and got dressed," Forman recalled. "One officer in particular, Don Kinney, really cared deeply about the sea otters and penguins. He started being their protector, feeding them and watching out for them. I don't know what I would have done without him."

Officer Kinney was a nineteen-year veteran of the NOPD. A sweet-natured street cop, he was average in every way except one: he brought his pet Moroccan cockatoo, Yogi, with him wherever he went. Yogi wasn't so much a mascot as he was an appendage. "Yogi had his own room, with a television on all the time," Kinney recalled. "He isn't much of a talker, although he'd say, 'I love you' or 'How are you?' But Yogi is the friendliest bird you ever met. He constantly wants me to rub his chest, to pay him a lot of attention, and I do."[51]

With Yogi on his shoulder, Officer Kinney holed up at the aquarium on Thursday, charging up police radio batteries in assembly-line fashion. He would also monitor radio text. Then, at a designated hour, he'd walk boxes of batteries across the street to Harrah's Casino, where they were handed out to fellow cops. Along with Officers Clayton Dunnaway and Tommy Green, he was keeping NOPD communications open. "But we were having problems with the aquarium generators," Kinney recalled. "The filters weren't right. We were getting 900 rpms instead of 1,800 rpms. We weren't keeping the fish and birds and animals cool enough. They were dying."[52]

Desperately, Officer Kinney tried to save the aquarium's wildlife. He fed the birds thawed mice and threw squid to the penguins. He made sure the alligators got slop and the sea turtles got released before they died. But he had only partial success. More than eight thousand fish died, despite his earnest efforts. "The value is in the millions," Forman recalled. "We started losing animals one by one. Don and others would dive into tanks and get out these eight-foot sharks and throw them in the Mississippi. The sharks all died. But Kinney did an incredible job and all the penguins lived."[53]

Throughout Thursday and Friday, Forman shuttled between the Audubon Zoo and the Aquarium of the Americas. "I had been giving people rides," Forman recalled, "but now I stopped. That was the saddest thing. The vast, vast majority of people, pushing grocery carts with their belongings, holding kids, were just good people looking for help. But a gun had been pulled on me so I had to stop."[54]

One evening, Forman drove with a police escort to Audubon Park and couldn't find a single zoo employee. About twelve people were supposed to be looking after the animals, making sure they were fed, and so the silence was eerie. Then Forman heard a noise coming from the swimming pool outside the zoo. "I went over there and there was a group bath going on," Forman recalled. "Everybody had gone swimming. They were barbecuing hamburgers. They had ice cream and they were swimming in the swimming pool. It was one of those incredible highs, strange and surreal, when I felt relieved, when I knew, 'God, everything is going to be all right.' "[55]

Philanthropist Julie Packer of the Packer Foundation sent a private plane to airlift the surviving animals from the Aquarium of the Americas to the Monterey Aquarium in California. A specially equipped, climate-controlled truck arrived to evacuate the penguins, seals, and otters. "As the penguins left, they all marched out the door in a straight line," Forman recalled. "The ordeal was over."[56]

Because Officer Kinney's own home had been totally destroyed, Forman allowed him to live at the aquarium for two months. Yogi was always right by his side. "Yogi was like a little kid," Kinney recalled. "Because of the trauma of Katrina he really didn't like being left for a minute. If I left him out of his cage, he would walk behind me like a human."[57]

Aquatic creatures of all kinds died in Louisiana, not just at the Aquarium of the Americas. The Coast Guard reported finding thousands of smashed fishing vessels in the coastal parishes. Katrina also decimated the enormous state seafood industry. For example, about 80 percent of the crawfish industry was wiped out by the dual hit of Katrina and Rita. Surging salt water annihilated their habitat. At least temporarily, a way of life was destroyed. Rice farmers in Louisiana had long made money by tossing cages in Atchafalaya Bay and pulling crawfish out of the mud. According to the *Times-Picayune,* about 5,000 mudbug-hunting jobs were lost due to the hurricanes. Most of the state's twenty crawfish processing plants were unable to reopen due to severe damage. The nation's major oyster beds lay just east of the Mississippi Delta, but following the churning of the seas, these reefs were dead. Before Katrina there were 2.3 million acres of oyster beds; by early September they were nearly all gone. "In some cases reefs that were there for forty, fifty years," oyster farmer Pete Vujnoich Jr. said, "are not there anymore."[58]

VI

On Thursday afternoon, even with the Superdome evacuating, an estimated 40,000 others were still stranded all over the Greater New Orleans area. Major Dalton Cunningham of the Salvation Army worried about the 200 people who had taken refuge at one of his group's buildings on South Claiborne Avenue in the city. Many of them were sick and all of them were hungry. On Thursday, Cunningham was told that help would probably not arrive for several days. "They said they're doing it by quadrant and we'll just have to take a number and get in line," Cunningham said, and then his thoughts returned to those in his care. "They are there without food. Their lives are threatened. I'm not even sure they'll be alive when we get there."[59]

At the Morial Convention Center, 20,000 tragedies were circling at once, one for each person inside or outside. Leaving his family encampment across the street in the Harrah's parking lot, Ivory Clark entered the madhouse of a building on Thursday. He was still looking for electricity, something to power the machine that controlled his aunt's asthma. He asked the police or National Guard for help. "They had generators," Clark recalled. "They admitted it to me, but refused to help Auntie."[60]

With a lot of time to kill sitting within view of the Convention Center, Clark's wife, Donna, scrawled observations and feelings in her journal. Some of her entries capture the hardships of those days, as well as her appreciation of her husband:

> We dozed on and off until mid-morning, awoke by fellow neighbors singing various gospel songs . . . they praised the Lord and praised the Lord until the troops came over and demand them to lower their voices or they would be arrested.
>
> The usual rituals, sleep, food, washing off with bottle water, trying not to smell if possible.
>
> Word kept coming the buses were coming. [No one] really got excited, simply because we had heard this same story before, over and over.
>
> Auntie was breathless. The humidity and the smoke from the smokers didn't help much. I got a couple of our neighbors in the garage to include my aunt in prayer, while others fanned her and sung hymns.
>
> My husband was truly a blessing to us all he never complained he did what a man was supposed to do. He protected us, stuck with us through it all and never once utter a negative word, always willing and doing a real man's duties as a husband, father, grandson-in-law, nephew-in-law, uncle. I truly thank God for him and may God continue to bless him in all ways.[61]

VII

While the Clarks and others at the Convention Center waited to be evacuated, a glimmer of hope appeared on Thursday in the form of Lieutenant General Russel Honore, who was the commander of the Department of Defense's Joint Task Force Katrina. He was able to announce in a briefing at the Pentagon that 7,400 troops would be in Louisiana by Thursday night, with an additional 6,000 in Mississippi. The buildup in Greater New Orleans was to be largely through the deployment of 1,400 National Guard soldiers trained in security, including the 300 from the Arkansas National Guard. Honore explained that troops would continue to arrive from the National Guard and the regular army until the total reached about 30,000. He gave the impression of a man not overwhelmed by the magnitude of the job at hand. For the time being, however, he was just another official spouting promises. In terms of improving the lot of hurricane victims, 7,400 soldiers on the ground were having little effect. Honore later acknowledged the master strategy that was the enemy in the situation. He told the *Times-Picayune* that "if he had been attacking the Gulf Coast, he would have used the same tactics as the deadly storm. Honore said Katrina's wrath amounted to a 'classic military maneuver,' using overwhelming power and the element of surprise to cripple communications and block transportation routes. And the storm covered its flanks by breaching levees and destroying railroads."[62]

As Honore arrived in New Orleans, pathologist Dr. Greg Henderson wandered around the Convention Center trying to help out wherever he could. He told people to sit in the shade and to drink from the few IV bags available because no needles were around. Walking around with a stethoscope and not much more, he was more priest than doctor. Calling the Convention Center "where hell opened its mouth," Henderson told Anderson Cooper of CNN that he saw approximately fifty dead bodies in hell's jaws, a number that was conflated. But nobody questions, however, that the streets of New Orleans were awash in bodies.[63]

Later in the day, General Honore flew to New Orleans and immediately began directing the soldiers—and anyone else who came into view. Honore had grown up on a scrubby farm in Lakeland, Louisiana, and had a role model in a relative, Charles Honore, who rose to the rank of major general in the U.S. Army. According to a cousin, the Honores were typical Cajun: "a little bit of French, a little bit of Indian, a little bit of Creole, a little bit of black, a little bit of white—just like a gumbo."[64] When Katrina hit, Russel Honore was serving as commander of the First Army at Fort McPherson in Georgia.

The moment he stepped off his Blackhawk chopper, he made his presence felt. "Put those damn weapons down," he snapped at soldiers he saw on a military vehicle, looking as though they were riding into open warfare. He admonished National Guard troops that they were not in Iraq, and that they needn't point weapons at people until provoked. He had the kind of cool strength for which New Orleans had been longing for days. Although Honore could not be everywhere at once, and in fact did little to change the overall situation on Thursday, his arrival in a city so sorely neglected was at least something. Barbara Starr, the Pentagon reporter for CNN, told Wolf Blitzer (on Friday) of traveling with Honore: "Up to 1,000 National Guard troops came into town earlier today. Those people in the Convention Center, thousands of them, are in absolutely dire straits and this has been General Honore's stated priority for the day, is [*sic*] to get that food convoy in with food, water, medical supplies and begin to medevac out those who are too sick or injured perhaps to stay. What I can tell you is, as we have traveled with the military today, General Honore has ordered, at the top of his lungs, every troop that he comes across to point their weapons down. He has repeatedly gone up to vehicles, gone up to National Guard troops standing sentry, even gone up to New Orleans PD and said, 'Please put your weapon point down. This is not Iraq.' Those are his words. He wants the profile here to very much be one of a humanitarian relief operation."

In the meantime, the Army Corps of Engineers and its contractors were trying a new strategy in the effort to plug the breached levee on the 17th Street Canal, so that the city could start the long process of pumping itself dry. While helicopters were still dropping sandbags into the 700-foot-long hole, Boh Brothers Construction began driving pilings across the width of the canal, so that a sort of dam could be constructed. That way, the water could be pumped out of the canal, at least up to the point at which the sandbags reached. That would facilitate the repair to the levee wall.[65]

The damming of the breached New Orleans levees became known as the "Super Sack." The company that worked with the U.S. Army Corps of Engineers was the Golden, Colorado, company Kaiser-Hill that had cleaned up the Rocky Flats nuclear weapons plant in Denver, one of the largest cleanups ever completed in the world. Kaiser-Hill was a joint venture by two preeminent firms, Kaiser Engineering and CH2M Hill. To dispose of nuclear waste, the company had designed special polypropylene sacks that could hold contaminated materials and soils. Coordinating with Lift-Liner vendors, MHF Logistical Solutions, and other companies, Kaiser-Hill sent 2,400 of these "super sacks" to the Corps of Engineers to be filled with sand and dropped in the breaches. The Corps loaded each bag with up to 16,000 pounds of Gulf South sand. They then had helicopters carefully lower them into the breached sections of the levees.[66] "I grew up around New Orleans," recalled Jerry Long, a vice president at Kaiser-Hill. "We recognized that robust bags used to ship radioactive soil were needed. You didn't want the wrong plastic bags that had chemicals in them. You didn't want fabrics that would disintegrate over time either. We had the product they needed. And we didn't hesitate, we shipped it to them in record time."[67]

VIII

While the breaches were being filled with Kaiser-Hill sandbags, City Councilwoman Jackie Clarkson was talking with Mayor Ray Nagin on Thursday on the fourth floor of the Hyatt. When would the breaches be repaired and the flooding subside was the million-dollar question. Also in the room were Chief Technology Officer Greg Meffert, Director of Homeland Security Terry Ebbert, and a few police officers. "Mr. Mayor," Clarkson said. "I've called as many people as I know to call, left every message, gotten in touch with everyone, gotten as many people out of town as I can. I'm out of a job. What can I do?"

Before Mayor Nagin could give Clarkson an assignment, another police officer burst into the room. "Let's go, Mr. Mayor," the officer said. "Gotta get you outta here." Nagin got up, grabbed Clarkson by the elbow, and headed for the stairwell, ready to climb up to his quarters on the twenty-seventh floor. After conferring privately with the officer for ten or twenty seconds, Nagin turned to Clarkson, filled with terror. "There's a couple of rabble-rousers leading the mob from the Superdome and they're trying to get through the Hyatt doors," Nagin told Clarkson. "They're coming after me."[68]

By this point in the Katrina saga, Nagin was paranoid about his security. He was cracking at the seams. Worried that the Superdome folks were going to get him, he had become all jitters. Perhaps he was consumed with guilt for not making an appearance at either the Superdome or the Convention Center. Or maybe he couldn't function properly without sleep. "Mr. Mayor," Clarkson said. "They're after you, not me. You're important and you're young. You get up those stairs. They're not after me. You go on. I'll think about it."[69]

Nagin didn't want no as an answer from Clarkson. He insisted that Clarkson come with him to the top bunker. "I thought, 'Do I want to die of a gunshot or a heart attack?'" Clarkson said. "I stayed right behind him up those steps. I decided at age seventy, I'd rather take my chances with a heart attack than a gun. The National Guard and the police held them off. There were only three or four rabble-rousers. The rest of them were very nice people. They just thought that we had resources in the Hyatt that they didn't have."[70]

Mayor Nagin spent the rest of Thursday in a paranoid state of mind. No rescue boat or Superdome speech or direct action for him. Unable to communicate with the outside world, terrified, he munched on peanuts and tried to listen to Garland Robinette. "The mayor had a wind-up radio and we would take turns winding it up," Clarkson recalled. "It was a present his wife had given him for Christmas. An old kind of battery radio to listen to WWL. He didn't even have a transistor. We kept looking out the window at all the people on the bridge between the Hyatt and the Superdome. They were getting water dropped for them. The police were out there. They were interacting with the National Guard. It was very calm. We were getting so hopeful. At that point we were listening on the radio. That was the first time we heard all the criticism that was on the radio."[71]

With Nagin and Clarkson was Sally Forman, the mayor's tireless communications director. During Katrina, she, in her own words, "acted more like the mayor's secretary."[72] A staunch defender of Nagin's erratic behavior during Katrina, Forman booked her boss on CNN, *60 Minutes,* and *Oprah*. She insisted he wasn't hiding in the Hyatt from August 29 to September 1. "We had no transistor radio," Forman recalled of City Hall's lack of communication. "Twice I stopped a woman sitting on the Hyatt floor with a transistor to glean information. We had no other way to know what was going on."[73]

Because they hadn't really heard the criticism being levied at the New Orleans city government by the major networks and the rest, Nagin, Forman, and Clarkson were all taken aback by what they heard on WWL. For days, these two strong-willed women had been urging Nagin to be proactive, to show his leadership stripes, to abandon the Hyatt hideout stance. Nagin *had* taken a boat trip to various flooded New Orleans neighborhoods, but he always made sure his clothes didn't get dirty. He had peeked into the Superdome a couple of times but always in a covert, spontaneous, glimpse-me-if-you-can fashion. Now, on Thursday, General Honore announced that federal troops were coming. They were arriving en masse. If Nagin did nothing to dispel the notion that he was AWOL from the first-responder drama, then his political career was probably finished. He would go down in history as one of worst U.S. mayors ever. "You have to get on the radio and tell them what happened," Clarkson urged Nagin, "that your city troops performed their jobs and are still doing the jobs and you've had no backup from the state or federal government and that's the problem and they need to get in here and you need to tell that story."

Since Nagin was incapable of taking the plunge of actually informing the public of what was happening, Forman and Clarkson took matters into their own hands. They took a satellite phone to a Hyatt room and convinced the mayor to dial up Garland Robinette. "That's when he did his famous irate speech and blasted the government and blasted the president," Clarkson recalled. "Sally and I were standing right at his side. We were going 'Go, go, go!' "[74]

All week long policemen, firemen, and medical personnel wanted the mayor to do something—anything. Forman later admitted that the mayor's problem was proper communication with the Superdome, which was connected to the Hyatt. So why didn't Nagin wade in the water, stand under a Superdome goalpost, and have a bullhorn moment or engage in boat rescues? Why stay sequestered in a hotel? Why abandon City Hall to Terry Ebbert? Why hide from the disaster? "Symbolically the mayor felt that they didn't need another political person in a rescue boat," Forman explained in his defense. "The best thing the mayor could do was run the operation on the ground. Politically it was perhaps a mistake. And he didn't have a bullhorn, so we made flyers to hand out at the Dome."[75]

Like Clarkson, Forman wanted the mayor to be bold, to let his blood vessels burst in a moment of authentic outrage at the slow federal response. His national image was sinking by the minute and he needed to pull a public relations rabbit out of his top hat. Time was of the essence. If he waited until Thursday night, then all the federal troops would be in town. He would lose his right to complain. His window of opportunity was dwindling away. "At the time he did the WWL interview, our biggest concern was the people," Forman said. "We were just hoping that locals were listening to the radio. I never thought it was going to have reverberations."

IX

Garland Robinette was the populist voice of passion in the midst of the frightening degeneration of New Orleans. Columnist Dave Walker of the *Times-Picayune* called him the "reluctant oracle of the post-Katrina apocalyptic air." Anyone who had a radio in the Greater New Orleans area was tuned into WWL, gaining bits of information, taking good advice, and something akin to the truth from the veteran newsman. At various times in Robinette's extended broadcast, people would call in and give a description of their situation wherever they were. "We stayed real," Robinette said. "Or tried to."

Robinette had become a local folk hero for having broadcast through

Katrina on Monday and having fled the floodwaters on Tuesday. After visiting his wife in Natchez and getting some sleep, he was back on the air from a Baton Rouge studio, living at a friend's house. Clear Channel had stepped into the fray and was broadcasting WWL live all over the world via the Internet. "I just got furious on the air," Robinette recalled. "I didn't care if I had a job or not. I kept saying on the air to the U.S. government, 'Where are you? Where the hell are you? This is day four and there's no help. What the hell is going on?'" With Clear Channel allowing Robinette free rein, the reluctant broadcaster started fielding calls from foreign countries. "Got a call from Australia. Got a call from England. All these listeners worldwide were asking, 'What can I do?' I said, 'This can't be happening in the United States.' I am reeling. I'm so mad. And the story I get from Sally Forman is that [she and Nagin] were in the Hyatt. She said, 'I'm going to call Garland's show.' So I'm really lettin' it fly, shaming the government. I'm in one of my rants saying 'People are dying on the streets! They're dying on the overpasses! They had no food!' Just ranting like that and suddenly on the other line was Sally Forman with Mayor Nagin. And he just started echoing what I'd been saying."[76]

Like everyone else in New Orleans, the mayor was angry and chose to let it show. He didn't sound like a mayor. He didn't even sound collected. And yet the way that Nagin lashed out, mimicking Robinette, was a perfect reflection of the mood of the city on Thursday. Knowing that federal troops were on the way gave him the opportunity to demand federal troops. That way his grandstanding words would be construed by the press to be decisive. It was the perfect, phony, cause-and-effect gambit.

Nagin began by speaking of his contact with President Bush:

> NAGIN: I told him we had an incredible crisis here and that his flying over in Air Force One does not do it justice. And that I have been all around this city, and I am very frustrated because we are not able to marshal resources and we're outmanned in just about every respect.
>
> You know the reason why the looters got out of control? Because we had most of our resources saving people, thousands of people that were stuck in attics, man, old ladies. . . .

And they don't have a clue what's going on down here. They flew down here one time two days after the doggone event was over with TV cameras, AP reporters, all kind of goddamn— Excuse my French, everybody in America, but I am pissed.

ROBINETTE: Did you say to the President of the United States, "I need the military in here"?

NAGIN: I said, "I need everything."

Now, I will tell you this—and I give the President some credit on this—he sent one John Wayne dude down here that can get some stuff done, and his name is General Honore. He came off the doggone chopper, and he started cussing and people started moving. And he's getting some stuff done. . . .

ROBINETTE: What do you need right now to get control of this situation?

NAGIN: I need reinforcements, I need troops, man. I need 500 buses, man. We ain't talking about— You know, one of the briefings we had, they were talking about getting public school bus drivers to come down here and bus people out here.

I'm like, "You got to be kidding me. This is a national disaster. Get every doggone Greyhound bus line in the country and get their asses moving to New Orleans."

They're thinking small, man. And this is a major, major, major deal. And I can't emphasize it enough, man. This is crazy.

I've got 15,000 to 20,000 people over at the Convention Center. It's bursting at the seams.

ROBINETTE: Do you believe that the President is seeing this, holding a news conference on it but can't do anything until Kathleen Blanco requested him to do it? And do you know whether or not she has made that request?

NAGIN: I have no idea what they're doing. But I will tell you this: You know, God is looking down on all this, and if they are not doing everything in their power to save people, they are going to pay the price. Because every day that we delay, people are dying and they're dying . . .

You know what really upsets me, Garland? We told everybody the importance of the 17th Street Canal issue. We said, "Please, please take care of this. We don't care what you do. Figure it out."

ROBINETTE: Who'd you say that to?

NAGIN: Everybody: the governor, Homeland Security, FEMA. You name it, we said it. . . .

ROBINETTE: If some of the public called and they're right, that there's a law that the President—that the federal government can't do anything without local or state requests, would you request martial law?

NAGIN: I've already called for martial law in the city of New Orleans. We did that a few days ago.

ROBINETTE: Did the governor do that, too?

NAGIN: I don't know. I don't think so.

But we called for martial law when we realized that the looting was getting out of control. And we redirected all of our police officers back to patrolling the streets. They were dead tired from saving people, but they worked all night because we thought this thing was going to blow wide open last night. And so we redirected all of our resources, and we held it under check.

I'm not sure if we can do that another night with the current resources. . . .

They're showing all these reports of people looting and doing all that weird stuff, and they *are* doing that, but people are desperate and they're trying to find food and water, the majority of them.

Now you got some knuckleheads out there, and they are taking advantage of this lawless situation. . . . But that's a small majority [sic] of the people. Most people are looking to try and survive.

And one of the things—nobody's talked about this—drugs flowed in and out of New Orleans and the surrounding metropolitan area so freely it was scary to me. That's why we were having the escalation in murders. People don't want to talk about this, but I'm going to talk about it.

You have drug addicts that are now walking around this city looking for a fix, and that's the reason why they were breaking into hospitals and drugstores. They're looking for something to take the edge off of their jones, if you will.

And right now, they don't have anything to take the edge off. And they've probably found guns. So what you're seeing is drug-starving crazy addicts, drug addicts, that are wrecking havoc. And we don't have the manpower to adequately deal with it. We can only target certain sections of the city and form a perimeter around them. . . .

Robinette pointed out that some people, aware that the federal government couldn't come in unless formally requested to do so, felt the response so far had been as good as could be expected. Nagin acknowledged that the remark he was about to make would probably get him in trouble:

NAGIN: But we authorized $8 billion to go to Iraq lickety-quick. After 9/11, we gave the President unprecedented powers lickety-quick to take care of New York and other places.

Now, you mean to tell me that a place where most of your oil is coming through, a place that is so unique when you mention New Orleans anywhere around the world, everybody's eyes light up— You mean to tell me that a place where you probably have thousands of people that have died and thousands more that are dying every day, that we can't figure out a way to authorize the resources that we need? Come on, man.

You know, I'm not one of those drug addicts. I am thinking very clearly. And I don't know whose problem it is. I don't know whether it's the governor's problem. I don't know whether it's the President's problem, but somebody needs to get their ass on a plane and sit down, the two of them, and figure this out right now.

Robinette wanted to know what the station could do to help.

NAGIN: Organize people to write letters and make calls to their congressmen, to the President, to the governor. Flood their doggone offices with requests to do something. This is ridiculous.

I don't want to see anybody do any more goddamn press conferences. Put a moratorium on press conferences. Don't do another press conference until the resources are in this city. And then come down to this city and stand with us when there are military trucks and troops that we can't even count.

Don't tell me 40,000 people are coming here. They're not here. It's too doggone late. Now get off your asses and do something, and let's fix the biggest goddamn crisis in the history of this country.

ROBINETTE: I'll say it right now, you're the only politician that's called and called for arms like this. And if—whatever it takes, the governor, President—whatever law precedent it takes, whatever it takes, I bet that the people listening to you are on your side.

NAGIN: Well, I hope so, Garland. I am just— I'm at the point now where it don't matter. People are dying. They don't have homes. They don't have jobs. The city of New Orleans will never be the same in this time.

ROBINETTE: We're both pretty speechless here.

NAGIN: Yeah, I don't know what to say. I got to go.

When Nagin hung up the telephone, he broke down crying. Had he just made a fool of himself? Had he succumbed to helpless anger and had a meltdown on the air? Or had that been his Rudy Giuliani moment? According to Clarkson, he sequestered himself in his bathroom for twenty or thirty minutes, despite Clarkson's and Forman's efforts to coax him out. "I was proud of him for delivering those words, sticking it to Blanco, FEMA, all of them," Clarkson said. "He did well." In the far-off distance you could hear snorting explosions and see columns of smoke. They were sequestered from the deluge, but couldn't escape its prolonged wrath. Both Clarkson and Forman were proud that Nagin told Bush and Blanco to "get off your asses and do something," and "let's fix the biggest goddamn crisis in the history of this country." With little knowledge of history, and unable to get accurate information, Nagin truly believed the flooding of New Orleans was bigger than Gettysburg or Pearl Harbor or 9/11. His sense of perspective was shattered. But he had at last found his voice (or at least *a* voice). His meltdown was a much appreciated tonic to the poststorm bureaucratic abyss of both FEMA and Homeland Security. "I think the world needed to hear it," Clarkson said. "I think the government needed to hear it. I'll tell you, two hours later, that general was on that highway, that convoy was on that highway."

Forman insisted that "get off your asses" wasn't meant as a direct punch at President Bush. It was just Mr. Mayor letting off some steam. "CNN showed up for a dub," Forman recalled. "We gave it to them. I never believed the mayor was speaking to the nation." That was a very naïve presumption, uttered defensively with a self-conscious air. Her immediate concern was the psychological state of Nagin. "I tried to get him to calm down," she said. "I kept saying, 'It's okay. It's okay.' He eventually stopped crying and wandered into another room. 'Let me be alone for a while,' he said. 'I need to be alone.'"

X

Michael Brown of FEMA was all over the television on Thursday, doing so many interviews that people watching began to wonder which business he was in: disaster response or television programming. One press briefing each day would have been understandable, leaving him the time to oversee the response. Instead, Brown was available for one-on-one interviews with all of the major networks and cable news channels. It was clear that Brown was being offered as the face of the disaster for the Bush administration. That day, however, it was not a kindly or empathetic face. Asked by Aaron Brown on CNN's *News Night* about the distress seen throughout the New Orleans area, especially by the poor populations, Michael Brown responded, "Unfortunately, that's going to be attributable to a lot of people who did not heed the advance warnings."[77]

FEMA was playing catch-up all week with the disaster that continually spun ahead of them. Contracts that should have been arranged before Katrina were still in the negotiation phase. Brown tried to fend off criticism of FEMA's response by pointing out that 30,000 troops were going to be on the ground in the affected areas within three days, that hospitals were being emptied, and that rescues were still under way. During the course of the *News Night* interview, viewers saw fresh footage of New Orleans. It no longer looked like a Third World country, but more like a biblical throwback. "Considering the dire circumstances that we have in New Orleans," Michael Brown said, "virtually a city that has been destroyed—things are going relatively well."[78] That was a preposterous statement, but he had become the preposterous face of the deluge.

The false, upbeat assessment from Brown, who had been forced by Chertoff to stay in Baton Rouge, was challenged by anyone who witnessed the Convention Center firsthand. "These people are being forced to live like animals," reported CNN's Chris Lawrence that same day. "And we are not talking a few families or a few hundred families. We are talking thousands and thousands of people just laid out over the entire street. Living in these horrible conditions, trash, feces, dirty food—I mean, in just the worst possible conditions. We saw a man right in front of us literally dying. Going into a seizure on the ground, people trying to prop his head up. They have no medicine, no way to evacuate him. . . . These people are saying, 'How much longer can we last? Where are the buses? Where is the plan? Where is the help?' People are saying, 'We don't have food. We don't have water.' Are you really just going to let us sit here and die like this? Because people are already starting to die right in downtown New Orleans."[79]

Behind the positive face of the director, FEMA employees were suddenly handing out no-bid assignments, costly for taxpayers and futile in terms of timely deliveries of vital necessities. And the director was playing catch-up, too. At two in the afternoon, Mayor Nagin had released an extraordinary special statement for an elected official in the United States: "This is a desperate SOS," it read. "Right now we are out of resources at the Convention Center and don't anticipate enough buses. We need buses. Currently the Convention Center is unsanitary and unsafe and we're running out of supplies."[80] The tour buses FEMA contracted were for the Superdome.

Incredibly, FEMA didn't become aware of anyone at the Convention Center until that day, Thursday. According to reports, the agency learned about the 20,000 people in and around the huge building by watching the TV news—and perhaps by hearing/reading Nagin's SOS. FEMA did know about a team of Swift Water Rescues airlifted to the scene by the state of California. The team had already rescued hundreds of people in New Orleans and neighboring Jefferson Parish. On Thursday, the Californians were ordered by FEMA to immediately cease operations because they hadn't received a security clearance from the federal agency.

Meanwhile, Brown continued to engage in bizarre e-mails to FEMA officials, many of them trying to *look* on top of things. "Please roll up your sleeves of your shirt . . . all shirts," press secretary Sharon Worthy instructed him. "Even the President rolled his sleeves to just below the elbow. In this crisis and on TV you need to look more hard-working. ROLL UP THE SLEEVES!"[81]

One of the most bizarre clashes of personality in American politics occurred when Brown did a spot check on devastated St. Bernard Parish. Later, in his testimony to a congressional committee investigating the federal, state, and local response to Katrina, Brown called Louisiana "dysfunctional," partly because of the state's unique system of parish presidents, sheriffs, and mayors, who were often political opponents. "I distinctly remember making a trip to St. Bernard Parish where I met Parish President [Junior] Rodriguez," Brown said. "A very vulgar, profanity-spewing encounter with him. Very entertaining. It was surprising to me. So after leaving this meeting with him, who I *think* is in charge of St. Bernard, somebody pulled me aside and said, 'Okay, you've done this, but you really have to go see Sheriff Jack Stephens.' "

Brown asked why this was necessary when he had just met the parish president. He was told that both men "think they're in charge of this parish." According to Brown, his security guards put him in a Humvee and made a circuitous trek down the back roads of the trashed-out parish. "I even joked with one of my aides, 'Hey, they're taking me to go dump me in the river somewhere. . . .' It was wild." Eventually, Brown was escorted down a trail to some forlorn structure straight out of *The Return of the Swamp Thing* and into the inner sanctum of Sheriff Stephens. "And we found him where he had commandeered someone's houseboat. I walked in and it was like a scene in a movie. I'd just been in St. Bernard with Rodriguez, where they had minimal supplies and we're having a discussion about what you need—how many MREs I can get you, 'you' meaning right there in your emergency operation center, so you guys can go do your work. I go to see this sheriff who is now living on this houseboat and I walk into this huge buffet. It was astonishing to me. And he thinks he's in charge of everything."

Brown had felt Rodriguez was "just one of these old guys who's just like a sailor—every other word is a cuss word." But he didn't take it personally. Now, as Sheriff Stephens launched into a litany of complaints and demands (while eating delicacies), Brown had time to analyze the man before him. According to Brown, Stephens was "a little more polished and, you know, a little more of a slick politician kind of guy who, with all these deputies all around, with these guns everywhere." Even though he grew up in rural Oklahoma, Brown had never experienced such back-road bumpkins. "I felt like I'm walking into a Mafia meeting somewhere." Aside from the lavish spread, Brown reported, there was "all the liquor you could drink. It was absolutely fascinating to me."[82] It was like an old bootleggers den, something from the prohibition days of Bonnie and Clyde.

Brown had suddenly entered Kingfish land. As Sheriff Stephens started offering up wisdom, and one of his policemen kept spitting tobacco juice into a Styrofoam cup, the uptight, refined, and polished Brown plotted an exit strategy. At the first lull in the sheriff's monologue, Brown wished the posse good luck in coping with the post-Katrina stresses and skedaddled off the houseboat.

Surrounding the feast was a mass despair. On the Sunday before Katrina, forty-seven-year-old Jimmy Pitre had picked up his mother, Joyce McGuier, from her house in Chalmette. He was forcing her to evacuate. He drove her to the Superdome, but she refused to join the long line. Her heart was bad. She couldn't handle the heat. He went to four New Orleans hospitals, which all rejected her. She didn't have a clear-cut medical illness, just heart palpitations. McGuier, a great scat singer, a jazz-blues devotee, and a straight-talking woman full of life, had been in poor health that summer. Jimmy drove her back to Chalmette and decided he would ride out the storm with her. "Brother," he later said. "I never left her side for a second during the storm."[83]

The Wall of Water lifted her house. In the span of ten minutes McGuier lost everything. With water cascading in, Pitre untied a boat, and with the help of his neighbor got his mother into it, safe from the rising flood. But it was too late. Joyce McGuier died of a heart attack. "Let me make one thing clear for the record," Pitre said emphatically, recalling the events of the morning. "She never got one mouthful of water in her lungs. Not a drop. No way did I let her drown. She didn't get water in her. I wouldn't let it happen. You understand?"[84]

Keeping the corpse of his mother in the boat, a defiant Pitre grappled with grief by becoming a hyperkinetic first responder. With no Wildlife and Fisheries or FEMA around, Pitre hoped to avenge the death of his mother. He pulled more than 300 Chalmations out of the floodwater. He took the frantic, displaced people to Chalmette High School. He wouldn't eat or drink for days. He rescued. He ran on crazed adrenaline, his mother in the boat for days. He prevented a rape of a young girl. "I didn't want nobody else to die," he said. "Nobody else around me was going to die."[85]

When Wildlife and Fisheries arrived, Pitre ended his efforts, which bordered on the maniacal. He slowly came to grips with what had happened. Neither FEMA nor the Red Cross ever gave a cent to Pitre for his mother's house. Nothing would bring his mother back. He lost everything he owned, including a truck, which he was forced to pay the notes on. There was no congratulatory letter from FEMA or the Red Cross, thanking him for saving hundreds of lives. He never appeared on ABC's *Nightline* or NBC's *Nightly News*. But when the deal went down, Pitre was the man. He piled desperate people onto the corpse of his mother and never complained or sniggered or condescended or rolled up his shirtsleeves. He acted. Pure and simple. It was a matter of the heart.

Five months before Katrina struck, Michael Brown *had* warned Michael Chertoff of just about *all* the breakdowns that occurred after Katrina. In a March 2005 white paper Brown argued that the Bush administration was not prepared for a major domestic natural disaster. He wasn't comfortable with FEMA being a part of Homeland Security.[86] However, Brown's efforts to persuade his boss to wake up failed. That Thursday Chertoff was interviewed by Robert Siegel of National Public Radio. Chertoff claimed to have no knowledge of the tens of thousands of people trapped at the Convention Center. In fact, the secretary of Homeland Security seemed remarkably uninterested, without the curiosity and compassion one might expect of any American in regard to a report on his fellow citizens.

> SIEGEL: We are hearing from our reporter—and he's on another line right now—thousands of people [are] at the Convention Center in New Orleans with no food, zero.
>
> CHERTOFF: As I say, I'm telling you that we are getting food and water to areas where people are staging. And, you know, the one thing about an episode like this is if you talk to someone and you get a rumor or you get someone's anecdotal version of something, I think it's dangerous to extrapolate it all over the place. The limitation here on getting food and water to people is the condition on the ground. And as soon as we can physically move through the ground with these assets, we're going to do that. So—
>
> SIEGEL: But, Mr. Secretary, when you say . . . we shouldn't listen to rumors, these are things coming from reporters who have not only covered many, many other hurricanes; they've covered wars and refugee camps. These aren't rumors. They're seeing thousands of people there.
>
> CHERTOFF: Actually I have not heard a report of thousands of people in the Convention Center who don't have food and water. I can tell you that I know specifically the Superdome, which was the designated staging area for a large number of evacuees, does have food and water. I know we have teams putting food and water out at other designated evacuation areas. So, you know, we've got plenty of food and water if we can get it out to people. And that is the effort we're undertaking.[87]

The arrival of Lt. General Russell Honore drew cheers from the Gulf South, which desperately needed straight talk from someone at the federal level. The three stars on his shoulders were the first tangible evidence that the Bush administration hadn't forgotten the region. One of the largest humanitarian relief operations in American history was finally under way. The sight of flatbed trucks filled with supplies and M-16s pointed skyward was seen by those still in the bowl as liberation forces. People shouted out, "Thank you, Jesus," and "Bring 'em on." Police Superintendent Eddie Compass, in a perfectly pressed police uniform, rode on the running board of a truck around the Convention Center shouting that food and water were on the way. "We got 20,000 people out of the Superdome," he told them, "and we're going to take care of you."[88]

As commander of the 1st Army, Honore knew how to lead troops. With a cigar in his mouth and a laserlike glint in his eye, he was not bad in leading civilians in the midst of a crisis, too. But with fires erupting all over the city, people dying at the Convention Center, and looters still on the prowl, Honore had his work cut out for him. "If you ever have 20,000 people come to supper, you know what I'm talking about," Honorc said. "If it's easy, it would have been done already. Our number one task is to deal with the concentration of people in New Orleans, as well as those that are isolated. And we're going to get after it. By and large these are families that are just waiting to get out of here. They are frustrated. I would be, too. I get frustrated at the cash register counter when the paper runs out."[89]

Chapter Fourteen

THE FRIDAY SHUFFLE AND SATURDAY RELIEF

> Finally, the undersigned can see no reason for a much longer stay in the area since, as we said before, ordinary facilities for receiving visitors are not available in the parsonage. Further activity, such as sniffing around for dead people on glaciers, might be a job for the police or the Scout movement, but is an unworthy task for the spiritual authorities of the country. Furthermore, I understand that I was only scheduled to spend this one day here in the west to complete my mission.
>
> —Halldór Laxness, *Under the Glacier*

I

At long last, on Friday September 2, President George W. Bush visited the Gulf Coast, the fifth day of the Hurricane Katrina crisis. His visit coincided with the arrival of General Russel Honore and massive amounts of food, water, and personnel in the region. In New Orleans, the Louisiana National Guard—under Governor Blanco's handpicked adjutant general,

Major General Bennett Landreneau—had taken control of the Superdome, loading people up on FEMA-contracted buses. The Blanco buses were being directed to the Convention Center. Yet so much was wrong in the region, that it will take years for anyone to feel dramatic improvement. President Bush must have known history would hold him largely responsible for the slow response. Before leaving Washington, he appeared in the Rose Garden to thank those people who were working to help the Gulf South. "The results are not acceptable," President Bush said. "I'm headed down there right now."[1]

It was a glimmer of the feisty George W. Bush, the one who could get things done, the blunt-spoken man who might put a scare into anyone, even the most complacent bureaucrat. But it was only a phrase in an otherwise vague, meandering statement. Behind the scenes, White House aides were uncertain that Bush actually understood what was going on, or not going on, in the Gulf South. In fact, he seemed to be the only person in America without an outraged opinion about the lackluster federal response to Katrina. As *Washington Post* columnist Dan Froomkin pointed out on September 6, "President Bush somehow missed the significance of what was happening on the Gulf Coast last week as he and his political guru, Karl Rove, flitted between Texas and California and, finally, Washington."[2]

Early Friday morning, the President boarded Air Force One for the three-hour flight to Mobile, Alabama, his first stop. He took a telephone call from U.N. Secretary-General Kofi Annan, who offered his "condolences and support" of the Katrina victims.[3] En route, he sat down to watch a compilation of news coverage recorded onto a DVD by White House Counselor Dan Bartlett. The night before the President's trip, according to reporting in *Newsweek*, "some White House staffers were watching the evening news and thought the president needed to see the horrific reports coming out of New Orleans." The magazine couldn't help wondering where Bush had been, what had kept him from watching the coverage for himself. "How could this be?" *Newsweek* asked. "How the president of the United States could have even less 'situational awareness,' as they say in the military, than the average American about the worst natural disaster in a century—is one of the more perplexing and troubling chapters in a story that . . . ranks as a national disgrace."[4] Later, President

Bush told NBC News's Brian Williams, "I don't see a lot of news. Every morning, I look at a newspaper. I can't say I've read every article in the newspaper. . . . I mean, I can tell you what the headlines are."[5]

Secretary of Homeland Security Michael Chertoff was onboard and so was Representative William Jefferson, the first African American to be elected to Congress in Louisiana since Reconstruction. Jefferson was a powerful member of the Ways and Means Committee. If Louisiana were to receive billions of dollars in federal aid, Jefferson would be a key conduit through which the cash would flow. Jefferson sat next to Secretary of Housing and Urban Development Alphonso Jackson on the plane. Both African-American leaders—one a Democrat, the other a Republican—urged President Bush to "reach out" to the black community. "Alphonso and I felt it was important for the President to put his feet on the ground in New Orleans," Jefferson said. "He was going to visit the 17th Street Canal, which was fine, but that was upscale and white. We thought he needed to meet the folks that were hurt. . . . But [the security advisors] thought it was too dangerous for him to be seen in a possibly angry crowd."[6]

Jefferson found President Bush to be in a tough, no-nonsense mood regarding Katrina. He was extremely focused. He didn't like hearing that relief efforts were slowed down. At one point during the day's travels along the hurricane-ravaged area of the Gulf Coast, President Bush saw a fire burning down below. "What's that?" Bush snapped. When Michael Chertoff explained that there were some isolated fires along the Gulf Coast, Bush exploded in anger. "Put the fire out, now!" he said. "I want that fire out."[7] Brown might have become the public whipping boy for Katrina, but President Bush was taking his frustration out on Chertoff.

During the hurricane, Alabama was hit hardest in its southwest corner, which juts like a lobster's claw into the Gulf of Mexico, forming Mobile Bay. The hurricane slammed into the area on the west side of Mobile Bay; the storm surge was forced up the bay and into the city of Mobile. Downtown streets had up to eleven feet of water.[8] The storm destroyed piers along the coast, but no section was as badly damaged as Dauphin Island, a barrier island dotted with homes. Much of it was uninhabitable after Katrina. "We got hammered," Mayor Jeff Collier of Dauphin Island said. "Boats were flying around like planes or something."[9] The death toll from the storm in Alabama, however, was low. Two people were killed in an automobile while driving during the storm. For most people in the state, Katrina was a trial, but not a catastrophe.

In Mobile, the President set up shop in a hangar at the Coast Guard Aviation Training Center at Mobile Regional Airport. Coast Guard Lieutenant Commander Jim Elliott had done a phenomenal job sending out cutters and helicopters to storm-ravaged areas immediately after Katrina passed. "We know how to join with other organizations to get the job done," Elliott recalled. "We were out the door as soon as the winds died down."[10]

"We had carefully coordinated our approach to rescuing four or five days in advance of Katrina," Walter Dickerson, director of Mobile County Emergency Management, recalled. "Police, Fire, Coast Guard . . . we made sure we had twenty-first-century communications together. Mobile County was the exact opposite of Orleans Parish."[11]

Waiting for the President in Mobile, full of bear hugs, was Haley Barbour. The two politicians were extremely tight. When President Bush was elected governor of Texas, Barbour was the bare-knuckled chairman of the Republican National Committee. They had forged a genuine liking for each other. All week long, while other Republicans jumped ship, criticizing Bush's Katrina performance. Barbour remained loyal, with not one iota of frustration aimed at Bush. Barbour was a disciple of Ronald Reagan's eleventh commandment: a Republican shall not criticize a fellow Republican. He knew President Bush to be a deeply compassionate man, one void of prejudice or hatred. "When he got to me, he cried," Barbour recalled. "Tears just ran down his cheeks. It made me cry."[12]

Michael Brown of FEMA was also on hand to meet President Bush, along with Alabama Governor Bob Riley. Their hugs were of the stiff, professional kind. Aside from members of the press, the only other people present were a few dozen Coast Guard officers, state officials, and local politicians. First, Bush was briefed on the situation. Governor Riley then spoke to the press, reporting that the town of Bayou Le Batre had lost 3,000 houses to the hurricane. Dauphin Island lost 60 percent of all homes. The governor described the state's initiatives to provide help to those affected. He spoke positively about the involvement of FEMA, even though he wasn't specific. The good feeling in Alabama made for a heartening beginning for the tour: Alabama had been hit, but not nearly to the extent of the other two states. Nonetheless, the "largely upbeat assessment" he'd been given probably put Bush in a good mood, and that wasn't necessarily a good thing. The journey had yet to begin, after all.

President Bush seemed relieved to enter into the congratulatory atmosphere in the hangar. It was something like a men's locker room, and he could give as good a halftime speech as anyone. Unfortunately, it wasn't halftime—not even for his tour. Mississippi, as he might come to see, was in far worse shape than Alabama. As Joe Scarborough of MSNBC later described Mississippi's Gulf Coast, the region was a "total loss . . . nuclear winter. As bad as it gets. Beyond war zones. There's rubble in war zones. There is nothing in Waveland."[13] And Louisiana, given the breached levees in New Orleans, was in worse shape than that.

In his remarks to the press, Bush started by citing the search and rescue work of the Coast Guard, but in oddly vague terms, almost as though it was presidential boilerplate:

> Well, first I want to say a few things. I am incredibly proud of our Coast Guard. We have got courageous people risking their lives to save life. And I want to thank the commanders and I want to thank the troops over there for representing the best of America.
>
> I want to congratulate the governors for being leaders. You didn't ask for this, when you swore in, but you're doing a heck of a job. And the federal government's job is big, and it's massive, and we're going to do it. Where it's not working right, we're going to make it right. Where it is working right, we're going to duplicate it elsewhere. We have a responsibility, at the federal level, to help save life, and that's the primary focus right now. Every life is precious, and so we're going to spend a lot of time saving lives, whether it be in New Orleans or on the coast of Mississippi.

A little later, the President did cite a specific case with which he was familiar. The fact that it concerned a millionaire rankled many of those Gulf South victims who were left without a penny by the storm.

> We've got a lot of rebuilding to do. First, we're going to save lives and stabilize the situation. And then we're going to help these communities rebuild. The good news is—and it's hard for some to see it now—that out of this chaos is going to come a fantastic Gulf Coast, like it was before. Out of the rubbles of Trent Lott's house—he's lost his entire house—there's going to be a fantastic house. And I'm looking forward to sitting on the porch.
>
> Out of New Orleans is going to come that great city again. That's what's going to happen. But now we're in the darkest days, and so we got a lot of work to do. And I'm down here to thank people. I'm down here to comfort people. I'm down here to let people know that we're going to work with the states and the local folks with a strategy to get this thing solved.

President Bush thanked the people of the three Gulf Coast states for the compassion they'd exhibited in the aftermath of Katrina, and he made reference to others integral to the recovery. "Again, I want to thank you all for . . ." President Bush said, with his voice trailing off in midsentence, as it tended to do, "and Brownie, you're doing a heck of a job. The FEMA director is working twenty-four— They're working twenty-four hours a day."

> Again, my attitude is, if it's not going exactly right, we're going to make it go exactly right. If there's problems, we're going to address the problems. And that's what I've come down to assure people of. And again, I want to thank everybody.
>
> And I'm not looking forward to this trip. I got a feel for it when I flew over before. It— For those who have not— Trying to conceive what we're talking about, it's as if the entire Gulf Coast were obliterated by a— The worst kind of weapon you can imagine. And now we're going to go try to comfort people in that part of the world.[14]

The President seemed as though he was suffering from jet lag—except that Washington was in the same time zone as Mobile. Sportily dressed, in a button-down shirt, he was quite open about the fact that he was on the

Gulf Coast to thank people, the kind of "to-do list" item that one keeps to oneself, unless one cannot think of another thing to say. That was President Bush's basic problem on Friday morning: he didn't actually have anything to say. Or maybe he was emotionally overwhelmed and was keeping a stiff upper lip. The DVD made by his aides had yet to sink in, apparently. It would seem fair to let him off the hook for one bad speech, but his words were crucially important that morning to everyone on the Gulf Coast and beyond. Saying that he was "down here to thank people" and then doing so did not represent the forward sweep of a leader, which was what the nation needed desperately at that critical juncture. *USA Today* commented that in the past, "President Bush has shown that he can be empathetic, sensitive and decisive. But those qualities eluded him for days after Hurricane Katrina, and the lapse could become a defining moment of his White House tenure."[15]

And so the President dutifully thanked people, but in general terms. Only one person was singled out. "Brownie, you're doing a heck of a job" became the catchphrase associated with President Bush's leadership in the Katrina disaster. Michael Brown was patently *not* doing a heck of a job, whether through his own failings or the lack of cooperation from others. The FEMA operation was itself an unmitigated disaster. The phrase became emblematic of the President's ignorance about the situation and his tendency to take a casual attitude toward it. It was inappropriate to use the nickname at that time, especially for a man struggling at his job. The only person in the world, in fact, who called the FEMA director "Brownie" was the President. As for his doing a "heck of a job," it's hard to imagine Dwight D. Eisenhower in the midst of D-Day referring to a subordinate in such a jocular way, or in congratulating him before the battle was through. "There was a time when FEMA understood that the correct approach to a crisis was to deploy to the affected area as many resources as possible," Senator Mary Landrieu said that afternoon, requesting that President Bush appoint a cabinet-level official to quarterback the Katrina response. "In order to resolve this dire situation we must return to the successful tactics of the past."[16]

The "Brownie" remark generated widespread criticism at the time. It remained in common parlance, an expression that served to describe anyone doing a horrendous job, especially when the speaker didn't care one way or the other. If a person was making a mess of fixing his or her own car, for example, the new American idiom was to say with a smile, "Brownie, you're doing a heck of a job." It wasn't much of a legacy for either man. "I was there for the Brownie comment," Jefferson recalled. "My take on it was that the President was following [Bob Riley's] lead, trying to put on a good partisan air."

Brown later claimed that the "Brownie" remark triggered malice toward him. Everybody who loathed President Bush saw the demolition of Brown as a way to get at the President. "That was my tipping point," Brown recalled, "because at that point he used the nickname with me, it caused the mainstream media to go, 'Who the hell is that? Who is this guy, this guy from FEMA?' "[17]

It was the *Time* story on September 8, 2005, that really put the new Katrina scapegoat on the hot seat. The article was brimming with allegations that Brown had padded his résumé and exaggerated his role in previous jobs. The FEMA director was, according to *Time*, touched by fraudulence.[18] Months after Katrina, Brown continued to claim that *Time* had smeared him to embarrass the Bush administration. This defiant anti-*Time* attitude was perhaps Brown's way of coping with becoming disdained by millions of Americans overnight. Brown said, "When someone asked my lawyer what was not true about the *Time* story, he said, 'Well, if you read from the very beginning to the end, the story is not true.' My résumé has gotten me confirmed by the Senate twice and a full-fledged FBI investigation [to secure] my top-secret clearances and so the only thing that I can conclude [is] that I'm a conservative Republican, but I'm not a right-wing nut. It really became apparent to me at that point that this was their way—*Time*'s or the reporter's way—to get to the administration, to get to the President."[19]

As the first week led into the second, resentment of FEMA only grew. Whether you were black or white, rich or poor, an evacuee or in the bowl, FEMA had become a dirty word. Like the Johnny Cash song "I've Been Everywhere," just name the Gulf Coast town—Pearlington, Waveland, Ocean Springs, Biloxi, Pascagoula, Dauphin Island—and they were unified around a single beef: FEMA. Now President Bush, who had been missing in action for four days, avoiding the smell of death, was saying that Michael Brown was doing a "heck of a job." It was too much for decent Gulf South folks to bear. Who was this guy Brown? Why was President Bush so chummy with him? The long knives were being pulled out. Following the *Time* article, Democrats and Republicans, storm victims and TV couch potatoes, all agreed: Brownie had to go. Everybody but Homeland Security Director Michael Chertoff, who on Friday stuck to the line that FEMA had done a "magnificent job." When pressed on how that was so, Chertoff answered, "You can't fly helicopters in a hurricane. You can't drive trucks in a hurricane."[20]

II

President Bush next flew to Biloxi, a "low chopper ride," as he said, which gave him a chance to see damage along the way. Air Force One, meanwhile, went on to New Orleans, the final stop of the Gulf Coast tour. Arriving in Biloxi in late morning, the President toured devastated areas in the company of Governor Haley Barbour and Biloxi Mayor A. J. Holloway. At a press conference afterward, President Bush was more detail-oriented than he had been in Mobile, although the troop deployments he mentioned in the region were out-of-date. ("We need to get troops," Bush said. "We had 1,200 troops arrive yesterday, I'm told. We're going to have 1,200 today." In fact, more than 10,000 troops made it to the coast on Friday.) He was also able to field questions in a serious way. The helicopter tour may well have sobered his attitude. Still, the President had only seen the devastation from above. His other senses weren't assaulted, as they were for those on the ground. That finally changed in Biloxi. After walking through the rubble, President Bush told the press that the damage was "worse than imaginable"[21] and added "I am satisfied with the response. I am not satisfied with all the results."[22]

To President Bush's credit, he met and hugged Katrina victims in Biloxi, even if they were prescreened by the Secret Service. "I don't have anything," a trembling Mississippi woman said, clearly shaken. "Well," President Bush told her, "they'll help you." "We have nowhere to go," she said. "I know," he said, trying to reassure her. "I know." "And I came here looking for clothes," she told the President. "Well," Bush said, "they'll get you some clothes."[23]

As they walked together through the blasted-out Biloxi neighborhood, arms around each other, it seemed that the President cared. It was the only moment on Bush's Gulf South trip in which he was, for a fleeting hour, shining the flashlight in his face, reassuring the Gulf South that the President *cared*. Unfortunately for the President, the Mobile "Brownie" comment superseded these more heartfelt Biloxi moments when Bush hugged and cried and commiserated with everybody he met.

Surrounding the Biloxi rubble photo-op were angry voices, though. "You can smell the death," said Phyllis Upshire-Davis, who lived outside Biloxi. She had nearly lost her life and had to identify the body of her neighbor, who had come to check on her just before the storm.[24] As of Friday morning, the death count in Harrison County alone was fifty-six. The total number of storm-related deaths in Mississippi was one hundred. Both numbers were still rising. More than a half-million people were without power in southern Mississippi, and restoration of electricity wasn't expected for three weeks. Some 17,000 people were staying in temporary shelters;[25] at least twenty times that number were staying in hotels or with relatives. Food was scarce along the coast. Reclaiming the neighborhood was extremely difficult, since roads were blocked with debris and draped with loose electrical wires. Traffic clogged whatever avenues were available. Morris Dees, founder of the Southern Poverty Law Center, discovered that driving from Montgomery, Alabama, to the coast, normally a three-and-a-half-hour trip, took eleven hours. "I filled up my motorcycle tank in Montgomery and headed down to Biloxi to check up on a niece," he recalled. "There would have been no way to make it in a car. Police from all over America clogged up the highway. And, man, the nails were everywhere. It was flat-tire city along the coast."[26]

Another problem facing the coast was the fact that it depended for immediate aid on areas to the north that were not ordinarily supposed to be gravely affected by hurricane damage. Hattiesburg and Jackson were not decimated by any means, but both cities had enough wind damage from Katrina to complicate efforts to send large-scale assistance to the coast. The final disaster on the Mississippi Gulf Coast was looting. Law enforcement officers had to be taken off rescue duty in Mississippi to maintain the peace. It was not captured on film, it was not covered much in the press, but Mississippi was, just like New Orleans, in danger of losing control of its neighborhoods, from Pass Christian to Ocean Springs.

Meanwhile, in front of the media that Friday, Governor Barbour told the President, "The federal government has been great."

"We have a responsibility to help clean up the mess," Bush replied.

"Y'all are helping," Barbour followed.[27] But up to that point, it just wasn't true.

During the first week, FEMA was nowhere to be seen along the coast, except in the form of about a half dozen medical teams. Hancock County, home to Bay St. Louis and Pass Christian, among other devastated towns, had yet to see a single FEMA official.[28] One would arrive on Saturday: one.

Nevertheless, Barbour insistently put a cheerful spin on his state's plight. After first calling Mississippi "Hiroshima," he was in a different frame of mind by Friday. He started calling his state's loss of 221 citizens and 68,000 homes across a 28,000-square-mile disaster zone "The Deal." Perhaps it was his country-boy pride that made him reluctant to sound dependent, whining, or defeated. Perhaps it was good politics to seem pleased, even if there was little to be pleased about. Like his old boss Ronald Reagan, Barbour knew that optimism and humor—even during dark hours—were better tonics than despair. More than any other politician in the aftermath of Katrina, Governor Barbour kept an upbeat, resilient attitude about "The Deal." He also used his extensive contacts to get provisions shipped into his state. "A burly former high school football player, Barbour prizes the rituals of good ol' boy bonding—the bawdiness, bellowing and back-slapping that lubricate so many of the friendships he collected," Mark Leibovich wrote in the *Washington Post*. "Which, again, might seem frivolous right now. Except that shared history comes in handy in times like these, and Barbour might have more shared history with more well-placed people than any governor in the country."[29] Whenever Barbour was pressed for a comment on the poor showing of FEMA and the White House, he would refer to the national government as "a good partner."[30] Governor Barbour's reluctance to complain—he was called the "stoic bubba"—was not shared by Robert Latham, director of the state Emergency Management Agency. Latham hadn't received anything he'd requested of FEMA. Mississippi had been abandoned. "I've lost my temper several times," he said. "It's time to throw the rule book out on this and do what's needed."[31]

Among those doing what was needed was Marsha Barbour, the governor's wife. She and her two adult children joined a Mississippi National Guard convoy at the end of the week and brought a pickup truck full of supplies to Gulfport. They used their pull in getting through roadblocks to run supplies into coastal cities and towns. Mrs. Barbour also rode with police on patrol against looters. The question along the coast was why FEMA couldn't get through. Meanwhile, Mississippi Senator Trent Lott was using his considerable connections to facilitate private supply efforts.[32]

Over and over again, FEMA actually stopped truckloads of supplies, water, or ice on some bureaucratic pretext or other.[33] Meanwhile, people stranded along the coast went without drinking water for days. On orders from FEMA, the *Bataan*, the Navy ship loaded with supplies and fitted with extensive medical facilities, had been redirected from New Orleans (just as one of its landing craft was preparing to reach the city) to the waters off Gulfport. Once again, however, it was stymied from extending any substantive help. The ship was to be some holding area for medical help. At noon on Friday, fifty-six physicians, nurses, and other medical workers were airlifted to the ship, where they had nothing at all to do.[34] Along the coast, their skills were desperately needed, but they were stranded onboard the *Bataan*, awaiting orders from FEMA.

At Keesler Air Force Base, the sprawling home of the Hurricane Hunters, most of the visiting airmen were being sent home, as soon as possible. The enormous medical center was emptied of patients and most personnel, leaving a skeleton staff to continue medical care in a huge tent turned into a modern hospital.[35] Those stationed permanently at the base were hard at work cleaning up the base and the surrounding community. The camp newspaper, which had been renamed the *Katrina Daily News*, was holding a contest to name the job of overcoming the ravages of the storm. "You survived it, you're living it; now name it," invited the editors.

With gallows humor, they immediately announced that some suggestions had already been rejected, including "Operation Deny Comfort" and "Operation Provide Gumbo."[36] The eventual winner may have lacked irony but it was spirited: "Operation Dragon Comeback."[37]

That Friday, as President Bush traveled from Alabama to Mississippi, Dr. Fredro Knight of Hancock Medical Center went in the opposite direction. After Katrina hit, he hadn't been able to reach his parents in Mobile until Thursday. "A Mississippi state trooper let me use his satellite phone," Knight recalled. "I got through and my mom answered. 'Hey, I'm alive,'" he declared. "She breathed a sigh of relief. Due to her recent stroke, however, she had a hard time speaking. 'Are you all right?' she asked. 'Are you hurt?'"[38]

"Yeah, I survived the storm, but it's been hell," Knight replied. "We're stitching up people around the clock. I don't know when I'll be home because I can't get out of here. My truck went underwater and there's no way in or out of Bay St. Louis. If I'm not home in two days, tell dad to call the National Guard to come in and fly me out. But I'm okay."

Knight's father, Fredro C. Knight Sr., had been a schoolteacher for ten years and then a principal for twenty-five. His mother had been an English teacher and choir director in Mobile. "My dad called the National Guard and they got me to Mobile," Knight recalled. "I sat with Mom and told her all my Katrina stories. Then it was back to the hospital in Bay St. Louis. I couldn't stay away from the patients for long."

Dr. Knight was able to spend the night in Alabama because relief had arrived at Hancock Medical Center in the form of a Disaster Medical Assistance Team. Funded by Homeland Security, DMAT was like a MASH unit. "They set up tents, they have doctors, they have nurses, they have pharmacists, they have respiratory therapists," Hal Leftwich, Hancock's CEO, explained. "They were all people who, in their own community, had geared for this particular service. They came in, a quasi-military unit, a fantastic group of people because they all wanted to help." Usually Hancock Medical Center saw 75 patients a day; with DMAT's assistance, they were able to tend to 1,100 patients daily. The Mississippi National Guard, for its part, set up a field hospital in front of the DMAT, equipping it with operating rooms and X-ray machines. "You'd be amazed how many medical support people came pouring in from all over," Leftwich recalled, mentioning units from North Carolina, California, and Virginia, among others.

By the time Dr. Fredro Knight returned from Mobile on Saturday, Hancock Medical Center had been transformed from a flooded-out wreck into the leading triage center in Mississippi. "We never slackened the pace," Knight said. "We never turned anybody away at our doors."[39]

Progress was nowhere near as noticeable on the rest of the Mississippi Gulf Coast. Just a few miles from the coast, small farms were still reeling from the storm. Chicken farms, with their flimsy coops, were especially hard hit. Typically, 90 percent of the chickens were killed, trapped inside the coops as the wind raced through. The survivors went hungry, although some farmers and animal lovers tried to save them (for humane reasons only—they were unsalable).[40] Likewise, in the Lower Ninth, a chicken processing plant not far from Jackson Barracks flooded, sending dead birds floating down Deleray Street. "What a health hazard," Reverend Willie Walker said. "The chicken stench was unbearable."

In the renewal of the Mississippi coast, goodwill sprung up within the business community. As competitive as commerce was in normal times, outside companies extended themselves to give locals a chance to right themselves after the storm. It was a reflection of the fact that all business was based on a partnership of one sort or another and when one side was hampered, the other side pulled harder. In the gambling industry, Harrah's Entertainment lost casinos in Gulfport and Biloxi as well as New Orleans. Within two days of the storm, it made a commitment to pay employees of the closed casinos a full salary for ninety days; it also made a $1 million contribution to a fund for employees in the region.[41] Sanderson Farms, a chicken processor based in Laurel, Mississippi, announced that it would pay its farmers the usual rate, even though all the chickens under contract had been killed or rendered unusable.[42]

The greatest favor granted to businesses along the coast, however, was the herculean effort of Mississippi Power.[43] For businesses, electricity was a requisite to opening. That Mississippi Power would beat its own estimate of four weeks to restore power, and have electricity back in most areas in only seventeen days, was an enormous benefit. A dismaying indication of just how tough life was during those seventeen days, was found in a Mississippi Power advisory directed at contractors and others involved in the cleanup: "Crews should not pile debris higher than 12 feet (from the ground) in their trucks. Transporting tall piles of trash runs a significant risk of snagging or coming in contact with power lines."[44]

Within a week of the storm, one in ten businesses in Biloxi would be operating again,[45] a small percentage, to be sure, but an initial push for the momentum necessary to bringing the coast back for good. "The people have got to understand that out of this rubble is going to come a new Biloxi, Mississippi," President Bush said on his tour, evoking the ghost of Hurricane Camille. "It's hard to envision it right now. When you're standing amidst all that rubble, it's hard to think of a new city."[46]

Elton Gary of Biloxi lived through the storm and the aftermath, feeling hungry sometimes and overheated all the time. To his chagrin, he was out of a job, since the wax factory where he worked had been destroyed when several railroad cars had blown through one of its walls. Nine days after the hurricane, when Gary pulled into his driveway, he noticed that his porch light was on—*it was on!* He sat in his car for a long time, looking at the light and letting it find its way to him again.[47]

III

From the disturbing quiet of Mississippi, President Bush traveled to the cacophony of Louisiana on Friday. The plan was for him to land at Baton Rouge and meet with state and local officials, including Governor Blanco and Mayor Nagin, before touring parts of the ravaged area. New Orleans looked as devastated as it had all week. The water had yet to recede. Crime prevailed. The stench was pervasive. Nevertheless, there were signs that the situation was finally taking a turn for the better. Five thousand troops rolled in on military vehicles loaded with supplies. Most of the soldiers were military police, who took positions at strategic points in the city and began to claim territory in the name of order. From a distance, the scene was reminiscent of the Allied troops entering Rome or Paris during World War II. They sat high in their vehicles, while the locals lined the streets, watching them pass. The response to this arriving army, though, was mixed. Many New Orleanians cheered, certain that water, food, and transportation would soon be available. A little girl waved at an approaching convoy, calling out, "Thank you, Mr. Army!"[48] Others had just about run out of patience. They swore at the trucks going by, venting five days' worth of anxiety and impatience. "I told a group of Army guys to get their rifles the hell out of my face," one New Orleans resident said on being confronted by the troops at last. "I had just lost my house. And they show up a week late with itchy fingers. F—— them."

Another good sign was that the Superdome was nearly empty. Buses had been collecting people for the previous twenty-four hours. Once the facility was entirely empty, the Convention Center would be next. It would take the entire weekend to pick up all of the willing evacuees. The next problem was deciding just where to take them. Over the weekend, many of those taken out of New Orleans were taken to the Louis Armstrong International Airport and then sent on to other destinations by airplane. Evacuees often didn't even know where they were going until they were seated on a plane, listening to the captain describe the itinerary. The state of Utah deserves special recognition for its open-armed acceptance policy. Under the leadership of Governor Jon Huntsman Jr., the state welcomed 582 evacuees, most of whom expressed a desire to find new homes, either in Utah or elsewhere. The state supported their efforts.

The worst aspect of the hurried evacuation was that FEMA, which was overseeing the understandably frantic effort, didn't keep track of who was going where.[49] In the golden age of information management, that was inexplicable. Package services were capable of telling a customer the location, to within a block, of the delivery truck carrying any certain package. A chain store can trace every box of diapers as it leaves the factory and then travels across the country to be stocked in a far-off store. While it's not as if Katrina evacuees should have been given bar codes, records ought to have been kept on each person. As it was, people were separated from their spouses, their elderly parents, and even their young children.

One week after the storm, the National Center for Missing and Exploited Children (NCMEC) set up a special Katrina hotline. Over the next thirty-six days, the line took 20,932 calls. People were separated, with few ways to locate loved ones. NCMEC posted names and, if available, photographs on its Web site. It registered 5,068 missing children—a disgracefully high number. While some children were unavoidably lost in the course of the storm, and some were safe with one parent, while the other fretted, the rescuers were often callous about separating family members. Eventually, NCMEC resolved 96 percent of its Katrina cases. The National Center for Missing Adults performed a similar service for the 12,514 adults reported as missing.[50] It eventually reunited all but a few dozen with those who were looking for them. Within the state, the Louisiana Department of Health and Hospitals set up its own clearinghouse for information on persons missing in the storm; it reunited 7,420 people with family and friends.

The staging area for the evacuation of approximately 60,000 people during the first days of September was the Louis Armstrong International Airport. And it was at the airport that Mayor Nagin and Governor Blanco were to meet with President Bush during his visit to New Orleans. Senators David Vitter and Mary Landrieu would be present, along with two congressmen, Piyush "Bobby" Jindal and William Jefferson, and the President's chief of staff, Andrew Card. With the rest of the airport in turmoil, the meeting was to take place on Air Force One.

Brian G. Lukas of WWL-TV was assigned to cover President Bush's stopover at the airport that Friday afternoon—or so he thought. He had left early for the airport in the station's van, wanting to get a prime filming location on the tarmac. What he encountered instead, however, was a roadblock operated by Louisiana state troopers, who denied him access to the airport. The White House didn't want local coverage of the Air Force One summit. Refusing to be turned back, Lukas pulled over and started working his cell phone, hoping his friends in Baton Rouge could intervene on his behalf. "They refused me entrance," Lukas recalled. "The President was trying to do a quick in-and-out of New Orleans, and wanted to control his image. The photograph of him staring out of Air Force One looking over the Gulf Coast had been a disaster. It made him look detached. I guess he didn't want a repeat."[51]

Lukas was sidelined on the periphery of the airport, behind a wire fence. He was about three hundred yards from Air Force One, within zoom-lens range. But even as the great powers gathered for their powwow, Lukas noticed something even more important, more historic: a human wave of displaced persons hobbling along the Airline Highway overpass, trying to walk out of New Orleans, unable to wait for buses any longer. Lukas had filmed the Beirut bombing in 1983 and the diseased slums of Jakarta and Brazil. Whenever there was a major earthquake or landslide in Central America, he would go and film the devastation. He once even searched the suburbs of Medellín in hope of catching shots of Colombian drug lords in their cocaine warehouses. In 1982, when Pan American flight 759 crashed into Kenner, he was the first cameraman on the scene, filming charred bodies being yanked out of the rubble. Nothing, however—not even his previous four days of Katrina-related footage—prepared him for the horror of the human misery he was now encountering on the Airline Highway overpass, in view of Air Force One. "They were walking together out of shelters," he recalled. "It was so gruesome, it was difficult to film it."

Sitting in the WWL truck, Lukas jotted notes about what he witnessed, never forgetting that President Bush was trying to hide from these poor people, because the Secret Service thought meeting inner-city African Americans was a security risk. "An elderly lady is slowly pushing her aged husband in a wheelchair," Lukas wrote. "His feet are swollen and bloodied. His bare feet had been scraping on the hot pavement. On the side of the road there are rocks and broken glass. His feet had glass embedded in them. I took one of my shirts and wrapped his feet then lifted his feet above the pavement with a board. I had only a few bottles of water and I gave it to them . . . they needed something to eat."

As the Airline Highway refugees approached the roadblock, he had carefully filmed the aged couple, zooming in on their anguish. "It seems that the world has abandoned them," Lukas wrote. "They were in a desperate situation. Finally aid came from a passing state trooper."

But the displaced people kept coming, erroneously believing they could get aid at the airport. They couldn't have been more wrong. President Bush had the airport locked down; security was not letting evacuees enter the premises. Why would all these state troopers be so cold and callous to these poor? They had billy clubs pulled for the most feeble human beings America could offer up. "Another group of people came over the overpass pushing grocery carts," Lukas wrote in his journal. "They were in desperate shape. In the cart they had what appeared to be two lifeless babies on plastic wrapping. Another cart had a small child covered with welts from mosquito bites. Other children were huddled in the carts and looked exhausted."[52]

President Bush had just completed a visit to the 17th Street Canal breach. When he arrived back by helicopter, water was handed out to the Airline Highway evacuees. "It was hot," Lukas wrote. "The heat was boiling up from the pavement; any water given to these exhausted people is welcomed with relish. This is tough to witness."[53]

Lukas was also assigned to a film story on the evacuees who were being helicoptered into the ariport. For hundreds of choppers, it was an ongoing circle run between the disaster zone and the airport. Wandering around the airport later, Lukas wrote: "Baggage carts carry the evacuees as they walk off, and in many cases, [are] carried off the helicopters. IV's are given on site . . . it is very hot and the tarmac is steaming. The emergency personnel try their best to shade the people from the sun with their hands. Litters are placed on the baggage cart to assist the victims. The elderly hold tightly to the medical staff . . . they are confused. This is an unimaginable scene. Inside the airport thousands of people are being attended to. Some have died, some are dying. There is an airline gate sectioned off by a white sheet where the dead and near-dead lay."

By all accounts, the emergency personnel at the airport were full of compassion. Nurses were washing patients, getting them clean, even trying to decontaminate them. Fleets of ambulances were coming and going with sirens flashing. "Doctors, nurses, soldiers, firefighters, and police are attending the victims of the storm," Lukas wrote. "There is a lot of crying . . . people are in pain. An elderly woman has just been brought into the morgue. There are so many elderly people. I cannot help but notice the swollen feet of many of the elderly . . . many cannot walk, they have to be carried. More victims keep coming in by helicopters and by ambulance. It is a constant stream."[54]

IV

The focus of the Air Force One meeting was an analysis of how more could be done to help the people of Louisiana. There was no one present whose political reputation had been enhanced by the first four days after Katrina, and they each knew it. Inasmuch as there was still time for someone to ride in as the knight in shining armor, the meeting was also about giving that person a boost, or an elbow.

When Mayor Nagin and the Audubon Institute's Ron Forman arrived at Air Force One, they were told to wait, since the President had gone to see the canal breach. An attendant on Air Force One, noticing Nagin's filthy clothes, asked if he would like to take a shower. "And he says, 'Yeah, I'd love to take a shower.' There's monogrammed towels and everything, all of the cosmetic perks," Forman recalled. What was truly important to Nagin was that his head was shaved and waxed just right for his photo-op with President Bush. Like a primping teenager, he just wouldn't get out of the shower. Guards rapped at the bathroom doors, telling the mayor, "You've got five minutes and then the President gets here." They knocked again. But Nagin feigned deafness, wanting to smooth his head just right. It was the kind of personal task you couldn't speed up too much. Finally, the attendant had had enough: they kicked the door hard and told Nagin to get out; the President was about to arrive. "Well, you know, for me [Air Force One] was a relief from everything I had been doing," Nagin recalled. "From the standpoint that I got a chance to take a shower, which I hadn't done in many days. For that, it was kind of nice . . . the President wasn't on there initially and this attendant showed me a room, where I could shower up. Tried to get me to rush and shower. I got out of there eventually."[55]

"Damn," he said to everybody as he walked out into the cabin of Air Force One. "I wasn't ready to get out of the shower! I was shaving my head and I was showering and, God, there was warm and hot water!"[56] As Forman joked, "He would have stayed in the shower and missed the meeting with Bush." There was something chillingly vain about Nagin's sense of judgment and his coiffing with the underclass suffering all around him.[57]

When the meeting began in the conference room near the middle of the plane, each elected official around the table told a story of incompetence or lack of cooperation on the part of FEMA. Representative Bobby Jindal told *Newsweek* that "the president just shook his head, as if he couldn't believe what he was hearing."[58] If nothing else, the President's reaction proved that Dan Bartlett and his other advisors had been right when they handed him a DVD of news coverage on the hurricane at the beginning of the day. It was inconceivable that the President hadn't heard such stories; they filled the newspapers, radio, and television reports nearly all week. Yet it was apparently all news to Bush. Inevitably, the looming question of troop strength came up early in the summit. Senator Vitter pressed the point, along with Mayor Nagin. On Wednesday, as discussed earlier, Vitter had relayed messages from Karl Rove's aides to Blanco's office, suggesting that the governor assign authority over National Guard troops to the President. Blanco could not see how that would help. Indeed, it might hinder efforts. According to the 1878 Posse Comitatus law, federal troops could not participate in law enforcement; federalized National Guardsmen would be that much less useful in the real world of New Orleans lawlessness. Anyway, she had no reason to step aside and allow George Bush to play the role of hero. He, for his part, was less than enthusiastic about bringing glory to her door, if he could possibly avoid it. People following the joust over the troops noted that the White House had not made any suggestions that Mississippi governor Barbour relinquish command of the National Guard in his state.[59] Blanco believed Bush's "federalizing" push in Louisiana was a "paper reorganization," part of the White House's spin effort to blame *her* for the post-Katrina mess in Louisiana.[60]

Everyone at the meeting was aware of the tension between the President and Governor Blanco. Mayor Nagin saw his role as that of a stick of dynamite that breaks the logjam. Senator Vitter recalled that Nagin lost his temper, slammed his hand on the table, and insisted that a chain of command needed to be established. He was trembling, eyes wide, as he seemed to be inching toward a nervous breakdown. A chain of command had been established, by the U.S. Constitution and that of Louisiana, but getting it to work seemed to be another story entirely. The discussion was overheated, "as blunt as you can get without the Secret Service getting involved," *Time* reported, quoting a participant.[61]

Nagin later gave his own version of what happened in an interview with Soledad O'Brien on CNN:

> I said, "Mr. President, Madam Governor, you two need to get together on the same page, because of the lack of coordination, people are dying in my city."
>
> . . . They both shook their head and said yes. I said, "Great." I said, "Everybody in this room is getting ready to leave." There was senators and his cabinet people, you name it, they were there. Generals. I said, "Everybody right now, we're leaving. These two people need to sit in a room together and make a doggone decision right now."
>
> The President looked at me. I think he was a little surprised. He said, "No, you guys stay here. We're going to another section of the plane, and we're going to make a decision."
>
> He called me in that office after that. And he said, "Mr. Mayor, I offered two options to the governor." I said—and I don't remember exactly what. There were two options. I was ready to move today. The governor said she needed twenty-four hours to make a decision.[62]

The President's offer concerned the federalization of the National Guard troops. At stake were the 13,268 National Guard troops already assigned to Louisiana. The contretemps did not actually affect any federal troops. Blanco's request of twenty-four hours to consider the President's plan may have been ill-advised, since she had no intention of ceding power, but it ultimately did not delay further deployments. Late that evening, the White House sent Blanco a form, ready for her signature, granting control of the troops to the federal government, under an arrangement that gave Lieutenant General Russel Honore command of both her National Guard and the regular Army troops, stipulating that he would take orders from each of them regarding their respective troops—except in the case of a dispute, which would be settled by a team of judges. And in the meantime, the President would prevail. It was a cockeyed proposal. Blanco didn't sign it, but, consistent with her nonconfrontational style, returned a message the next morning acknowledging Bush's choice of Honore as commander of the Army troops in Louisiana. She then called the White House to speak with the President and was told by Andrew Card that he was on his way to the Rose Garden to announce that with the governor's permission, he was taking over command of the National Guard in Louisiana. She disabused Card of that notion immediately.[63] For the time being, Governor Blanco had won a battle for state's rights, long a tradition with Southern politicians of both parties.

As of Friday, Blanco's Louisiana National Guard was about to be enhanced by 30,000 fresh troops. The cavalry was indeed on the way. They had one essential assignment left to complete: the evacuation of the Convention Center. Lieutenant Colonel McLaughlin was just finishing the evacuation of the Superdome when he received the orders. Here's what he recorded in his diary on September 2:

> We then get the mission to seize control of the Convention Center—it will be a combat assault to wrest control from the thugs/criminal element entrenched there amongst those who are simply trying to survive. Not originally meant to be an evacuation/shelter site, it was coopted by a large crowd in the many thousands of families, old, sick, and bad guys. We launch our mission at noon and quickly cordon off the area from the aquarium along the river levee and the Riverwalk, to the bridge area and we set up 9 food and water lanes. I accompany a team to sweep the Riverwalk, we capture looters including one who had like a rat in a trap gotten into a cubicle he could not get out of (one of those with the steel chain security fences that roll down). The entire Riverwalk is looted, every shop and restaurant broken into, merchandise strewn everywhere and more senseless damage, the famous lobster tank at Anthony's shattered with about 60 dead lobsters on the floor. In my estimation, 80% of the crowd in the Convention Center area is either wearing looted clothing (brand-new tennis shoes, shirts, hats with the tags still on) or carrying shopping bags/pushing carts filled with looted items. I found one guy with a brand-new North Face whitewater rafting drybag filled to the brim with looted items: 6 pairs of tennis shoes, 20 or so shirts, 8 ballcaps, 20 watches

still in their boxes, about 40 Gameboys or X-box videogames, all brand-new—we turned him over to NOPD. Many, many people looted merchandise that was not critical to survival, unless one contends that DVDs, CDs, and 4 new pairs of tennis shoes, plus looted electronics will ensure survival. In my estimation, most looting was not for food and drink, which were readily available for free, it was simply opportunistic. I was stunned at the amount of looted stuff piled high in boxes, plastic storage bins, and shopping carts in people's possession, and the massive amounts of looted stuff simply strewn about abandoned. Any reasonable American seeing what I saw in even a 30-minute span would vote for a shoot-to-kill directive to deter looting—it was criminal, despicable, and embarrassing to see people show no shame in openly carrying around looted items. In fact, many looters pranced about and flaunted the looted goods almost as if to brag "look what I stole." It was a long hot day of patrols, stopping obvious looters, confiscating looted liquor. I broke at least 500 bottles of looted liquor, never thought I'd smash a $250 bottle of cognac on purpose, we were usually lied to about the origins of the looted goods ("I found this on the sidewalk," "It's everybody's stuff now," "My nephew gave this to me," "I brought it all from home")—it was depressing and you simply cannot imagine how hard it was not to smash someone in the face with a rifle butt after getting such a stupid answer. We fed everyone that evening and ran patrols all night—it was a calm and serene night. Very few NOPD officers in the area and we had just heard that approximately 500 or 600 NOPD officers were AWOL, you can imagine how we felt. We save their city and the CBD (Central Business District)—the few NOPD cars I saw simply drove through the crowds, and did not stop and offer help or info—in direct contrast to our efforts in working the crowds, letting them express their frustrations, giving them info (mainly about the bus staging efforts), and keeping the peace. Of my entire career, this took the most patience, the most intense effort to keep an equilibrium, and a determined effort to not simply beat the crap out of people who were openly stealing—I kept reminding myself that there were some good decent people in the crowd. We did rescue and shelter many tourists caught by the storm, a nice Spanish family from Barcelona, a couple from Russia, and many others. For the most part, the crowd was very angry with Mayor Nagin and with the NOPD; many remarked that we were the only ones giving them aid and info. There were tons of urban myths/rumors amok—any rumor you can conceive of.[64]

Clashes of egos occurred between the Louisiana National Guard—which had been trying to control the anarchy since Katrina—and the new, fresh federal reinforcements under General Russel Honore. According to McLaughlin, the U.S. Army acted as though the Louisiana Guard had screwed up and that the 30,000 additional forces dispatched to the Gulf South were going to have to mop up their mess. But in truth, by Friday, the Louisiana National Guard *had* evacuated the Dome and was in the process of clearing out the Convention Center. It hadn't been pretty. But they had managed to seize control of the situation without the U.S. Army or the Marines. A symbolic showdown occurred on Friday between Colonel David Aycock of the Louisiana National Guard and General Honore of the U.S. First Army.

The Louisiana Guard had brought in Heavy Expanded Mobility Tactical Trucks (HEMTT), carrying pallets of water and MREs. "The idea was to have the feeding lanes orderly," McLaughlin recalled. "Because we had 24,000 people to feed, we ordered some forklifts to help with the distribution process." Just before the forklifts arrived, General Honore suddenly showed up in the parking lot with three television cameras in tow. He was chewing his trademark cigar, barking orders at everybody, trying to play "the John Wayne dude" Mayor Nagin so admired. Honore ordered David Aycock of the Louisiana National Guard to "dump his loads," to drop the water pallets to the ground from the trucks so that evacuees could take what they needed. "You see, we didn't want to just dump the pallets because we were dealing with water," McLaughlin recalled. "You don't want the bottles to break. The MREs would probably survive. But if you just dropped them on the pavement anarchy would break out. We wanted to keep the distribution lanes as orderly as possible." Colonel Aycock said calmly, "Sir, we've sent some people to get forklifts."[65]

At that Honore went ballistic, cursing up a storm. "Get the fucking

HEMTTs and dump all that shit," he barked. "What kinda fool are you?" Without question, General Honore was right. When people are dying of thirst, you don't make them queue up in lines as though they were waiting to see a rock 'n' roll concert. You dump the water, fast. Yet it was also true that the Louisiana National Guard had been holding Fort New Orleans while President Bush tragically hesitated in sending in the federal reinforcements. General Honore could have pulled Colonel Aycock aside for a brief chat, telling him why he wanted the water dumped. Instead he humiliated a Louisiana National Guard officer in front of his men. "I know what military leadership is," McLaughlin said. "If you're going to correct somebody, especially a full colonel, you take them to the side and say it quietly. But to bawl out a full colonel in front of three media people, for a sound bite? I was sick in the pit of my stomach. What General Honore was saying was 'I'm here. I'm in control. You locals are worthless.'"[66]

For the same reason, the implication of the meeting on Air Force One was troubling to Governor Blanco. While General Honore was tacitly exerting control over her National Guard, he did not command them and she wanted to keep it that way. To that end, she was trying to defend Louisiana's legal rights with the President of the United States. She wrote an open letter to President Bush requesting 40,000 additional troops. She coyly insisted that if New Orleans was to be rescued, the President's "personal involvement" must be to "ensure the immediate delivery of federal assets needed to save lives that are in jeopardy hour by hour." While Blanco was making this proclamation, the Bush administration was already seizing control of New Orleans and blaming her—and the NOPD and the Louisiana National Guard—for the post-Katrina bedlam.

There are many versions of what exactly President Bush and Governor Blanco said as they spoke quietly on Air Force One. Bush had already gotten an agreement from Nagin to federalize the troops, though the mayor had no say in such matters. Blanco, of course, did have a legal stake in the disposition of the troops, and she remained skeptical of the notion of a federal takeover. She saw it as simply the Bush administration trying to scapegoat Louisiana for everything that went wrong. She felt the blame should have been firmly pointed at President Bush's slow response, Secretary Chertoff's disinterest, and Director Brown's fecklessness. That Friday evening, President Bush made a frontal attack on Governor Blanco, dismissing her "open letter" and raising the stakes. "Around midnight, the power struggle was ratcheted up when the Bush administration sent Blanco a legal memorandum proposing that the governor request that the federal government take charge of the evacuation of New Orleans," wrote Michael Eric Dyson in *Come Hell or High Water: Hurricane Katrina and the Color of Disaster*. "The Bush administration sought control of the police and National Guard forces that reported to Blanco. After talking throughout the night with administration officials, Blanco and other Louisiana officials rejected the offer for fear that it would amount to a federal declaration of martial law."[67]

What was their tug-of-war all about? Much of it concerned public perception. President Bush *had* the authority to federalize the Louisiana National Guard. He could have invoked the Insurrection Act of 1807, which gives the President the right to "suppress insurrections," a particularly touchy issue in the Deep South. Since World War II it had been invoked three times, race relations being intrinsic to each case: in 1957 in Little Rock, and 1963 in Oxford, both revolving around desegregation, and in 1992 during the Rodney King race riots in Los Angeles. Nicholas Lemann wrote in *The New Yorker*, "The Insurrection Act of 1807 outlines the script that the administration evidently wanted Governor Blanco to follow: a governor asks the President to federalize local law enforcement in order to suppress an insurrection; the President issues a proclamation ordering the 'insurgents to disperse'; they don't; the cavalry rides to the rescue. . . . But the President has the option of sending in troops without being asked when the law isn't being enforced or the rights of a class of people are being denied—which was clearly the case in New Orleans, not just because crime was rampant, but because so many people were trapped in hellish conditions."[68]

Governor Blanco was flabbergasted by Nagin's performance on Air Force One: taking the long shower, pounding his fist, using cuss words with abandon. He was unprofessional and, she feared, unglued. He seemed to have made a prearranged pact with Andy Card, acting jovial with him like he was already firmly in the Bush camp. He was praising President Bush ad nauseam, despite the fact that the day before he had criticized the federal response on the radio. He simply told everybody in Louisiana what they wanted to hear—except Blanco. "We always sensed that he was used as a tool, as a buffer," Blanco said in a December 2005 interview. "It was bizarre, you know, that little [get your asses down here] remark he made on national TV and I saw it again recently when I was looking at the copies of my public comments. When we met on Air Force One, Nagin was falling apart. He was near nervous breakdown."[69]

Quite naturally, Nagin saw himself in a different light. His version of the Air Force One meeting with Bush cast himself as the balanced, fair-minded voice of reason. Even though he had denounced Bush on WWL the evening before, Nagin insisted that the President knew that all those remarks were just straight talk. "But he was well aware of it," Nagin told CBS's *60 Minutes*. "And I pulled him aside with the governor. I said, 'Look. That was uncharacteristic of me. But consider being in my shoes. What would you have done? And if I said anything disrespectful to the office of the President or the governor, I apologize. But tell me, What we gonna do now?' "[70]

According to Nagin, President Bush accepted his quasi-apology. "The President basically said, 'Mr. Mayor, I know we could've done a better job, and . . . we're gonna fix it.' " According to Nagin, Bush insisted that the New Orleans mayor be deadly honest with him, without pulling any punches. "He said to me, 'I think I've been hearing a lot of stuff that . . . may not be true,' " Nagin recalled. " 'I wanna hear from you. Tell me the truth, and I will help you.' And I looked in his eyes, and he meant it. And when he meant it, I told him the truth." Nagin claimed Bush was "brutally honest" and "we talked turkey."[71]

Even though Blanco didn't fully respect Nagin and thought him a man of bad faith, she telephoned him after the meeting and inquired about his mental health. "I called him that night and said 'Ray, you need to get out of town. You need to go sleep somewhere.' What was going on was that he was locked up on the twenty-fourth [*sic*] floor of the Hyatt and anyone who wanted to see him had to climb twenty-four [*sic*] flights of stairs and he was afraid to come out. But he'd come out once a day or something like that and go make some crazy remarks to the media and then go hide. He was, like, going through a near nervous breakdown." Whether Blanco's advice was heeded or not, the mayor did soon leave the Hyatt. The governor recalled, "Well, then he left for five days! In the heat of everything that was going on, he's screaming about no leadership and he's a total void. I never made any public comments about that and still would have a hard time. . . . I don't want to talk about him. It doesn't serve me well to talk about him, best I don't talk about him."[72]

The period to which Governor Blanco was referring was Wednesday, September 7, through Monday, September 12. While the streets of New Orleans were still flooded, Mayor Nagin left for Dallas and leased a house. He exaggerated that his own New Orleans home on Park Island Drive had been severely damaged (it wasn't). Although Nagin didn't deny spending almost an entire week in Dallas, he was perplexed as to why Blanco and others were upset over the hiatus. "Why would a governor of the state of Louisiana be ticked about that?" he asked. "I don't get that. I mean, I took care of my city as best I could. I got it organized. I got rescues. I didn't leave [for Dallas] until that last bus left New Orleans. . . . Why does that upset somebody?" Like Blanco, Police Chief Eddie Compass was stunned that Nagin simply abandoned New Orleans at its darkest hour. In NOPD circles, Nagin's handle had been "New Orleans One." Since the police were furious that Nagin had leased a house in Texas while New Orleans was 80 percent underwater, he became known in NOPD circles as "Dallas One."[73] A few police officers made signs saying "Dallas One" as a protest, posting them around the Wal-Mart and Tchoupitoulas Street. "I pulled the signs down," Compass recalled. "I knew what it meant. I told the guys it just wasn't good for the city."[74] Meanwhile, Nagin found the time on September 5 to tell NBC News, "It wouldn't be unreasonable to have 10,000 deaths."[75] It was the most irresponsible statement any politician made during the Great Deluge. Just imagine the pain such a dire forecast brought to displaced people frantically trying to find a loved one. A leader doesn't hype the casualty toll. It shatters morale and increases panic. Ironically, even with Nagin exaggerating the numbers, many assumed he was, in fact, lowballing the real figure.

There were 1,351 deaths attributable to the hurricane: 2 in Alabama,[76] 2 in Georgia, 16 in Florida,[77] 228 in Mississippi,[78] and 1,103 in Louisiana.[79] The numbers edged upward in late September as more bodies were found.

Officials had to make judgment calls to determine whether particular deaths would have occurred in due course or were the direct result of Katrina, the evacuation, or the stress of the aftermath—it could be very hard to say. Don Moreau, operations chief in one of the regional coroner's offices assigned to work on Katrina deaths in Louisiana, gave the example of a man with a terminal disease who died as he was being evacuated under trying conditions. The coroner was supposed to decide whether it was a "Katrina related" death. "Short of 1-800-ASK-GOD," Moreau said, "I don't know how to determine that."[80]

In the December interview, Governor Blanco cited Nagin's exaggeration of the death toll as pandering to the press; he was ready to tell anybody what they wanted to hear. "I felt like those Nagin statements made it more difficult for everybody," Blanco said. "When he said '10,000 dead,' then the media would be asking me, every time I went up, 'How many people do you think died?' And I would have to respond, 'We have no way of knowing that right now. We have no confirmation. We're still in lifesaving mode.' "

When pushed to explain Nagin's motivation for such an exaggeration, Blanco, accentuating the near nervous breakdown, offered a plausible answer: "I think I can guess what happened to Nagin, but he'd have to tell his part, is that after a time, you have a tendency to want to satisfy people when they ask you the same question, day in and day out. He probably just threw it out, like estimates of 10,000. I never did that because I felt like I had no way of knowing it. We still had to do lifesaving missions before we started on body recovering and then the reports would have to be done. I just know and understand the dangers of overstating in the big national and international media and I just try to use facts as much as possible without speculating. They love you to speculate. Then you own the number. I think Ray is saying now that he threw the number out as a possibility and then it became his number. I didn't let them tag me with a number."[81]

The events that followed in the wake of Hurricane Katrina were spun into legends even as they were happening. Rumors were folded into the news cycle and repeated as fact before they could be corroborated or checked. Stories of rampant murders at the Superdome were later discounted: there were no murders at the Superdome. There was a tale of twenty-two people found in Jefferson Parish, tied together by a rope, drowned in the storm's fury: it wasn't true, even though it was reported in newspapers all over the country. The most upsetting case of rumor-mongering, however, would come from a surprising source. On the Tuesday after the storm, Police Superintendent Eddie Compass met Oprah Winfrey, who was visiting the city of New Orleans on a mission. On her September 9 show she said, "As this catastrophe unfolded, I watched, like all of you, and I felt helpless, and I wanted to do something. So I picked up the phone and I called some of my friends and said, 'Let's go down there and see what we can do.'" While in New Orleans, finding "stories you haven't heard," Oprah did get a true scoop. In talking with Compass, she heard his emotional account of a friend and fellow officer who committed suicide. She asked him if there were still dead people in the houses. "Oh, God, thousands," Compass said.

Compass's prediction that thousands of dead bodies would be found in homes was far off. It was perhaps an understandable bit of unintentional hyperbole. Unfortunately, the police superintendent didn't stop there.

"Inside the Superdome," Oprah Winfrey said to her audience, "he had seen horrors that will haunt him the rest of his life."

"We had little babies in there," Compass said. "Some of the little babies getting raped . . ."[82]

In the South, "babies" can refer to children of any age and at least one child rape was confirmed at the Convention Center (which Compass ought not to have confused with the Superdome). Compass had no right, however, to exaggerate whatever reports he may have heard into "little babies getting raped." He was losing his grip on reality, and within two weeks Nagin forced him to resign.

"Hindsight is easy," Compass later said in an interview. "I screwed up during Katrina. Communication was very bad. There was so much misinformation. Marlon Defillo was handling communications for the police; Marlon and I came to the conclusion that 'no comment' was worse than talking to the media. It was damned if I do and damned if I don't. Terry Ebbert told me to do the interviews because Nagin wasn't in shape to." In a 2006 interview, Nagin made it seem that Compass was the one unable to cope with "the pressure he was under."

Compass's babies-getting-raped exaggeration grew out of deep concerns he had for his own family. "My wife was eight months pregnant," Compass recalled. "And it was reported [falsely] that my twenty-four-year-old daughter was raped at the Ritz-Carlton. The communications were horrible. The reports I was getting of pedophiles raping small kids made me sick. My whole life is about helping the kids. So I broke down crying with Dr. Phil and Oprah and all that. In hindsight, I shouldn't have reported rapes—I gave too much information. Some of it was wrong information. I admit I made mistakes. But I was working twenty hours a day and trying to show the media that the police department had nothing to hide. We would report what we knew."

Compass was at City Hall when he received an e-mail from Nagin basically firing him. "He asked me to come up with a thirty-five- to forty-day exit strategy," Compass recalled. "I was in my room with Colonel Ebbert and Marlon Defillo when I read the e-mail. It's sad when your boss, and supposed friend, treats you like that. Jesus Christ! I just couldn't believe it, you know. Those guys really had to calm me down."[83]

V

On Friday, while President Bush held his Air Force One summit, Lieutenant Commander Jimmy Duckworth was still in "all hands on deck" high alert at the Coast Guard's temporary headquarters in Alexandria, Louisiana. Every hour, he was sending Coast Guard boats into the bowl of New Orleans. Meanwhile, Coast Guard helicopters, under the leadership of Lieutenant Shelly Decker (Air Desk) evacuated nuns out of the Ursuline Academy on Nashville Avenue and medical personnel from Memorial Hospital on Napoleon Avenue. On average, the Coast Guard rescued 5,500 people a year; by Friday it had already surpassed that mark. One junior Coast Guard pilot, in fact, saved more than 50 people in a twenty-four-hour period. With President Bush in the area, "everybody under the sun," according to Duckworth, was pitching in with boat and helicopter rescues. "DOD had started to come," he recalled. "Air Marine, private patrolling-type helicopters, every description in the air, flying people everywhere."[84]

The sky was filled with Hueys, Blackhawks, Chinooks, and Apaches. But it was the Coast Guard that led the way in the air and in the water after Katrina. It was because they were doing what they had been trained to do on a daily basis—save lives. By September 7 they had rescued 24,135 people by boat and helicopter, plus another 9,400 from eleven hospitals.[85]

Duckworth knew what these new federal troops were thinking—that television "did not do justice to the destruction and devastation that we saw." Although Duckworth was, for the most part, anchored to Alexandria, he had taken a Vietnam-era artillery observation plane on a mission earlier in the week, hovering over Orleans, Jefferson, and St. Bernard parishes at about 1,000 feet. The antiquated plane had a propeller in both the front and the back. He was particularly worried about the people still stranded in Chalmette. He wanted to deploy LCACs—100-foot-long and 50-to-65-foot-wide behemoths—into the parishes, but he wasn't sure they could fit over the levees. Instead of rescuing ten people at a time the LCACs could bring in at least sixty people per launch. "My plan was to go down to Lower St. Bernard to see if we could get the LCACs up the Violet Canal because, I thought, if we could get them up the canal we could get them onto St. Bernard Highway and we could get into the neighborhoods and rescue thousands," Duckworth recalled. "I asked my pilot to fly over the Mississippi River–Gulf Outlet and follow it down so I could find the Violet Canal and follow it back. And we did that. I could not believe the devastation to the MRGO levee. It was just basically not even there. Just a fucking ant pile. The Bayou Bienvenu Flood Control Structure just looked like it had been stomped by a giant, just completely in disarray."[86]

Upon flying toward the river from the intersection of the MRGO and the Violet Canal, Duckworth was stunned by the horrific vastness of the four serious oil spills, that of Murphy Oil being the largest. A Coast Guard pollution control group worried about the potential health risks from the benzene. It came to the conclusion that the four oil spills combined were on the order of approximately three-quarters the size of the notorious Exxon-Valdez spill that occurred in Alaska in March 1989. "All of a sudden my eyes started burning, and I thought there was a hydraulic leak in the plane or something," Duckworth recalled. "It was just really, really bad vapors, and I looked down and saw the oil that was down there in the Murphy tank. There was oil everywhere and what was happening was vapors from the oil baking in the sun came up one thousand feet, so it was hard to keep your eyes open at that."[87]

But Duckworth was too busy to worry about benzene exposure. He had FEMA to deal with. Task Force Katrina, under General Honore's command, wanted the Mississippi River opened ASAP. That would take Coast Guard/FEMA cooperation. Duckworth was game, but he quickly learned that FEMA's primary problem was its complete ignorance of Louisiana geography. "I worked a lot with FEMA during the outset and the biggest challenge that I saw for FEMA was having an appreciation for the nuances of New Orleans and the surrounding area," Duckworth said. "The fact that the river doesn't necessarily run north–south was a big deal. The East Bank can be the North Bank. Where is this in relation to that? Oh, it's not just the Port of New Orleans, it's the Port of Plaquemine, St. Bernard–New Orleans, Port of South Louisiana. FEMA didn't seem to have full appreciation for that, for locale. And I found myself and some of my fellow operators who are from New Orleans, we found ourselves acting as translators for a lot of FEMA people to help get them to understand what it's going to take to do something, what are the logistics involved in getting from this point to that point. You can't drive there, you're going to have to go by boat, etc. But, to be fair, there was no way FEMA could have absorbed the magnitude of the destruction. I'm not a Michael Brown apologist, but I will say that had Douglas MacArthur been in charge of the preparation and response for Katrina, it would have made an ass out of him."[88]

On Friday, Duckworth cracked a smile for the first time since Monday. Things were looking up: the National Guard reinforcements, President Bush's Gulf South tour, the U.S. Army 82nd Airborne Division en route, the FEMA and Blanco buses, the Kaiser-Hill company sandbags, the fact that Fats Domino was found alive. The light rays were starting to shine through.

Meanwhile, the community of Alexandria just kept rolling out the Southern hospitality. A citizens' group was barbecuing nightly dinners for the Coast Guard, with casserole dishes appearing around the clock. The police department was using its squad cars as taxis for the Coast Guard, shuttling some officers back and forth to Baton Rouge. All the city restaurants had a "Feed the Coast Guard for Free" program. Schoolkids plastered walls with signs reading "Thank You" and "Well Done." The Alexandria International Airport became a staging area, allowing tons of equipment and matériel to be flown in. American flags appeared on front lawns. "Looking back on it," Duckworth said, "one of the things that really helped out the crew was when you're walking off watch and you're all hot and tired and there's a little lady walking in with a cake. That happened over and over."

Throughout the Great Deluge, Lieutenant Commander Duckworth kept detailed notes of what the Coast Guard was doing. His journal was filled with time charts, mission plans, basic math, telephone numbers, accomplished goals, and persistent problems. On Friday Duckworth made a note that he later pointed to as a key to the government's troubling response during the first week after Katrina. The note read:

- Communications Difficulties
- Transportation Challenges
- Mobility of STA and AIRSTA
- Many of our crew have lost everything
- Impact to?—Damage to Harvey Lock + Bridges
- Numerous sunk vsls + Barges
- Damage to Mineral + Oil Infrastructure
- Ugly Civil Unrest
- Safety and Health Concerns
- 1,000s of Bodies Decomposing
- Good Citizens Are Suffering for Acts of Some
- Unknown # Trapped in Buildings
- C-130 Is Poor!
- 12 Survivors on Spencer 9 Are Filipino
- 3 Mine Sweepers from Navy?[89]

Almost every page of Duckworth's journal pointed out problems that needed solving. When pressed to explain, however, who or what was to blame for the post-Katrina search, rescue, and recovery quagmire, he balked, saying he wasn't a "woulda, coulda, shoulda kinda guy." He particularly refused to criticize anything that President Bush or Governor Blanco did; they were, to his mind, his bosses. He respected both of them. But then, not wanting to dodge the question entirely, he turned to page fourteen of the journal and said one word, "bureaucracy." He then pointed to something he wrote that Friday which, just weeks after Katrina, remained incomprehensible in its stilted red-tape language. "However, FEMA rep FIAT CA DEP is Requiring Request for Type II Incident MGMT TEAM for Assessment for Establishment of DOD Supported Base Camp." When asked what this meant, Duckworth shrugged. "That's the point. The Coast Guard was successful because Captain Paskewich had gotten our assets out of New Orleans before the storm, we situated ourselves in Alexandria, and we threw away the playbook," he said. "We took all comers and didn't wait for Type II Incident Management Teams. We winged it. We entered the game and stayed in the game until our job was done." And then, he added, "Too much bureaucracy can be a big, big problem in a catastrophe."

VI

Like Lieutenant Commander Duckworth, Mississippi Governor Haley Barbour—the Yazoo City/Zeiglerville booster—believed that stoic optimism and a can-do spirit was the best way out of the disaster. A proponent of this philosophy was Bay St. Louis Mayor Eddie Favre. Each day after Katrina when on the Gulf Coast a lumber store reopened or a fast-food restaurant started clearing away debris, Mississippians started talking about progress. "We're just not a whiny people," Favre said. "But we're compassionate beyond belief. Even though our town was largely gone, we took care of each other. We didn't count on the feds. We knew it was up to us and eventually the lights would go on."[90] Everybody *believed* the lights would shine. That's the incredible thing about Mississippi Gulf Coast optimism. "When the lights came on, that was a blessing from God," Eddie Bigelow told the Associated Press. "Every day is a little better. It's like giant steps, if you saw this place [Gulfport] last Tuesday [after Katrina hit]."[91]

The Mississippi Gulf Coast optimism was contagious. If Bruce Springsteen was the Rock 'n' Roll Laureate of New Jersey and John Mellencamp ruled Indiana, then surely Captain Jimmy Buffett was the voice of the Gulf South. Born on Christmas Day 1946 in Pascagoula, Mississippi, the son of a naval architect, Buffett was raised in Mobile, Alabama. All up and down Highway U.S. 90—which ran from Jacksonville Beach, Florida, to Van Horn, Texas—he had kinfolk living along the sand dunes of Bay St. Louis, Pascagoula, Florida, Pass Christian, Gulf Shores, and even the off-shore islands. A natural-born explorer, Buffett liked nosing around the alternative U.S. 90 roads, like the Old Spanish Trail and South Wayside. "Even as a kid, New Orleans was my focal point," Buffett said. "I considered home anywhere along that ribbon, where French culture existed."[92]

Buffett's 1975 hit, "Margaritaville," and its follow-up, "Cheeseburger in Paradise," became virtual anthems for the Gulf South tourist industry. Captain Buffett, usually clad in Hawaiian shirts and cutoffs, baseball cap in place, always looked ready for the beachball good times. Defying genre, mixing country and western with calypso, he launched the Parrothead Movement; his sold-out concerts were like Mardi Gras celebrations, where booze flowed freely and outlandish carnival outfits were worn. Before long he opened Margaritaville Cafés in Key West and New Orleans (by 2005 there were seven more); this was followed by a series of *New York Times* bestselling novels, like *Tales from Margaritaville* and *Where Is Joe Merchant?* "Everything was going great and then Katrina came and, man, I just couldn't think," Buffett recalled. "When those winds hit, and those levees broke, a helluva lot of culture went out of the northern edge of the Caribbean."

Buffett was actually in Key West the week before Katrina had hit Florida. August was the month he liked to scuba dive. "It was so damn hot," Buffett recalled. "The usually cool water felt like bathwater. There was nothing scientific about it. But I remember thinking, and saying to people, 'The next storm in the Gulf of Mexico is goin' to be a bad one. Real bad.' I'd dived for decades in August and the water was never that warm."[93]

When Katrina made landfall in Buras, Louisiana, Buffett was playing in Indianapolis. But whenever he wasn't singing and strumming, he was glued to the television. Not only did he have relatives to worry about, but he owned houses on Mobile Bay. Then there was his restaurant on Decatur Street in New Orleans. "I didn't think I had a role as a first responder," he said, "but I got down there as quickly as possible."

Immediately, Buffett offered all his New Orleans employees an advance in salary of four weeks, hotel lodging, and jobs at his other Margaritaville restaurants.[94] Hurricanes had always been a motif at Buffett's Caribbean-themed eateries, where lightning, thunder, and a spinning hurricane cloud were used as side effects during the musical revues.[95] His new post-Katrina gambit was "Blue Dog Relief"; all of his restaurants would promote levee 5 protection of New Orleans by selling George Rodriguez Blue Dog prints with the slogan "To Stay Alive You Need Levee 5."[96]

As an entertainer beloved in the Gulf South he didn't want to denounce President Bush for the slow federal response. That was not his style.

Shortly after the Indianapolis gigs, Buffett made his way to Mobile Bay. Reports of the New Orleans levee breaks were dominating the news, but Buffett considered Mississippi the ground zero of Katrina. Standing in Pascagoula, Mississippi, seeing rubble, made him feel angry—real angry. But who was he mad at? Buffett was profoundly moved by a *Times-Picayune* op-ed called "Open Letter to the President":

> Dear Mr. President:
>
> We heard you loud and clear Friday when you visited our devastated city and the Gulf Coast and said, "What is not working, we're going to make it right." Please forgive us if we wait to see proof of your promise before believing you. But we have good reason for our skepticism.
>
> Bienville built New Orleans where he built it for one main reason: It's accessible. The city between the Mississippi River and Lake Pontchartrain was easy to reach in 1718. How much easier it is to access in 2005 now that there are interstates and bridges, airports and helipads, cruise ships, barges, buses and diesel-powered trucks.
>
> Despite the city's multiple points of entry, our nation's bureaucrats spent days after last week's hurricane wringing their hands, lamenting

the fact that they could neither rescue the city's stranded victims nor bring them food, water and medical supplies.

Meanwhile there were journalists, including some who work for the *Times-Picayune*, going in and out of the city via the Crescent City Connection. On Thursday morning, that crew saw a caravan of 13 Wal-Mart tractor trailers headed into town to bring food, water and supplies to a dying city. Television reporters were doing live reports from downtown New Orleans streets. Harry Connick Jr. brought in some aid Thursday, and his efforts were the focus of a *Today* show story Friday morning.

Yet, the people trained to protect our nation, the people whose job it is to quickly bring in aid were absent. Those who should have been deploying troops were singing a sad song about how our city was impossible to reach.

We're angry, Mr. President, and we'll be angry long after our beloved city and surrounding parishes have been pumped dry. Our people deserved rescuing. Many who could have been were not. That's to the government's shame.

Mayor Ray Nagin did the right thing Sunday when he allowed those with no other alternative to seek shelter from the storm inside the Louisiana Superdome. We still don't know what the death toll is, but one thing is certain: Had the Superdome not been opened, the city's death toll would have been higher. The toll may even have been exponentially higher.

It was clear to us by late morning Monday that many people inside the Superdome would not be returning home. It should have been clear to our government, Mr. President. So why weren't they evacuated out of the city immediately? We learned seven years ago, when Hurricane Georges threatened, that the Dome isn't suitable as a long-term shelter. So what did state and national officials think would happen to tens of thousands of people trapped inside with no air conditioning, overflowing toilets and dwindling amounts of food, water and other essentials?

State Rep. Karen Carter was right Friday when she said the city didn't have but two urgent needs: "Buses! And gas!" Every official at the Federal Emergency Management Agency should be fired, Director

> Michael Brown especially. In a nationally televised interview Thursday night, he said his agency hadn't known until that day that thousands of storm victims were stranded at the Ernest N. Morial Convention Center. He gave another nationally televised interview the next morning and said, "We've provided food to the people at the Convention Center so that they've gotten at least one, if not two meals, every single day."
>
> Lies don't get more bald-faced than that, Mr. President. Yet, when you met with Mr. Brown Friday morning, you told him, "You're doing a heck of a job."
>
> That's unbelievable.
>
> There were thousands of people at the Convention Center because the riverfront is high ground. The fact that so many people had reached there on foot is proof that rescue vehicles could have gotten there, too.
>
> We, who are from New Orleans, are no less American than those who live on the Great Plains or along the Atlantic Seaboard. We're no less important than those from the Pacific Northwest or Appalachia. Our people deserved to be rescued. No expense should have been spared. No excuses should have been voiced. Especially not one as preposterous as the claim that New Orleans couldn't be reached.
>
> Mr. President, we sincerely hope you fulfill your promise to make our beloved communities work right once again. When you do, we will be the first to applaud.[97]

The newspaper had articulated Buffett's attitude in a nutshell. The Mississippi country boys like himself would join the New Orleans rising. Bourbon Street was still his touchstone. They hadn't lost New Orleans, Biloxi, or Mobile, after all. The spit-in-your-face spirit was alive and well. The Gulf South, including New Orleans, was going to rise again. Piano legend Eddie Bo would still be singing "Hook & Sling" and the fried shrimp po'-boys would be better than ever at Trapani's in Bay St. Louis. The oyster beds would come back; they always did, and the Original Oyster House would be serving half-shells with crackers before you could say "Spanish Fort." Ever since he went through a Dylan phase in his twenties,

Buffett had stayed away from protest anthems. He considered himself a song-and-dance man, not Baudelaire.

But just seeing his aunt's Pascagoula home destroyed was hard to take. The simple place had been a depository for precious memories, the station where his boyhood imagination soared. Somehow, all of his sailor's dreams came back to this house. He had even written a song, titled "The Pascagoula Run," about his teenage nights at the Stateline bar, along the waterfront where there were "pinball machines and Cajun queens." The chorus went:

> It's time to see the world
> It's time to kiss a girl
> It's time to cross the wild meridian
> Grab your bag and take a chance
> Time to learn a Cajun dance
> Kid, you're going to see the morning sun
> On the Pascagoula run[98]

So Captain Buffett sat down and did what he did best; compose songs that Gulf Coasters could relate to. He wrote with his drummer friend Matt Betton, who had a gorgeous melody just waiting for Buffett's salty lyrics—an off-center, post-Katrina coping song titled "Move On." While Emeril Lagasse was downsizing, and keeping Delmonico closed, Buffett opened his New Orleans Margaritaville Café pronto. So what if he lost money? The show must go on. Next, he signed up to perform at the New Orleans Jazz and Heritage Festival, claiming he was going to get "his ass on stage dancing" when Fats Domino took to the piano.[99]

Somewhere in Buffett's sea-breezy spirit were the seeds of the New Orleans rebuild. At the time of the 150th Mardi Gras, Buffett was still so popular in the Gulf South that he could have run for governor of Florida, Alabama, Mississippi, or Louisiana and won. He wasn't a first responder or a member of a blue-ribbon commission, but he had the Mike Fink roarer's attitude, the feisty, fiery, indomitable, clumsy, shrewd joie de vivre of the Gulf South on his guitar fret. Buffet headed to his Key West recording studio in February 2006 and laid down tracks for his new album,

Party at the End of the World. He went for a swim on the beach. The water wasn't hot anymore. The dolphins were back, wiggling like hula hoops. He felt refreshed. Pretty soon the Gulf South would be comin' on strong. When asked, however, if he would consider running for governor of Florida, Mississippi, Alabama, or Louisiana—all states where he owns homes—he declined. "I probably couldn't pass the drug test," he said. "So I'll just keep singin' my songs."[100]

VII

Even as stranded Biloxians were desperate for rescue and relief, some entrepreneurial Gulf Coasters were chanting "rebuild" by Saturday, September 3. But there was an opportunist strain to their voices that did not necessarily put them on the same side as everyone else. Marlin Torguson, a native of northern Minnesota, got his start in the casino racket helping Native American tribes get permits to operate slot machines. In 1990, when the Mississippi legislature approved casino gambling (on water) he headed for the Gulf Coast, purchasing a large tract of land from jazz clarinetist Pete Fountain. Before long he had opened two Casino Magic gambling complexes—one in Bay St. Louis, the other in Biloxi. By 1999 he had made a fortune and could live wherever he wanted. He chose Jacksonville, Florida. "I'm a born chicken," he confessed with a crooked grin. "That is when it comes to hurricanes."[101]

Watching the storm on TV from Florida, Torguson grew excited. Katrina *could* be seen as an opportunity. Sure, the cleanup would cost billions. But the Biloxi strip could now be rebuilt properly. He was ecstatic a month later when Governor Barbour declared that Mississippi casinos could be land-based. That was a windfall for casino investors. And, as luck would have it, Torguson had the blueprints and financing for a $500 million megacasino complex called Bacaran Bay—a name he picked out of thin air. The complex would have 740 hotel rooms and condos nestled along a little waterway near the Gulf. But it wouldn't be on the water. Those days were over, thanks to Katrina. There would be a huge convention center, forty bowling lanes, two wedding chapels, and a stadium-style music venue, where country and western legends could perform. "We're also going to have a major blues club and a dog hotel with groomers," Torguson enthused shortly after Katrina. "The Mississippi Gulf Coast is going to be better than before. You understand? Land-based, real buildings not on water. The dead and debris and mops will be gone before you know and we'll be back."

That's how salesmen dreamers like Torguson saw the Katrina disaster, even though hundreds of corpses had yet to be identified and frightened people still screamed for water and medical care at the Superdome. The winning ticket was to polish your shoes and make a real estate deal. The important thing now was political contacts. Did you give money to Haley Barbour's 2003 campaign for governor? Had you made sure never to double-cross former Louisiana Senator Trent Lott? Could you get a subcontract deal out of Halliburton? Did you know Steve Wynn? Would you be willing to have a bank account in the Cayman Islands? For decades, the Longshoreman's Union had made it cost-prohibitive to hire union workers. Now, freed from such constraints, and with big labor in disarray, cheap Mexican labor could pour into Mississippi and Louisiana and build Gallerias and Astrodomes and Six Flags. Big Texas and Las Vegas money was on its way—unstoppable. Uncork the champagne bottles. "You've gotta understand," Torguson enthused. "The Biloxi Airport brought in 850,000 passengers a year. The new one—which will be finished mid-2006—will turnstile 2.2 million a year. See what I mean? Biloxi is going to have twenty to twenty-five new casinos in the next five years."[102] The old Gulf Coast was gone, swept away by the hurricane and the human tornadoes that came afterward.

VIII

As the first week of the Katrina disaster came to an end, attention was focused on those who were still left in the blighted cities and towns of the Gulf South. But an astonishing 1.7 million of the people who had once lived in those neighborhoods were already gone, scattered throughout the United States. Many had moved in with friends or relatives, believing that the storm would be a temporary interlude, and that they could return home in a matter of days. By the end of the week, it was clear that, for most people, other arrangements would have to be made. Fortunately, communities all over America were eager to help evacuees, in any way possible.

Armantine Verdin and her son, Xavier, who were stranded on a shrimp boat in St. Bernard Parish for two days, finally walked along a levee to a ferry, which dropped them on the West Bank of the Mississippi, where a National Guard truck scooped them up and dropped them off under a bridge in New Orleans, where a crowd was awaiting evacuation. Eventually, a bus took them to the Reliant Arena in Houston. All the while, Monique Michelle Verdin, Armantine's granddaughter, was frantically making telephone calls to find them. She described what happened next. "After four days of floating, being herded, walking, riding, and waiting, my ninety-year-old grandmother and seventy-two-year-old mentally handicapped uncle, now alone to fend for themselves, were loaded onto a bus and dropped off at the Astrodome in Houston." She went to fetch them there, and a friend who lived near Baton Rouge offered them a place to stay in a trailer home, near a golf course. Armantine was told she could not return to her home in Chalmette in upper St. Bernard Parish until the following summer. "From what I understand," Monique said, "upper St. Bernard is a toxic waste site. Thousands of barrels of oil have spilled in the middle of suburbia. My grandmother desperately wants to go home. But home to what? And when? My grandmother needs a place to live where she can continue to plant her favored butter beans in the earth, breathe clean air, listen to the birds, and catch rainwater in a cistern. Simple survival."

Instead, they were displaced and making the best of it. "It is quite surreal!" Monique said. "My grandmother and Xavier have never seen a golf course except on TV. We watch the old white men as entertainment at the fourteenth hole."[103] They were among the lucky ones.

The story of Armantine Verdin makes it a little bit easier to understand the frustration of David Moore, a retired teacher from Akron, Ohio, who took the initiative in mid-September to go to Texas shelters for Katrina refugees and offer good housing up north. He couldn't understand at first why no one took him up on his offer. The fact that Akron had a reputation for slightly colder weather than New Orleans may have been a factor. But as Moore found out, the pull of home was even stronger than that. By the time he arrived, the people still at shelters didn't have anything left, except the tie to home. The strong, the smart, and the motivated were long gone by mid-September. They knew how to take care of themselves even when, as Monique Verdin said of her grandmother's situation, the "world is upside down." Those who were left in the shelters two weeks after the storm did not know how to survive except in the confines of their home turf. "I cried every night when I went back to the hotel," Moore said of his visit to Houston. "The people that wanted to get out, they got out. These were the hard core."[104]

AMERICANS GAVE well over $1 billion in the first two months after Katrina to help people whose lives were torn apart by the storm and its aftermath. The *Chronicle of Philanthropy* listed the top dozen recipients of generosity (as of early November):

American Red Cross Chapter Headquarters	$1,300,000,000
Salvation Army USA	$275,000,000
Bush-Clinton Fund for Katrina Relief	$100,000,000
Catholic Charities	$84,000,000
Habitat for Humanity	$53,600,000
United Way	$36,000,000
Samaritan's Purse	$33,000,000
United Jewish Charities	$23,600,000
America's Second Harvest	$22,700,000
Baton Rouge Area Foundation	$20,600,000
Southern Baptist Convention Disaster Relief	$20,100,000
Humane Society of the United States	$18,000,000[105]

The Bush-Clinton Fund for Katrina Relief had been initiated on Thursday, September 1, in a press event in the Oval Office at the White House.[106] The President's father, George H. W. Bush, the forty-first president, and

Bill Clinton, the forty-second, headed the fund-raising project, just as they had a similar, private-sector effort in the aftermath of the tsunami in Southeast Asia less than a year before. The partnership of the former political opponents had come as a welcome surprise when President Bush appointed them as humanitarian ambassadors in the tsunami disaster. In the wake of Katrina, the concept was slightly shopworn, in terms of its shock value, but the two presidents were still an effective team when it came to bringing in money. Only one day after the launch of the fund, Bill Clinton received a telephone call from Lee Scott, chairman of Wal-Mart, headquartered in Clinton's native Arkansas, pledging a whopping $17 million.[107] The fund was slow in dispersing the funds, waiting until early December to make its first contribution. It made a $40 million grant to help colleges rebuild after the storm; the rest of the money went to hurricane relief projects established by the respective governors, and to faith-based charities that were working on the ground in helping those affected rebuild their lives.[108]

The Red Cross, by far the recipient of the most money, was also the target of the most criticism. In case of a national disaster, the Red Cross is not a mere private institution, but a recognized part of the federal response. It is supposed to open and operate shelters, depending on FEMA for supplies. After Hurricane Katrina, the Red Cross did not have enough shelters, by a wide margin, or ready means by which to expand its efforts. In the absence of Red Cross shelters, local governments and private groups opened various spaces for storm refugees, but the Red Cross did not know how many of these there were. The organization had been caught utterly off guard. It dispensed money to people who patently hadn't been affected by the storm and remained oblivious to many who were.[109] FEMA made a poor partner in the post-Katrina effort, continually losing critical paperwork or simply denying requests with no explanation. Even while receiving positive answers from FEMA management, the Red Cross was stymied by what it called the federal agency's "mushy middle."[110] The Katrina response was twenty times larger than any previous American Red Cross effort, encompassing 220,000 volunteers, yet it was an effort that could have been much better. "Katrina," concluded the House Select Committee to Investigate the Preparation for and Response to Katrina, "was too much for the Red Cross."[111] That would be understandable, except that millions of Americans entrusted the group with their contributions, in hopes of materially helping the victims.

Then there was the unrecorded charity, from person to person. "I met a fellow from Vermont, a truck driver," Haley Barbour recalled, speaking of a visit he made to a shelter in Mississippi. "He and sixteen other truck drivers had driven down from Vermont to deliver seventeen trailers of food to Gulfport. I couldn't believe it . . . seventeen tractor trailers all the way from Vermont. Then he told me it was his third trip."[112]

Doris M. Jones, age eighty-seven, had lived in New Orleans for thirty years, having worked as an accountant at a hotel. In retirement, she resided in a small apartment complex, enjoying her friends there and the way they took care of one another. The complex weathered the storm in good shape, but one day about a week afterward, the authorities came in and told the residents they had fifteen minutes in which to pack their belongings and leave in the Army trucks awaiting them outside. It was part of the mandatory evacuation proclaimed right after the storm, but put into practice only after the homeless were safely moved out. "The first thing I did was dump my underwear drawer in a bag," Miss Jones said later. "Then I took all my jewelry and perfumes. I left there in my bedroom slippers, my underwear, and my jacket. It was horrible the way they treated us." Eventually, she was sent to Camp Edwards, an Army base on Cape Cod, all alone, without any relatives in the world and without even her friends from the complex. And also without the belongings she treasured most. "I loved my clothes," she said, "all color-coordinated and hanging up nice. Now they're all gone." As someone interested in fashion, she mentioned to some of the people at the Army camp that when she thought of Massachusetts, she thought of Filene's Basement, the bargain source for haute couture in downtown Boston. Word seemed to get out about that.

A few days later, Filene's Basement sent a town car to pick Miss Jones up at the base and bring her into the store for a VIP tour. She had the best bargain anyone ever had at the store: anything she wanted, for free. "This is to me a lifetime dream come true," she said. Employees lined up to greet her. A group at the bottom of the escalator—on the way to women's wear—cheered her and held a sign reading, "Welcome, Doris."[113] There were a few weeks in September and October 2005 when communities across America could not do enough for Katrina evacuees like Doris Jones. Everyone can imagine what it is to lose all sense of home. The only remedy in those early days was the word on the sign "Welcome."

Patrick Wooten, the roofer who had literally beaten back the looters eyeing his Algiers home, ended up at Camp Edwards as well. When he and his family arrived, they were dirty and tired, having been at Louis Armstrong International Airport for several days while awaiting evacuation. At the airport, "we had to sleep on the floor with the dogs and everything," Wooten said. "The dogs peeing all on the floor, so you know that's how you made your clothes smell." After the three-hour flight to Cape Cod, the passengers were met on the tarmac by a line of people: Red Cross volunteers, physicians, ministers, and others from the community. These Cape Codders were so happy to see the evacuees, safe and sound at last, that they naturally opened their arms to hug each passenger. Patrick Wooten recalled, "You come out there and getting off the plane, I said, 'Oh, man, I really don't want to do this to these people, touch them. I'm stinky.' And so they're overlooking that filth, you know what I'm saying, man? 'Come here and give me a hug.' I say, 'Okay, here you go, but I'm stinky.' They ain't worried about how stinky I was. They said, 'Come here.' I said, 'Oh, man, this is love.' "

IX

Michael Prevost, the heroic Newman High School administrator, had decided it was time to leave the Lakeview area on Friday. President Bush had visited the 17th Street Canal, new rescue boats were arriving en masse, and he was plumb exhausted. It was time to stop canoeing. On the previous day, Thursday, he had gotten through to his fourteen-year-old daughter, Maddine, in Virginia. "Daddy, where are you?" she asked. "In New Orleans, honey," he answered. Her commonsense response was, "Why?" Prevost decided she was right—enough was enough. The next day, he packed his canoe, paddled past the London Avenue Canal and through

Lakeview, checking on a few friends' houses on the way. With Chelsea, his dog, still his loyal companion, he eventually made it to Metairie, getting off at the I-10/610 split. Two ambulances were there and they brought Prevost and his dog to the Causeway/I-10 split. He was appalled at the sea of humanity congregating at the crossroads. "Just like Mogadishu or something," he said. "It was chaos. When they dropped us off I knew why I had been reluctant to go to the Superdome."[114]

A Louisiana state trooper immediately confronted Prevost. "You can't take your dog on the bus," the trooper said. "You see that mass of people, that's the line and no dogs." An SPCA rescue truck was nearby, so he took Chelsea over to them. He would have to part with his dog. They said good-bye, in a brokenhearted way, but Prevost assumed they would be reconnected soon. Few buses ever showed up. People were screaming in the oppressive heat. Garbage and flies were everywhere. Prevost decided he would start walking down Causeway Boulevard. Anything was better than hanging out with this angry crowd. Out of the corner of his eye he saw a new Ford Explorer. He walked over and started talking to the driver, who was a roofer. He was heading to Ochsner Hospital in New Orleans to help fix their Katrina damage. As the truck window went down he could feel the air conditioner blast out at him. "I said to him I could use a ride, anywhere out of here," Prevost recalled. "I told him to just drop me off on Airline Highway. He looked at me and said, 'You're all right aren't you? You don't have a gun?' He was very apprehensive. Earlier in the week his roofing crew had been looted and roughed up by a mob. 'Look, I'll take you to Ochsner, I have to do a little business there and then I'll take you to Baton Rouge.' "

To Prevost this was miraculous. While heading West on I-10 later that afternoon the roofer asked Prevost where he wanted to be dropped off in Baton Rouge. Prevost gave him his sister's address. The roofer put the address into his GPS direction finder and drove right to the location. When they arrived, Prevost's sister, worried that he had died, was pulling plywood off the windows. A homecoming was under way. In a single afternoon Prevost had gone from paddling in New Orleans floodwaters to having a Greek supper in an air-conditioned house with his sister's family in Baton Rouge. "It's not a happy ending though," Prevost said, near tears.

"I was never able to find Chelsea. Nobody at the SPCA knew what happened to her. I hired pet detectives, the whole bit, but no luck. They've just got to change that law about no pets on buses. It's caused a lot of grief."[115]

Dr. Ruth Berggren, the AIDS specialist at Big Charity, had arranged for the evacuation of her Nine West unit, but not before letting the national media know what had happened to New Orleans's hospitals after the hurricane. For example, when the shots of snipers were heard in the area around the hospital, its neighbor Tulane Hospital, and the Saratoga parking garage (its roof had already served as a helipad during the evacuation of Tulane), Berggren let CNN and the other major news outlets know. Berggren's old friend Libby Goff, now living in Texas, had arranged for a private jet to bring up to forty people from Charity back to Dallas.

Meanwhile, Berggren's story and Goff's lobbying persuaded CNN's editorial director, Richard Griffiths, to send his chief medical correspondent, Dr. Sanjay Gupta, to the scene. "A lot of these doctors came in because of the hurricane," Gupta told CNN viewers on Friday, September 2. "They knew they might be needed. They've been here since Saturday. They plan on staying until every patient has been taken out. . . . They've been very, very diligent about taking care of these patients, which is remarkable. I think a lot of lives that otherwise would have been lost have been saved by these doctors, who have not slept in several days, have very little food and water themselves, and are operating under the most remote conditions really possible."[116]

After Gupta's helicopter deposited him at Big Charity, it was clear the sick patients there were desperate to be evacuated. CNN could hardly just do its on-scene report and decamp, leaving everybody stranded. "We couldn't ethically just fly out our crew," Griffiths said. "So I started guilting companies in the South to send in their helicopters." Suddenly CNN was chartering a helicopter to make repeated Big Charity evacuations and finding another seven to do the same.

"What happened was a miscarriage of justice in America," Griffiths recalled. "And we were impassioned to do something from Atlanta. It was frustrating just watching from our control room. But we had an idea of the big picture. So if we could do our small part at Big Charity, great."[117]

On Friday afternoon Dr. Berggren, like a cattle drover, lined up the staff of Nine West, the ward of AIDS and tuberculosis patients, along the downstairs wall of the hospital. She was a clipboard queen, determined to keep her part of the evacuation orderly. Because CNN had taken on Big Charity as a corporate mission, the world was wondering what would happen to Nine West. "So we're all standing there in this dark hall and the panic level is rising, everyone's sweating," Berggren recalled. "They're bringing down violent psychiatric patients (whom they had sedated and wrapped with duct tape to the gurney), so they're coming out like mummies, wild-eyed. The people that are transferring them are shoving these gurneys down the hall real fast. So we're all flattened up against the wall and there's people with bullhorns yelling, 'Move to the left! Move to the right!' No matter how close you got to the wall. That moment was so unearthly, unlike anything I had ever experienced."[118]

The Texas Wildlife and Fisheries "cowboys" arrived at the emergency-room loading dock with flat-bottom boats. All of the eighteen Nine West patients, fourteen with AIDS and four with tuberculosis, were helped into the boats. Each patient's medical records had been carefully sealed in a plastic zippered folder and hung via a chain around the neck. "It was essential that whatever shelter they ended up in knew their history," Berggren said. "Now, at long last, we were all parting." Some of these Nine West patients were brought to evacuation buses; others, unfortunately, were dropped off at the Superdome. "When I learned they weren't taken immediately out of town as promised, I was livid," Berggren said. "But they all eventually made it to a safe place."

With all of Big Charity's patients evacuated, Dr. Breggren was ready to evacuate her Nine West staff, which included nurses, ward clerks, and cleaning specialists. At four in the afternoon, Dr. Berggren and eighteen others, including the twelve-year-old son of one of her nurses, boarded four or five Louisiana Wildlife and Fisheries boats. With Berggren's husband, Dr. Tyler Curiel, as their guide, they headed to the Saratoga parking garage. They felt they were entering the unknown, clambering into glass-bottom boats on a jungle cruise. Like a mother hen, Dr. Berggren took control of the Nine Westers, making sure they evacuated together. "They started screaming with the bullhorns that they wanted five females on the boat and we thought that was really strange," Dr. Berggren recalled. "We had never heard about any gender separation, so we said 'Uh-uh.' The Wildlife and Fisheries rescuers just couldn't understand why none of our nurses would step forward. But they eventually let us on the boats as a cohesive unit." In an article Dr. Berggren wrote for the *New England Journal of Medicine*, she complained about "rough game wardens," who were "oblivious to our requests to travel together."[119]

When the boats pulled up to the Saratoga garage, Curiel sprinted to the roof to let the National Guard know the Nine West group had arrived. He met with disappointment on the rooftop: there were no helicopters. They waited anxiously for half an hour while game wardens prowled the rooftop with guns. There were still no copters. They all walked downstairs to Loyola Avenue and eventually found two school buses to take them to the airport.

Roadblocks prevented the buses from driving down I-10; the police told them that the airport was too full, with more than 45,000 people trying to escape New Orleans. "They kept trying to get us to turn around," the doctor recalled. "I was still wearing my badge, and every time I had to talk to a police officer, I would just get pushy and shove my ID in their faces." After talking her way through two such roadblocks, the pair of buses were stopped a third time. "There was this very red-faced, I think maybe drunk policeman, at least he was very sleep-deprived and slurring his words and determined to turn us around," Dr. Berggren recalled. "He didn't want to let us go any further."[120]

Just as her frustration level peaked, she looked at her cell phone, which was serviceable for the first time since Katrina. She dialed Libby Goff. "You've got to convince this policeman that we've got a private plane waiting for us," Dr. Berggren told her friend. "I can't get through to him." A brazen, "let me at him" Libby agreed. Hopping off the bus, Dr. Berggren, told the officer that "Texas" wanted to talk to him, and she handed over the cell phone. A polite, fast-talking Libby pulled some mumbo jumbo about how Governor Perry, Ross Perot Jr., and other Lone Star heavyweights were waiting in Dallas for the arrival of these VIP Nine Westers and he, the roadblock cop, was screwing up the entire gambit. "I was never rude," Goff recalled. "But somehow I found the right words. His whole demeanor changed. He backed off and said, 'I'm sorry, I don't want to prevent anybody from having an evacuation.' "[121]

The passengers cheered. At last, Nine West was at the departure location, the Signature Aviation Terminal. This private hangar was a ghost town: totally abandoned, not a security guard or private plane in sight. Stacked up in boxes on the floor were thousands of heated MREs. Why weren't they being distributed? Who was keeping them stashed here? What in the hell was FEMA up to? She never found out the answers to those questions. All of the staff sat down and rested on the cool concrete. Unfortunately, Berggren's cell phone was once again dead. The last word was the private plane from Chicago would arrive at 10 P.M., so they waited, eating MREs and dozing off. Hours passed—there was no plane at ten, nor at one, three, or five in the morning. Finally, at dawn, Berggren got a signal on her cell. It was Goff, who had been trying to reach her all night; the jet was coming at 10 A.M.—not P.M.—and its tail number was 933. "When the plane finally arrived at 10:15 Saturday morning, the nurses all said they were going to name their next child 933," Dr. Berggren recalled. "We were treated like royalty on this plane, with food and drink and wipes. I'd been telling Libby all of our woes. She had prepped our pilots about what provisions we most needed; so they bought wipes. The nurses, a lot of them had never been on an airplane. They were thunderstruck. They got giddy and one of them started drinking."[122]

When Nine West landed in Meacham International Airport in Fort Worth, there was even a welcoming committee of old neighbors from when Berggren had lived in the area and others. In flight the doctors, nurses, and staff of Nine West had agreed to kiss the ground when they landed. Instead, when they disembarked, everybody started crying. Dr. Berggren was the only one with enough composure to follow through with their plan. "I kissed the ground," she recalled. "Yes, I did. Everybody else spent the next twenty minutes crying. It was an incredible catharsis. We had all kept it together for a week, but as soon as we got on the ground in Fort Worth, we just all fell apart."[123]

Some of the nurses and staff were flown to Houston or San Antonio, but most stayed on in Dallas. An automobile caravan chauffeured Dr. Berggren and the others to three homes in University Park. Libby Goff, who had arranged the rescue, had made a huge banner and draped it across her house. It read, "We're So Glad You're Here!" She had also stocked her living room with shopping bags full of deodorant, toothpaste, underwear, T-shirts, and everything else they needed. Dr. Mike Harris, another friend of Dr. Berggren, was on hand to disseminate medication. The deluge was over. "Beautiful sight of complete strangers talking, hugging, holding, crying," Goff wrote in her diary about the event. "Father Powers is called to aid with counseling. No questions asked, he is there in minutes. Weary put to bed, strangers crying with strangers, holding them in their arms."[124]

Months after her evacuation, Dr. Berggren was still extremely emotional about their Texas-size welcome. A prismatic wave of memory gave her chills, in a good way. Crying to her, however, was a sin of sentiment. She tried to stay stoic. "We were so embraced that I have to fight back tears," she said. "Just think about it. The message to these humble people from Charity, who had really just been doing their best to survive, was that they were of great value. As one of the Charity nurses said, 'To get to Dallas and feel like you were worth something. It redeems my faith not just in America, but in the whole concept of humanity.' "[125]

Chapter Fifteen

GETAWAY (OR X MARKS THE SPOT)

I have little knowledge of the Bible. It seems to me, though, that when God made the rain fall for forty days and nights, he fully understood that Noah would rebuild human society after the Great Flood ended. If Noah had been a lazy man, or a hysterical man given to despair, then there would have been great consternation in God's heaven. Fortunately, Noah had the needed will and ability, so the deluge played its part within God's plan for man, without playing the tyrant beyond God's expectations. Did God, too, count on a built-in harmony of "balancing out"? (And if so, does God not seem rather vicious?)

—Kenzaburo Oe, *Hiroshima Notes*

I

Standing outside the Superdome, waiting for an MRE, Diane Johnson thought about dying. The man-made misery was worse than the storm. And she wondered about whether her brothers and sisters at

Noah's Ark Church survived the deluge. Her fortitude was vanishing under the unrelenting sun. It hurt to breathe. "Here I was waiting for a sip of water, being held up; I couldn't walk." Stripped of her medications, absentminded due to the dizzy spells, Johnson was taxing her high threshold for pain. You name the ailment—from a dropping bladder to a damaged liver—she was afflicted by it. Somehow, through the miracle of modern medicine, she had created a prescription pill regiment for herself that worked. Every day, under doctor's orders, she gobbled pills and capsules. While waiting to be evacuated from the Superdome, however, sitting in a wheelchair, it wasn't the medication she was missing most. "I wanted my electric chair," she said. "When we swam out of our house [in the Lower Ninth], there was no way to take it along."[1]

An ambulance finally picked up Johnson at the Superdome and rushed her to a hospital in Baton Rouge. Her husband was at her side. The ill Johnson knew that her time was growing short. Funny thing about death, she thought. Most people think you shrivel up and pass. Not always so. She was like a balloon, getting bigger by the minute: bloated neck and bloated eyes, bloated stomach and bloated behind. So be it. She was a Christian woman. Devoted. Soon all the drudgery would be over. She was headed to the Maker. She was on the ethereal borderline between life and death. "When the spittle dries up," she later said, laughing, "you know you're in trouble."[2] But on Saturday, at least, Johnson's New Orleans nightmare was over. Baton Rouge was taking care of her.

Ivory Clark and his eight dependents were also queued up in MRE lines at the Convention Center that Saturday morning. Water was being distributed next to a huge outdoor Wyland mural depicting underwater life: humpback whales, schools of dolphins, and a stingray frolicking in the blue ocean. Things were finally looking up. The Louisiana National Guardsmen were marching around, securing even the transplanted palm trees that lined the median on Convention Center Boulevard. Nearly all of the nearby shops had been looted. Helicopters were hovering low, blowing the street garbage in every direction, unceremoniously dropping twenty-four-bottle cases of water, which burst out of their packs, forcing those who were thirsty to scramble as if on an Easter egg hunt. Clark, the gatherer, was able to snag six or seven bottles. "My real concern was Auntie," he recalled. "She was on her last legs. We had to get her medical attention fast. For a ninety-one-year-old, Sedonna Green was hanging tough. I kept rubbing my brass cross, keeping my fingers crossed."[3]

Lieutenant Colonel Bernard McLaughlin of the Louisiana National Guard was helping to orchestrate evacuations from the Convention Center that Saturday. Because the center had never meant to shelter 20,000 people, everything was being done in ad hoc fashion. People jammed the entrances to the center like a dense herd of human flesh. Everybody wanted out. Pushing was the norm. The Hilton parking lot had been transformed into "feeding lanes," the spot where Ivory Clark was getting MREs and plotting his next course of action. Approximately ten bus lanes had been established along Tchoupitoulas Avenue. McLaughlin's biggest chore was setting up a security perimeter around the entire center and kicking looters out of the Riverwalk Marketplace. Besides his own Guardsmen, McLaughlin, still in regulation uniform, was assisted by seventeen or eighteen Louisiana State Police Alcohol and Tobacco agents. He had with him a Beretta M-9 9mm pistol and fifty rounds of ammunition.[4] "What shocked me was when I got to the Convention Center, I expected to see a lot of NOPD," McLaughlin recalled, "but I realized very quickly that there were very few."[5]

The insensitive, brutish, sloppy police work McLaughlin saw "in the game" made him livid. "I will never forget this and it shocked me," McLaughlin said. "On Saturday I'm with first-rate Louisiana State Police and we're trying to answer questions for the crowd. They had a bunch of vehicles and we were like a security corridor to be able to react to things, because people were filtering in and out, answering questions left and right. I saw marked New Orleans [police] units driving through the crowds. I'm talking about elbow-to-elbow people, their lights flashing, their windows rolled up. They weren't stopping and that's when I started seeing these Cadillac Escalades driving through and I was wondering where they were getting those vehicles, not realizing till later that they got them from the [Sewall] dealership. But I was unimpressed with how NOPD handled that because we could have used some of them there with us. I know and understand that some of them had lost homes and their families were gone and stuff. But a lot of us were in the same boat. I left my home in Lake Charles and was down there. I lost a home two weeks later in Hurricane Rita. My point being, we're down there thinking that we're facing the same risk, we're professionals. We expected them to do what we did in the National Guard, which is to act like Katrina was their primary mission, everything else, family, possessions, that comes second, and I was disappointed in them."[6]

But why would the NOPD just drive by throngs of stranded people in stolen cars? For what purpose? What were they trying to prove? "I don't know," McLaughlin said. "I thought it cold and impersonal. This was their city. Those were their people. And their driving through with lights flashing was an obvious 'Get out of our way!' They'd hit the siren now and then. It was interesting because I commented to some of the state police, who were as critical as I was, like what was the deal with these guys? Did they not understand basic policing? Everybody should understand basic policing: Get out, stop, help, circulate with the crowds. Maybe they were going somewhere. But it seemed to us like they were driving through to look at the situation, not stopping to help and try to deal with it."[7]

Since Katrina hit, he hadn't gotten much sleep. Even with his eyes half-closed, however, he had the look of an army field officer who didn't suffer fools lightly. On Friday night at the Convention Center, he slept right next to the feeding lanes on a little grassy mound. He was working twenty hours a day. "I hadn't had a bath in a week," he recalled. "I'm stinking. I'm filthy dirty. I took my boots, kind of made a little pillow out of my boots. I was about to get a couple hours' sleep. And I'm thinking, 'I can't believe this! I'm a lawyer with a great law practice and I'm literally homeless, sleeping in New Orleans, next to the CC.' I was thinking that it couldn't be real, it was the Comedy of the Absurd or something."[8]

The security task-force commander for the Convention Center was Lieutenant Colonel Jacques Thibodeaux, a jack-of-all-trades who did a remarkable job keeping the morale of the Louisiana National Guard high all week. His main order was: Talk to the people. . . . They don't bite. As leader of Louisiana's Special Reaction Team (SRT), he was in charge of evacuating the approximately 20,000 from the Convention Center. He felt sorry for the destitute. He didn't like that Blackwater troops had arrived in

New Orleans to secure buildings; they had no authority and were confusing people. Thibodeaux, who was also a deputy U.S. marshal, was a great believer in listening to complaints, wearing compassion on your sleeve, and creating an openhearted environment. Social inequity, however, was not his evacuation mandate. To him the Convention Center was divided into just two kinds of displaced people: armed and unarmed. He wanted all the weapons collected by Saturday morning before the helicopters and buses started evacuating. The best way to know who had a gun or knife was for somebody to provide a tip.

Most Louisiana National Guardsmen—along with those from Arkansas and California, who had also arrived at the Convention Center—believed that alcohol was the real culprit of the mass social unrest at both the Superdome and the Convention Center. On Saturday morning McLaughlin and other Guardsmen went on "booze-busting patrol," an essential element of crowd control in general. He was dumbfounded that about one-tenth of the people at the center were openly walking around with liquor. It was an unwholesome, quarrelsome trend. The National Guard didn't need drunkards getting dehydrated in the sun. He saw bottles of Tanqueray, Smirnoff, Jack Daniel's, and Cuervo. These were clearly looted, most from the nearby Cajun Market.

At a Loews hotel four blocks from the Convention Center, McLaughlin encountered a group of families camping out. He promised them the evacuation buses were en route. A burly fiftyish woman came up to him to lodge a complaint. " 'Sir,' she said, 'we got some people next door set up a bar and they're selling liquor to anybody for a buck.' " A diligent McLaughlin went over to check it out. Sure enough, a man in his thirties had brought a stolen bar into the Loews lobby and was mixing drinks. "I don't know where they got it, but they took it from somewhere," McLaughlin recalled. "A real bar, a long, dark wood bar with the liquor bottles that had the metal pourers still on them. They had two pitchers of water and two pitchers of orange juice and I'll never forget when I walked in this guy's wiping some double old-fashioned glasses." McLaughlin asked him, " 'What are you doing here?' 'I got a bar,' the guy said. 'Do you want a drink?' 'No, I don't think so!' " McLaughlin snapped back. " 'Dude, you're shut down.' "

An argument ensued. The bartender insisted he was just being entrepreneurial; why not make a little cash while waiting for the buses and helicopters? he groused, but he was outmatched. "'Let me tell you something,'" McLaughlin told him. "'This is looting. You looted that. I could arrest you right now. That's a fifteen-year felony. That's a three-year mandatory minimum sentence!'" McLaughlin had had enough and he was getting a little nervous. He was a lone white official in an all-black crowd, playing the prohibitionist. "'Look,'" McLaughlin said, "'do me one favor. Turn around and grab the wall.'" The bartender was oddly compliant. McLaughlin started hurling the bottles at a nearby wall, one at a time. "I then proceeded to smash thirty bottles in about ninety seconds," he recalled. "Everything shattered. They were out of business."[9]

II

Lieutenant Colonel McLaughlin wasn't the only one perplexed by the gruntish arrogance of the NOPD surrounding the Convention Center. Professor Lance Hill, executive director of the Southern Institute for Education and Research at Tulane University, had heard John Burnett of NPR's *All Things Considered* report on the mass confusion at the Convention Center. As a fifty-four-year-old do-gooder, he cringed with sadness. How could the city not provide bottled water? Why wasn't food being distributed? Since his Uptown house hadn't been damaged in the storm, he decided to go help the stranded. Pulling out a can of spray-paint from his garage, he wrote "AID" across his Ford Taurus next to primitive red crosses. He emptied his refrigerator and went into grocery stores, filling his car with water. Then he headed to the Convention Center. He wasn't a NOLA Homeboy, he wasn't trying to rescue anyone. All he wanted to do was deliver water. "On my way there, I saw a bloated corpse," Hill recalled. "I later found out her name was Vera Briones Smith. She had been there for days. The police refused to let people cover her up or touch her. I remember thinking, 'If that's how they're treating the dead, who knows how they are treating the living.'"[10]

When he pulled up at Hall D of the Convention Center, the stranded were overjoyed. The plastic bottles of Kentwood he handed out were guzzled and the bags of chips ripped into. "People started placing orders with me," he said. "Mothers wanted Pedialyte, milk." He promised everybody he'd be back. He spent all of Friday dropping off provisions. He encountered no problems save a teenager banging on the hood of his car, asking him about buses. But then a blockade policeman stopped Hill on one of his runs. "What are you doing?" the officer asked. "Bringing water to the folks," he responded. "No," the officer said. "If you do that, those guys will rush you." Hill explained that he'd been bringing water regularly and had encountered no problems. After a lecture, he was allowed to drop off water, but not to interact with the people. "Don't come back," the officer told him. "No more."

Quite naturally, with people dehydrated, desperate for water, Hill continued his "AID" deliveries. But the police began harassing him, interfering with his drop-offs. At one juncture, as he was driving near the Uptown overpass, a pickup truck with two police officers in it motioned for Hill to pull over. He just ignored them. He wasn't going through the third degree again. "Then they fired three or four shots over the bow of my car," Hill recalled. "Then I stopped. Eventually, they let me go. Didn't arrest me; just told me to slow down."[11]

Helicopters continued to fly along the Mississippi River. The entrenched New Orleanians who hadn't evacuated sighed with relief. The choppers were, if nothing else, a sign of outside civilization. Photographer David Spielman, whose shop was in The Rink just a half-block from Commander's Palace, reported to his Uptown-Garden District friends that eating landmarks such as Tipitina's and Clancy's and Casamento's had escaped the floodwaters; their culinary future looked bright. Since Katrina struck, the fifty-five-year-old Spielman kept a finely written diary of his solitary experiences wandering around the Uptown–Garden District. "The streets are being cleaned little by little," Spielman wrote. "National Guard everywhere, pouring into the city faster than the water, the only traffic on the street are the Humvees and military trucks. Checkpoints everywhere, guard camps being set up in Audubon Park, Children's Hospital, school and fire houses. Helicopters overhead day and night like dragonflies hovering over a pond."

A marathon runner, Spielman, didn't forsake his daily jog even during the worst days of the Great Deluge. In his diary he wrote of runs he made at dawn.

> My runs in the park are strange as it is only the birds, squirrels, and waterfowl there with me. The camped Guard are amused by me running, must be betting this old guy's nuts to be out doing this, after all this. Life in the Monastery is strange, so quiet, yet the presence of the Sisters is there. Each morning I go into their chapel and say a prayer for them and us. I keep a candle burning for them in front of the altar. I then ring the bell in the hall outside the chapel three times reminding God that they might not be here, but please don't forget them. Still no sign of the police, their absence is very noticeable. NOPD seems to mean No Police Department. I get checked often at roadblocks, my station wagon, gray and diminishing hair, I think, helps in reassuring the soldiers I'm not a looter. If it weren't for the Guard this place would be taken apart by looters.[12]

But mainly what Spielman saw cropping up all over New Orleans were *X*'s on houses. *X* marked the spot all over New Orleans, and the Gulf South in general, following Katrina. Cans of DayGlo spray paint were handed out by FEMA to second and third responders. The primary goal was to mark in bright red or orange which buildings had been searched and whether bodies were found. The FEMA schematic—published verbatim in the *Times-Picayune*—was as follows:

a) A simple slash designates a "hasty search," while an X marks a more thorough primary or secondary search.
b) The date is marked at the top of the grid.
c) The initials of the search squad are painted to the left.
d) To the right are notations for hazards: gas and water leaks, downed wires, rats or dead animals.[13]

Deciphering these spray-painted signs on New Orleans homes was like engaging in an ancient ritual. One thought of Exodus 12:7–13: "They shall take of the blood, and strike [it] on the two side posts and on the upper door post of the houses. . . . And the blood shall be to you for a token upon the houses where ye [are]: and when I see the blood, I will pass over you, and the plague shall not be upon you to destroy [you]." In a reversal of God's promise to protect, the symbols staining the door posts or walls of houses often meant death. If you knew how to read the symbols correctly, it told how the flooding unfolded in the bowl. In 1665, London suffered an outbreak of bubonic plague, also known as the Black Death, in which 15 percent of the city's population perished. In an attempt to control the plague, households that were afflicted were quarantined, a red cross painted on their door next to the phrase "Lord have mercy on us."

There was, however, no *X* on the door of Memorial Medical Center of New Orleans when Tony Zumbado, accompanied by Josh Holm, found the forty-five dead there on September 6. During the first post-Katrina week, medical staff at Memorial Medical Center managed to evacuate seventy patients, largely on Acadian Ambulance helicopters. Then the employees left and the center was empty, really. All the evacuees told horror stories of dank hallways, terrible isolation, and fear of disease.

For days Zumbado and Holm wanted to get inside Memorial, to check if there were any survivors in the 317-bed Uptown hospital. But they never made it. Every time they came within a quarter-mile of the hospital, they were called on to rescue someone stranded in the Central City Garden District neighborhood. The two NBC freelancers saved dozens of storm victims, pulling them off floating plywood and plastic garbage cans and into their pirogue. On September 6, Zumbado and Holm hardened their resolve and refused to be distracted until Memorial Medical Center loomed in front of them like an ugly Oz. They were both seized with fear, hearing unintelligible sounds coming from the massive hospital, faint noises crying out in the otherwise silent street. Or so they thought. Something was clearly amiss. You could sense it. Sizing up their situation, Zumbado decided it was best to wade around the hospital's perimeter looking for an easy entrance. If worse came to worst, they would smash a window and crawl inside. "Sure enough, a door was unlocked," Holm recalled. "We were, at that moment, more fearful of remembering how to get back to the canoe."[14]

The Memorial lobby was U-shaped and smelled atrocious, forcing them to blink away tears. Glancing around, they saw oxygen tanks, syringes, and garbage scattered about. Clearly, the lobby had been used as a makeshift triage area in the hours after Katrina. Using their headlamps as a light source, they scanned the wreckage, craning left to right in search of bodies. They heard a faint bell, a muted wind chime coming from somewhere. Unless Memorial was haunted, they weren't alone. "I thought I was going insane," Holm recalled. "I kept hearing a jingling bell."[15] As they crept down a corridor, the noise got louder. Suddenly, right in front of them, an orange-and-gray cat appeared, hollow-eyed, its collar tag the source of the mysterious sound. "I said in my head one hundred times, 'Let's get out of here,'" Holm recollected. "But I knew that to verbalize it wouldn't go over well with Tony."

Moments later, they came across their first dead patient, an old black man in bed with a grimace on his face, his white hospital gown torn half off, his contorted hands grasping at demons, rigor mortis locking them in place. "That was the first dead person I've ever seen the eyes of," Holm said. "The corpses in the water were facedown. It looked like he died the most painful death he could."[16]

Farther down the corridor, they heard another noise. It grew louder as they approached. Was somebody alive? Had a doctor stayed behind? Zumbado opened the door and they both jumped back. WWL was playing on a transistor radio, which was next to a dead woman soaked in blood. Blood was splattered everywhere, as if somebody had slaughtered a goat. Both Zumbado and Holm almost vomited from the stench, their bodies perspiring as if they had a fever. "From that moment onward we started running into dead bodies that were just lying in the hallways, covered up with sheets," Zumbado recalled. "And we started videotaping and counting and opening up sheets to see if they were female, if they were elderly, that kind of stuff and we were videotaping that. Then we went to the emergency room, which was flooded, and there was a gentleman left on the emergency room table. There were dogs in there, barking. Yes, there were a couple of dead dogs, a cat. It was horrific. It was horrible and the smell, you know from the poisoned water coming up. We knew that the higher we got, the more bodies we'd probably find. I was overwhelmed with the first floor and the people we videotaped, so we decided to go up to the second floor."[17]

As they approached the stairwell, the sight of a mangy black cat, only half alive, hissing on a chair left to pry the door open, caused their hearts to leap violently. "Two big green eyes glaring from a black cat on a black chair in a black hole of a hospital," Holm later described the scene. "I wanted out. I just couldn't handle anymore. But Tony just wouldn't turn back."[18]

The second-floor hallway was full of dead people covered up, mostly special needs patients. Zumbado kept lifting off the flimsy white sheets and filming the mortified flesh. Many of the hallway dead had a pillow placed under their heads and were still hooked up to air tanks, as if a nurse had been tending to them until the last moment. Whether that last moment had come because the patient died or the caretaker fled was impossible for Zumbado to say. "I tried to videotape whatever ID was left next to the body," Zumbado recalled. "They had left medical records next to the bodies, so that if someone came in at least they would know who the people were. So, as I videotaped that, as we're walking down the hallway, I saw a sign on the door of the hospital chapel saying Do Not Enter, and it was written on yellow legal-pad paper."

Zumbado and Holm, their headlamps shining in each other's eyes, knew something horrible was up. Taking a deep breath, Zumbado made sure his PD-150 camera was working, then he headed toward the handwritten sign. Slowly he pushed the door open and was overwhelmed by the heavy, humid smell of decomposition. Dead bodies, nineteen in all, were laid out in the hospital chapel, each of them blanketed with a thin cotton sheet. "The stench just came and slapped me in the facc and I turned around and just vomited," Zumbado recalled. He fell back into the hall. "I just wailed over the second-floor balcony and we were both sick for about five minutes and coughing." On the altar he had spotted an open Bible, as though somebody had performed last rites. "It seemed to me that somebody had made a last-ditch effort to maybe give them the respect and a memorial, knowing they were going to die."[19]

Most normal people would have fled the nightmare, sprinted back down the stairs, hopped in the pirogue, and paddled back to St. Charles

Avenue. But there was nothing normal about Zumbado. After all, he was a hurricane chaser, one with a macho Cuban streak, a veteran videographer with a long history of contained recklessness and amazing diligence. They had ventured this far into the house of ghouls; they had a journalistic obligation to forge onward. After World War II, General Dwight D. Eisenhower had ordered Army Air Corps Motion Picture teams to film the Nazi concentration camps. Emaciated bodies of Jews and Gypsies, and other so-called undesirables had been piled up like cords of wood next to the ovens. Eisenhower didn't just demand footage of the Holocaust, he forced German villagers to queue up and witness what the Third Reich had done to thousands of innocent people. A prescient Eisenhower knew that if the atrocities weren't documented, there would be Holocaust deniers, bankrupted charlatans who claimed the extermination of almost six million Jews hadn't happened.

Zumbado knew the same would be true of Katrina. He could rush back to Camp NBC, tell Heather Allan and Brian Williams about Memorial Medical Center, but his words wouldn't do the chapel and its victims justice. They would, of course, believe him. They were his winter soldiers, colleagues he could count on. But over time, when the crisis faded, a Katrina revisionism would settle in. Regency Press would publish a book saying that President Bush hadn't really been slow to respond, that there hadn't been any snipers, and that the purity of distilled water had been available at the Convention Center. He could already hear the revisionists, blaming the media for overreacting, claiming the Gretna Bridge Incident and the Cadillac-looting NOPD never happened. So through Katrina, even while saving people, Zumbado used his camera like a weapon, shooting everything in sight. He didn't care whether the *Today* show or *Hardball* thought it was too gruesome to air. He really didn't care whether they aired his footage at all. He was making a documentary record. And when the Katrina deniers poked their heads out of their holes, NBC would be able to hit them with the footage of the bodies. Let them try to deny the corpses in the deep freezer at the Convention Center. Let them dare to say the dozens of floating corpses he had recorded on film in Central City were illusions. Let them try to claim that the dead in the Memorial chapel weren't real.

One reason Zumbado was determined to film *everything* in Memorial, regardless of its TV value, was that back in April 2000 he was the first cameraman in when the INS agents seized six-year-old Cuban Elián González. "The INS kicked and beat me, grabbed my camera, made me turn it off," Zumbado recalled. "They later lied that they had roughed me up. And because I had turned my camera off, as ordered, I had no footage of what they did. They later tried to deny their actions. From that moment onward I decided to film everything. No more shutting off. Let the truth be seen."[20]

When Zumbado's stomach stopped convulsing, when he no longer felt the need for smelling salts, he turned around, as if facing a blaze of gunfire, and pushed his way through the chapel door. He counted each body, out loud, one to nineteen. He zoomed in on the gold cross lording over the macabre, makeshift morgue. In a flashback he remembered a few days before, when he was working side by side with the NOLA Homeboys, rescuing people from the deluge, how he saw a group of nurses yelling, "Help! Come get us!" out of Memorial's windows. He remembered thinking at least they were safe, on high ground. What hell must they have been going through? Who gave these nineteen their last rites? What verse of the Bible had a doctor or nurse chosen to read? He didn't know if these bodies would ever be embraced with a good-bye, a prayer, or a song, if anybody would ever bless them before they were buried in the copper-red Louisiana mud. What he did know, however, was that this wasn't the time to get emotional. He dug deep inside, hanging on to the deep professionalism that he had crafted after decades in the disaster business. "Josh and I are deeply Catholic," Zumbado recalled. "Wherever we are, no matter where, we go to church on Sunday. So we held hands, bowed our heads, and said a prayer for the dead."[21]

Their prayer, however, was brief. They couldn't breathe. Zumbado had been half-holding his breath while filming and praying and he was starting to turn ashen gray. He kept remembering Eisenhower and Buchenwald. Quickly, he ripped off the sheets to film the faces of the dead. He'd heard that Jefferson Parish had sent a refrigerated truck to New Orleans to bring the corpses out of the city to a FEMA camp in St. Gabriel, Louisiana. These were the lucky bodies about to be toe-tagged. He wanted to make sure that the families of the Memorial patients had closure, that their loved ones didn't become unidentifiable corpses in a FEMA morgue. His lungs and eyes, however, just couldn't take it anymore. After filming five or six faces, he bolted out the chapel door, coughing and hacking uncontrollably. He had done his best. "Let's get out of here," he told Holm. "Come on!" With their headlamps in place, they practically slid down the stairs past the black cat and quickened their pace as they hopscotched over the first-floor corpses. Eventually they burst out of the metal doors into the sunlight. Soaked with sweat, muttering expletives over and over again, they found their canoe. "Are you okay?" Zumbado asked, clasping Holm on the back of the neck. "Yeah," a shaken Holm said. "Yeah. I'll be okay. Just give me a minute."

That evening Zumbado's footage was the lead story on NBC and MSNBC. The Memorial chapel and St. Rita's Nursing Home became the symbols of Katrina deaths. There was no *X* spray-painted on the block-long Memorial hospital, no hieroglyphics written in the bottom right-hand corner. Just forty-five corpses and accusations that some of the dead may have been euthanized by Memorial Medical Center doctors and nurses. That issue would have to be decided by the courts. "All I knew," Zumbado said, "was that days later, I went back by Memorial and the smell of death was in the air outside. The stench was so powerful that it made its way out of the hospital, just cut through the fortresslike walls."[22]

III

Five days after Zumbado and Holm had reported on Memorial Medical Center the corpses were still there. They just continued to rot. Nobody, in fact, knew who was responsible for their removal. The State of Louisiana, without proper officials to assign to the task, turned to FEMA to help. "Among the many failures in government planning revealed by Hurricane Katrina, one was particularly striking: No one, it seems, figured out ahead of time who was going to pick up the dead," Renae Merle and Griff Witte of the *Washington Post* explained in mid-October 2005. "When the storm swept through the Gulf of Mexico six weeks ago and left hundreds of bodies to decompose in homes and streets, Louisiana officials looked to the Federal Emergency Management Agency for help in removing them. But since cities and localities had historically recovered bodies from mass casualties, FEMA says, it had made no arrangements."[23]

People from all over the world were aghast at how the dead were simply left to decompose. It became the warped journalistic cause célèbre. Every editor wanted a corpse picture. More than a week after Katrina, FEMA hired Houston-based Kenyon International Services Inc., body removal specialists, to collect the deceased. It didn't take Kenyon long to learn that FEMA was in the midst of a bureaucratic breakdown. Every time Kenyon sent a mobile morgue to Louisiana, FEMA tried to slow down their body-recovery procedures. It was almost as if FEMA *wanted* to leave corpses in the street. Governor Blanco, siding with Kenyon, lambasted FEMA. "While recovery of bodies is a FEMA responsibility," she said "no one has been able to break through the bureaucracy to get this important job done."[24]

For a city known for festive jazz funerals, the image of corpses on sidewalks was a blow to the very soul of New Orleans. Although on the Saturday after Katrina a guitarist in the French Quarter was singing "Mr. Bojangles" and "Rocky Raccoon" in Jackson Square, most of the music was gone from the city. But slowly the jazz funerals would come back. It was just a matter of time. The long-defunct Onward Brass Band would play again.

Fittingly, it was the late Austin Leslie, whose fried chicken, as Kim Stevenson wrote in the *New York Times*, "was considered the gold standard even by the South's most persnickety chefs,"[25] who received the first honors when the jazz funeral resumed. Leslie died several weeks after Katrina, as a result of the stress. Like Diane Johnson, Leslie had spent two days stuck in an attic waiting to be rescued. He was eventually evacuated to Atlanta, hoping to get back to the deep-fryer as soon as possible. But in late September he died, burning up with fever. He wasn't considered one of the Katrina dead, but the storm surely hastened his demise. On October 9, just after 2 P.M., according to the *Times*, on the corner of North Broad and St. Bernard, a jazz funeral was held for Leslie, beginning at Pampy's

Creole Kitchen. Gralen B. Banks, a member of the Black Man and Labor and a Hyatt building supervisor, said of the jazz funeral: "This is the first opportunity we had to show the whole spirit of New Orleans. And we're not going to pass it up for love or money."[26]

IV

While there was no official moment when the Great Deluge ended, when the buses finally emptied the Convention Center on Saturday, the dark veil of Katrina started to lift. "We set up the bus lanes and in just six hours, from 9 A.M. to 3 P.M., we closed the entire area," McLaughlin recalled in his field diary. "There are some very sad stories late Saturday. I took custody of a fifteen-year-old child, carrying a backpack, with a borderline IQ—he was a nice-looking, gentle black kid, who had been abandoned by his parents at his home and was alone for six days until he was found by an ambulance crew—I personally put him on a bus. What kind of parent abandons a special-needs child? Of everything, this angers me the most."[27]

One father who never lost track of his brood was Ivory Clark. When the buses pulled up on Convention Center Boulevard, Clark seized the moment. While everybody else headed for the first three or four buses, he wheeled Auntie toward the end of the convoy, the rest of his family chasing after him. By this time, Ivory's body felt broken. Somehow, from the New Orleans Grand Palace Hotel to Charity Hospital to the Superdome to the Convention Center, he had kept his morale up. He had even hummed old Marvin Gaye and James Brown songs to make the young ones laugh. But above all else, he thought about football: just pretended he needed to drive Auntie's wheelchair into an imaginary end-zone. He had the notion of getting his family on one of the last buses. This way they could board with dignity, no elbows flying. Eventually, he wheeled Auntie straight into a Louisiana National Guard barricade. Breathless, sweating profusely, Clark tried to convey his predicament to the Guardsman standing there as if protecting Buckingham Palace: how his auntie could barely breathe due to chronic asthma, how ninety two-year-old Sedona Green was dehydrated and crippled with heat exhaustion. "Can't we get on one of these buses?"

Clark asked. "We've got to stay together as a family." The Guardsman glanced over the two elders in the Clark clan, seeing that indeed they clearly needed medical attention. "Well," he said. "Go get them ready. Come on. You're going in the sky."[28]

At that moment on Saturday, dozens of helicopters started landing at the I-10 end of the Convention Center, where Clark and his family were gathered. They were directed into helicopter lanes. The National Guard was trying to medevac the sick and the elderly first, but Clark had been adopted by this one Guardsman who promised to get his family out of New Orleans as a single unit. Clark never got his name, but he was their savior. Winking at Clark, patting his son, Gerald, on the back, bringing them water. It was as if this Guardsman, from somewhere in Louisiana, realized that this was the very reason he had joined the National Guard in the first place: to help fine upstanding families like the Clarks. A helicopter ground controller kept shouting out low numbers: "two" or "three" or "five." That designated how many available spots there were on a chopper. When "eight" was shouted out, Clark's new Guardsman friend burst out laughing, giving the double thumbs-up, anxious to help them get aboard. None of the eight, including Clark, had ever been on a helicopter before. They were instructed to hunch downward, to cut into the wind being generated by the propeller blades. "We all acted like we were carrying a football," Clark recalled. "We were determined not to get hit, heads down." Without any hitches, they all boarded the helicopter. Their Guardsman friend made sure they were all buckled in. The helicopter held around thirty evacuees. Everybody was apprehensive, unsure of where they were going. But anywhere in the United States had to be better than sleeping on oil slicks in Harrah's garage. Upon liftoff, Clark gazed out in disbelief. It was the first time he saw how badly New Orleans had flooded. Nobody had yet told him whether his home at 8534 Edinburgh Street had flooded. He looked in that direction. His section of Holygrove was under seven or eight feet of water. For days he had endured, not once shedding a tear. With everybody now safe and the realization that his home was a Katrina casualty, he reached his breaking point. He sobbed. Lake Pontchartrain had literally poured into all the restaurants and homes in Lakeview he knew so well. He spotted horses trapped on a little piece of dry land and people waving at the sky from I-10. "I thought New Orleans had turned to hell," Clark said. "Every second I felt better about gettin' the hell out of the cursed place."[29]

The Clark family landed at the Armstrong International Airport and were loaded onto a flatbed Jeep. A couple of forms were filled out and then Auntie was rushed to the second-floor triage center. Boxes of Lance snacks were being passed out and there was air-conditioning. She was given respiratory treatment and nitroglycerin pills, was washed and provided with clean clothes. "Boy, oh, boy, when Auntie came out, when they were done with her, she was a new lady," Clark recalled. "All clean and beautiful. No more wheezin'."[30]

Some airport official, seeing Ivory Clark and family waiting for Auntie to get released, asked Clark if he would mind sweeping up a section of the airport. Clark was a chef, not a janitor, but he leapt at the opportunity to help out. Grabbing a broom, Clark and his son, Gerald, began sweeping up, picking up trash and dumping it into plastic garbage bags. By the time they were done, the place looked spick-and-span. Clark was rewarded with a bag of Zapp's chips and an ice-cold Coca-Cola, his favorite drink. "I knew everything was okay when I was given that Coke," Clark said. "I thought 'Aaahh . . . civilization.'"[31]

The airport official, impressed with the job Clark had done, promised to get his family on a Baton Rouge–bound bus within the hour. She was as good as her word. The only snafu that developed was that wheelchairs weren't allowed on the bus. So Clark and his son picked up Auntie as if she were a giant log: one held on to her shoulders, the other to her legs. The entire family got on board. "Eventually we got goin'," Clark said. "All I could do was hug and kiss Donna. I was so happy to be alive, to have my beautiful wife. I wasn't sure how I got where I was, but I was free from that mess."[32]

Upon arriving at LSU's Pete Maravich Assembly Center, the Clark family was given first-rate medical attention. Everybody got eye tests, had their blood pressure taken, and were even given appropriate prescriptions. "Nobody in the hurricane did better than those people in Baton Rouge," Clark said. "They were so kind, treated us real well."

On Saturday the Cajun Navy, led by Andy Buisson, returned from Lake Charles to New Orleans. (Sara Roberts was too ill to make this second rescue trip.) Once again they slept around the Convention Center and Harrah's Casino. Virtually nobody was left in the grand halls. The evacuation had been successful. "During the night, a contingent of mules galloped past, chased by police. A cool front came through providing relief," Buisson wrote in his diary. "The 42-floor Sheraton lit up briefly, amidst downtown darkness, a sign of hope to those who were awake to see it. Much different on this Sunday morning compared to four days before when the Dome, Convention Center, once overflowing with anxious evacuees, was now occupied only by mounds of stinking debris and garbage next to unopened cases of bottled water. The City was silent but for the passing helicopters that filled the sky. The radio carried stories of snipers and criminals. Some told of bands of thugs roaming the streets reducing law enforcement to defending a single building from the rooftop. We saw none of that, but knew we were not seeing the entire City."[33]

But it was the arrival of the famed U.S. Army's 82nd Airborne Division—known in history for its heroism during D-Day—that truly quieted the week-long crisis. When the President finally signed deployment orders for the regular military on Saturday morning, the 82nd could finally move, which was exactly what it was trained to do. At any given time one-third of the division's available troops were ready to leave immediately for assignment anywhere in the world. On Saturday, September 3, the "go order" came at 10:05 A.M. (EST)[34] and by 4:00 in the afternoon, Pope Air Force Base, located in North Carolina near the 82nd's home at Fort Bragg, was a busy place. As the division chaplain later recalled, troops from the 82nd were sorting themselves out near the runways at Pope: some were leaving for Afghanistan and some for Iraq.[35] And approximately 3,600 were headed for hard duty closer to home, in Louisiana. By the time the troops climbed onto the transport planes, though, their commander, Major General William Caldwell IV, was already headed for New Orleans to oversee the Great Deluge operation personally.

Caldwell was primed for action. He later admitted that it had been "disheartening" for his division to wait day after day for the call to act.[36] The paratroopers in their maroon berets were anxious to go to Louisiana; they'd even rehearsed for it during the week. "They feel very honored they can help other Americans," Caldwell said.[37] As soon as Caldwell arrived in Orleans Parish he met with Lieutenant General Russel Honore, who assigned the 82nd to search-and-rescue activity in greater New Orleans, along with humanitarian missions whenever the need arose, which was often. The 82nd set up headquarters at the Louis Armstrong International Airport, roughing it in the primitive conditions that prevailed there. To the degree that comforts were available, they were reserved for the ill and elderly. The airport had been practically deserted Friday morning, but by Saturday it was inundated with evacuees, medical personnel, and patients. The paratroopers soon found that there was little for them to do in terms of keeping the peace, that the facility had adopted its own weary sense of order. The 82nd was instrumental though in helping to organize the process of evacuation. Moreover, the troops fanned out over greater New Orleans performing house-by-house searches for people in need of help. The first problem was that the city was still largely underwater. The division, arriving with only four boats, soon requisitioned eighty-two more. "We eventually became the 82nd Waterborne Division," Major General Caldwell commented wryly.[38] The 82nd was joined in Louisiana by troops from the First Cavalry Division from Fort Hood, Texas. In addition, Marines from Camp Pendleton near San Diego, and from Camp Lejeune in North Carolina were sent to Mississippi.

In the city of New Orleans, regular military troops carried automatic rifles, but typically their weapons were not loaded. "We're just trick-or-treating," a sergeant with the 82nd explained to Dan Baum of *The New Yorker*. "If I saw someone going in that store right there, I couldn't do anything but radio it in."[39] Members of the Army who arrived in New Orleans were not allowed to be on police duty, though they did free up National Guardsmen and law enforcement officers for peacekeeping. The 27,000 National Guardsmen deployed to the region took up that chore with vehemence, and so did the many police sent in from cities all over the region. In fact, one of Lieutenant General Honore's jobs was to keep the peacekeepers from coming on too strong. In any case, by Saturday, New Orleans had been wrested from the chaos of the preceding week. A kind of calm settled over the streets, not a calm of relaxation or relief, but a weary calm, resigned to the fact that the future was going to be painful but at least it would not be as ugly as the previous week. The Army was on hand to make sure of that much. "Once the 82nd Airborne Division arrived the cockroaches that had caused the problems ran for cover," a relieved Terry Ebbert said. "They weren't going to challenge these guys. They instantaneously stabilized the command and control. They had their own radio system operating throughout the city. By nightfall on Saturday New Orleans was very stable. The crisis was over."[40]

V

On Monday, September 5, the Army Corps of Engineers had finally managed to close the breach in the 17th Street Canal. The other levee repairs had already been completed, and so that evening, after a desperate week, the city's great pumps started working again.[41] Within weeks, the city was drained. The result was not much of an improvement. For miles on end, the buildings were uninhabitable, the life that had once coursed through them all gone. At the time, ABC News and the *Washington Post* commissioned a poll, asking a sampling of Americans whether New Orleans should be rebuilt in its entirety or limited to the neighborhoods situated above sea level (that is to say, the richer neighborhoods). The result of the poll was close. Forty-nine percent thought it should be rebuilt the way it had been; 43 percent thought the low-lying neighborhoods should be abandoned; 9 percent had no opinion.[42] The wealthier neighborhoods flickered back to life first. The people who had been evacuated from the low-income sections were generally ambivalent about returning. "I don't know if I want to go back," said Corey Jones, a twenty-five-year-old truck driver with three children. "They lied to us. We got played like fools."[43]

Reverend Bill Shanks, a New Orleans minister, liked his city better in the aftermath. He considered New Orleans to be free of its many sins, as he saw them, and the pride it had taken in them, from witchcraft to free love to open homosexuality and abortion. "God simply, I believe, in His mercy," said Reverend Shanks, "purged all that stuff out of there."[44] An Internet letter that circulated in the aftermath put it more succinctly. "I always thought New Orleans was a toilet," the writer asserted. "It finally got flushed." The idea that Katrina had come as retribution was more common that it should have been. Irresponsibly, adding insult to injury, Mayor Nagin promulgated Reverend Sharks's vengeful God theory. "Surely God is mad at America," Nagin said in a public speech on January 16, 2006, to celebrate Martin Luther King Jr. Day. "He sent us hurricane after hurricane and it has destroyed and put stress on this country." He went even further, pretending he had had an imaginary conversation with Dr. King, who told him that New Orleans should be a "Chocolate City."[45] Suddenly "Ray speak" had turned into rank demagoguery, a mayor luring citizens—at a time when they craved healing the most—into racial politics.

The question of whether the political response to Katrina was implicitly racist was debated from the first days. That the local government was ill-prepared and the federal government uncaring was obvious during that critical first week. Many of the problems and attitudes at every level predated Hurricane Katrina by years. The difference was that right after the storm, city and state leaders were doing their best with whatever they had. Leaders at the federal level, on the other hand, meaning Secretary Chertoff and President Bush, shirked the Gulf Coast until pressured to act, days late. And so it follows that local mistakes were committed before anyone knew the racial makeup of the victims. The lag at the federal level started after it was obvious who was affected the most. The fact that the federal response *could* have been better, starting at the moment the hurricane struck, begs the questions: Under what circumstances could it have been better? If the victims were white? If they were rich? If they had not been members of a voting bloc that the Republican Party had a motive to disperse? All of those factors offered explanations to receptive minds. The one that rings truest, though, is that cronyism riddled FEMA and its contractors in the Bush administration, making incompetence and not racism the key to the response. As Lieutenant Commander Duckworth noted, the bureaucracy "was to blame."

President George W. Bush was ultimately responsible for the ineffective response from August 29 to September 2. If Michael Chertoff was haughty and aloof, then Bush should have changed all that with a phone call on Monday night. He should have lit a fire—a bonfire—under him and made sure that Homeland Security didn't do another blasted thing until the homeland and all who lived in it were secure once again. The President could have moved mountains but was sadly aloof himself, as the storm and its spinoffs passed before him. His approval rating sank to the lowest level of his presidency up to that time; a Zogby international poll had it at 41 percent on September 8. Politically, he was wounded, and not merely from his own poor performance. At the end of the first week after Katrina, Bush tried in every way possible to pressure Governor Blanco into ceding control of troops in her state, along with, effectively, responsibility for the course of the response. It was the sort of political fight that Bush was used to winning, but Blanco, for her part, stood up to the President. As a practical matter, few had done that before. Bush, operating with a majority in both houses of Congress and what he seemed to regard as a mandate stemming from the September 11, 2001, terrorist attacks, had been able to ignore opposition at most other critical junctures of his presidency. Blanco gambled everything in refusing to give the President the chance to take charge; she no doubt felt that, due to the work of her office, the momentum of the disaster response was about to change as the first weekend after Katrina arrived. It was a battle largely hidden from the public, but in winning a battle royale with the President, Blanco changed the second term of George Bush, leaving him open to other attempts to curtail the sweeping power he assumed for himself. In the span of one week in late summer 2005, the United States was changed, and not just on the battered coastline along the Gulf of Mexico. The country could always bounce back from a natural disaster, and the hurricane was a natural disaster. But the Great Deluge was a disaster that the country brought on itself.

VI

After about six days of nonstop rescues, Reverend Willie Walker finally made his way to Noah's Ark Church. Water had virtually washed away his files and library. Nothing much was left of the interior. The pews had been churned into kindling, as if run through with a buzz saw. The

Bibles, some of them rare, were all mush. The carpet was soaked and smelled like microwaved cottage cheese. His organ had no top and his electric keyboard was a tangle of meaningless wires. Somehow Walker salvaged his podium and his specially made sign, which read: "PEACE for the Weary / LOVE and ENCOURAGEMENT for the Hurt / SPIRITUAL GROWTH in CHRIST for the Lost / SHELTER for the Needy / During the Time of Storms, We Will Stand Strong."[46] His church was nonexistent, except in his heart. His sign would go on the altar of his new church someday: a reminder of the Great Deluge.

Katrina had taken quite a toll on his congregation. Desperately, he tried to reach Demetriam Williams and family, the Simmonds and Trumble families, Elouise Washington, and Ernestine Carter. He was especially concerned about Diane Johnson, worried about her high blood pressure and sickle-cell anemia. Eventually, he reached her in Orlando, where, with the help of the Foundation of Hope, she had been evacuated to. Along with her husband and son, she had been bused to Lafayette, given medical attention at the Cajun Dome, and then shipped to Orlando. Pastor Gus Davies of the Northland A Church Distributed of Longwood, Florida, was looking after her; Walker was grateful. She was battered; sorrow had mummified her spirit. "She was still thankful to be alive," Reverend Walker said. "Barely. She told me about her ordeal swimming through the attic, stuck at the Convention Center, the evacuation to Lafayette, life in the Cajun Dome." He told her to come back to New Orleans, that Noah's Ark Church was sunk but it would come back. He was setting up shop at Evening Star Missionary B.C. on Hickory Street, just a few blocks from the Mississippi River. He was now preaching in Pigeontown, where some of the Section 8ers who had looted Oak Street had lived.

There will never be exact statistics about the damage and death toll inflicted by Katrina. Were a looter shot by the NOPD or an old man who died of a heart attack victims? But the rough ballpark statistics the State of Louisiana put out, not including the Mississippi-Alabama coast, are nevertheless illuminating. More than 200,000 homes were destroyed while another 45,000 were deemed unlivable. Add to that the 15,000 apartments washed away and you get the scope of the housing crisis. Even months after the storm, Louisiana had 400,000 displaced residents. Both St. Bernard

Parish and Plaquemines Parish were 90 percent obliterated. The human death toll, while not comparable to the great Galveston hurricane of 1900, was still high at approximately 1,300 deaths. "So many people died of Katrina-related stresses," Walker said. "Norman Robinson of WDSU properly started announcing these deaths as due to Katrina stresses."[47]

Because Diane Johnson's Lower Ninth Ward home was uninhabitable, Walker was trying to find her a FEMA trailer. He visited her duplex on Tricou Street to report to her whether any of her belongings were salvageable. There was her Port-o-potty for handicapped people, two paintings, and, though missing wheels, her maroon Pronto electric wheelchair. The joystick was in fine shape. He also found a couple of photographs of her, cracked and mud-caked, but salvageable. He was about to call Diane in Orlando with the mixed news, then he got a telephone call. Johnson had just died. On her death certificate the cause was listed as pulmonary failure/high blood pressure.[48] Her body was being shipped back to New Orleans for burial. "Katrina had overwhelmed her," Walker recalled. "She couldn't get her blood pressure regulated. And she was depressed. You can't stay negative all the time; the stress will kill you."[49]

Reverend Walker and the Johnson family took solace in knowing that Diane was at one with Jesus at the time of her death. She was not an outcast or a castaway. She was pressed toward the Rock that was Jesus. Since childhood, sickle-cell anemia had been her thorn, but she used the ailment as a badge of honor. Like Paul, it made her humble. She understood that sometimes the worst enemies are within ourselves. Many of her Lower Ninth Ward friends were hatching conspiracy theories about the Industrial Canal breach. They said that white developers anxious for their land had dynamited the levee, just like they did back in 1927. A distant relative claimed that barge *ING 4727* was purposely unleashed on the Lower Ninth Ward like a missile, the white power structure wanting to rid the area of "niggers." Johnson nodded away conspiracy theories as the devil's work. "Katrina was beyond us," Johnson said in an interview from Orlando shortly before her death. "I'm too tired to worry about yesterday. I pray both in and out of season. I don't let go of Jesus' unchanging hand just because of a hurricane." She loved everybody at Northland A Church Distributed, her church in Orlando, particularly Reverend Davies, who hailed from Sierre Leone and took special care of her. But Orlando was sterile to her compared with the kinetic Lower Ninth. She missed Soft and Sassy Beauty Salon, where once a month she had had her hair done. And she longed for vegetables from Pham Grocery, only a half-block from the house. She was given a new Pronto, but it wasn't as smooth-running as her old one. But most of all she kept praying that the old oak tree across Tricou Street was standing tall like it always was. Indeed it was.

When pressed to talk about Katrina and racism, Johnson shrugged it off. "We've been bamboozled," she said. "But I've turned it over to Jesus." Even as she was dying in Orlando—and it was clear Katrina had torn her health asunder—she never lost her faith. She kept talking to Reverend Walker about heaven, hoping to find it. At Noah's Ark Church she had brushed up against the tassel of Jesus; now, in death, she would get to hug the son of the Maker. "An incredible lady" was how thirty-two-year-old Laura Young, a Northland A Church Distributed administrator, described her. "She said 'thank you' more than anybody else I've ever met."[50]

Reverend Walker presided over her funeral in Pigeontown at Evening Star Missionary Baptist Church. Her body lay in an open casket, with children coming up and rubbing her hair, touching her cheeks. She had on a light blue shirt, the kind you buy off the rack at Target for $3.99. Screams and groans and moans were heard. Handkerchiefs were placed over mouths and rubbed across foreheads. Popsicle-stick fans with Johnson's picture on one side of the flapping cardboard were handed out. The local NAACP paid for them. A bad organist hammered out the chords of songs like "Amazing Grace" and "Since I Got Over." A red rosary was placed around her neck. Nobody had blessed it. Because all her heirlooms had been washed away, there weren't too many mementos to lay on her body. There was a stained photograph of her as a child. Like all children, she was a pure angel. And then there was a snapshot of her with her husband, Darryl, taken in one of those Woolworth's camera booths in the 1960s. She lay in a silver coffin, with interesting moldings on the sides. Truth be told, however, it was the cheapest in the mortician's catalog, just a step up from the pauper's pine box. Her son, Willie, had picked it out. He was there in the little chapel, dressed in a blue cotton shirt, no necktie, screaming out in anguish. After the viewing, he fell to his knees in front of his mama's body, trying to stop the coffin from being closed shut. There was heaving, sobbing, fanning, and organ pounding. A faint breeze, almost undetectable, came blowing through the back doors.

When Reverend Walker took the podium, he was a pillar of strength. His beard was trimmed and an affirmative aura beamed from his infectious smile. Everybody who gathered for the funeral had two things in common: they loved Diane Johnson and they had survived Hurricane Katrina. This was no time to talk about the lethal ineptitude of Bush, Brown, Chertoff, Nagin, or Blanco. Instead of mourning Johnson's death or Katrina bashing, Walker spoke of how lucky Diane Johnson was to be in heaven. No more chains. No more floodwater. No more sickness. No more post-Katrina stress. "She made it through her Katrina tribulations," Walker intoned. "Her home may have been debris. Mud may have overwhelmed her household. Her wardrobe may have been lost in the flood. But she is now dressed in white robes, ready to meet the Maker. She is no longer a displaced person or refugee or shelter victim. She is now in a clean place, without dirty hands."

All of the congregation was in tears, fanning themselves. Many of Johnson's family and friends had lost everything in the Great Deluge, but they somehow had managed to put on Sunday clothes for the memorial service. They weren't in the diaspora anymore. They were home, in New Orleans, ready to start the rebuilding even if they were now living in Florida or Texas or Arizona or Massachusetts. Walker quoted powerful words from Revelation 7:14–17: "These are they which came of great tribulation, and have washed their robes, and made them white in the blood of the Lamb. Therefore are they before the throne of God, and serve him day and night in his temple: and he that sitteth on the throne shall dwell among them. They shall hunger no more, neither thirst any more; neither shall the sun light on them, nor any heat. For the Lamb which is in the midst of the throne shall feed them, and shall lead them unto living fountains of water: and God shall wipe away all tears from their eyes."

Reverend Walker was trying to convey a message of endurance. Like drugs or alcohol or adultery or greed, Katrina had to be overcome. Recovery was in order. Assuaging guilt was necessary. If only he had made Diane

Johnson leave that Saturday before the storm. If only Mayor Nagin had evacuated the poor and the sick. If only the U.S. Army Corps of Engineers had constructed better levees. Such nagging thoughts weren't going to bring her back. You couldn't rebuild your new life on the back of the deluge. The new life began in forgiveness. Including forgiving yourself. "It's over," Reverend Walker said. "Hallelujah! Rejoice! It's over. No more Katrina issues on this side of life."

Timeline

All times are Central Standard Time.

Saturday, August 27, 2005

5:00 A.M. Hurricane Katrina is in the Gulf of Mexico, about 165 miles west of Key West, Florida, and 435 miles southeast of the Mississippi River Delta (24.4 N by 84.4 W). It is gathering strength at a forward speed of just 7 mph. Winds are blowing steadily at 115 mph. Hurricane force winds can be felt 40 miles from the eye. Katrina is a Category 3 storm.

9:00 A.M. Plaquemines and St. Charles parishes in Louisiana declare mandatory evacuation.

10:00 A.M. FEMA director Michael Brown appears on CNN to encourage residents of southeastern Louisiana to leave as soon as possible for safety inland. The National Hurricane Center (NHC) issues a hurricane watch for metropolitan New Orleans and warns that Katrina could make landfall as a Category 5 storm.

12:00 noon The Louisiana National Guard calls 4,000 troops into service, representing practically all available troops. Three thousand more of the state's Guard soldiers are in Iraq, along with high-water vehicles and other heavy equipment.

1:00 P.M. Airlines begin to close down operations at Louis Armstrong International Airport. Alabama governor Bob Riley orders evacuation of the southernmost areas of Mobile and Baldwin countries.

1:30 P.M. Governor Kathleen Blanco and local officials go on television advising people to evacuate.

4:00 P.M. Contra-flow plan goes into effect in southeast Louisiana, allowing traffic to move outward on nearly all lanes of the major highways in the vicinity. The NHC's hurricane watch is extended as far east as the Alabama-Florida border.

5:00 P.M. Governor Blanco and Mayor C. Ray Nagin appear at a press conference to warn residents of the storm. Nagin declares a state of emergency in New Orleans, but

stops short of calling for a mandatory evacuation. Tulane University suspends orientation activities for the fall term, telling students and others to leave the region.
7:25–8:00 P.M. Max Mayfield, director of the NHC, takes the initiative, calling officials in Alabama, Louisiana, and Mississippi to warn them of the severity of the coming storm.
10:00 P.M. The NHC raises the hurricane watch to a warning, predicting that the entire central Gulf Coast will be in danger.

Sunday, August 28

As of Sunday morning, the Coast Guard is at work closing ports and waterways in the hurricane's predicted path. Personnel, vessels, and aircraft are positioned in staging areas to be ready for the aftermath of the storm.

12:40 A.M. Katrina is rated a Category 4 hurricane.
1:00 A.M. The eye of the hurricane is located in the Gulf, 310 miles south-southeast of the Mississippi River Delta (25.1 N by 86.8 W), gathering power as it crawls along at 8 mph.
6:00 A.M. Hancock County, Mississippi's westernmost coastal county, orders a mandatory evacuation for all residents.
6:15 A.M. The National Oceanic and Atmospheric Administration reports that Katrina is now a Category 5 storm.
7:00 A.M. Gulf Coast residents wake up to the news that Katrina is a Category 5 hurricane, with winds blowing steadily at 160 mph. The eye is located in the Gulf, 250 miles south-southeast of the Mississippi River Delta (25.7 N by 87.7 W), moving at 12 mph.
8:00 A.M. The Superdome opens as a shelter of last resort. Jackson County, the easternmost coastal county in Mississippi, invokes mandatory evacuation orders for all residents south of U.S. Route 90 and in other low-lying areas.
9:25 A.M. President George W. Bush calls Governor Blanco, advising that she and Mayor Nagin order a mandatory evacuation. Plans are in place to do so imminently.
9:30 A.M. Mayor Nagin orders a mandatory evacuation of New Orleans. The storm is due to come ashore in approximately fifteen hours.
10:00 A.M. President Bush speaks with FEMA Director Michael Brown and Governors Kathleen Blanco (Louisiana), Haley Barbour (Mississippi), Jeb Bush (Florida), and Bob Riley (Alabama). Officials in Harrison County, Mississippi's central coastal county, order mandatory evacuations for those in threatened zones.
10:11 A.M. The National Weather Service issues a dire warning of the probable impact of Katrina: "Most of the area will be uninhabitable for weeks . . . perhaps longer."
11:00 A.M. Michael Brown arrives in Baton Rouge.
3:00 P.M. Cars are leaving Greater New Orleans at the rate of 18,000 per hour. Even so, about 112,000 people in New Orleans do not own a car or have access to one.

4:00 P.M. Highways remain clogged in the areas north and west of New Orleans. About 80 percent of the city's population of 485,000 has evacuated, leaving about 100,000 people.

4:15 P.M. President Bush, Michael Brown, and Michael Chertoff participate in an electronic briefing conducted by Max Mayfield, who warned of the danger of destruction and flooding in the wake of Katrina. Michael Brown also tried to prepare federal leaders for the magnitude of the coming disaster: "We're going to need everything that we can possibly muster, not only in this state and in the region, but the nation, to respond to this event." Both Bush and Chertoff will repeatedly claim during the week that they were taken by surprise by the damage wrought by the hurricane.

4:40 P.M. The videoconference ends and top federal officials return to their own pursuits. Congressman Thomas Davis III (R-Virginia) later notes: "The president is still at his ranch, the vice president is still fly-fishing in Wyoming, the president's chief of staff is in Maine. In retrospect, don't you think it would have been better to pull together? They should have had better leadership. It is disengagement."

5:00 P.M. The first sign of the hurricane: rains from the outermost fringe begin to fall along the coast from southern Louisiana to Biloxi and Gulfport on the central Mississippi Coast.

6:00 P.M. A curfew goes into effect in New Orleans. Approximately 10,000 people are in the Superdome; an unknown number are waiting in houses and other buildings all over the region. Seventeen of Mississippi's twenty-seven casinos are closed. Many had shut their doors even earlier, at 2 A.M., to allow employees to evacuate.

8:30 P.M. The last train to leave New Orleans before the hurricane departs with many empty cars.

9:00 P.M. Rains from Katrina's outer bands come down in New Orleans.

10:30 P.M. The last of the people seeking refuge in the Superdome are searched and allowed in. Between 8,000 and 9,000 citizens are in the stands, about 600 are in a temporary medical facility. About 300 officials, medical workers, and staff members oversee the crowd, with security provided by 550 National Guard troops.

Monday, August 29

12:00 midnight The storm surge begins to press up onto Mississippi's western coastline, with beachside residents in Waveland reporting a foot of water in their homes.

2:00 A.M. The eye of the hurricane is passing 130 miles south-southeast of New Orleans (28.2 N by 89.6 W). Winds are blowing steadily at almost 155 mph. Katrina is a Category 4 storm, moving slowly at a rate of 12 mph. A slow-moving hurricane is much more destructive than one that passes quickly over inhabited areas.

3:00 A.M. The 17th Street Canal begins to suffer a breach, according to National Guard reports.

4:00 A.M. The eye of the hurricane is 90 miles south-southeast of New Orleans and 120 miles south-southwest of Biloxi (28.8 N by 89.6 W). Winds are blowing steadily at almost 150 mph. Katrina is a Category 4 storm. The storm surge begins to arrive at the central part of the Mississippi Coast.

5:00 A.M. A civilian calls the Army Corps of Engineers to report that the 17th Street Canal is breached.

5:15 A.M. Greater New Orleans loses electric power. At the Superdome, auxiliary generators provide dim light but no air-conditioning.

5:20 A.M. The Biloxi/Gulfport Regional Airport records gusts of 78 mph, which indicate that hurricane-force winds have arrived, according to the National Weather Service.

6:00 A.M. The eye of the hurricane is 70 miles south-southeast of New Orleans, between the mouth of the Mississippi River and Grand Isle. Rain is falling at the rate 1 inch per hour. All of southeastern Mississippi is pounded by the wind and rain. The storm is 95 miles south-southwest of Biloxi (29.1 N by 89.6 W). Winds are blowing steadily at almost 145 mph. Katrina is a Category 4 storm, with a forward speed of 15 mph. The National Weather Service station at the Biloxi Gulfport Airport is no longer functioning.

6:10 A.M. The eye of Katrina makes its landfall near Buras, Louisiana, as a Category 4 hurricane. Pressure in the center of the storm is 920 mb, the third lowest for a hurricane at landfall.

6:30 A.M. Buras has been destroyed. Nearly every building in lower Plaquemines Parish, which escorts the Mississippi River into the Gulf, is obliterated.

7:00 A.M. High tide along the Mississippi Coast—but the water is still rising, as a result of the storm surge. U.S. Route 90—2 miles from the beachfront—is dotted with boats thrown ashore by the growing waves.

7:10 A.M. The storm has reached Pascagoula, the biggest city along the eastern part of the Mississippi Coast; winds of 118 mph are recorded.

7:11 A.M. Telephone service fails in southeastern Louisiana, as landlines and most cell phone towers are affected by the storm.

7:50 A.M. A massive storm surge sends water over the Mississippi River–Gulf Outlet (MRGO) and the Industrial Canal, causing immediate flooding in St. Bernard Parish and eastern neighborhoods of New Orleans. Water levels in most areas are 10 to 15 feet. In some places, the water is so deep that police officers in boats have to steer carefully around streetlights. Ninety-five percent of the parish is underwater.

8:00 A.M. The eye of the hurricane is passing 40 miles to the southeast of New Orleans (29.7 N by 89.6 W), with a forward speed of only 15 mph. Winds are blowing steadily at almost 135 mph; the hurricane has grown in width with hurricane-force winds at least 100 miles from the eye. Katrina is a Category 4 storm, affecting all of southeast Louisiana. Winds and rain are also beginning to tear into the Mississippi Coast.

Mayor Nagin appears on NBC's *Today* show and, in the midst of an otherwise upbeat report, asserts that water is coming through the levees in places: "We will have significant flooding, it is just a matter of how much."

The Waveland Police Department building is pounded by water.

8:14 A.M. The Industrial Canal is breached, flooding the Lower Ninth Ward.

8:15 A.M. Water rises so quickly at St. Rita's Nursing Facility in St. Bernard Parish that it overtakes bedridden patients. Thirty-five will die during the next half hour.

The water inside the Waveland police headquarters has risen 3 feet in fifteen minutes. The twenty-seven people make their way and end up swimming for something solid to which they can cling.

8:30 A.M. Hurricane Katrina pushes a 25-foot storm surge into the flat, unprotected Mississippi Coast.

9:00 A.M. Two holes open up in the Superdome roof. The 8-mile-long bridge connecting New Orleans to Slidell along Interstate 10 is impassable, a crumpled wreck.

9:30 A.M. In the Lower Ninth Ward of New Orleans, the water is 6 to 8 feet deep.

9:45 A.M. The storm surge along the central part of the Mississippi Coast is now 28 feet deep.

10:00 A.M. Hurricane Katrina makes a second landfall near Pearlington, Mississippi, at the border with Louisiana. The National Hurricane Center discontinues all hurricane watches. The eye of Katrina is moving ashore at the Louisiana-Mississippi border (30.2 N by 89.6 W). Winds are blowing steadily at almost 125 mph. Hurricane-force winds are felt 125 miles from the eye. Katrina is a Category 3 storm, moving at about 17 mph.

10:30 A.M. President Bush declares emergency disasters in three states: Louisiana, Mississippi, and Alabama.

10:40 A.M. The Waveland police are still clinging to the upper branches of bushes and trees, trying to survive in 10 feet of water. They will be rescued shortly.

10:45 A.M. (approximate) President Bush calls Secretary Chertoff from Air Force One, but does not discuss the Gulf Coast, then in the throes of the hurricane. He has questions about immigration policy.

11:00 A.M. Brown issues a memo ordering 1,000 FEMA employees to the Gulf Coast. He gives them two days to arrive. Word reaches the Emergency Operations Center in New Orleans of the breach in the 17th Street Canal levee. At the Hancock County Emergency Operations Center, located inland in Bay St. Louis, Mississippi, the water is 3 feet deep. The building is 27 feet above sea level. Few other buildings in the vicinity are on such high ground.

12:00 noon The eye of the hurricane is located 40 miles south-southwest of Hattiesburg, Mississippi (30.8 N by 89.6 W), increasing its pace to 17 mph. Winds are blowing steadily at 105 mph. Katrina is a Category 2 storm.

The storm surge rises from the open sea, across Lake Pontchartrain, and finally over the communities on the north shore. In towns such as Lacombe and Slidell, the water is 15 to 20 feet deep.

President Bush makes a speech in Arizona concerning Medicare reform. Michael Brown participates in a videoconference that includes Governor Blanco and White House Deputy Chief of Staff Joe Hagin. Brown states that he has spoken with President Bush twice during the morning. In response to a question from Hagin, Blanco says that she has not heard that any levees have breached, but that that could change at any time.

12:15 P.M. The hurricane begins to take leave of the Mississippi coast. Barometric pressure rises so quickly that people report that their ears popped. The wind changes direction as the storm passes to the north. The same force that had been pushing water from the sea onto the land, causing the storm surge, begins to send it back in the other direction. The storm surge slowly drains from the coast.

1:00 P.M. Mayor Nagin announces the breach of the 17th Street Canal but does not elaborate on the implication.

1:00 P.M. (approximate) The Louisiana Department of Wildlife and Fisheries is the first official rescue group to begin operations in New Orleans, arriving in an initial convoy of seventy vehicles, with boats attached. Looting begins in various parts of New Orleans: Mid-City (sporting goods warehouse), Tremé (grocery store), and Uptown (hardware store).

2:00 P.M. The eye of the hurricane is 20 miles west-southwest of Hattiesburg, Mississippi (31.4 N by 89.6 W). Winds are blowing steadily at almost 95 mph. Katrina is a Category 1 storm. U.S. Route 90 in Mississippi is covered by 7 feet of water in Harrison County, as the storm surge has yet to recede completely. People have been floating in the water or clinging to treetops since midmorning.

3:00 P.M. Four feet of water is reported in the Lakeview section of New Orleans. With the storm proceeding to the north, the water trapped with the storm surge in Lake Pontchartrain shifts south, pressing against the levees and floodwalls protecting the city of New Orleans. The extra pressure widens the breached levees.

4:00 P.M. The eye of the hurricane is located 30 miles northwest of Laurel, Mississippi (31.9 N by 89.6 W), moving at 18 mph. Winds are blowing steadily at 75 mph. Katrina is a Category 1 storm. The London Avenue Canal levees are breached in two places. Water is gushing into New Orleans.

5:00 P.M. About 1,000 people rescued from flooded neighborhoods have been dropped off at the Ernest Morial Convention Center in New Orleans and told that buses would arrive to pick them up. The buses never arrive. Officials and workers inside the center, numbering about forty, tell the evacuees that there is no provision for sheltering people: no food, water, or other necessities. The doors are locked.

6:45 P.M. The Coast Guard is conducting rescues with its initial fleet of seven helicopters.

8:00 P.M. Governor Blanco speaks with President Bush to impress upon him the destruction caused by Katrina: "Mr. President, we need your help. We need everything you've got." Secretary of Defense Rumsfeld is attending a San Diego Padres baseball game, at the invitation of the team owner.

9:27 P.M. Michael Chertoff's chief of staff, John F. Wood, receives firsthand description of the levee breaks and the "extensive flooding" in New Orleans, courtesy of FEMA official Marty Bahamonde.

9:30 P.M. (approximate) President Bush turns in for the night, without taking any action on the Katrina disaster or Governor Blanco's requests for assistance.

10:30 P.M. Homeland Security in Washington receives a message regarding Bahamonde's firsthand description of the disastrous flooding in New Orleans.

11:00 P.M. White House receives firsthand report of the levee breaks and the dangerous, extensive flooding in New Orleans.
11:05 P.M. Deputy Homeland Secretary Michael Jackson receives e-mail report on Bahamonde's account of New Orleans disaster.
12:00 midnight (approximate) People outside the Convention Center break a door and force their way in, unlocking the other doors.

Tuesday, August 30

As of Tuesday morning, people with nowhere else to go flock to the Convention Center; by the end of the day, there will be 20,000 inside.

Patients and staff members are stranded in New Orleans hospitals, all but one of which are without power, and conditions are deteriorating. Six hundred are stuck in Charity Hospital (250 patients and 350 others), 1,200 in Tulane University Medical Center (160 patients, 1,040 others—and 76 pets), 2,000 in Truro Infirmary (250 patients and 1,750 others), 650 in Methodist Hospital (150 patients and 500 others), and 2,060 in Memorial Medical Center (260 patients and 1,800 others).

Eight oil refineries in Louisiana have ceased operation. They account for 8 percent of America's refining capacity. Most offshore drilling operations are also shut down; 89 percent of production has been suspended.

1:30 A.M. The boiler room at Tulane University Hospital, which houses the emergency generators, fills with water. Within three hours, all power is out.
3:00 A.M. When the power fails at Methodist Hospital, volunteers, including some children, replace ventilator machines by hand-pumping air into the lungs of ailing patients, hour after hour.
6:00 A.M. The area around the Superdome and New Orleans City Hall, which had been dry on Monday, is under 3 feet of steadily rising water.
7:00 A.M. President Bush, in San Diego, is told of the severity of the crisis along the Gulf Coast. He is advised to end his six-week vacation early and he agrees. After finishing his appearances in California, he will return to his ranch in Texas and then return to Washington on Wednesday. In later statements, he maintains that as of Tuesday morning, he was informed that New Orleans had "dodged the bullet."
8:00 A.M. The number of National Guard troops on the ground in the disaster zones of their respective states is nearly 3,800 in Louisiana, more than 1,900 in Mississippi, and 800 in Alabama. In addition, Alabama has sent specialists to assist Mississippi.
9:00 A.M. (approximate) Michael Chertoff leaves Washington for Atlanta, where he will attend a conference on avian flu. He later claims that he believed New Orleans to be safe, having read some early, positive reports in the morning newspapers. He does not explain why he ignored the dire descriptions delivered to his staff the night before.
10:00 A.M. Mayor Nagin announces that efforts to use sandbags to block the breach in the 17th Street Canal levee have not been successful. The breach has grown to about 200 feet. Looting is reported all over New Orleans.

10:53 A.M. Mayor Nagin decrees a mandatory evacuation for the city and empowers officers to take residents away, even against their will.
11:00 A.M. President Bush makes remarks on the anniversary of V-J Day at the naval air station North Island in San Diego, California.
11:30 A.M. In Louisiana and Mississippi, 1.1 million homes and other buildings are without power, according to Entergy.
12:00 noon Fires are seen around New Orleans. The fire department struggles to reach them, let alone fight them. In Harrison County, Mississippi, 240 of the state's National Guard troops have arrived for recovery work.
1:00 P.M. (approximate) The Louisiana Department of Wildlife and Fisheries is operating more than 200 flat-bottomed boats in Greater New Orleans. "We'd load a boat with people, run to the nearest high ground or road, unload them, and go back out," said a warden.
1:30 P.M. Louis Armstrong International Airport is functioning again. Unscheduled flights for relief efforts use the facility.
2:00 P.M. A casino official in Mississippi petitions the state legislature to enact changes in the regulations specifying that gambling is legal only on floating vessels; he wants inland casinos to be legalized.
3:00 P.M. Governor Blanco, Senator Landrieu, Senator Vitter, and other officials report on conditions in southeast Louisiana. Water is still pouring into the city and streets are blocked with up to 10 feet of water.
4:07 P.M. House Speaker Dennis Hastert suggests that Senator Trent Lott (R-Mississippi) call on President Bush to visit the Mississippi Coast, saying that "the people of Mississippi are flat on their backs. They're going to need your help."
7:00 P.M. Secretary Chertoff designates the Katrina destruction an "Incident of National Significance."

Wednesday, August 31

As of Wednesday morning, 78,000 people in Mississippi are living in shelters.

Waters have stopped rising in New Orleans, as levels in the city and the lake reached parity.

The population inside the Superdome has grown to 26,000 and officials have locked the doors.

The temperature inside Methodist Hospital, which has been without power for more than twenty-four hours, is 106 degrees.

The Coast Guard is continuing rescue operations with 4,000 personnel, 37 aircraft, 15 cutters, and 63 small boats.

8:00 A.M. Two hospitals, Charity and University, lose their generators, from either flooding or lack of fuel. There is no light, air-conditioning, or power to run medical equipment. President Bush participates in a videoconference on the subject of Katrina; FEMA Director Mike Brown reports on flooding in Greater New Orleans and the breached levees.

9:00 A.M. (approximate) Governor Blanco tries repeatedly to get a call through to the President; when he finally takes her call, she requests 40,000 troops.

10:00 A.M. Governor Blanco makes a joint announcement with FEMA that plans have been laid to evacuate residents remaining in Greater New Orleans to the Astrodome in Houston. Air Force One flies low over the Gulf Coast, so that President Bush can see the damage. Joint Task Force Katrina, the Pentagon's command center for disaster response, is organized at Camp Shelby, Mississippi. The person in charge is Lieutenant General Russel Honore.

11:00 A.M. With the Superdome closed to new refugees, people with nowhere to go are now heading toward the Convention Center.

11:30 A.M. Air Force One is seen in the skies over the Gulf Coast. President Bush sees the devastation of the Gulf Coast from the air, ordering the pilot of his official jetliner to fly low over the region.

12:00 noon The water stops rising in New Orleans—the levels in the city and the lake have reached parity.

12:00 noon (approximate) Secretary of Health and Human Services Michael Leavitt declares that a public health emergency exists in the states affected by Katrina.

12:15 P.M. Senator David Vitter (R-Louisiana) conveys a message to Governor Blanco from Bush political advisor Karl Rove, advising her that the White House wanted to federalize the evacuation of New Orleans.

2:00 P.M. (approximate) Evacuation of the Superdome begins with first priority given to the 700 seriously ill and disabled people housed in the stadium. Officials finish moving them out in late evening.

2:20 P.M. Governor Blanco telephones President Bush, informing him that federalization of the evacuation and the Louisiana National Guard will not be necessary. She asks him to announce when federal troops will arrive in the disaster zone.

3:00 P.M. President Bush convenes a task force at the White House to discuss ways to improve the response. It meets for an hour.

4:00 P.M. About 200 refugees from the flooded city of New Orleans are met on the Crescent City Bridge to the town of Gretna by police officers brandishing guns and shouting obscenities. The police chief, Arthur Lawson, had given orders to refuse entry to those needing help or a dry place to camp. "My obligation is to the people of Jefferson Parish," he said.

4:11 P.M. President Bush addresses the nation in his first speech devoted to Hurricane Katrina.

7:00 P.M. Martial law is declared in New Orleans. Mayor Nagin pulls the police force off rescue detail and orders officers to focus entirely on controlling the looting, which has become rampant.

Thursday, September 1

As of Thursday morning, an estimated 4,000 people are stranded on the Interstate 10 overpass in New Orleans, many are elderly and sick and have been without their prescription drugs for days.

The Convention Center houses 15,000 to 25,000 people—estimates are difficult to make—who are without supplies.

Around the United States, the price of regular gasoline shoots past $3 per gallon, purportedly as a result of the disruption in refining in Louisiana.

Baton Rouge has replaced New Orleans as the biggest city in Louisiana.

2:00 A.M. The first of thousands of evacuees arrive in Houston and are welcomed into the Astrodome.

6:12 A.M. In an interview on *Good Morning America,* Bush inexplicably says, "I don't think anybody anticipated the breach of the levees."

7:00 A.M. In an interview on National Public Radio, Homeland Security Secretary Michael Chertoff calls the reports of thousands of people stranded in and around the Convention Center in New Orleans "rumors," and states, "Actually, I have not heard a report of thousands of people in the Convention Center who don't have food and water."

8:00 A.M. According to a report in the morning's Arlington Heights, Illinois *Daily-Herald,* House Speaker Dennis Hastert says "it doesn't make sense" to rebuild New Orleans in the same low-lying area. "First of all your heart goes out to the people, the loss of their homes," he told the editorial board, "but there are some real tough questions to ask about how you go about rebuilding this city."

10:00 A.M. (approximate) A detail of eighty-eight policemen attempt to take control of the Convention Center but are turned away.

11:00 A.M. (approximate) Dr. Norman McSwain, chief of trauma surgery at Charity Hospital, calls the Associated Press to report that the institution has no food, barely any fresh water, and very limited power. It is being ignored by rescuers and was not on any rescue plan.

1:44 P.M. President Bush appears in the Oval Office to introduce two presidential predecessors, his father, George H. W. Bush, and Bill Clinton, as co-chairmen of the Hurricane Katrina Relief Fund. One of Bush's associates describes him on this afternoon as "angry, tired, grumpy."

2:00 P.M. Mayor Nagin delivers an ultimatum to the rest of the country, on CNN. Calling his statement "a desperate SOS," he insists that "I need reinforcements. I need troops, man. I need five hundred buses, man. This is a national disaster. I've talked directly with the President. I've talked to the head of the Homeland Security. I've talked to everybody under the sun." The Army Corps of Engineers reports that water levels in Lake Pontchartrain are falling, meaning that the city is also draining somewhat. The Corps also says that progress is finally being made in the effort to plug the 17th Street Canal levee.

3:00 P.M. (approximate) Evacuations of two local hospitals are suspended due to attacks from sniper fire.

8:00 P.M. Mayor Nagin calls in to the radio program of Garland Robinette and complains vociferously about the federal response. In comments crackling with slang and obscenities, the mayor blasts federal authorities, including the President. The interview ends with both Robinette and Nagin unable to talk, for all the raw emotion they feel, along with everyone else in Greater New Orleans.
9:00 P.M. Governor Blanco says that Speaker Hastert's comments on the rebuilding of New Orleans amounted to an effort to "kick us when we're down. I demand an immediate apology."
10:00 P.M. In Washington, the Senate convenes a special session to vote on a $10.5 billion emergency relief bill for the disaster zone. It is called H.R. 3645, the "Emergency Supplemental Appropriations Act to Meet Immediate Needs Arising from the Consequences of Hurricane Katrina, 2005." The bill passes.
11:00 P.M. Officials in Houston deem that the Astrodome is filled to a workable capacity after 11,375 people have been admitted. They close the doors and begin sending new arrivals from Louisiana to other shelters in Texas.

Friday, September 2

As of Friday morning, Louis Armstrong International Airport has been set up as a medical center and is ready to receive patients from area hospitals and others needing help.

The number of National Guard troops in the disaster zone has risen to 19,500; 6,500 are in New Orleans.

The Louisiana SPCA is working to save the thousands of pets left behind in Greater New Orleans.

2:00 A.M. The last staff members are evacuated from Memorial Medical Center in New Orleans. Forty-five bodies remain behind.
4:35 A.M. Those still in New Orleans are jarred by several loud explosions at a chemical warehouse.
4:55 A.M. Alan Gould, a man who has been living inside the Convention Center for three days, tells CNN that with no food, water, or help from the outside, people are suffering and many are dying. He calls the situation "genocide." Told that thousands of troops are expected over the next few days, Gould tells CNN, "Okay, so we got to sit by and watch four or five more elderly sick people die or some—another baby die, or whatever, while they're making up their mind to come in?"
6:20 A.M. The head of Emergency Operations for New Orleans, former Marine officer Terry Ebbert, expresses the bottled-up frustration of people all over the country: "This is a national disgrace. FEMA has been here three days, yet there is no command and control. We can send massive amounts of aid to tsunami victims, but we can't bail out the city of New Orleans."
9:00 A.M. (approximate) A well-coordinated force of 1,000 National Guard troops storms the Convention Center to take control. It is an enormous building and a time-consuming task. The troops have 200,000 meals with them.

10:30 A.M. Army Chinook helicopters evacuate the last people from Tulane University Hospital.
10:35 A.M. In Mobile at the start of a tour of the Gulf Coast, President Bush praises Brown: "Brownie, you're doing a heck of a job." He seems utterly out of touch with FEMA's ineffectual response to the disaster.
12:00 noon Convoys of army trucks arrive in New Orleans, with supplies, food, and water. Lieutenant General Russel Honore is in command. In Washington, the House of Representatives passes the $10.5 billion relief bill approved by the Senate the night before.
12:30 P.M. In Biloxi, Bush says, "I am satisfied with the response. I am not satisfied with all the results."
1:00 P.M. Evacuation of the Superdome continues at a slow pace; those at the Convention Center continue to wait for buses.
2:07 P.M. The United Nations announces plans to send aid to the Gulf Coast. Individual nations from Venezuela to Canada to Germany have already started sending personnel and matériel to the disaster zone.
2:30 P.M. President Bush visits the 17th Street Canal levee in New Orleans and has his picture taken in front of the effort to patch the breach. Huge machinery is being used to lower building materials into the gap.
3:00 P.M. (approximate) With the Astrodome deemed full, having accepted more than 11,000 evacuees, the city of Houston opens the Reliant Center, which will accept another 11,000 people from the hurricane zone.
4:00 P.M. President Bush meets with Governor Blanco, Mayor Nagin, and others aboard Air Force One. Nagin almost loses control as he demands that Bush and Blanco work out a chain of command for the deployment of military personnel.

The hospitals in New Orleans are finally evacuated.

5:01 P.M. President Bush makes a five-minute speech at Louis Armstrong International Airport promising that New Orleans will be rebuilt. Some people consider his flippancy on the occasion inappropriate. Referring to New Orleans, he said, "I believe the town where I used to come from Houston, Texas, to enjoy myself—occasionally too much—will be that very same town, that it will be a better place to come to."
8:30 P.M. (approximate) Amtrak trains are being used to take evacuees out of New Orleans.
9:40 P.M. On the telethon "A Concert for America," produced by NBC, singer Kanye West goes off-script to complain about the specter of racism in the federal response to the Katrina victims, saying, "You know, it's been five days [without aid], because most of the people are black. . . . George Bush doesn't care about black people."
11:20 P.M. Governor Blanco receives a fax from Bush Chief of Staff Andrew Card, indicating that she need only sign an attached letter requesting that the federal government assume control of the rescue and recovery in Louisiana, including oversight of National Guard troops.

SATURDAY, SEPTEMBER 3

As of Saturday morning, the state of Texas has taken in 220,000 evacuees from Louisiana, not counting those staying with relatives and friends. The number staying in the state's ninety-seven official shelters is about 120,000, with about 100,000 in hotels.

Katrina has impacted 90,000 square miles—an area equivalent to the size of Great Britain.

2:15 A.M. With the evacuation of the Superdome almost complete—2,000 people are left—the National Guard abruptly stops the operation because buses have been ordered to stop going to the stadium. No one at the Superdome knows the reason.

7:30 A.M. President Bush prepares to announce that the federal government is taking over control of the Louisiana National Guard and other rescue functions. He has signed the $10.5 billion emergency spending bill passed by Congress.

7:56 A.M. Governor Blanco faxes a letter to Andrew Card at the White House, refusing the federal government's attempt to assume control.

8:00 A.M. President Bush changes his announcement to cover only the deployment of 7,000 active-duty troops, which would, he says, arrive in the Gulf Coast over the next three days; the troops are from the 82nd Airborne, the First Cavalry, and the Marines First and Second Expeditionary Forces.

9:00 A.M. Governor Blanco makes a statement describing a new attitude in New Orleans, now that 13,000 National Guard troops are on the ground there. She predicts that people in the Convention Center will be moved out that day.

9:30 A.M. Michael Brown announces that 1.9 million meals (army rations) and 6.7 million bottles of water are on hand in the disaster zone.

12:00 noon Buses arrive at the Convention Center to take people to safety and comfort elsewhere.

1:45 P.M. Helicopters take the last evacuees off the Interstate 10 overpass in New Orleans.

3:00 P.M. James Lee Witt, the former director of FEMA, has been hired by Governor Blanco to help the state of Louisiana organize its relief efforts. Michael Brown's response to the hiring: "That is absolutely the right thing to do. He will make a huge difference."

4:00 P.M. The New Orleans *Times-Picayune* reports that the streets of New Orleans are "getting safer by the minute."

5:00 P.M. Loyola University announces that the fall semester has been canceled. Tulane and other colleges in New Orleans have already made the same decision.

5:47 P.M. The Superdome is empty.

9:50 P.M. Louisiana State Police announces that the Convention Center is empty.

Notes

Chapter 1: Ignoring the Inevitable

1. Interview with Laura Maloney, January 5, 2005.
2. SPCA National Mission Statement, SPCA Archive, New York, N.Y.
3. Press releases, July 7 and August 27, 2005, Louisiana SPCA.
4. Bulletin, "Hurricane Katrina Advisory Number 16," National Hurricane Center, Miami, Fl., August 27, 2005, 5 A.M.
5. Joseph B. Treaster, "Hurricane Drenches Florida and Leaves Seven Behind," *New York Times*, August 27, 2005.
6. Interview with Kathy Boulte, October 18, 2005.
7. Interview with Laura Maloney, December 15, 2005.
8. Agency Group 05, "NRC Continues to Monitor Nuclear Plants Affected by Hurricane Katrina," Nuclear Regulatory Commission, August 30, 2005.
9. "New Orleans Braces for 'The Big One,'" CNN News (online), August 28, 2005.
10. "Katrina Evacuation Directives," New Orleans *Times-Picayune*, August 27, 2005.
11. Interview with Benny Rousselle, November 29, 2005.
12. John Smith Kendall, *History of New Orleans* (1922).
13. Ibid.
14. W. Adolphe Roberts, *Lake Pontchartrain* (Indianapolis: Bobbs-Merrill, 1946), p. 28.
15. Ibid., p. 33.
16. Jim Fraiser, *The French Quarter of New Orleans* (Jackson: University Press of Mississippi, 2003), p. 7.
17. Philomena Hauck, *Bienville: Father of Louisiana* (Lafayette: University of Southwestern Louisiana Press, 1998).
18. Fraiser, *French Quarter of New Orleans,* pp. 7–28.
19. Interview with Park Moore, November 29, 2005.

20. America's Wetland Resource Center, "The Official Numbers," updated online January 21, 2006.

21. Elizabeth Kolbert, "Watermark," *The New Yorker,* February 27, 2006.

22. Mike Dunne, *America's Wetland: Louisiana's Vanishing Coast* (Baton Rouge: Louisiana State University Press, 2005).

23. Mark Fischetti, "Drowning New Orleans," *Scientific American*, October 2001, pp. 76–85.

24. John W. Sutherlin to author, October 23, 2005.

25. Theodore Roosevelt, *A Booklover's Holiday in the Open* (New York: Charles Scribner's Sons, 1916), pp. 376–77.

26. Interview with Daryl Malek, November 7, 2005.

27. C. L. Lockwood and Rhea Gray, *Marsh Mission: Capturing the Vanishing Wetlands* (Baton Rouge: Louisiana State University Press, 2005), p. 9.

28. Sierra Club memo.

29. Office of Senator Mary Landrieu, press release, May 16, 2005.

30. Press release, University of New Orleans, January 19, 2000.

31. U.S. Army Corps of Engineers, *Defending New Orleans*, October 2005.

32. Dunne, *America's Wetland*, p. 73.

33. Interview with King Milling, January 4, 2006.

34. Roy Blount Jr., "Feet on the Sheet," *Louisiana Cultural Vistas*, Fall 2005, p. 72.

35. Elizabeth Kolbert, "Storm Warnings," *The New Yorker*, September 19, 2005, p. 35.

36. John McPhee, *The Control of Nature* (New York: Farrar, Straus and Giroux, 1989).

37. Fischetti, "Drowning New Orleans."

38. John McQuaid and Mark Schleifstein, "Washing Away," New Orleans *Times-Picayune*, June 2002.

39. Joel K. Bourne Jr., "Gone with the Water," *National Geographic*, October 2004, pp. 88–89.

40. Eric Berger, "Keeping Its Head Above Water," *Houston Chronicle*, December 1, 2001.

41. Interview with Eric Berger, October 20, 2005.

42. John M. Barry, *Rising Tide: The Great Mississippi Flood of 1927 and How It Changed America* (New York: Simon & Schuster, 1998).

43. Ken Gewertz, "The Lessons from Katrina," *Harvard University Gazette*, December 12, 2005.

44. "Assessing the Katrina Plan Failure," BBC News, October 6, 2005. See also Jack Williams, "Hurricane Scale Invented to Communicate Storm Danger," *USA Today*, May 17, 2005.

45. "The Saffir-Simpson Scale," Atlantic Oceanographic and Meteorological Laboratory, Miami, Fl., 2005.

46. David Longshore, *Encyclopedia of Hurricanes, Typhoons, and Cyclones* (New York: Checkmark Books, 2000), pp. 174–75.

47. "Anatomy of a Disaster," *U.S. News & World Report*, September 26, 2005.

48. Michael Lewis, "Wading Toward Home," *New York Times Magazine*, October 9, 2005.

49. Innovative Emergency Management, "Southeast Louisiana Catastrophic Hurricane Functional Plan," September 20, 2004, p. 1.
50. Ibid., p. 20.
51. Interview with Clancy Dubos, January 5, 2006.
52. John McQuaid, " 'Hurricane Pam' Exercise Offered Glimpse of Katrina Misery," *Boston Globe*, September 9, 2005.
53. "City of New Orleans Comprehensive Emergency Management Plan."
54. Audrey Hudson and James G. Lakely, "New Orleans Ignored Its Own Plan," *Washington Times*, September 9, 2005.
55. Press release, "Mayor Nagin Urges Citizens to Prepare for Hurricane Katrina," City of New Orleans, August 27, 2005.
56. Patrick Fagan, "Katrina: Correcting Culpability and Fixing Fallacies" (unpublished, report provided to author, November 2, 2005).
57. Chris Rose, "Hollywood South, Baby!" New Orleans *Times-Picayune*, August 27, 2005.
58. Gwen Filosa, "Nagin Says He'll Oppose Building Moratorium," New Orleans *Times-Picayune,* January 22, 2006.
59. Tim Padgett, "Can New Orleans Do Better?" *Time*, October 24, 2005 p. 36.
60. Jeremy Alford and Allen Johnson Jr., "Katrina's Quiet Gift," *Gambit Weekly*, December 27, 2005, p. 12.
61. Bruce Nolan, "Katrina Takes Aim," New Orleans *Times-Picayune*, August 28, 2005, p. 1.
62. "Suffering and Semantics," nola.com, September 9, 2005.
63. Christy Oglesby, " 'My Momma's Body Is on the Roof,' " CNN, October 17, 2005, transcript.
64. Gordon Russell, "Nagin Orders First-Ever Mandatory Evacuation of New Orleans," New Orleans *Times-Picayune*, August 28, 2005, p. 1.
65. Interview with Terry Ebbert, March 8, 2006.
66. Interview with Tonya Brown, September 30, 2005.
67. Interview with Jackie Clarkson, November 18, 2005.
68. Interview with Kathy A. Lawes-Reed, December 8, 2005.
69. Margaret Mitchell, *Gone with the Wind* (New York: Warner Books, 1993), p. 842.
70. "The Lost City," *Newsweek*, September 12, 2005, p. 48.
71. Fraiser, *French Quarter*, p. 206.
72. Rich Thomaselli, "N.O.'s 5B Tourism in Tatters," *Advertising Age*, September 2, 2005, p. 1.
73. Anne Rice, "Do You Know What It Means to Miss New Orleans," *New York Times*, September 4, 2005.
74. Fact sheet, Port of New Orleans, December 2005.
75. A. J. Mistretta, "Rebuilding Energy," *Biz New Orleans*, June 1, 2004.
76. U.S. Census Bureau, State and County Quick Facts, 2000.
77. Ibid.
78. Michael Grunwald, "The Steady Buildup to a City's Chaos," *Washington Post*, September 11, 2005.

79. Maureen Dowd, "Sex, Envy, Proximity," *New York Times*, October 15, 2005.
80. Jim Reed's letter of August 24, 2005, quoted in Jim Reed and Mike Theiss, *Hurricane Katrina: Through the Eyes of Storm Chaser* (Helena, Mont.: Farcountry Press, 2005), p. 4.
81. Interview with Dubos.
82. Interview with Julie Silvers, October 7, 2005.
83. Keith O'Brien and Bryan Bender, "Chronology of Errors," *Boston Globe*, September 11, 2005.

Chapter 2: Shouts and Whispers

1. Proclamation No. 48 KBB 2005, "State of Emergency—Hurricane Katrina," State of Louisiana Executive Department.
2. Kathleen Blanco, letter to George W. Bush, State of Louisiana Executive Department, August 27, 2005.
3. Ibid.
4. Quoted in Kerry Emanuel, *Divine Wind: The History and Science of Hurricanes* (New York: Oxford University Press, 2005), p. 174.
5. Bob Williams, "Governor Blanco and Mayor Nagin Failed Their Constituents," *Wall Street Journal*, September 19, 2005.
6. "Statement on Federal Emergency Assistance for Louisiana," Office of the Press Secretary, the White House, August 27, 2005.
7. Press release, "Homeland Security Prepping for Dangerous Hurricane Katrina," Department of Homeland Security, August 28, 2005.
8. Interview with Bob Mann, December 21, 2005.
9. Interview with Cedric Richmond, January 6, 2005.
10. Kathleen Blanco, "Response to U.S. Senate Committee on Homeland Security and Governmental Affairs Document and Information Request" (submitted December 2, 2005).
11. Interview with Cedric Richmond, December 11, 2005.
12. "Exhibits Across South Also Struggling," *Hollywood Reporter*, August 30, 2005.
13. Interview with Nick Mueller, November 28, 2005.
14. Interview with Marc Morial, September 28, 2005.
15. Jacques Morial, interview with Jessica Maruri, October 28, 2005.
16. Interview with Oliver Thomas, January 6, 2005.
17. Evan Thomas and Arian Campo-Flores, "The Battle to Rebuild," *Newsweek*, October 3, 2005.
18. Michael Eric Dyson, *Come Hell or High Water* (New York: Basic Civitas Books, 2006), p. 8.
19. Nik Cohn, *Triksta: Life and Death and New Orleans Rap* (New York: Alfred A. Knopf, 2005). Quotations are from the Web sites of Juvenile and Lil' Wayne.
20. Interview with Rafael C. Goyeneche III, December 28, 2005.
21. Interview with Warren Riley, December 5, 2005.

22. Ibid.
23. Ibid.
24. Office of the Governor of Louisiana, "Response to U.S. Senate Committee on Homeland Security and Governmental Affairs Document and Information Request Dated October 7, 2005, and to the House of Representatives Select Committee to Investigate the Preparation for and Response to Hurricane Katrina," December 2, 2005.
25. Clancy Dubos, "First Responders: New Orleanians of the Year," *Gambit Weekly*, January 3, 2006.
26. Ruth Berggren, "Unexpected Necessities—Inside Charity Hospital," *The New England Journal of Medicine* 353, October 13, 2005.
27. Interview with Nick Felton, January 6, 2006.
28. Erika Bolstead, "New Orleans Advises Residents to Evacuate as Katrina Draws Near," *Miami Herald*, August 27, 2005.
29. Bulletin, "Hurricane Katrina Advisory Number 18," National Hurricane Center, Miami, Fl., August 27, 2005.
30. Bruce Nolan, "Katrina Takes Aim," New Orleans *Times-Picayune*, August 28, 2005, p. 1.
31. Mark Schleifstein, "Hurricane Center Director Warns New Orleans: This Is Really Scary," New Orleans *Times-Picayune*, August 27, 2005.
32. Interview with Kathleen Blanco, December 21, 2005.
33. "FEMA Chief Taken Off On-Site Efforts," Fox News, September 10, 2005.
34. Walter Maestri, interview with Daniel Zwerdling, *All Things Considered*, National Public Radio, September 9, 2005.
35. Ray Nagin, *Anderson Cooper 360,* CNN, January 20, 2006, transcript.
36. Bolstead, "New Orleans Advises Residents to Evacuate."
37. Interview with Janine Butscher, September 27, 2005.
38. Interview with Judy Oudt, November 30, 2005.
39. Mary Vuong, "Mark a Bad Season's End with Your Own Hurricane," *Houston Chronicle*, November 30, 2005.
40. Andrew Travers, "Ridgewood Native's Heart Aches for New Orleans Home," *Ridgewood News*, September 16, 2005.
41. Interview with Andy Ambrose, October 6, 2005.
42. Interview with David Vitter, December 21, 2005.
43. Interview with Ronald Mack Jr., December 8, 2005.
44. Interview with Kenny Bourque, September 16, 2005.
45. Kevin Moran, "Big Easy Gets Busy as Katrina Takes Aim," *Houston Chronicle*, August 28, 2005.
46. Michael Barnett blog, August 28, 2005.
47. Brenda Norrell, "A Pointe-au-Chien Family Describes Survival After Katrina," *Indian Country Today*, September 28, 2005.
48. Interview with Benjamin Johnson, December 21, 2005.
49. Ray Nagin, interview with Stone Phillips, *Dateline,* NBC, September 11, 2005.

50. Ray Nagin, *Anderson Cooper 360*, CNN, January 20, 2006, transcript.
51. Interview with Joe Donchess, December 13, 2005.
52. "Three Nursing Home Residents Die in Evacuation," New Orleans *Times-Picayune*, August 29, 2005.
53. Membership List 2005, Louisiana Nursing Home Association, Baton Rouge, La.
54. Interview with Donchess.
55. Paul Rioux, "Doomed Nursing Home Had Offer to Bus Transport," New Orleans *Times-Picayune*, September 9, 2005.
56. Doug Simpson, "Nursing Home Owners Charged with Negligent Homicide in 34 Flood Related Deaths," Associated Press, September 13, 2005.
57. Nicole Gelinas, "Who's Killing New Orleans?" *New Orleans City Journal*, Autumn 2003.
58. Zora Neale Hurston, *Their Eyes Were Watching God* (New York: Harper Perennial, 1998), p. 158.
59. Interview with Reverend Willie Walker, October 3, 2005.
60. Ibid.
61. Ibid.

Chapter 3: Storm vs. Shoreline

1. Joseph Conrad, *Typhoon and Other Stories* (New York: Random House, 1991), p. 189.
2. "Hurricane Hunters: GPS in the Eyewall," *Geospatial Solutions*, October 2005, p. 13.
3. "Anatomy of a Disaster," *U.S. News & World Report*, September 26, 2005, p. 29.
4. Quoted in "Katrina," *National Geographic*, September 2005 (special edition), p. 33.
5. Martin Merzer, "As Hurricanes Intensify, Potential Damage Increases Exponentially," *Miami Herald*, September 28, 2005.
6. National Weather Service data.
7. Quoted in Kerry Emanuel, *Divine Wind: The History and Science of Hurricanes* (New York: Oxford University Press, 2005), p. 136.
8. Herman Melville, *Moby-Dick*.
9. Jeffrey Kluger, "Global Warming: The Culprit," *Time*, October 3, 2005, p. 43.
10. William M. Gray and Philip J. Klotzbach, *Forecast of Atlantic Hurricane Activity for October 2005 and Seasonal Update Through September* (Fort Collins, Col.: Colorado State University, 2005).
11. Al Lewis, "Column," *Denver Post*, August 30, 2005.
12. Interview with William Gray, October 25, 2005.
13. "Anatomy of a Disaster," p. 29.
14. "Eye to Eye with Hurricane Katrina," Hurricane Hunter reports, Kessler Air Force Base, Biloxi, Miss.
15. Kluger, "Global Warming," p. 46.
16. Interview with Terry Jonson, November 16, 2005.
17. Interview with Marc Morial, October 17, 2005.

18. Interview with Diane Johnson, November 24, 2005.
19. Bulletin, "Hurricane Katrina Advisory Number 23," National Hurricane Center, Miami, Fl., 10 A.M., August 28, 2005.
20. Weather message, 10:11 A.M., National Weather Service, New Orleans, August 28, 2005.
21. Brian Williams, "Daily Nightly" blog, September 5, 2005.
22. Brian Friel, "The Forecaster of Disaster," *National Journal*, September 24, 2005, p. 2946; interview with Robert Ricks, October 24, 2005.
23. Robert Ricks, interview with Brian Williams, September 15, 2005, NBC News transcript.
24. "Timeline: Hurricane Katrina Actions, Non-Actions," *Executive Intelligence Review*, September 16, 2005.
25. Bruce Nolan, "Katrina Takes Aim," New Orleans *Times-Picayune*, August 28, 2005, p. 1.
26. Emanuel, *Divine Wind*, pp. 151–52.
27. Bill Harper, "Mississippi Braces for Major Hit from Katrina," *USA Today,* August 28, 2005.
28. Interview with Greg Iverson, October 26, 2005.
29. Ibid.
30. Ibid.
31. Interview with Charlie West, December 14, 2005.
32. C. Ray Nagin, "New Orleans Mandatory Evacuation Declaration," August 28, 2005, transcript.
33. Jan Moller, "Blanco Tours New Orleans," New Orleans *Times-Picayune*, August 31, 2005.
34. Joe Gyan Jr. and Scott Dyer, "N.O. Mayor Endorses Jindal, Defies Party," *Baton Rouge Advocate*, November 4, 2003, p. 2.
35. Interview with C. Ray Nagin, March 15, 2006.
36. Will Sentell, "Blanco Wins in Historic Vote; Woman Elected Governor," *Baton Rouge Advocate*, November 16, 2003, p. 1.
37. James Dao, "After the Storm, She Tries to Mend State, and Career," *New York Times*, December 29, 2005.
38. Tyler Bridges, "Blanco's Bid," *Gambit Weekly*, December 7, 2004.
39. Edward Alden, Joshua Chaffin, Christopher Swann, and Holly Yeager, *Financial Times*, September 10, 2005.
40. Interview with Reverend Willie Walker, October 3, 2005.
41. Evan Thomas, "How Bush Blew It," *Newsweek*, September 19, 2005.
42. Robert Block, Amy Schatz, Gary Fields, and Christopher Cooper, "Power Failure; Behind Poor Katrina Response, a Long Chain of Weak Links," *Wall Street Journal,* September 6, 2005.
43. Interview with Oliver Armstrong, October 27, 2005.
44. Susan B. Glasser and Michael Grunwald, "The Steady Buildup to the City's Chaos," *Washington Post*, September 11, 2005.

45. Interview with Armstrong.
46. Interview with Marc Morial, September 20, 2005.
47. Author interview with Collette Creppell, October 24, 2005.
48. Ibid.
49. Marc Caputo, "Category 5 Storm Barrels Toward New Orleans," *Miami Herald*, August 28, 2005.
50. "New Orleans Empties Out Amid Fear, Prayers," Associated Press, August 30, 2005.
51. "Orleans Watering Hole Stays Open," *San Francisco Chronicle*, September 9, 2005.
52. Walter "Wolfman" Washington, interview with Jessica Maruri, October 23, 2005.
53. Interview with Chris Owens, December 31, 2005.
54. Interview with Ivory Clark, November 25, 2005.
55. Ibid.
56. Interview with Donna Clark, December 7, 2005.
57. Transcript, FEMA videoconference August 28, 2005, Associated Press, pp. 1–2.
58. Ibid., pp. 6–7.
59 Ibid., pp. 14–15.
60. Press briefing by Scott McClellan, Office of the Press Secretary, the White House, March 6, 2006.
61. Chris Matthews, interview with Haley Barbour, *Hardball*, MSNBC, March 2, 2006.
62. Steve Bowman, Lawrence Kapp, Amy Belasco, "Hurricane Katrina: DOD Disaster Response" (Washington, D.C.: Congressional Research Service, September 19, 2005), p. 16.
63. "President Discusses Hurricane Katrina; Congratulates Iraqis on Draft Constitution," White House, August 28, 2005, transcript.
64. Jake Calamusa, interview with Bill O'Reilly, *The O'Reilly Factor*, August 29, 2005.
65. Anna Badkhen, "Race Against Time," *San Francisco Chronicle*, August 31, 2005.
66. Calamusa interview, *The O'Reilly Factor*.
67. Interview with Clark.
68. Interview with Sarah Jaffe, October 19, 2005.
69. Ibid.
70. Statement, "Federal Emergency Assistance for Mississippi," White House, August 29, 2005; statement, "Federal Emergency Assistance for Alabama," White House, August 29, 2005.
71. Drew Brown, "States Have Enough Troops," Knight Ridder, August 29, 2005.
72. "Hurricane City: Dauphin Island, Alabama's History with Tropical Storms," radio broadcast.
73. Gary Mitchell, "Alabama Coast Evacuating Early for Katrina," *Decatur Daily*, August 29, 2005.
74. Interview with Janet McQueen, November 22, 2005.
75. Interview with Hal Leftwich, December 6, 2005.
76. Interview with Fredro Knight, December 7, 2005.
77. Ibid.
78. Ibid.

79. Interview with Sydney Saucier, December 8, 2005.
80. Interview with Angie Gambino, December 6, 2005.
81. Interview with Saucier.
82. Interview with Nagin, March 15, 2006.
83. "The Security Czar," *Gambit Weekly*, March 3, 2005.

Chaper 4: The Winds Come to Louisiana

1. Interview with Joaquin Zumbado, November 14, 2005.
2. Interview with Josh Holm, November 18, 2005.
3. Interview with Zumbado.
4. Interview with Heather Allan, December 16, 2005.
5. Interview with Zumbado.
6. Interview with Holm.
7. Office of the Governor of Louisiana, "Response to U.S. Senate Committee on Homeland Security and Governmental Affairs Document and Information Request Dated October 7, 2005, and to the House of Representatives Select Committee to Investigate the Preparation for and Response to Hurricane Katrina," December 2, 2005.
8. Interview with Marsha Evans, October 25, 2005.
9. Dwight Landreneau, "Comments," *Louisiana Conservationist*, November/December 2005, pp. 1–5.
10. Adam Einick, "The Call of Duty: LDWF Search and Rescue," *Louisiana Conservationist*, November/December 2005, p. 5.
11. Interview with Tim Bayard, December 22, 2005.
12. Interview with Jeff Goldblatt, December 30, 2005.
13. Ibid.
14. Ibid.
15. Interview with Garland Robinette, December 27, 2005.
16. Jaquetta White, "Firm Shut Down Oil Rigs in Gulf, Removed Workers," New Orleans *Times-Picayune*, August 28, 2005.
17. Interview with Cody Nicholas, December 7, 2005.
18. Ibid.
19. Interview with Jimmy Duckworth, November 21, 2005.
20. Kane quoted in Stephen Ambrose and Douglas Brinkley, *The Mississippi River and the Making of a Nation* (Washington, D.C.: National Geographic Society, 2003), pp. 5–25.
21. Interview with Duckworth.
22. "Katrina Intensifies into a Powerful Hurricane, Strikes Northern Gulf Coast," NASA report, August 30, 2005.
23. Peter Whoriskey and Guy Gugliotta, "Storm Thrashes Gulf Coast," *Washington Post*, August 30, 2005.
24. Interview with Eddie Favre, September 16, 2005.
25. National Hurricane Center advisory, August 30, 2005.

26. Keith O'Brien and Bryan Bender, "Chronology of Errors," *Boston Globe*, September 11, 2005.
27. Press release, U.S. Army Corps of Engineers, Vicksburg, Miss., August 30, 2005.
28. Brian Hayes, "Natural and Unnatural Disasters," *American Scientist*, November/December 2005.
29. Michelle Krupa, "Workers' Lives Were Priority," New Orleans *Times-Picayune*, October 11, 2005.
30. Carl Quintanilla, "Manned Pumps Evacuated Before Katrina Hit," NBC News, October 18, 2005.
31. Bob Ross, "Second-Guessing Is Not Broussard's Focus," New Orleans *Times-Picayune*, October 9, 2005.
32. Michelle Krupa, "Leaders Battle over Pump Decision," New Orleans *Times-Picayune*, October 13, 2005.
33. Interview with Zumbado.
34. Ibid.
35. Jennifer Broome, "Personal Account of Hurricane Katrina," WOIA–San Antonio (NBC), posted September 16, 2005.
36. Interview with Judy Frank, November 7, 2005.
37. Tim Jones, "Plaquemines, First Parish Hit by Katrina, May Be Last to Recover," *Chicago Tribune*, October 16, 2005.
38. "Summary of Hurricane Katrina," NCAA Satellite and Information Service, 2005.
39. Tommy Tomlinson, "For Now, Survivor Plans to Stay Behind with His Dogs," Knight Ridder, September 6, 2005.
40. Lee Hancock, "French Quarter Comes Through Relatively Unscathed," *Dallas Morning News*, August 29, 2005.
41. Interview with Chris Owens, December 31, 2005.
42. Bill Taylor, "Battered Stadium Protects Sick, Poor," *Toronto Star*, August 30, 2005.
43. Mary Foster, "Super Storm Rips Piece off Superdome's Roof," Associated Press, August 29, 2005.
44. Interview with Oliver Thomas, January 6, 2006.
45. Interview with Jackie Clarkson, November 18, 2005.
46. Interview with Gralen Banks, March 6, 2006.
47. Ibid.
48. Interview with Warren Riley, December 5, 2005.
49. "Shooter Get 65-Year Sentence," New Orleans *Times-Picayune*, December 13, 2005.
50. Interview with Riley.
51. Broome, "Personal Account of Hurricane Katrina."
52. Raymond B. Seed et al., Preliminary Report on the Performance of the New Orleans Levee Systems in Hurricane Katrina on August 29, 2005 (Berkeley: University of California Press, 2005), pp. 2–5–8.
53. David Cuthbert, "Katrina's Lost Lives," New Orleans *Times-Picayune*, January 8, 2006.

54. Ceci Connolly and Manuel Roig-Franzia, "Katrina Proved Deadly in Every Section of New Orleans," *Washington Post*, October 23, 2005.
55. Interview with Michael Prevost, October 26, 2005.
56. U.S. Census Bureau, *Census 2000 Sample Characteristics (SF3)*; from a compilation by the Greater New Orleans Community Data Center.
57. Interview with Prevost, December 28, 2005.

Chapter 5: What Was the Mississippi Gulf Coast

1. Interview with Eddie Favre, November 7, 2005.
2. "Hurricane Katrina Advisory Number 27," National Hurricane Center, Miami, Fl., August 29, 2005.
3. Dahleen Glanton, "Biloxi Resident Emerges from Hiding to Find Her Town in Shambles," *Chicago Tribune*, August 30, 2005.
4. Geoff Pender and Michael Newsome, "Katrina Wreaks Havoc on Mississippi," Biloxi *Sun-Herald*, August 29, 2005.
5. Leslie Williams, "A Week in the Ruins of Mississippi," New Orleans *Times-Picayune*, September 8, 2005.
6. "In Quotes," *Katrina Daily News*, Keesler Air Force Base, September 5, 2005.
7. "CE Personnel Brave Katrina's Winds in Big Rescue," *Katrina Daily News*, Keesler Air Force Base, September 5, 2005.
8. Jim Garamone, "Katrina: Riding Through the Monster," *American Forces Press*, September 12, 2005.
9. Anita Lee, "Grim Work of Recovering Bodies Begins in Biloxi," Biloxi *Sun-Herald*, September 7, 2005.
10. Julie Goodman, "Katrina: The Recovery," Jackson *Clarion-Ledger*, October 2, 2005.
11. Kelly Knauer, *Hurricane Katrina: The Storm That Changed America* (New York: Time Books, 2005), p. 4.
12. Interview with Frank Griffith, January 4, 2006.
13. Interview with Laura Stepro, January 7, 2006.
14. Interview with Michael Prendergast, December 6, 2005.
15. Interview with Glen Volkman, December 3, 2005.
16. Interview with Israel Neff, December 3, 2005.
17. Interview with Stepro.
18. Interview with Michael Veglia, December 7, 2005.
19. Interview with Alex Coomer, December 8, 2005.
20. Interview with Ronnie Williams, December 7, 2005.
21. Interview with Veglia, December 28, 2005.
22. Interview with Hardy Jackson, November 21, 2005.
23. " 'Take Care of the Kids and the Grandkids,' " CNN.com, August 30, 2005.
24. Interview with Jackson.
25. Tommy Tomlinson, "For Now, Survivor Plans to Stay Behind with His Dogs," Knight Ridder, September 6, 2005.

26. Lee Hancock, "French Quarter Comes Through Relatively Unscathed," *Dallas Morning News*, August 29, 2005.
27. Interview with Nick Breazeale, October 27, 2005.
28. Ibid.
29. Greg Iverson, "My Story," *Kiplinger*, November 2005.
30. Gary Rivlin, "Bright Spot on Gulf as Casinos Rush to Rebuild," *New York Times*, December 14, 2005.
31. Quoted in "Katrina" *National Geographic*, September 2005 (special edition), p. 20.
32. Cathy Booth Thomas, "It's Worse Than You Think," *Time*, November 28, 2005, p. 35.
33. Interview with Eddie Favre, December 21, 2005.
34. Interview with Ron Vanney, December 23, 2005.
35. Interview with David Stepro, December 29, 2005.
36. Interview with Angie Gambino, December 7, 2005.
37. Interview with Favre, December 21, 2005.
38. Interview with Dara Adano, November 4, 2005.
39. Interview with Eddie Favre, November 12, 2005.
40. George Pawlaczyk and Malcolm Garcia, "Loved Ones Grapple with Confusion, Red Tape in Search for Missing," Knight Ridder, September 27, 2005.
41. John Christofferen, "Fountain Says Storms Won't Keep Him Down," Associated Press, November 6, 2005.
42. "Weary Fountain Ran Dry," New Orleans *Times-Picayune,* March 8, 2006.
43. Interview with Hal Leftwich, December 7, 2005.
44. Interview with Sydney Saucier, December 8, 2005.
45. Interview with Jo Ann Garcia, January 8, 2006.
46. Ibid.
47. Interview with Gambino.
48. Ibid.
49. Interview with Leftwich.
50. Interview with Fredro Knight, January 9, 2006.
51. Interview with Janet McQueen, November 21, 2005.
52. "Hospital Manages to Keep Doors Open," *Houston Chronicle*, September 3, 2005.
53. "Hancock Medical Center Could Take Six Months to Fully Reopen," The Jackson Channel.com, November 7, 2005.
54. Tom Wilemon, "Biloxians Coping, Remembering," Biloxi *Sun-Herald*, September 1, 2005.
55. Brian MacQuarrie, "Storm Reduces Mississippi Coastal Resort Region to Rubble, Ruins," *Boston Globe*, August 31, 2005.
56. Ibid.
57. Tonya Walker, interview with Renee Montaigne, *Morning Edition*, National Public Radio, September 27, 2005.
58. Interview with Veglia, December 7, 2005.
59. Ben Nelms, "TV Face of Katrina Tragedy Loses Wife, Finds Hope with His Palmetto Family," *The Citizen News* (Fayetteville, Ga.), September 16, 2005.

60. Shaila Dewan and Abby Goodnaugh, "Rotting Food, Dirty Water, and Heat Add to Problems," *New York Times*, September 2, 2005.

61. Robert Little, "Life and Death Make Everything Else Unimportant," *Baltimore Sun,* August 21, 2005.

62. Lewis Lapham, "Notebook: Hurricane Katrina," *Harper's*, November 1, 2005.

63. Jerry Mitchell, Chris Joyner, and John Fusquay, "Flooding, Wreckage, Death Sweep Miss.," Jackson *Clarion-Ledger*, August 30, 2005.

64. Gary Fineout, "Biloxi, Gulfport Resemble War-Torn Cities in Hurricane's Wake," *Miami Herald*, August 30, 2005.

65. Richard Ford, "A City Beyond the Reach of Empathy," *New York Times*, September 4, 2005.

66. Interview with Greg Iverson, October 6, 2005.

67. Mitchell, Joyner, and Fusquay, "Flooding, Wreckage, Death Sweep Miss."

68. Jim Butler, "War Zone: Scenes from the Day After Katrina," Biloxi *Sun-Herald*, August 31, 2005.

69. Interview with Frank Griffith, January 3, 2006.

70. Interview with Charlie West, December 14, 2005.

71. Robert Little, "Storm Turns Beachfront into Junkyard," *Baltimore Sun,* August 31, 2005.

72. Tim Reid, "Aquarium Dolphins Found Alive," *The Times* (London), September 15, 2005.

73. Coyt Bailey, interview with Melissa Block, *All Things Considered*, National Public Radio, August 30, 2005.

74. Don Hammack, Anita Lee, and Scott Dodd, "Gulf Coast Begins to Assess Devastation from 'Our Tsunami,'" Biloxi *Sun-Herald*, August 30, 2005.

75. Tony Gnoffo, "In Pascagoula, Katrina Claims a Neighborhood," *Philadelphia Inquirer*, August 31, 2005.

76. Interview with Trent Lott, November 29, 2005.

77. Ibid.

78. "Lott Sues to Collect for Katrina Damage," *Houston Chronicle*, December 18, 2005.

79. Pender and Newsome, "Katrina Wreaks Havoc on Mississippi."

80. Bobby Cleveland, "Fishing Boats Stuck in Back Bay," Jackson *Clarion-Ledger*, September 8, 2005.

81. Interview with Favre, December 21, 2005.

82. Chris Joyner, "Katrina: The Recovery," Jackson *Clarion-Ledger*, October 13, 2005.

83. Interview with Favre, December 21, 2005.

84. Chris Joyner, "Mountain of Debris Daunting Problem," Jackson *Clarion-Ledger*, September 22, 2005.

85. Stan Tiner, "Our Worst Fears . . . ," Biloxi *Sun-Herald*, August 30, 2005.

86. Alex Leavy, "In Gulfport Evacuees Huddle at a School, out of Options" *St. Petersburg Times*, September 4, 2005.

87. *Scarborough Country*, September 1, 2005, MSNBC transcript.

Chapter 6: The Busted Levee Blues

1. Interview with Jim Amoss, January 10, 2006.
2. Ibid.
3. Ibid.
4. Frank Harris, *The Bomb* (New York: Mitchell Kennedy, 1909), p. 63.
5. Interview with Amoss, January 5, 2006.
6. Interview with Doug MacCash, January 10, 2006.
7. Interview with James O. Byrne, January 18, 2006.
8. Interview with MacCash.
9. Ibid.
10. Matt Scallan, "For Yacht Club, the Race Must Go On," New Orleans *Times-Picayune*, November 27, 2005.
11. Interview with Byrne.
12. Interviews with Byrne and MacCash.
13. Interview with Byrne.
14. Interview with MacCash.
15. Interview with Byrne.
16. Interview with Amoss, January 10, 2006.
17. Intermediate Advisory 25B, National Hurricane Center, Miami, Fl., August 29, 2005, 2 A.M.
18. "National Hurricane Center Report," August 29, 2005.
19. Mary Foster, "Katrina Rips Piece Off Superdome's Roof," Associated Press, August 29, 2005.
20. Anne Bochell Konigsmark, "Storm Drain: What Next for Dome," *USA Today*, December 29, 2005.
21. Interview with Brian Williams, January 10, 2006.
22. "Rumors of Death Greatly Exaggerated," New Orleans *Times-Picayune*, September 26, 2005.
23. Intermediate Advisory 26A, National Hurricane Center, August 29, 2005, 6 A.M.
24. Interview with Byrne, January 13, 2006.
25. Ann Carns, "Long Before Flood, New Orleans System Was Prime for Leaks," *Wall Street Journal*, September 25, 2005.
26. Ibid.
27. "Residents Say Levee Leaked Months Before Katrina," *Morning Edition*, National Public Radio, November 22, 2005.
28. Bob Marshall, "Levee Leaks Reported to S & WB a Year Ago," New Orleans *Times-Picayune*, November 18, 2005.
29. Ibid.
30. "Government's Failures Doomed Many," *Seattle Times*, September 11, 2005.
31. John McQuaid, "Katrina Trapped City in Double Disasters," New Orleans *Times-Picayune*, September 7, 2005.
32. Raymond B. Seed, Testimony, Senate Committee on Homeland Security and Governmental Affairs, November 2, 2005.

33. Graham Bulton, "Like Santa Claus," *Forbes*, October 9, 1995.
34. Interview with Blaine Kern, November 6, 2005.
35. Martin Miller, "A Long Road to Recovery," *Los Angeles Times*, October 12, 2005.
36. Interview with Kern, December 17, 2005.
37. Interview with Joaquin Zumbado, October 23, 2005.
38. Ibid.
39. Interview with Josh Holm, November 16, 2005.
40. Interview with Richard L. Stalder, December 30, 2005.
41. Interview with Zumbado.
42. Dan Baum, "Deluged; When Katrina Hit, Where Were the Police?" *The New Yorker*, January 1, 2006.
43. Interview with Heather Allan, December 16, 2005.
44. Anderson Cooper, CNN, September 2, 2005, transcript.
45. Eric Boehlert, "Covering Katrina," *Rolling Stone*, October 6, 2005, p. 45.
46. Brett Martel, "Angry God Sent Storms, Mayor of New Orleans Says," *Chicago Tribune*, January 17, 2006.
47. Interview with Allan.
48. Cormac McCarthy. *No Country for Old Men* (New York: Alfred A. Knopf, 2005), p. 216.
49. Interview with Tim Bayard, December 22, 2005.
50. Nik Cohn, "The Day the Music Died," *The Guardian*, January 15, 2006.
51. Interview with Warren Riley, December 7, 2005.
52. Ibid.
53. Interview with Jimmy Duckworth, November 21, 2005.
54. Ibid.
55. Press release, "Coast Guard Stages Resources for Post-Hurricane Response," U.S. Coast Guard, August 28, 2005.
56. Interview with Michael O'Dowd, December 18, 2005.
57. Interview with Duckworth.
58. Amanda Ripley, "How the Coast Guard Gets It Right," *Time*, October 23, 2005.
59. Stephen Barr, "Coast Guard's Response to Katrina a Silver Lining in the Storm," *Washington Post*, September 6, 2005.
60. Ripley, "How the Coast Guard Gets It Right."
61. "Guardsman Recounter Storm's Fury and Reserves at Jackson Barracks," New Orleans *Times-Picayune*, September 16, 2005.
62. Interview with Terry Ebbert, March 7, 2005.
63. George W. Bush, interview with Brian Williams, December 7, 2005.
64. Interview with Boisie Bolinger, October 3, 2005.
65. Interview with Renee Marcus, December 14, 2005.
66. Interview with Ebbert, March 9, 2005.
67. Maureen Dowd, "Sex, Envy, Proximity," *New York Times,* October 15, 2005.
68. Jeremy Manier and Michael Hawthorne, "Path Puts Off City's Day of Reckoning," *Chicago Tribune*, August 30, 2005.

69. Interview with Ebbert, March 6, 2005.
70. Raymond B. Seed et al., "Performance Preliminary Report on the New Orleans Levee Systems in Hurricane Katrina," University of California at Berkeley, 2005, sections 1–4 to 1–6 (re August 29, 2005).
71. Interview with Gene Alonzo.
72. Laura Parker, "What Really Happened at St. Rita's?" *USA Today*, November 28, 2005.
73. Interview with Nita Hutter, December 16, 2005.
74. Ibid.
75. Dean E. Murphy, "Storm Puts Focus on Other Disasters in Waiting," *New York Times*, November 15, 2005.
76. Interview with Jimmy Deleray, October 17, 2005.
77. Michael Barnett blog, August 30, 2005.
78. Interview with Kathleen Blanco, December 21, 2005.
79. Ibid.
80. Interview with Byrne, January 18, 2006.

Chapter 7: "I've Been FEMA-ed"

1. Interview with Ralph Blumenthal, January 20, 2006.
2. Penny Brown Roberts, Michelle Millhollon, and Joe Gyan Jr., "Big Blow," *Baton Rouge Advocate*, August 30, 2005.
3. Interview with Blumenthal.
4. Interview with Jeremy Alford, January 24, 2006.
5. Ralph Blumenthal, "First Priority: Signs of Life," New York Times News Service, August 31, 2005.
6. Ibid.
7. Interview with Blumenthal.
8. Daryl Lang, "This Is Not a Normal Assignment," *Photo District News*, September 1, 2005.
9. Marty J. Bahamonde, testimony, Senate Committee on Homeland Security and Governmental Affairs, October 20, 2005.
10. "Guam Storm Victims Are Housed in Hotels," *New York Times*, September 1, 1992.
11. Interview with Marty Bahamonde, November 7, 2005.
12. Ibid.
13. Marty Bahamonde, e-mail to David Passey, August 18, 2005, 4:46 P.M.
14. Marty Bahamonde, e-mail to David Passey et al., Bahamonde testimony, Senate Committee on Homeland Security and Governmental Affairs, October 20, 2005, exhibit 12.
15. "High Winds Punch Holes in Superdome Roof," Associated Press, August 30, 2005.
16. Ibid.
17. Brian Williams, "Three Months Later, a Return to the Superdome," November 29, 2005, MSNBC transcript.
18. John Burnett, "More Stories Emerge of Rapes in Post-Katrina Chaos," National Public Radio, December 21, 2005.

19. Interview with Judy Benitez, October 20, 2005.
20. Natalie Rule, e-mail to Marty Bahamonde and David Passey, Bahamonde testimony, Senate Committee on Homeland Security and Governmental Affairs, October 20, 2005, exhibit 10.
21. Interview with Marty Bahamonde, January 22, 2006.
22. Ibid.
23. Ibid.
24. Interview with Gordon Russell, January 23, 2006.
25. Interview with Bahamonde.
26. Interview with Alford.
27. *Late Show with David Letterman*, CBS, September 13, 2005, transcript.
28. Bahamonde testimony, Senate Committee on Homeland Security and Governmental Affairs, October 20, 2005.
29. Daren Fonda and Rita Healy, "How Reliable Is Brown's Résumé?" *Time*, September 8, 2005.
30. *The Early Show*, CBS, September 9, 2005.
31. Fonda and Healy, "How Reliable Is Brown's Résumé?"
32. Molly Ivins, "Another Bungling by the Bushies," *Charleston Gazette*, September 10, 2005.
33. Fonda and Healy, "How Reliable Is Brown's Résumé?"
34. Jimmy Carter, Town Hall meeting, American University, September 19, 2005, transcript.
35. Angie C. Mack, "A Crisis Agency in Crisis," *U.S. News & World Report*, September 19, 2005, p. 36.
36. Date from National Climatic Data Center, National Oceanic and Atmospheric Administration.
37. South Florida *Sun-Sentinel* clippings, May 2004.
38. Larry Eichel, "What Went Wrong," Knight Ridder, September 11, 2005.
39. Eric Boehlert, "The Politics of Hurricane Relief," salon.com, September 5, 2005.
40. A. J. Rodriguez, "wish list," August 29, 2005, supplied by Mitch Landrieu.
41. Leo Bosner, interview on *Frontline*, September 28, 2005, PBS transcript.
42. Dru Orja Jay, "FEMA Turned Away Aid," *Français International News*, September 6, 2005.
43. Robert Block, Amy Schatz, Gary Fields, and Christopher Cooper, "Power Failure: Behind Poor Katrina Response, a Long Chain of Weak Links," *Wall Street Journal*, September 6, 2005.
44. Bosner, interview on *Frontline*.
45. Interview with David A. Ross, December 3, 2005.
46. Interview with Warren Riley, December 5, 2005.
47. Bill Steigerwald, "Government Can Take Lesson from Wal-Mart," Pittsburgh *Tribune-Review*, October 14, 2005.
48. Interview with Eddie Favre, October 17, 2005.
49. "Companies Pitch In," CNN, September 15, 2005, transcript.

50. Interview with Sarah Jaffe, January 30, 2006.
51. Press release, "First Responders Urged Not to Respond to Hurricane Impact Areas," FEMA, August 29, 2005.
52. Aaron Broussard, *Meet the Press*, NBC, September 4, 2005.
53. Interview with Keith Calhoun, November 16, 2005.
54. Patricia J. Williams, "The View from Lott's Porch," *The Nation*, September 26, 2005.
55. Thomas Frank, "Roots May Save Lower 9 Ward," *USA Today*, December 6, 2005.
56. "Fats Domino Found OK in New Orleans," CNN, September 1, 2005, transcript.
57. Rick Coleman, *Blue Monday* (New York: Da Capo, 2006), p. x.
58. Interview with Diane Johnson, November 23, 2005.
59. Interview with Diane Johnson, December 20, 2005.
60. Interview with Charmaine Neville, December 20, 2005.
61. "Anatomy of a Disaster," *U.S. News & World Report*, September 26, 2005.
62. Kathleen Blanco, interview with Anderson Cooper, January 19, 2006, CNN transcript.
63. "Brown's Claims Not Fully Supplanted by FEMA Hurricane Response Plan," September 27, 2005, ABC News transcript.
64. Bruce Alpert, "FEMA Chief Dawdled," New Orleans *Times-Picayune*, November 3, 2005.
65. Bahamonde testimony, Senate Committee on Homeland Security and Governmental Affairs, October 20, 2005.
66. David D. Kirkpatrick and Scott Shane, "Storm and Crisis: The Former Administrator, Ex-FEMA Chief Tells of Frustration and Chaos," *New York Times*, September 15, 2005.
67. Rick Klein, "Ex-FEMA Chief Spreads the Blame," *Boston Globe*, September 28, 2005.
68. Kirkpatrick and Shane, "Ex-FEMA Chief Tells of Frustration and Chaos."
69. Spencer Hsu, "After the Storm, Chertoff Vows to Reshape DHS," *Washington Post*, November 14, 2005.
70. Evan Thomas, "How Bush Blew It," *Newsweek*, September 19, 2005.
71. Alison Young, Shannon McCaffrey, and Seth Borenstein, "As New Orleans Flooded, Chertoff Discussed Avian Flu in Atlanta," Knight Ridder, September 15, 2005.
72. Interview with Charles Foti Jr., January 22, 2006.
73. *The O'Reilly Factor*, Fox, August 29, 2005.
74. Suzanne Rodgers, interview on *Paula Zahn Now*, CNN, August 29, 2005.
75. Jonathan S. Landay, Alison Young, and Shannon McCaffery, "Chertoff Delayed Federal Response, Memo Shows," Knight Ridder, September 13, 2005.
76. Michael Brown to Michael Chertoff, memo, Department of Homeland Security, August 29, 2005.
77. Jeff Crouere, "President Bush's 'Brownie' Is Black Mark on Louisiana History," BayouBuzz.com, November 4, 2005.
78. Randolph E. Schmidt, "Red Cross Says Number of People in Shelters Growing," Associated Press, August 31, 2005.
79. "Hurricane Katrina Aftermath: American Red Cross Relief Efforts," Reuters, September 14, 2005.

80. http://haircut.net/auction.
81. 2005 Meteorite Community Charity Raffle, http://www.aerolite.org/meteorite-raffle.htm, September 17, 2005.
82. Anthony Tan and Hedwig Kröner, "Latest Cycling News for September 5, 2005; Armstrong Helps Katrina Victims," http://cyclingnews.com, September 5, 2005.
83. Peter Whoriskey and Sam Coates, "Looting, Fires, and a Second Evacuation," *Washington Post*, August 31, 2005.
84. Cecilia M. Vega, "As Bodies Recovered, Reporters Are Told 'No Photos, No Stories,'" *San Francisco Chronicle*, September 13, 2005.
85. Interview with Russell.
86. Interview with Lamar Montgomery, January 22, 2006.
87. Interview with Jimmy Duckworth, November 21, 2005.
88. Bill Walsh, "Federal Report Predicted Cataclysm," New Orleans *Times-Picayune*, January 24, 2006.
89. Interview with Ivory Clark, November 25, 2005.

Chapter 8: Water Rising

1. Interview with Kathleen Blanco, December 19, 2005.
2. Interview with Raymond Blanco, December 19, 2005.
3. Interview with Bob Mann, December 19, 2005.
4. Interview with Mary Landrieu, January 26, 2006.
5. Ibid.
6. Interview with Philip L. Capitano, January 27, 2006.
7. Interview with Eddie Compass, March 13, 2006.
8. "The Education of Eddie Compass," *Gambit Weekly*, May 28, 2002.
9. Walt Philbin, "Despite Murder Rate, Crime Down," New Orleans *Times-Picayune*, August 20, 2005.
10. Interview with Eddie Compass.
11. Interview with Mary Landrieu.
12. Interview with Kathleen Blanco.
13. Arthur Davis, *Design for Life*, unpublished manuscript supplied by Arthur Davis.
14. Ibid.
15. Interview with Kathleen Blanco.
16. Interview with Mary Landrieu.
17. Interview with Kathleen Blanco.
18. Interview with Raymond Blanco.
19. Interview with Kathleen Blanco.
20. Ibid.
21. Interview with Ty Bromell, January 27, 2006.
22. Ibid.
23. Interview with Dwight Brasher, January 26, 2006.
24. Interview with Bromell.
25. Interview with Angele Davis, January 27, 2006.

26. Interview with Kathleen Blanco.
27. Ibid.
28. Ibid.
29. Mary Curtius, "E-Mails Reveal Fussing over Blanco Image," *Los Angeles Times*, December 12, 2005.
30. Interview with Kathleen Blanco.
31. Ibid.
32. Interview with Bromell.
33. Interview with Davis.
34. Ibid.
35. Leonard Kleinpeter, letter to author, March 13, 2006.
36. Interview with Jim Amoss, January 10, 2006.
37. Interview with Garland Robinette, December 27, 2005.
38. Ibid.
39. Interview with Sam Jones, December 28, 2005.
40. 2005-246 LDWF Search and Rescue Missions, September 8, 2005.
41. John Burnett, "A Confederacy of New Orleans Characters," National Public Radio, October 1, 2005.
42. John Pope, "Storm Puts Dent in West Nile Virus," New Orleans *Times-Picayune*, January 23, 2006.
43. Interview with Jeff Goldblatt, December 30, 2005.
44. Interview with Compass.
45. Interview with Jimmy Duckworth, November 21, 2005.
46. Interview with Kathleen Blanco.
47. Interview with Mann.
48. Interview with Michael Knight, December 8, 2005.
49. Ibid.
50. Interview with Dyan French Cole, March 25, 2006.
51. Trymaine D. Lee, "Momma's Mission," New Orleans *Times-Picayune*, September 18, 2005.
52. Interview with Rick Matthieu, March 13, 2006.
53. Interview with French Cole.
54. Mama D, testimony before U.S. Congress, December 5, 2005, transcript.
55. Interview with Jimmy Deleray, October 17, 2005.
56. Ibid.
57. Ibid.
58. Interview with Mitch Landrieu, December 2, 2005.
59. Interview with Harry Lee, February 7, 2006.
60. Ibid.
61. Ibid.
62. Ibid.
63. Ibid.
64. Ibid.
65. Interview with Brian Williams, March 13, 2006.

66. Ibid.
67. Ibid.
68. Ibid.
69. Interview with Brian Williams, January 10, 2006.
70. Interview with Jean Harper, March 3, 2006.
71. Albertine Arseneau, interview on *All Things Considered*, National Public Radio, August 30, 2005.
72. "City a Woeful Scene," New Orleans *Times-Picayune*, online edition, August 30, 2005.
73. Ibid.
74. Stephen J. Hedges, "Guard Units Stretched Thin," *Chicago Tribune*, September 17, 2005.
75. Pete Simons, interview on *The Situation Room*, CNN, August 30, 2005.
76. Ibid.
77. Press release, "Coast Guard Continues Katrina Rescues," U.S. Coast Guard, August 30, 2005.
78. Interview with Jimmy Duckworth, December 8, 2005.
79. Raphael Obermann, blog, searchwarp.com, August 29, 2005.
80. Vince Laforet, "Triage," Dispatches: Notes from the Field, blog, digitaljournalist .org, September 2, 2005.
81. Interview with Duckworth, December 8, 2005.
82. Ibid.
83. Interview with Mary Fortune and Brandi Idris, November 21, 2005.
84. Interview with Duckworth, December 8, 2005.
85. Jim Washington, "Five Local Navy Ships to Help on Gulf Coast," Norfolk *Virginia-Pilot*, August 31, 2005.
86. Joe Gyan Jr., "Navy Ships Assist in Range of Activities," *Baton Rouge Advocate*, September 9, 2005.
87. Maggie Fox, "When Katrina Got Tough, Nurses Got Inventive," Reuters, September 15, 2005.
88. Lou Dolinar, "Katrina: The Untold Story," *The Nation*, October 24, 2005, p. 38.
89. Jesse Jackson, interview on *Anderson Cooper 360*, CNN, September 2, 2005, transcript.
90. Peter Whoriskey and Guy Gugliotta, "Evacuation and the Recriminations," *Washington Post*, September 4, 2005.
91. Aaron Neville, interview on *Your World with Neil Cavuto*, Fox News Channel, November 23, 2005, transcript.

Chapter 9: City Without Answers

1. Michael Chertoff, interview by Tim Russert, *Meet the Press*, NBC, September 4, 2005.
2. Joseph B. Treaster and Kate Zernike, "Hurricane Katrina: The Overview; Hurricane Slams into Gulf Coast; Dozens Are Dead," *New York Times*, August 30, 2005.
3. Peter Whoriskey and Sam Coates, "Amid the Devastation, Some Feel Relief," *Washington Post*, August 30, 2005.
4. Peter Whoriskey and Guy Gugliotta, "Storm Thrashes Gulf Coast," *Washington Post*, August 30, 2005.

5. Doug MacCash and James O'Byrne, "Levee Breach Floods Lakeview, Mid-City, Carrollton, Gentilly, City Park," New Orleans *Times-Picayune*, online edition, August 30, 2005.
6. Interview with Charlie Melancon, December 18, 2005.
7. Kerry Emanuel, *Divine Wind: The History and Science of Hurricanes* (New York: Oxford University Press, 2005).
8. Robert Novak, "Lawyers vs. Katrina," CNN.com, September 8, 2005.
9. Interview with Robert Novak, January 31, 2006.
10. Joseph Conrad, *Heart of Darkness* (New York: W.W. Norton, 1988), p. 250.
11. Interview with Novak.
12. Bill Walsh, "Squabbles Hindered Rescue Efforts," New Orleans *Times-Picayune*, January 31, 2006.
13. Leo Bosner, interview on *Morning Edition*, National Public Radio, September 16, 2005.
14. Lara Jakes Jordan, "After Storm FEMA Refused Boats, Trucks," New Orleans *Times-Picayune*, January 30, 2006.
15. Ibid.
16. "Joan Baez Performs at Crawford Anti-War Vigil," Associated Press, August 21, 2005.
17. Michael A. Fletcher, "Cindy Sheehan's Pitched Battle," *Washington Post*, August 13, 2005.
18. Elaine Quijano, "Soldier's Mom Digs in Near Bush Ranch," CNN.com, August 7, 2005.
19. 2006 State of the Union, post speech commentary, Bob Schieffer, CBS News, January 31, 2006.
20. Evan Thomas, "How Bush Blew It," *Newsweek*, September 19, 2005.
21. Ibid.
22. Lloyd Grove, "Deep Fishing for Solution to Katrina," New York *Daily News*, September 7, 2005.
23. Ibid.
24. Julian Malone, "Chertoff Blames Slow Katrina Response on Failure to Prepare," Cox News Service, October 22, 2005.
25. White House press release, "President Commemorates 60th Anniversary of V-J Day," August 30, 2005.
26. Howard Fineman, "A Storm-Tossed Boss," *Newsweek*, September 19, 2005, p. 38.
27. "Katrina Timeline; Misdirected Aid," *All Things Considered*, National Public Radio, September 9, 2005.
28. Clancy Dubos, "The Madness of C. Ray," *Gambit Weekly*, January 24, 2006.
29. Interview with Joseph B. Treaster, January 28, 2006.
30. Gordon Russell, "Nagin: Mistakes Were Made at All Levels," New Orleans *Times-Picayune*, September 11, 2005.
31. William Kalec, "Taking Off the Gloves," New Orleans *Times-Picayune*, January 29, 2006.
32. Tom Planchet, "Katrina Blog Part II," WWL-TV local news Web site, August 30, 2005.
33. Bob Marshall, "Levee System Along River Held Its Ground in Storm," New Orleans *Times-Picayune*, January 23, 2006.

34. Anne Rochell Konigsmark, "Amid Ruins, 'Island' of Normalcy in the Big Easy," *USA Today*, December 19, 2005.
35. FEMA news conference, *The Situation Room*, CNN, August 30, 2005.
36. Ibid.
37. John Zarrella, *The Situation Room*, August 30, 2005, CNN transcript.
38. Dan Shea, "Underwater," New Orleans *Times-Picayune*, online edition, August 30, 2005.
39. CNN Reports, *Katrina: State of Emergency* (Kansas City: Andrews McMeel, 2005), p. 32.
40. Greg Henderson, e-mail to Charlie Henderson, August 30, 2005, supplied by Greg Henderson.
41. Leslie Williams, "Eastern N.O. Residents Call for MR-GO to Close," New Orleans *Times-Picayune*, January 29, 2005.
42. Interview with Chad Clark, December 8, 2005.
43. Ibid.
44. Ibid.
45. Ibid.
46. Ibid.
47. Paul Rioux and Manuel Torres, "Receding Floodwaters Reveal Devastation of St. Bernard," New Orleans *Times-Picayune*, online edition, September 2, 2005.
48. J. Michael Brown, interview with Shannon Gibney, September 2005, "New Orleans Survivor Story."
49. Kathleen Blanco press conference, August 30, 2005, CNN transcript.
50. J. Michael Brown, "New Orleans Survivor Story."
51. Interview with Mitch Landrieu, December 2, 2005.
52. Robert Palmer, *Deep Blues* (New York: Penguin, 1981), pp. 74–75.
53. Interview with Mitch Landrieu.
54. Jennifer Brown, "Oil Spills Foul Louisiana Neighborhoods," *USA Today*, October 20, 2005.
55. Interview with Clark.
56. Interview with Clark.
57. Stephen J. Hedges, "Crew of Navy Ship Ready to Play Larger Role in Relief Effort," *Chicago Tribune*, September 3, 2005.
58. Interview with Mary Landrieu, January 26, 2006.
59. Sandy Davis, "Katrina Leaves Slidell a 'Nightmare'," *Baton Rouge Advocate*, August 31, 2005.
60. Chris Kirkham, "The Scene from South Slidell," New Orleans *Times-Picayune*, August 31, 2005.
61. "Texan Blues Star Brown Dies at 81," BBC News, September 11, 2005.
62. Planchet, "Katrina Blog Part II."
63. Kathleen Blanco press conference, August 30, 2005.
64. Interview with Terry Ebbert, March 17, 2006.
65. Interview with Jackie Clarkson, December 8, 2005.
66. Ibid.
67. "City a Woeful Scene," New Orleans *Times-Picayune*, online edition, August 30, 2005.

68. Mike Perlstein and Brian Thevenot, "Even a Cop Joins in the Looting," New Orleans *Times-Picayune*, online edition, August 30, 2005.
69. Planchet, "Katrina Blog Part II."
70. Perlstein and Thevenot, "Even a Cop Joins in the Looting."
71. Greg Henderson, e-mail to Charlie Henderson, August 30, 2005.
72. James Varney, "N.O. Cops Reported to Take Cadillacs from Dealership," New Orleans *Times-Picayune*, September 29, 2005.
73. "Ex-Cop Indicted in Stolen Truck Case," New Orleans *Times-Picayune*, January 21, 2006.
74. Interview with David Vitter, December 21, 2005.
75. Al Thomas note left at Marriott Courtyard, September 2, 2005, courtesy of Mimi Fast.
76. Patrick Wooten, interview on *All Things Considered*, National Public Radio, September 16, 2005.
77. Miriam Hill and Nicholas Spangler, "No Evidence Backs Up Report of Rescue Helicopters Being Fired Upon," Knight Ridder, October 2, 2005.
78. Interview with Eddie Jordan, December 29, 2005.
79. "Anatomy of a Disaster," *U.S. News & World Report*, September 26, 2005.
80. Interview with Warren Riley, December 5, 2005.
81. Ibid.
82. Matthew Magee, "Dispatches from the Edge of Civilization," *Sunday Herald*, September 18, 2005.
83. Holbrook Mohr, interview on *Day to Day*, National Public Radio, August 31, 2005.
84. Haley Barbour, interview on *Today*, August 30, 2005.
85. Scott Dodd, "Katrina's Fury Irrevocably Changed Mississippi Coast," Knight Ridder, August 30, 2005.
86. Susan M. Moyer, *Katrina: Stories of Rescue, Revolution, and Rebuilding in the Eye of the Storm* (Champaign, Ill.: Spotlight Press, 2005).
87. Tony Gnoffo, "In Pascagoula, Katrina Claims a Neighborhood," *Philadelphia Inquirer*, August 31, 2005.
88. Chris Joyner, "Katrina: The Recovery," Jackson *Clarion-Ledger*, October 13, 2005.
89. Timothy R. Brown, "Senator Trent Lott Urges President Bush to Visit Mississippi," Associated Press, August 31, 2005.
90. James Janega and Angela Rozas, "U.S. Recalls FEMA Chief," *Chicago Tribune*, September 10, 2005.
91. Jonathan S. Landay, Alison Young, and Shannon McCaffrey, "Chertoff Delayed Federal Response, Memo Shows," Knight Ridder, September 13, 2005.
92. Planchet, "Katrina Blog Part II."
93. Bill Taylor, "In the Big Easy, Despair Is Rising with the Waters," *Toronto Star*, August 31, 2005.
94. "Government's Failures Doomed Many," *Seattle Times*, September 11, 2005.
95. Ibid.
96. John O'Neil and Maria Newman, "Brown Asserts He Alerted White House Quickly on Katrina," *New York Times*, February 10, 2006.

97. Press release, "Guard, NORTHCOM Responds to Hurricane Aftermath," Agency Group 9, Department of Defense, August 30, 2005.
98. Drew Brown, "U.S. Military to Assist Relief Effort with Aircraft, Logistical Support," Knight Ridder, August 30, 2005.
99. David D. Kirkpatrick and Scott Shane, "Ex-FEMA Chief Tells of Frustration and Chaos," *New York Times*, September 15, 2005.

Chapter 10: The Smell of Death

1. Interview with Sara Roberts, February 1, 2006.
2. Interview with Sam Jones, December 28, 2005.
3. Interview with Ronny Lovett, February 2, 2006.
4. Interview with Sara Roberts, February 6, 2006.
5. Ibid.
6. Interview with Lovett.
7. André Buisson, "R & R in the City," August 31 to September 6, 2005, unpublished diary.
8. Interview with Roberts, February 6, 2006.
9. Interview with André Buisson, February 7, 2006.
10. Interview with Roberts, February 6, 2006.
11. Buisson, "R & R in the City."
12. Interview with Roberts, February 6, 2006.
13. Interview with Lovett.
14. Interview with Roberts, February 6, 2006.
15. Buisson, "R & R in the City."
16. Ibid.
17. Interview with Reverend Willie Walker, October 3, 2005.
18. Interview with Reverend Willie Walker, January 26, 2006.
19. Bob Herbert, "Evacuees Find Humiliation, Not Help," New Orleans *Times-Picayune,* September 15, 2005.
20. Interview with Walker, October 3, 2005.
21. Robert Block, "Documents Reveal Extent of Fumble on Storm," *Wall Street Journal,* September 13, 2005.
22. Lawrence McLeary, interview on *All Things Considered,* National Public Radio, September 9, 2005.
23. "A Full Blackout for 2.1 Million Gulf Coast Homes," *Toronto Star,* August 31, 2005.
24. Liane Hansen, "Power Woes Add to New Orleans Anguish," *Weekend Sunday Edition,* National Public Radio, December 4, 2005.
25. John Schwartz, "Pumping: Behind the First Roar of Machinery to Drain the City," *New York Times,* September 8, 2005.
26. Paul Davidson, "Power Company Gets Lights on Fast," *USA Today,* September 7, 2005.
27. "Entergy Corp.'s New Orleans Subsidiary Files for Bankruptcy," *Transmission and Distribution World,* October 2005.

28. Rick Bush, "This Is Our Fight," *Transmission and Distribution World*, October 1, 2005.
29. Erika Bolstead and Nicholas Spangler, "After Weeks of Turmoil, New Orleans' Police Chief Steps Down," Knight Ridder, September 27, 2005.
30. Interview with Eddie Compass, February 25, 2006.
31. Joseph B. Treaster and Christopher Drew, "City to Offer Free Trips to Las Vegas for Officers," *New York Times,* September 5, 2005.
32. Josephine Ervin, "Pastor Gives Solace at Church," Shenikah Glory Full Gospel Baptist Church Web site newsletter, December 2005.
33. Interview with Charmaine Neville, February 8, 2005.
34. Office of the Governor of Louisiana, "Response to U.S. Senate Committee on Homeland Security and Governmental Affairs Document and Information Request Dated October 7, 2005, and to the House of Representatives Select Committee to Investigate the Preparation for and Response to Hurricane Katrina," December 2, 2005.
35. Sidney Barthelemy, CNN interview quoted on *All Things Considered,* National Public Radio, August 31, 2005.
36. Office of the Governor of Louisiana, "Response to U.S. Senate Committee."
37. Marty Bahamonde, testimony, Senate Committee on Homeland Security and Governmental Affairs, October 20, 2005.
38. Interview with Terry Ebbert, March 8, 2006.
39. "Press Gaggle with Scott McClellan," August 31, 2005, White House Press Office, transcript.
40. Interview with Kathleen Blanco, December 19, 2005.
41. Ibid.
42. Marty Bahamonde, e-mail to Cindy Taylor et al., August 31, 2005, 2:44 P.M.
43. Interview with Michael Brown, February 4, 2006.
44. Andrew Zajac and Andrew Martin, "Offer of Buses Fell Between the Cracks," *Chicago Tribune,* September 23, 2005.
45. "Katrina's Big Contracts Go to Companies in Political Loop," Associated Press, October 19, 2005.
46. Interview with Tony Zumbado, February 3, 2006.
47. Ibid.
48. Ibid.
49. Interview with Tony Zumbado, December 1, 2005.
50. Tony Zumbado, MSNBC interview, September 1, 2005, transcript.
51. Interview with Josh Holm, February 5, 2006.
52. Keith Spera, "Desperation, Death on Road to Safety," New Orleans *Times-Picayune,* August 31, 2005.
53. Interview with Ivory Clark, December 26, 2005.
54. Ibid.
55. Jennifer Loven, "Bush Views Devastation of Hurricane Katrina from Air Force One," *Billings Gazette,* August 31, 2005.
56. "Press Gaggle with Scott McClellan," August 31, 2005.

57. Peter Baker, "Vacation Ends, and Crisis Management Begins," *Washington Post,* September 1, 2005.
58. "Press Gaggle with Scott McClellan," August 31, 2005. White House Press Office transcript.
59. "Analysis: President Bush Surveys Hurricane Damage," *Morning Edition,* National Public Radio, September 1, 2005.
60. Howard Fineman, "A Storm-Tossed Boss," *Newsweek,* September 19, 2005.
61. "Anatomy of a Disaster," *U.S. News & World Report,* September 26, 2005.

Chapter 11: Blindness

1. Kris Axtman, "Lost-and-Found Families," *Christian Science Monitor,* September 15, 2005.
2. CNN News Group, *Katrina: State of Emergency,* CNN Reports, October 2005.
3. "National Response Plan: Local/Federal Response Strategies & Coordination Structures," Department of Homeland Security Web site, December 2005.
4. Ibid.
5. José Saramago, *Blindness,* trans. Giovanni Pontiero (San Diego: Harcourt Brace, 1997).
6. Patrick Water, "Morial Likens Katrina to September 11," New Orleans *Times-Picayune,* February 3, 2006.
7. Interview with Michael Brown, February 4, 2006.
8. Michael D. Brown, testimony, Senate Homeland Security and Governmental Affairs Committee, February 10, 2006; "An Early Alarm from New Orleans," *New York Times,* February 10, 2006.
9. "Preliminary Observations Regarding Preparedness and Response to Hurricanes Katrina and Rita," Government Accountability Office, February 1, 2006.
10. Spencer Hsu, Jody Warrick, and Rob Stein, "Documents Highlight Bush-Blanco Standoff," *Washington Post,* December 5, 2005.
11. Interview with Kathleen Blanco, December 19, 2005.
12. Office of the Governor of Louisiana, "Response to U.S. Senate Committee on Homeland Security and Governmental Affairs Document and Information Request Dated October 7, 2005, and to the House of Representatives Select Committee to Investigate the Preparation for and Response to Hurricane Katrina," December 2, 2005.
13. Interview with Blanco.
14. Interview with Bob Mann, December 19, 2005.
15. "The 2000 Campaign: Transcript of Debate Between Vice President Gore and Governor Bush," *New York Times,* October 4, 2000.
16. Katrina Woznicki, "Navy Hospital Ship Comfort to Head for New Orleans," *MedPage Today,* August 31, 2005.
17. Lynn Armitage, "Homegrown Heroes," *OC Metro: Business Lifestyle Magazine,* October 27, 2005.
18. Press release, Orange County Urban Search and Rescue California Task Force 5 Deployed to Assist with Hurricane Katrina, August 31, 2005.

19. Armitage, "Homegrown Heroes."
20. Don Thompson and Robert Jablon, "Governor: Katrina Devastation 'Challenges Limits of Despair,'" Associated Press, September 2, 2005.
21. Ann Scott Tyson, "Strain of Iraq: War Means the Relief Burden Will Have to Be Shared," *Washington Post,* August 31, 2005.
22. Robert Block, Amy Schatz, and Gary Fields, "Power Failure: Behind Poor Katrina Response," *Wall Street Journal,* September 6, 2005.
23. Interview with Bernard H. McLaughlin, February 10, 2006.
24. Bernard H. McLaughlin, e-mail to author, February 9, 2006.
25. Ibid.
26. Interview with McLaughlin.
27. General MRE information from Federation of American Scientists, www.fas.org.
28. Bernard H. McLaughlin, diary, August 31, 2005, supplied by McLaughlin.
29. "Katrina Timeline (Unexecuted Plans)," *All Things Considered,* National Public Radio, September 9, 2005.
30. Press release, "Governor Richardson Orders 200 New Mexico National Guard to Aid in Hurricane Katrina Relief," Office of the Governor, New Mexico, August 30, 2005.
31. Michael Isikoff and Mark Hosenbell, "Red Tape," *Newsweek,* September 14, 2005.
32. Jim Lee with Victoria Tester, "After the Storm: A Grant County National Guardsman Tells What It Was Like to Be on the Scene in Louisiana After Hurricane Katrina," Silver City, (N. Mex.) *Desert Exposure,* November 2005.
33. Anne Hall and Doug Struck, "At Nursing Home, Katrina Dealt Only the One Blow," *Washington Post,* September 23, 2005.
34. Ibid.
35. Terry Ebbert, "Katrina Takes Aim," panel at Tulane University, February 12, 2006.
36. Interview with Mary Landrieu, January 26, 2006.
37. Belinda Rathbone, *Walker Evans* (Boston: Houghton Mifflin Company, 1995), p. 96.
38. Manuel Roig-Franzia, "Ghost Town: Down at the End of the World, Houses Walk and the Dead Rise Up," *Washington Post,* September 21, 2005.
39. Kathy Anderson, "The Ways of Water," New Orleans *Times-Picayune,* December 21, 2005.
40. Raymond Seed, testimony, Committee on Homeland Security and Governmental Affairs, November 2, 2005.
41. Richard Ford, "A City Beyond the Reach of Empathy," *New York Times*, September 4, 2005.
42. Interview with Kristina Ford, October 17, 2005.
43. Interview with Ben Jaffe, February 11, 2006.
44. William A. Fagaly, *Tools of Her Ministry: The Art of Sister Gertrude* (New York: Rizzoli, 2004), p. 5.
45. Ibid.
46. Interview with Jaffe.
47. Ibid.
48. David Fricke, "Native Sons," *Rolling Stone,* October 6, 2005, p. 127.

49. Nekesa Mumbi Moody, "Amid the Horror, a Lament for New Orleans Musical Heritage," Associated Press, September 2, 2005.
50. Interview with Mark Broyard, December 7, 2005.
51. Ibid.
52. Interview with Roger Guenveur Smith, December 2005.
53. Interview with Broyard.
54. Jacqueline Beaumont, "Texas Families Await Word from Relatives in Hurricane-Ravaged New Orleans," Beaumont (Tex.) *Enterprise*, August 30, 2005.
55. Blaine Harden and Shankar Vedantam, "Many Displaced by Katrina Find Shelter in Relatives' Homes," *Washington Post,* September 8, 2005.
56. Ron Thibodeaux, "Ailing Refugees Find Care at LSU Arena," New Orleans *Times-Picayune,* September 1, 2005.
57. Interview with Sherry Watters, February 9, 2006.
58. "LSU QB Shares Home with Fats Domino," Associated Press, September 4, 2005.
59. "Fats Domino Returns Home to New Orleans," Reuters, October 19, 2005.
60. Interview with Bill White, February 8, 2006.
61. Bill Murphy, Rad Sallee, Salatheia Bryant, Melissa Phillip, Peggy O'Hare, Dale Lezon, and R. G. Ratcliffe, "Houston: Buses Bring Thousands from Superdome to Astrodome," *Houston Chronicle,* September 1, 2005.
62. Interview with White.
63. "Press Conference with Officials from Homeland Security, the Environmental Protection Agency, and the Departments of Health and Human Services, Energy, Transportation, and Defense," Department of Homeland Security, August 31, 2005.
64. George Rush and Joanna Malloy, "As South Drowns, Rice Soaks in N.Y.," New York *Daily News,* September 1, 2005.
65. Lisa Rosetta, "Frustrated: Fire Crews to Hand Out Fliers for FEMA," *Salt Lake Tribune,* September 1, 2005.
66. Interview with Terry Ebbert, March 8, 2005.
67. John T. Edge, *Fried Chicken: An American Story* (New York: Putnam, 2004).
68. Kim Severson, "Austin Leslie, 71, Dies; Famed for Fried Chicken," *New York Times,* September 30, 2005.
69. "Anatomy of a Disaster," *U.S. News & World Report,* September 26, 2005.
70. Interview with Leonard Kleinpeter, March 13, 2006.
71. George W. Bush, text of speech released by Department of Homeland Security, August 31, 2005.
72. "Waiting for a Leader," *New York Times,* Op-Ed, September 1, 2005.
73. Robert D. McFadden and Ralph Blumenthal, "Hurricane Katrina: The Overview," *New York Times,* September 1, 2005.
74. Interview with Jeff Goldblatt, December 28, 2005.
75. Interview with Jeff Goldblatt, December 13, 2005.

Chapter 12: The Intense Irrationality of a Thursday

1. George W. Bush, interview with Diane Sawyer, *Good Morning America,* ABC, September 1, 2005.

2. Elisabeth Bumiller, "Democrats and Others Criticize White House's Response to Disaster," *New York Times,* September 2, 2005.

3. Joby Warrick, "White House Got Early Warning on Katrina," *Washington Post,* January 24, 2006.

4. Bumiller, "Democrats and Others Criticize."

5. Interview with Catrina Williams, February 16, 2006.

6. Ibid.

7. Ibid.

8. "Six Tales of Courage," *Time,* December 19, 2005.

9. Press release, "Harmony Hills Student Honored for Saving Lives," North East Independent School District, San Antonio, Texas, January 24, 2006.

10. Interview with Williams.

11. Leslie Turk, "Cries for Help," Lafayette (La.) *Independent,* September 7, 2005.

12. Interview with Williams.

13. Ibid.

14. Quoted in *Executive Intelligence Review,* September 30, 2005.

15. Interview with Richard Zuschlag, January 31, 2006.

16. Patrick J. Sauer, "The Anti-FEMA," *Inc. Magazine,* December 2005.

17. Interview with Zuschlag.

18. Sauer, "The Anti-FEMA."

19. Interview with Zuschlag.

20. Interview with James Cardiff, March 12, 2006.

21. Ibid.

22. Ibid.

23. Ibid.

24. Interview with Henry McEnery III, March 12, 2006.

25. Ibid.

26. Interview with Cardiff.

27. Interview with Zuschlag.

28. Ceci Connolly, "Frustration Grows in Days Stranded on Interstate 10," *Washington Post,* September 9, 2005.

29. Meghan Gordon and Sheila Grissett, "I-10 Becomes Refugee Camp," New Orleans *Times-Picayune,* September 3, 2005.

30. Ellen Barry, "Stranded on Highway and Left to Their Fate," *Irish Times,* September 2, 2005.

31. Connolly, "Frustration Grows in Days Stranded on Interstate 10."

32. Jocelyn Noveck, "Use of Word 'Refugee' to Describe Katrina's Displaced Disputed," Associated Press, September 6, 2005.

33. Michael Eric Dyson, *Come Hell or High Water: Hurricane Katrina and the Color of Disaster* (New York: Basic Civitas Books, 2006), p. 1.

34. Michael Kranish, "Bush Vows Probe of 'What Went Wrong,'" *Boston Globe,* September 7, 2005.

35. Gerard Baker, "Space, Food, Medicine, Protection: It's Better Here in Barbara's Hall of Plenty," *The Times* (London), September 16, 2005.

36. Jack Shafer, "Don't Refloat," slate.com, September 7, 2005.
37. Brian G. Lukas, "Covering Hurricane Katrina," personal journal, September 1–4, 2005.
38. Gordon and Grissett, "I-10 Becomes Refugee Camp."
39. *Oprah,* September 7, 2005, transcript.
40. Chip Johnson, "Police Made Their Storm Misery Worse," *San Francisco Chronicle,* September 9, 2005.
41. Ibid.
42. Ed Bradley, *60 Minutes,* CBS, December 18, 2005.
43. Interview with Sheperd Smith, February 1, 2006.
44. Rob Nelson, "Charred Oakwood Determined to Reopen," New Orleans *Times-Picayune,* October 19, 2005.
45. Interview with Smith.
46. Ibid.
47. Steve Inskeep, "Evacuees Were Turned Away at Gretna, Louisiana," *Morning Edition,* National Public Radio, September 20, 2005.
48. Brock N. Meeks, "Gretna Mayor Defends Bridge Blockade," MSNBC.com, September 22, 2005.
49. John Burnett reporting, *All Things Considered,* National Public Radio, September 15, 2005.
50. Interview with Natalie Rand, December 28, 2005.
51. Janet McConnaughey, "Man Listed as Hurricane Katrina Victim Was Shot by Police," Associated Press, October 10, 2005.
52. Janet McConnaughey, "La Crosse Man's Brother Killed by Police in New Orleans," Associated Press, October 9, 2005.
53. Interview with Conia Sherman, December 28, 2005.
54. Denise Moore, interview with Ira Moore, "After the Flood," *This American Life,* WBEZ radio, September 9, 2005.
55. Burnett, *All Things Considered.*
56. Wil Haygood and Ann Scott Tyson, "'It Was as If All of Us Were Already Pronounced Dead,'" *Washington Post,* September 15, 2005.
57. "A Failure of Initiative: Final Report of the Select Bipartisan Committee to Investigate the Preparation for and Response to Hurricane Katrina," U.S. House of Representatives, February 2006, p. 28.
58. Haygood and Tyson, "'Already Pronounced Dead.'"
59. Interview with Ivory Clark, February 16, 2006.
60. Ibid.
61. Ibid.
62. Reed Abelson and Alan Feuer, "10,000 Patients and Staff Members Await Evacuation from Barely Functional Hospitals," *New York Times,* September 1, 2005.
63. Neil Osterweil, "Hospitals Evacuate in Wake of Rising Waters from Katrina," *MedPage Today,* August 30, 2005.
64. Chris Adams, Thomas Fitzgerald, and Steven Thomas, "Hospital Workers Describe Effort to Save Patients," Knight Ridder, September 13, 2005.

65. Carrie Kahn, "New Orleans Hospital Staff Discussed Mercy Killings," *All Things Considered,* National Public Radio, September 16, 2006.
66. Jim Montgomery, "The Hospital Spirit Survives Hurricane Katrina's Wrath," *American Hospital Association,* September 19, 2005.
67. Interview with Christopher Wormuth, December 29, 2005.
68. Madeline Brand, "One Week Later, Catching Up with Dr. Randy Roig," *Day to Day,* National Public Radio, September 7, 2005.
69. Madeline Bar Diaz, "Hospital Crisis: Harrowing Five Days Until Rescue," Fort Lauderdale *Sun Sentinel,* September 7, 2005.
70. Adam Nossiter, "Dispute over Historic Hospital for the Poor Pits Doctors Against the State," *New York Times,* December 17, 2005.
71. "Two New Orleans Hospitals Plead for Help," Associated Press, September 1, 2005.
72. Interview with Ruth Berggren, January 29, 2006.
73. Ibid.
74. Ibid.
75. Ibid.
76. Interview with Peter DeBlieux, January 29, 2006.
77. Marilynn Marchione, "Most Troubled New Orleans Hospital Finally Gets Some Help," Associated Press, September 2, 2005.
78. Steve Sternberg, "For City's Historic Hospital, Help Is Needed 'in a Hurry,' *USA Today,* September 2, 2005.
79. Interview with Tyler Curiel, February 18, 2006.
80. Interview with Berggren.
81. Interview with Peter DeBlieux, October 25, 2006.
82. Interview with Libby Goff, February 19, 2006.
83. Matt Scallen, "Airport Medical Teams Treat Thousands," New Orleans *Times-Picayune,* September 9, 2005.
84. Jay Root and Matt Stevens, "Health-Care Workers Struggle to Care for the Sick and Elderly," Knight Ridder, September 2, 2005.
85. Interview with Larry Dixon, February 18, 2006.
86. Ibid.
87. Ibid.
88. Interview with Arthur "June" Thomas, February 18, 2006.
89. Ibid.
90. Interview with Dixon.
91. Ibid.
92. Ibid.

Chapter 13: "It's Our Time Now"

1. Interview with Bruce Foret, February 26, 2006.
2. Ibid.
3. Ibid.
4. Interview with George Morales, February 27, 2006.

5. Ibid.
6. Interview with Foret.
7. Ibid.
8. Ibid.
9. Interview with Calvin Fayard, September 28, 2005.
10. Interview with Ivory Clark, March 2, 2006.
11. Interview with Hank Staples, March 1, 2006.
12. Interview with Foret.
13. Ibid.
14. Interview with Jeff Amann, March 1, 2006.
15. Ibid.
16. Ibid.
17. Interview with Foret.
18. Ibid.
19. Interview with Warren Riley, December 5, 2005.
20. Interview with Harry Lee, February 27, 2006.
21. Allen G. Breed, "From Catastrophe to Chaos; Gunfire, Corpses Left in the Open, and a Slow Exodus Out," Associated Press, September 1, 2005.
22. "Frustration Boils; Mayor Nagin, Blanco Irate About Delays," *Baton Rouge Advocate,* September 2, 2005.
23. Ibid.
24. Ibid.
25. Ibid.
26. Paul Salopek, "A Stressful Toll on Haggard Cops," *Chicago Tribune,* September 8, 2005.
27. Trymaine Lee, "Seventh District Struggles to Come to Terms with Officers Suicide," New Orleans *Times-Picayune,* December 18, 2005.
28. Interview with Eddie Compass, February 25, 2006.
29. Interview with Riley.
30. Lee, "Seventh District Struggles to Come to Terms with Officers Suicide."
31. Interview with Riley.
32. Ibid.
33. Ibid.
34. Bernard H. McLaughlin, diary, September 1–3, 2005, supplied by McLaughlin.
35. Interview with Terry Ebbert, March 21, 2006.
36. Interview with Bill White, February 6, 2006.
37. Ibid.
38. Josh Peter, "Taking the Wheel," New Orleans *Times-Picayune*, October 16, 2005.
39. Interview with Judge Eckels, January 26, 2006.
40. Ibid.
41. Mary Foster, "Refugees Getting Showers and Meals in Houston," Associated Press, September 1, 2005.
42. Interview with Laura Maloney, December 13, 2005.
43. Interview with Denise Okojo, December 14, 2005.

44. Gloria Dauphin, "Service Dog Rescued and Reunited," LSPCA Web site, October 7, 2005.
45. Interview with Maloney, December 13, 2005.
46. Interview with Laura Maloney, March 2, 2006.
47. Interview with Ron Forman, January 10, 2006.
48. Ibid.
49. Ibid.
50. Ibid.
51. Interview with Don Kinney, February 25, 2006.
52. Ibid.
53. Interview with Ron Forman.
54. Ibid.
55. Ibid.
56. Ibid.
57. Interview with Kinney.
58. Matthew Brown, "Oyster Fishers Snag La. Jobs," New Orleans *Times-Picayune*, February 6, 2006.
59. Sam Coates and Dan Eggen, "A City of Despair and Lawlessness," *Washington Post*, September 2, 2005.
60. Interview with Ivory Clark, February 16, 2006.
61. Donna Johnson-Clark, Katrina diary, September 1–2, 2006, supplied by Johnson-Clark.
62. Mary Swerczek, "General Likens Hurricane to Well-Planned Attack; He Pledges Help Until State Can Take Over," New Orleans *Times-Picayune*, September 14, 2005.
63. Greg Henderson, interview with Anderson Cooper, CNN, September 9, 2005.
64. Dave Montgomery, "Man in Charge of Military Along Gulf Is No-Nonsense, All Army," Knight Ridder, September 11, 2005.
65. Mark Schleifstein, "Surge, Breach, and a 26-Foot-Deep Gorge," New Orleans *Times-Picayune*, September 1, 2005.
66. Kaiser-Hill press release, September 8, 2005.
67. Interview with Jerry Long, March 2, 2006.
68. Interview with Jackie Clarkson, December 8, 2005.
69. Interview with Clarkson.
70. Ibid.
71. Ibid.
72. Interview with Sally Forman, January 30, 2006.
73. Ibid.
74. Interview with Clarkson.
75. Interview with Sally Forman.
76. Interview with Garland Robinette, March 2, 2006.
77. "FEMA Chief: Victims Bear Some Responsibility," CNN.com, September 1, 2005.
78. Ibid.

79. "White House Press Briefing on Disaster Efforts; FEMA Halts Rescue Efforts in New Orleans as Gunfire Endangers Workers," CNN, September 1, 2005, transcript.
80. Coates and Eggen, "A City of Despair and Lawlessness."
81. Al Kamen, "A Worthy Move from FEMA," *Washington Post*, November 14, 2005.
82. Interview with Michael Brown, February 4, 2006.
83. Interview with Jimmy Pitre, March 2, 2006.
84. Ibid.
85. Ibid.
86. Memorandum from Michael D. Brown to Michael Chertoff, subject: Component Head Meeting, March 2005.
87. Michael Chertoff, interview with Robert Seigel, *All Things Considered,* National Public Radio, September 1, 2005.
88. Allen G. Breed, "National Guardsmen Arrive in New Orleans," Associated Press, September 2, 2005.
89. Lieutenant General Russel Honore, "A John Wayne Dude," CNN, September 3, 2005, transcript.

Chapter 14: The Friday Shuffle and Saturday Relief

1. Jennifer Loven, "Bush Says Relief Efforts Not Acceptable" Associated Press, September 2, 2005.
2. Dan Froomkin, "Dealing with Political Disaster," *Washington Post,* September 6, 2005.
3. Scott McClellan, "Press Gaggle with Scott McClellan: Aboard Air Force One en Route to Alabama, Mississippi, and Louisiana," September 2, 2005.
4. Evan Thomas et al., "Deadly Mistakes," *Newsweek,* September 19, 2005.
5. *NBC Nightly News,* December 12, 2005, transcript.
6. Interview with William Jefferson, February 24, 2006.
7. Ibid.
8. Jeb Schrenk, "Storm Smacks Alabama Cost," *Mobile Register,* August 30, 2005.
9. Interview with Jeff Collier, February 17, 2006.
10. Stephen Barr, "Coast Guard Response Is Katrina Silver Lining in the Storm," *Washington Post,* September 10, 2005.
11. Interview with Walter Dickerson, January 18, 2006.
12. Mark Leibovich, "Buddy Story," *Washington Post*, October 10, 2005.
13. Joe Scarborough, "Mississippi Rising," MSNBC.com blog, November 29, 2005.
14. "President Arrives in Alabama, Briefed on Hurricane Katrina," Office of the Press Secretary, White House, September 3, 2005.
15. Judy Keen and Richard Benedetto, "A Compassionate Bush Was Absent Right After Katrina," *USA Today,* September 9, 2005.
16. "Relief Convoy Finally Rolls In," *Milwaukee Journal,* September 3, 2005.
17. Rachel Donadio, "The Gladwell Effect," *New York Times,* February 5, 2006.
18. Interview with Michael Brown, February 2, 2006.

19. Daren Fonda and Rita Healy, "How Reliable Is Brown's Réumé," *Time,* September 8, 2005.

20. "New Orleans Mayor Lashes Out at Feds," CNN, September 2, 2006.

21. Loven, "Bush Says Relief Efforts Not Acceptable."

22. *Lou Dobbs Tonight,* CNN, September 2, 2005, transcript.

23. Mike Williams and Ken Herman, "Help at Last: Federal Response Takes Flak as Bush Tours Region," *Atlanta Journal-Constitution,* September 3, 2005.

24. Dahleen Glanton, "'You Can Smell the Death,'" *Chicago Tribune,* September 2, 2005.

25. Joshua Cogswell, "Little Aid, Lots of Misery," Jackson *Clarion-Ledger,* September 5, 2005.

26. Interview with Morris Dees, February 28, 2006.

27. Joe Atkins, "Barbour Praised Bush Too Quickly," Jackson *Clarion-Ledger,* November 6, 2005.

28. Geoff Fender, "More Action, Less Talk Required for Recovery," Biloxi *Sun-Herald,* September 5, 2005.

29. Mark Leibovich, "Buddy Story."

30. Jill Lawrence, "Governors Handle Crisis in Own Ways," *USA Today,* September 12, 2005.

31. Cogswell, "Little Aid, Lots of Misery."

32. Sally Jenkins, "By Hook or Crook, Surviving Storm," *Washington Post,* September 19, 2005.

33. Keith Naughton and Mark Hosenball, "Cash and Cat' 5 Chaos," *Newsweek,* September 26, 2005.

34. Stephen J. Hedges, "Crew of Navy Ship Ready to Play Larger Role in Relief Effort," *Chicago Tribune,* September 3, 2005.

35. Michael Newsom, "Keesler Medical Staff Making Do in Tent," *Biloxi Sun-Herald,* September 18, 2005.

36. "Name This Operation Contest," *Katrina Daily News,* Keesler Air Force Base, September 1, 2005.

37. Photo caption, *Katrina Daily News,* Keesler Air Force Base, September 15, 2005.

38. Interview with Fredro Knight, February 28, 2006.

39. Ibid.

40. Sheila Kast, "Mississippi Agricultural Operations Reel from Storm," *Weekend Edition,* National Public Radio.

41. Suzette Parmley, "Mississippi's Gaming Industry Devastated," *Philadelphia Inquirer,* August 31, 2005.

42. Barry Schlacter, "Timber, Poultry Hurt Worst by Katrina," *Fort Worth Star-Telegram,* September 15, 2005.

43. Barnaby J. Feder, "The Power Grid; Utility Workers Come from Afar to Help Their Brethren Start Restoring Service," *New York Times,* September 1, 2005.

44. Press release, "Mississippi Power Restoration Update," Mississippi Power, September 15, 2005.

45. Dennis Cauchon, "Some Areas Inch Toward Normality," *USA Today,* September 9, 2005.

46. Press release, "President Tours Biloxi, Mississippi Hurricane Damaged Neighborhoods," September 2, 2005, www.whitehouse.gov.
47. Cauchon, "Some Areas Inch Toward Normality."
48. "Frustration Boils Mayor Nagin, Blanco Irate About Delays," *Baton Rouge Advocate,* September 1, 2005.
49. Interview with Brian G. Lukas, February 18, 2006.
50. Press release, "Katrina/Rita Missing Persons Hotline," National Center for Missing and Exploited Children, February 6, 2006.
51. Interview with Brian G. Lukas, February 18, 2006.
52. Brian G. Lukas, "Covering Hurricane Katrina," personal journal, September 2, 2005, supplied by Lukas.
53. Ibid.
54. Ibid.
55. Interview with C. Ray Nagin, March 15, 2006.
56. Interview with Ron Forman, January 10, 2006.
57. Ibid.
58. Thomas et al., "Deadly Mistakes."
59. Karen Tumulty, "The Governor," *Time,* September 19, 2005.
60. Robert Travis Scott, "Politics Delayed Troops Dispatch to N.O.," *Times-Picayune,* December 11, 2005.
61. Thomas et al., "Deadly Mistakes."
62. C. Ray Nagin, interview with Soledad O'Brien, *American Morning,* CNN, September 5, 2005.
63. Robert Travis Scott, "Politics Delayed Troops Dispatch to N.O.," New Orleans *Times-Picayune,* December 11, 2005.
64. Bernard H. McLaughlin, diary, September 2, 2005, supplied by McLaughlin.
65. Interview with Bernard McLaughlin, February 10, 2006.
66. Ibid.
67. Michael Eric Dyson, *Come Hell or High Water: Hurricane Katrina and the Color of Disaster,* (New York: Basic Civitas Books, 2006), pp. 100–107.
68. Nicholas Lemann, "The Talk of the Town; Insurrection," *The New Yorker,* September 26, 2005.
69. Interview with Kathleen Blanco, December 19, 2005.
70. Scott Pelley (correspondent), "Katrina Response Sparks Outrage," *60 Minutes,* CBS, September 5, 2005.
71. Ibid. Also CNN transcript; and Dyson, *Come Hell or High Water,* p. 102.
72. Interview with Blanco.
73. Interview with Eddie Compass, February 25, 2006.
74. Ibid.
75. *Today*, NBC, September 5, 2005, transcript.
76. Bruce Baughman, director, Alabama State Emergency Management Agency, testimony, House Government Reform Committee, November 9, 2005.
77. "Reported Katrina Deaths, State by State," Associated Press, September 17, 2005.

78. Maurice Possley and John McCormick, "Officials Struggle with Katrina-Related Deaths," *Chicago Tribune*, November 1, 2005.

79. "January 30, 2006: Updated Number of Deceased Victims Recovered Following Hurricane Katrina," Bureau of Media and Communications, Louisiana Department of Health and Hospitals, January 30, 2006.

80. Possley and McCormick, "Officials Struggle with Katrina-Related Deaths."

81. Interview with Blanco.

82. *Oprah,* September 9, 2005, transcript.

83. Interview with Compass.

84. Interview with Jimmy Duckworth, March 3, 2006.

85. "Hurricane Katrina; The U.S. Coast Guard at Its Best," dedication, published by Faircourt, Tampa, Florida.

86. Interview with Duckworth.

87. Ibid.

88. Ibid.

89. Jimmy Duckworth, U.S. Coast Guard journal, August 29 to October 1, 2005 (private papers of Duckworth).

90. Interview with Eddie Favre, November 5, 2005.

91. David Rouse, "Small Signs of Normality," Associated Press, September 9, 2005.

92. Interview with Jimmy Buffett, February 15, 2006.

93. Ibid.

94. Mike Monson, "Katrina Evacuees Calling C-U Home," *News-Gazette* (Illinois), September 8, 2005.

95. Dawn Bryant, "Hurricane Show Returns to Margaritaville," (Myrtle Beach, South Carolina) *Sun News*, November 20, 2005.

96. http://margaritaville.com.

97. "Open Letter to the President," New Orleans *Times-Picayune*, September 4, 2005.

98. Jimmy Buffett, "Off to See the Lizard," song.

99. Interview with Buffett.

100. Ibid.

101. Interview with Marlin Torguson, October 26, 2005.

102. Ibid.

103. Brenda Norrell, "A Pointe-au-Chien Family Describes Survival After Katrina," *Indian Country Today*, September 28, 2005.

104. David Giffels, "Resettling in Area Is Tough Sell," *Akron Beacon Journal*, September 25, 2005.

105. Brennen Jensen and Nicole Lewis, "Hurricane Donations: How Much Has Been Raised and How It Will Be Spent," *Chronicle of Philanthropy*, November 10, 2005.

106. G. Robert Hillman, "Bush Seeks Counsel of Two Ex-Presidents," *Dallas Morning News,* September 1, 2005.

107. Elizabeth Williamson, "Contributions Near $100 Million," *Washington Post,* September 2, 2005.

108. "A Message from Former Presidents George H.W. Bush and William J. Clinton,"

letter of December 7, 2005, posted on Bush-Clinton Katrina Fund Web site, bush clintonkatrinafund.org.

109. Interview with Michael Prevost, December 28, 2005.

110. Sanjay Gupta, *American Morning*, CNN, September 2, 2005.

111. "A Failure of Initiative: Final Report of the Select Bipartisan Committee to Investigate the Preparation for and Response to Hurricane Katrina," U.S. House of Representatives, February 2006, p. 343.

112. Haley Barbour, testimony, Committee on Homeland Security and Governmental Affairs, U.S. Senate, February 2, 2005.

113. Megan Tench, "Filene's Basement Spree Is Evacuee's 'Dream Come True,'" *Boston Globe*, September 24, 2005.

114. Interview with Prevost.

115. Ibid.

116. Sanjay Gupta, "City of New Orleans Falling Deeper into Chaos," *American Morning,* CNN, September 2, 2005.

117. Interview with Richard Griffiths, February 28, 2006.

118. Interview with Ruth Berggren, January 29, 2006.

119. Ruth Berggren, "Unexpected Necessities—Inside Charity Hospital," *New England Journal of Medicine* 353 (October 13, 2005): 1550–53.

120. Interview with Berggren.

121. Interview with Libby Goff, February 21, 2006.

122. Interview with Berggren.

123. Ibid.

124. Libby Goff, diary, August 29–September 3, supplied by Goff.

125. Interview with Berggren.

Chapter 15: Getaway (or X Marks the Spot)

1. Interview with Diane Johnson, December 3, 2006.
2. Ibid.
3. Interview with Ivory Clark, February 6, 2006.
4. Interview with Bernard McLaughlin, March 7, 2006.
5. Ibid.
6. Ibid.
7. Ibid.
8. Ibid.
9. Interview with McLaughlin.
10. Interview with Lance Hill, March 8, 2006.
11. Ibid.
12. David Spielman, diary, September 3, 2005.
13. Michael Perlstein, "For Tales of Life and Death, the Writing's on the Walls," New Orleans *Times-Picayune*, September 17, 2005.
14. Interview with Josh Holm, November 18, 2005.
15. Ibid.

16. Ibid.
17. Interview with Tony Zumbado, November 14, 2005.
18. Interview with Holm.
19. Interview with Zumbado, November 14, 2005.
20. Interview with Tony Zumbado, March 8, 2006.
21. Ibid.
22. Interview with Zumbado, November 14, 2005.
23. Renae Merle and Griff Witte, "Lack of Contracts Hampered FEMA," *Washington Post*, October 10, 2005.
24. Ibid.
25. Kim Severson, "Austin Leslie, Dies, 71; Famed for Fried Chicken," *New York Times*, September 30, 2005.
26. Shaila Dewan, "With the Jazz Funeral's Return, the Spirit of New Orleans Rises," *New York Times*, October 10, 2005.
27. Bernard H. McLaughlin, diary, September 3, 2005, supplied by McLaughlin.
28. Interview with Clark.
29. Ibid.
30. Ibid.
31. Ibid.
32. Ibid.
33. André Buisson, diary, September 4, 2005.
34. William Caldwell IV, interview with Jed Babbin, Radioblogger.com, September 9, 2005.
35. Jay Price, "Katrina Notebook," *Raleigh News & Observer*, September 5, 2005.
36. David Wood, "The General in Charge Is a Man in a Hurry," Newhouse News Service, September 5, 2005.
37. Gerry J. Gilmore, "82nd Airborne Division Becomes 'Waterborne' in New Orleans," American Forces Press Service, September 21, 2005.
38. Ibid.
39. Dan Baum, "Deluged," *The New Yorker*, January 9, 2006.
40. Interview with Terry Ebbert, March 8, 2006.
41. John Schwartz, "Pumping: Behind the First Roar of Machinery to Drain the City," *New York Times*, September 8, 2005.
42. Deborah Acomb, "Hurricane Katrina," *National Journal*, September 10, 2005.
43. Matt Sedensky, "Final Chapter in Storm Exodus Is Epic Bus Journey," Associated Press, September 3, 2005.
44. Tom Walker, "Hidden Hands Behind Katrina," *New Statesman*, October 3, 2005.
45. Brett Martel, "Angry God Sent Storms, Mayor of New Orleans Says," Associated Press, January 17, 2006.
46. Noah's Ark Church sign, personal papers of Reverend Willie Walker, Kenner, Louisiana.
47. Interview with Willie Walker, March 5, 2006.
48. Interview with Gus Davies, March 7, 2006.
49. Interview with Walker.
50. Interview with Laura Young, March 7, 2006.

Acknowledgments

When Pulitzer Prize–winning historian David McCullough wrote *The Johnstown Flood,* his first book, he included an appendix honoring the 2,209 killed in the notorious Pennsylvania dam break of May 31, 1889. It was a roll call of remembrance. Long ago, I learned you never go wrong following David's elegant lead. He is the master of graciousness and the thank-you note. I wanted to emulate him in this regard, but unlike the Johnstown flood, it is not yet possible to include an accurate list of victims of Katrina. Just this morning I opened up my local newspaper, the *Times-Picayune,* to read a B-4 obituary-page story about two Hurricane Katrina victims whose corpses were just pulled out of the Lower Ninth Ward rubble yesterday. Today is March 17, 2006. Over six months have passed since Katrina roared ashore. On August 30, 2005, the morning after Katrina made landfall, the Homeland Security Operations Center sent out an urgent 6:00 A.M. bulletin claiming that "it could take months to dewater" New Orleans. They were right. What Homeland Security didn't anticipate, however, is that decomposed bodies would still be collected from the 5400 block of Prieur Street (among wooden debris) or the 2900 block of Higgins Boulevard (in a kitchen) six months after that flood bulletin was released. "I thought I'd read it all," Reverend Billy Graham said, as he toured the Lower Ninth in March 2006, "but it doesn't compare to what you see in just a few minutes' tour of this area."

At this writing, the Louisiana State Health Department—such as it is—tallied 1,080 Katrina dead; the U.S. Army Corps of Engineers claimed that in Orleans Parish alone there were approximately 400 people missing and presumed dead. While these statistics are amorphous, the demographic who suffered the most in the Great Deluge were not. They were the poor, ill or critically ill elderly. Not "African Americans" or "the unemployed" or "women of color." "Old folks" or "senior citizens" suffered the most. Women like Diane Johnson—who died of post-Katrina stress but is not included in the official death toll—are our country's true invisible people. As the old bluegrass song goes, Johnson was just "Old and in the Way." Even National Guardsmen during the Great Deluge seemed uninformed that the elderly dehydrate faster and are unable to be ripped off of their medication regiments in a willy-nilly fashion. "Katrina shows how hard it is for younger people in charge of the story, the social response or the rescue effort to put themselves in the shoes of the vulnerable elderly," Margarette M. Gullette perceptively wrote in the *Louisiana Weekly*, a New Orleans–based African-American newspaper in operation since 1925. "That is part of ageism: not hatred but ignorance, indifference, and the failure to imagine oneself as older and in need of care."

For those first post-Katrina weeks I lived with my family (plus a dog and two cats) out of the Omni Hotel in Houston. The hotel lobby had the whacked-out, mad feeling of a train station at wartime. Desperate evacuees, lucky to have rooms in Houston, were urgently working their cell phones, crowded around the lobby's large-screen TV, watching with bated breath the minute-by-minute news of the deluge, trying to stay grounded. We were like cossacks coming to despoil the czar's palace, stripping the Omni of its bathrobes and short-circuiting the business center's Xerox machines. So a salute is in order to the Omni employees who treated all the frantic Katrina drifters with great patience, kindness, and understanding. They were the epitome of professionalism.

A few weeks later, suffering from claustrophobia, my wife and I leased an apartment at the Royalton, a high-rise that ran out of room due to

Louisianans looking for four- to six-month leases. Much of this book was written in the Royalton's lobby, next to the free soda machine. I became a de facto member of the graveyard shift, chatting with the manager Natalie Santibanez, concierges Irenio Bantolina and Eddie Longoria, courtesy officer Onyeaga Nkemakolam, maintenance man Abraham Ibrahim, porter Carlos Villalobos, concierge Sergio Vasquez, doormen Reggie Lang, John Miller, and Gregory Porter, and valets Zaher Alsaleh, Emmanouil Charalampidis, Amin Eltaha, Juan Gamez, Frank Gleuni, Hassan Kassem, Julian Manzano, Aziz Sawadago, Kevin Starks, and Kenneth Tims, among others. Their diversionary banter about Houston gangs, real estate prices, Enron, the Mexican-American border—anything but Katrina—was most welcome.

Now, here is the buried lead of the acknowledgments. Upon arriving in Houston as an evacuee, the president of Rice University, David Leebron, and his wife, Ping, without *any* bureaucratic red tape offered me research quarters at the top of Fondren Library. Every amenity—a parking space, library access, electronic equipment, telephones—was placed at my disposal to write about the Great Deluge. My conduit to Leebron, Melissa Kean, Rice University's official historian, became my hall suitemate. She is a walking encyclopedia of all things Houston, a funny, vibrant, kind person, and an unbelievable asset to the university.

When Katrina hit I was working on a book for William Morrow about Theodore Roosevelt and conservation. In early September, shortly after the deluge subsided, my longtime editor, friend, and sparring partner, Claire Wachtel, suggested that I write about Katrina instead of TR. "You've got to do it," she kept saying. She was right. No editor could have worked more diligently on my behalf. She is a formidable talent and a believer in my work. I consider her family.

Others at Morrow deserve special thanks. Kim Lewis, executive managing editor, miraculously turned my manuscript into pages in an eclipsed week. Michael Morrison, president and group publisher, was always ready with advice and encouragement. He really believed in the book. Lisa Gallagher, senior vice president and publisher of William Morrow, was responsible for handling the various media requests and launch strategies that plagued us from the outset. Likewise Ben Bruton—from

Nebraska—always kept me informed of the reality, not the shadow, of what William Morrow had planned for this book. When Claire was absent, her assistant, Lauretta Charlton, as well as marketing manager Kevin Callahan, stepped into the fray, unafraid to get bloodied up. I'm grateful to them both. Sean Griffin likewise assisted with this book in numerous ways.

Because this book deals with mass negligence it had to be vetted carefully. This arduous task was performed by attorney Ellis Levine, who went over every line with a fine-tooth comb. One weekend we were on the telephone together for eight hours. By Monday I was able to sleep soundly. The rest of the Morrow team was vice president and associate general counsel Beth Silfin, design manager Betty Lew, production director Susan Kosko, production manager Derek Gullino, and designer Brian Mulligan.

My longtime agent, Lisa Bankoff of ICM, once again did a terrific job of looking after my publishing interests. Together we decided that some of the proceeds from this book will go to the Historic New Orleans Collection and St. Augustine Roman Catholic Church in New Orleans.

So much for the New York crowd. This book was written in Houston, New Orleans, and Baton Rouge, on the run and in places like Katz Delicatessen, Mother's Restaurant, and the St. Charles Tavern. (All right . . . and a lot of Starbuckses). And my SWAT team, an unlikely confluence of friends/Internet researchers, were always at my side. Just days upon arriving in Houston I got a telephone call from Emma Juniper, a twenty-year-old Clark University student and former personal assistant to the late Hunter S. Thompson. Armed with her white laptop—a Gonzo sticker on the top—she was full of grief concerning the plight of Katrina victims. She wanted to volunteer with the Red Cross and asked if she should. I said yes. Next thing I knew she had flown to Denver, rented a seven-person van, picked up her stepmother, Kristen Hannum, and twenty-one-year-old sister, Kira, and started driving to Texas. Using her computer, she was matching up displaced families scattered around Colorado, Oklahoma, and North Texas with their friends and relatives in the Astrodome shelter. Emma, I soon learned, was so proficient at this, that I hired her away from the Red Cross. My wife, Anne, and I rented her an apartment next to ours in Houston. We adopted her. And let me tell you, that young woman can scan cyberspace. Because I'm computer illiterate, her skills were desperately needed to fact-check and retrieve articles. She also made me feel old. A big new rage in college is called "couch surfing," where Generation Z-ers drift from city to city, sleeping for free on fellow surfers' sofas. It's the whole Kerouac thing done via the Internet. Due to her couch surfing, I got to meet an eclectic mix of smart young people. I learned a lot.

My dear friend Julie Fenster helped me in myriad ways. A fellow contributing editor at *American Heritage,* she was instrumental in framing my vision for the book. Every couple days we would talk by telephone and she would read my chapters with a critical eye. A first-rate historian, Fenster is also a great Internet researcher, and she constantly found blogs and stories for me to consider incorporating into the text. I don't know anybody smarter. Unfortunately on December 14, Fenster's mother, Ruth, died of cancer. She was a great lady. It was quite a setback to Julie but, even under those grim circumstances, she continued to push me forward in writing this book. Dozens of times, when I was in despair or exhausted, she said, "You've got to stay with it." Again, she is like family.

Then there was twenty-four-year-old Jessica Maruri, who made coffee drinks at the Starbucks on Maple Street in New Orleans. I began most of my pre-Katrina mornings in this shop, reading the *New York Times* and writing on yellow legal pads. I got to know Jessica—a 2003 graduate of Brandeis University—and found her savvy, bookish, and underappreciated. Like our family, when New Orleans flooded she had evacuated to Houston, finding employment at the Starbucks at Rice Village. I hired her away for this project. She became an extremely valuable tape transcriber, Internet researcher, and appointment secretary.

The fourth wheel of this operation was twenty-five-year-old Andrew Travers, the assistant director of the Roosevelt Center at Tulane University. Flushed out of New Orleans by the flood, he landed at his parents' house in Ridgewood, New Jersey. He ran the Roosevelt Center's non-Katrina-related work from that outpost for four months (George and Rita Travers deserve a special nod for allowing him to take over their home office.) In January he came back to New Orleans and opened up our campus headquarters. Working double-time, he ran the office during the day and at night came over to my Jefferson Avenue house and helped me pull this manuscript together. He is the best.

While writing this book I taught three classes at Tulane University: "History of the Cold War," "Theodore Roosevelt and the American Conservation Movement," and "Emergence of the Modern United States, 1917–1945." Occasionally during lectures I would deviate from a topic and talk about Katrina. I first tested out some of this prose on them, reading aloud about Reverend Willie Walker at the Carrolton Avenue boat launch and Tony Zumbado filming Memorial Medical Center. Their feedback, at a few critical junctures, was important. Tulane University President Scott Cowen and Provost Lester Lefton were also supportive despite the enormous pressure of keeping the school afloat in the flood. My friends and colleagues in theTulane history department supported me in numerous ways. They include Jim Boyden (especially), George Bernstein, Kenneth Harl, Richard Latner, Colin MacLachlan, Linda Pollock, Lawrence Powell (especially), Susan Schroeder, Randy Sparks, Richard Teichgraeber, Roseanne Adderley, Rachel Devlin, James Hood, F. Thomas Luongo, Samuel Ramer, Gertrude Yeager, Emily Clark, Neeti Nair, Jennifer Neighbors, Marline Otte, George Trumbull, Justin Wolfe, Terrence Fitzmorris, David Goldstein, Lance Hill, Edwin Lyon, and Bruce Raeburn.

Another auxiliary member of our *Great Deluge* SWAT team was New York photographer Lindsay Brice, who is originally from Valdosta, Georgia. We've been friends for fifteen years. Many of her excellent photographs are published here for the first time. She ended up in New Orleans as a freelancer, sleeping in her car. Heather Allan of NBC, worried about her safety, allowed her to infiltrate Camp NBC, to claim a patch of trailer-floor carpet as her bed. Someday an entire book of her Katrina photographs will be published. She also served as photo editor for this book, selecting, with my guidance and her expert eye, the appropriate images.

The reason I was able to travel around the post-Katrina Gulf South was a result of having press passes. When Katrina approached, *Rolling Stone* (Jann Wenner, Will Dana) immediately activated my credentials. Later *Vanity Fair* (Graydon Carter, David Friend) did the same. At various times NBC News, CNN, and Fox News Channel also allowed me to join up with their forces. It made all the difference in the world.

I personally conducted all of the interviews in this book. A number of key people helped me facilitate them, including Andy Ambrose, Jim

Amoss, Judy Benitez, Ruth Berggren, Ralph Blumenthal, Irene Briganti, Ty Bromell, Tuwan Brown, Africa Brumfield, Andy Buisson, Dyan "Mama D" French Cole, Rick Coleman, Joe Donches, Ron and Judy Drez, Clancy Dubos, Jimmy Duckworth, Terry Ebbert, Mimi Fast, Ben Fee, Caroline Feldman, Daren Fonda, Bruce Foret, Sally Forman, Charles Foti Jr., Tami Frazier, Nate Fredman, Lawrence Geller, Libby Goff, Jeff Goldblatt, Rafal C. Goyeneche, Tammy Haddad, Jean Harper, Caroline Heldman (Mama D filmmaker), Ben and Sarah Jaffe, Darryl Johnson, Benjamin Karp, Melissa Kean, Anura and Ravien Kulpath, Hal Leftwich, Marc Maglieri, Bob Mann, Rick Matthieu, Janet McQueen, Bernard McLaughlin, Maria Midence, Matt Moseley, Julie Oppenheimer, Warren Riley, Sara Roberts, Garland Robinette, Norm Robinson, Pharisa Robinson, Melanie Roussel, Theresa Simmons, E. Frank Snellings, Oliver Thomas, Anita Thompson, Juan Thompson, Reverend Willie Walker, Christopher Wormuth, Richard Zuschlag, and Tony Zumbado.

Special thanks to attorney George Tobia for grappling with my legal uncertainties, Kristi and John Schiller for Texas-style hospitality during my four months in Houston, and Meredith Cullen for friendship in the trenches. My wife's friends Kelley and Harper Trammell of Houston welcomed us to our new surroundings. Film documentarian Stephen Rue helped me facilitate a couple of crucial interviews. Frances and Calvin Fayard allowed me to crash at their New Orleans home during the first weeks of the Great Deluge. And Rini, Wally, and the Marcus family deserve a special salute for riding out Katrina with us at One River Place.

And to Sean, Robin, Dylan, and Hopper . . . the family . . . for helping us escape Katrina and for putting us up during Hurricane Rita. Sean's heroics in New Orleans, such as diving into swamp water to save the life of a schizophrenic woman are unforgettable.

Once again, my wife Anne, daughter Benton, and son Johnny are my joys. I love you guys. Likewise, Mom and Dad.

Grateful acknowledgment is made to reprint the following:

"Brownsville Girl" by Bob Dylan and Sam Shepard. Copyright © 1986 Special Rider Music. All rights reserved. International copyright secured. Reprinted by permission.

"Change in the Weather" by John Fogerty. All rights administered courtesy of John Fogerty. All rights reserved. Used by permission.

"Circulate the Rhythm" by Robert Hunter. Copyright © 1991 Robert Hunter. All rights administered by Ice Nine Publishing Company. All rights reserved. Used by permission.

"Everything Is Broken" by Bob Dylan. Copyright © 1989 Special Rider Music. All rights reserved. International copyright secured. Reprinted by permission.

"The Future" by Leonard Cohen. Copyright © 1992 Stranger Music Inc. All rights administered by Sony/ATV Music Publishing, 8 Music Square West, Nashville, TN 37203. All rights reserved. Used by permission.

"Asiniig" by Louise Erdrich. Copyright © 2003 Louise Erdrich. All rights administered by HarperCollins Publishers, 10 East 53rd Street, New York, NY 10022. All rights reserved. Used by permission.

"Hurricane" from *Collected poems, 1917–1982* by Archibald MacLeish. Copyright © 1985 by the Estate of Archibald MacLeish. Used by permission of Houghton Mifflin Company. All rights reserved.

"The Maker" by Daniel Lanois. All rights administered courtesy of Daniel Lanois. All rights reserved. Used by permission.

"Pascagoula Run" by Jimmy Buffett. All rights administered courtesy of Jimmy Buffett. All rights reserved. Used by permission.

"Tryin' to Get to Heaven" by Bob Dylan. Copyright © 1997 Special Rider Music. All rights reserved. International copyright secured. Reprinted by permission.

"Welcome to Jamrock," words and music by Damien Marley, Stephen Marley, and Ini Kamoze. Copyright © 2005 EMI April Music Inc., ZNS Music Publishing, Biddah Muzik, Inc., and Nine Sounds, Inc. All rights for ZNS Music Publishing controlled and administered by EMI April Music Inc. All rights reserved. International copyright secured. Used by permission.

INDEX